Texts and
Monographs
in Physics

W. Beiglböck
M. Goldhaber
E. H. Lieb
W. Thirring
*Series Editors*

Arno Böhm

# Quantum Mechanics

Springer-Verlag
New York   Heidelberg   Berlin

Arno Böhm

Department of Physics
Center for Particle Theory
The University of Texas at Austin
Austin, Texas 78712
USA

*Editors:*

Wolf Beiglböck

Institut für Angewandte Mathematik
Universität Heidelberg
Im Neuenheimer Feld 5
D-6900 Heidelberg 1
Federal Republic of Germany

Maurice Goldhaber

Department of Physics
Brookhaven National Laboratory
Associated Universities, Inc.
Upton, NY 11973
USA

Elliott H. Lieb

Department of Physics
Joseph Henry Laboratories
Princeton University
P.O. Box 708
Princeton, NJ 08540
USA

Walter Thirring

Institut für Theoretische Physik
   der Universität Wien
Boltzmanngasse 5
A-1090 Wien
Austria

With 105 Figures

ISBN 0-387-08862-8 Springer-Verlag New York
ISBN 3-540-08862-8 Springer-Verlag Berlin Heidelberg

**Library of Congress Cataloging in Publication Data**

Böhm, Arno, 1936–
  Quantum mechanics.

  (Texts and monographs in physics)
  Bibliography: p.
  Includes index.
  1. Quantum theory.   I.  Title.
QC174.12.B63        530.1'2        78-17453

Printed in the United States of America.

9 8 7 6 5 4 3 2 1

To Sharifa

# Preface

This book was written as a text, although many may consider it a monograph.

As a text it has been used several times in both the one-year graduate quantum-mechanics course and (in its shortened version) in a senior quantum mechanics course that I taught at the University of Texas at Austin. It is self-contained and does not require any prior knowledge of quantum mechanics. It also introduces the mathematical language of quantum mechanics, starting with the definitions, and attempts to teach this language by using it. Therefore, it can, in principle, be read without prior knowledge of the theory of linear operators and linear spaces, though some familiarity with linear algebra would be helpful. Prerequisites are knowledge of calculus and of vector algebra and analysis. Also used in a few places are some elementary facts of Fourier analysis and differential equations. Most physical examples are taken from the fields of atomic and molecular physics, as it is these fields that are best known to students at the stage when they learn quantum mechanics.

This book may be considered a monograph because the presentation here is different from the usual treatment in many standard textbooks on quantum mechanics. It is not that a "different kind" of quantum mechanics is presented here; this is conventional quantum mechanics ("Copenhagen interpretation"). However, in contrast to what one finds in the standard books, quantum mechanics is more than the overemphasized wave-particle dualism presented in the familiar mathematics of differential equations. "This latter dualism is only part of a more general pluralism" (Wigner) because, besides momentum and position, there is a plurality of other observables not commuting with position and momentum. As there is no principle that brings into prominence the position and momentum operators, a general formalism of quantum mechanics, in which every observable receives the emphasis it deserves for the particular problem being considered, is not only preferable but often much more practical. The lesson that I believe can be learned from the situation in particle theory is that more is needed than just the solutions of differential equations and there exist algebraic relations other than just the canonical commutation relation. In atomic and molecular physics the use of these general algebraic methods of quantum mechanics may be merely of practical advantage but not necessary, but in particle physics they seem to be essential. It is this general form of quantum theory that is presented here.

I have attempted to present the whole range from the fundamental assumptions to the experimental numbers. To do this in the limited space available required compromises. My choice may, to a certain extent, reflect my personal taste. But it was mainly influenced by what I thought was needed for modern physics and by what I found, or did not find, in the standard text-books. Detailed discussions of the Schrödinger differential equation for the hydrogen atom and other potentials can be found in many good books.[1] On the other hand, the description of the vibrational and rotational spectra of molecules are hardly treated in any textbooks of quantum mechanics, though they serve as simple examples for the important procedure of quantum-mechanical model building. Also, "elementary particles are much more similar to molecules than to atoms" (Heisenberg). So I have treated the former rather briefly and devoted considerable space to the latter.

Groups have not been explicitly made use of in this book. However, the reader familiar with this subject will see that group theory is behind most of the statements that have been cast here in terms of algebras of observables.

This is a physics book, and though mathematics has been used extensively, I have not endeavored to make the presentation mathematically rigorous. Most of the mathematics I believe to be rigorously justifiable within the framework of the rigged Hilbert space, which—in contrast to von Neumann's formation but in accord with Dirac's presentation[2]—is assumed to be the underlying mathematical structure. Except in the mathematical inserts, which are given in open brackets [M:  ], the reader will not even be made aware of these mathematical details.

The mathematical inserts are of two kinds. The first kind provides the mathematics needed, and the second kind indicates the underlying mathematical justification. The whole first chapter is a mathematical insert of the first kind. As presented here, the mathematics can only be appreciated in its applications. This suggests the pedagogical advice that the reader should not attempt to read the book in a linear fashion, one fully understood page after another, but that he should be content to obtain a superficial under-

---

[1] It has, usually, also been adequately treated in an undergraduate course.

[2] See, for instance, Dirac (1958), Jauch (1958), Ludwig (1954), and von Neumann (1932).

standing of a subject at first reading and then return to it later for a deeper understanding.

Quantum mechanics starts with Chapter II, where the most essential basic assumptions (axioms) of quantum mechanics are made plausible from the example of the harmonic oscillator as realized by the diatomic molecule. Further basic assumptions are introduced in later chapters when the scope of the theory is extended. These basic assumptions are not to be understood as mathematical axioms from which everything can be derived without using further judgment and creativity. An axiomatic approach of this kind does not appear to be possible in physics. The basic assumptions are to be considered as a concise way of formulating the quintessence of many experimental facts.

The book consists of two clearly distinct parts, Chapters II–XI and Chapters XIV–XXI, with two intermediate chapters, Chapters XII and XIII. The first part is more elementary in presentation, though more fundamental in subject matter, and gives a more approximate description because it treats all systems as stationary. Chapter XII introduces the basic assumption of time development. Chapter XIII is just an application of previously developed concepts and attempts to illustrate the characteristic features in which quantum mechanics differs from the classical theories. The second part, which starts with Chapter XIV, treats scattering and decaying systems. The presentation there is much more advanced.

Chapter XIV gives a derivation of the cross section under very general conditions, and Section XIV.5 may be the most difficult section of the book. Section XIV.5 and Chapter XV may be omitted in first reading if the reader can accept the results without derivation. Starting with Chapter XVI, applications are restricted to rationally symmetric systems without spin. Two different points of view—one in which the Hamiltonian time development is assumed to exist, and the other making use of the $S$-matrix—are treated in a parallel fashion. The required analyticity of the $S$-matrix is deduced from causality. One of the main features of the presentation is to treat discrete and continuous spectra from the same point of view. For this the rigged Hilbert space is needed, which provides not only a mathematical simplification but also a description which is closer to physics. In the last chapter, the rigged Hilbert space is used to describe a decaying system by eigenvectors of the energy operator with complex energy. This establishes the link between the $S$-matrix description of a resonance at a pole and the usual description of states by vectors in a linear space, and is another example of the advantages that the rigged Hilbert space provides.

Written at the height of the atomistic point of view—this book tries to expose also the complementary way of understanding, the holistic view. [See also Heisenberg's last lecture published in Physics Today **29**, 32 (1976).] Though never mentioned explicitly except in the brief Epilogue, the duality between atomism and holism is the recurring theme throughout the book.

*Austin, Texas*                                                                 Arno Böhm

# Acknowledgments

I am indebted to so many for their help, encouragement, and advice that I am hesitant to list individuals for fear of leaving out some who have made important contributions.

First there are those from whose teaching and writings I learned the subject of this book: G. Ludwig, "Grundlagen der Quantenmechanik." M. L. Goldberger and K. M. Watson, "Collision Theory." The principal mathematical source was I. M. Gelfand and collaborators "Generalized Functions." My principle sources for the experimental facts were G. Herzberg "Molecular Spectra and Molecular Structure" and H. S. W. Massey, E. H. S. Burhop "Electronic and Ionic Impact Phenomena." Many other sources were used and are acknowledged in the appropriate places and in the bibliography.

Of my colleagues: J. Ehlers supplied the initial inspiration and made many suggestions. A. Garcia, E. C. G. Sudarshan, and J. Werle have read part of the manuscript and provided helpful criticism and encouragement. The connection between causality and analyticity was discussed with N. G. van Kampen, who suggested the present form of Section XVIII.4. G. Bialkowski explained the phase shift analysis and helped with the writing of Section XVIII.8. R. H. Dalitz, C. J. Goebel, and H. A. Weidenmüller read Chapter XX and explained the problems connected with the eigenphase representation. L. Horwitz read Chapters XVIII, XX, and XXI and suggested important improvements. L. O'Raifeataigh read and improved Chapter XIX. The manuscript of Chapter XXI was discussed in the Spring of 1976 with B. Nagel. H. Baumgärtel explained some of the mathematical problems in Chapter XXI. The subject of Chapters XVIII and XXI was discussed with L. Fonda, G. C. Ghirardi, A. Grecos, and I. Prigogine, whose explanations were very helpful for the writing of the final version. H. C. Corben and F. Rohrlich examined Section IX.3a. L. Frommhold has given numerous explanations of the experimental facts throughout the writing of the book and A. Zaidi helped with the experimental data from nuclear physics. H. Stapp has given valuable advice and suggested many improvements, particularly for the Epilogue. The brief historical remarks proved more vexing than expected, and for the final version I received advice from W. Yourgrau, H. Rechenberg, E. Wolf, and from E. P. Wigner, who also suggested some other improvements.

The students to whom I am most indebted are G. Grunberg, who helped with substantial parts of the manuscript and eliminated many errors, and P. Moylan, who arranged the index. Also, S. Nadkarni and R. B. Tesse corrected and prepared parts of the manuscript.

Support from ERDA, the Center for Particle Theory at the University of Texas, and the NSF through its International Office during the writing of the book is gratefully acknowledged. The Texas University Research Institute provided a grant for the preparation of the manuscript, for which I am grateful to Vice-President Lieb. Finally, I wish to express my gratitude to the University of Texas, the Physics Department and Chairman T. A. Griffy for their support.

The help I recieved from the people at Springer, particularly W. Beiglböck, V. Borsodi, and Jeff Robbins, exceeded all my expectations.

# Contents

# CHAPTER I

# Mathematical Preliminaries

The mathematical language of quantum mechanics is introduced in this chapter. It does not contain much more than the basic definitions. A proper understanding of this material will come through its application in the following chapters.

## I.1 The Mathematical Language of Quantum Mechanics

To formulate Newtonian mechanics, the mathematical language of differential and integral calculus was developed. Though one can get some kind of understanding of velocity, acceleration, etc., without differential calculus (in particular for special cases), the real meanings of these physical notions in their full generality become clear only after one is familiar with the idea of the derivative. On the other hand, though, the abstract mathematical definitions of calculus become familiar to us only if we visualize them in terms of their physical realizations. Nowadays, no one would attempt to understand classical mechanics without knowing calculus.

Quantum mechanics, too, has its mathematical language, whose development went parallel to the development of quantum mechanics and whose creation in its full generality was inspired by the needs of quantum physics. This is the language of linear spaces, linear operators, associative algebras, etc., which has meanwhile grown into one of the main branches of mathematics—linear algebra and functional analysis. Although one might obtain some sort of understanding of quantum physics without knowing its mathematical language, the precise and deep meaning of the physical notions in their full generality will not reveal themselves to anyone who does not understand its mathematical language.

Therefore I shall start the quantum-mechanics course with some of the vocabulary and grammar of this language. I shall not try to be mathematically rigorous, since one can still communicate in a language that one does not speak completely correctly. I shall also not give all the mathematics that is needed at the beginning, and you need not be worried if you do not understand everything right away; one learns a language best by using it. I shall give at the beginning only as much mathematics as is necessary to make the initial statements about physics. We shall then have to learn new mathematical notions whenever they arise, while we proceed with the development of the physical ideas.

Before we start to study the mathematical structures that are employed in quantum mechanics, we should make the following observation: A mathematical structure is not something real—it only exists in our mind and is created by our mind (though often inspired by outside influences). It is obtained by taking a set of objects and equipping this set with a structure, by defining relations between these objects. Modern mathematics distinguishes three basic kinds of structures: algebraic, topological, and ordering. The mathematical structures we use are complicated combinations of these three. For example, the real numbers have an algebraic structure given by the usual laws of addition and multiplication, they have a topological structure given by the meaning of the usual limiting process for an infinite series of numbers, and they have an ordering structure given by the relations expressed by $<$.

We shall use predominantly algebraic structures, although in order to speak the mathematical language of quantum mechanics correctly, topological structures are essential. We shall start with the definition of a linear space, then introduce linear operators and give the definition of an associative algebra. That will provide us with enough vocabulary and grammar to enable us to start communicating physics, in the process of which the meaning of these mathematical structures will be filled with substance.

## I.2  Linear Spaces, Scalar Product

A set $R$ of elements $\phi, \psi, \chi, \ldots$ is called a *linear space* if the sum $\phi + \psi$ of any two elements $\phi, \psi \in R$ and the product $a\phi$ of any element $\phi \in R$ with any complex number $a$ are defined and have the following properties:

$$\text{if } \phi, \psi \in R, \text{ then } \phi + \psi \in R. \tag{2.1a}$$

$$\phi + \psi = \psi + \phi. \tag{2.1b}$$

$$(\phi + \psi) + \chi = \phi + (\psi + \chi). \tag{2.1c}$$

There exists in $R$ a "zero" element $0$ such that $\phi + 0 = \phi$ for all $\phi \in R$.   (2.1d)

$$\text{if } \phi \in R, \text{ then } a\phi \in R. \tag{2.1e}$$

$$a(b\phi) = (ab)\phi. \tag{2.1f}$$

$$1 \cdot \phi = \phi. \tag{2.1g}$$

$0 \cdot \phi = 0$ (the number zero appears on the left, the zero element on the right),   (2.1h)

$$a(\phi + \psi) = a\phi + a\psi, \tag{2.1i}$$

$$(a + b)\phi = a\phi + b\phi. \tag{2.1j}$$

The element $(-1)\phi$ is then usually denoted by $-\phi$; by properties 2.1g, k, h),

$$\phi + (-\phi) = (1 + (-1))\phi = 0\phi = 0.$$

The elements $\phi, \psi, \chi$ of the space $R$ are called *vectors*. A set $M$ in the linear space $R$ is called a *subspace* of $R$ if $M$ is a linear space under the same definitions of the operations of addition and multiplication by a number as given for $R$, i.e., if it follows from $\phi, \psi \in M$ that $a\phi \in M$ and $\phi + \psi \in M$.

An expression of the form $a_1\phi_1 + a_2\phi_2 + \cdots + a_n\phi_n$ is called a *linear combination* of the vectors $\phi_1, \phi_2, \ldots, \phi_n$; the vectors $\phi_1, \phi_2, \ldots, \phi_n$ are said to be *linearly dependent* if there exist numbers $a_1, a_2, \ldots, a_n$, not all zero, for which $a_1\phi_1 + a_2\phi_2 + \cdots + a_n\phi_n = 0$. If the equation $a_1\phi_1 + a_2\phi_2 + \cdots + a_n\phi_n = 0$ holds only for $a_1 = a_2 = \cdots = a_n = 0$, then the vectors $\phi_1, \phi_2, \ldots, \phi_n$ are called *linearly independent*. A space $R$ is said to be *finite-dimensional* and, more precisely, *n-dimensional* if there are $n$ and not more than $n$ linearly independent vectors in $R$. If the number of linearly independent vectors in $R$ is arbitrarily great, then $R$ is said to be *infinite-dimensional*. Every system of $n$ linearly independent vectors in an $n$-dimensional space $R$ is called a *basis* for $R$.

Every vector $\phi$ of an $n$-dimensional space $R$ can be uniquely represented in the form $\phi = a_1 e_1 + \cdots + a_n e_n$, where $e_1, e_2, \ldots, e_n$ is a basis for $R$. The numbers $a_1, \ldots, a_n$ are called the *coordinates* of the vector $\phi$ *relative to the basis* $e_1, \ldots, e_n$.

Evidently, when vectors are added, their corresponding coordinates relative to a fixed basis are added, and when a vector is multiplied by any number, all the coordinates are multiplied by that number.

A linear space is called a *scalar product space* (or *Euclidean space*), if in it a function $(\phi, \psi)$ is defined, having the following properties:

$$(\phi, \phi) \geq 0 \quad \text{and} \quad (\phi, \phi) = 0 \text{ iff } \phi = 0. \tag{2.2a}$$

$$(\psi, \phi) = \overline{(\phi, \psi)} \tag{2.2b}$$

(the bar denotes complex conjugate).

$$(\phi, a\psi) = a(\phi, \psi) \tag{2.2c}$$

($a \in C$, the set of complex numbers).

$$(\phi_1 + \phi_2, \psi) = (\phi_1, \psi) + (\phi_2, \psi). \tag{2.2d}$$

This function is called the *scalar product* of the elements $\phi$ and $\psi$. Such a scalar product can be introduced in every finite-dimensional space; for example, if $e_1, \ldots, e_n$ is a basis for $R$ and $\phi = a_1 e_1 + \cdots + a_n e_n$, $\psi = b_1 e_1 + \cdots + b_n e_n$, then, putting

$$(\phi, \psi) = \bar{a}_1 b_1 + \cdots + \bar{a}_n b_n, \tag{2.3}$$

we get a function $(\phi, \psi)$ satisfying the conditions (2.2).

With the scalar product one defines the *norm* $\|\phi\|$ of a vector $\phi \in R$ by

$$\|\phi\| = \sqrt{(\phi, \phi)}. \tag{2.4}$$

A complex-valued function $h(\phi, \psi)$ of two vector arguments is a *Hermitian form* if it satisfies

$$h(\phi, \psi) = \overline{h(\psi, \phi)}, \tag{2.5b}$$

$$h(\phi, a\psi) = ah(\phi, \psi) \qquad (a \in \mathbb{C}), \tag{2.5c}$$

$$h(\phi_1 + \phi_2, \psi) = h(\phi_1, \psi) + h(\phi_2, \psi). \tag{2.5d}$$

If in addition $h$ satisfies

$$h(\phi, \phi) \geq 0 \tag{2.5a}$$

for every vector $\phi$, then $h$ is said to be a *positive Hermitian form*. A positive Hermitian form is called *positive definite* if

$$\text{from} \quad h(\phi, \phi) = 0 \quad \text{follows} \quad \phi = 0 \text{ for every vector } \phi. \tag{2.6}$$

[Since the definition (2.5), (2.6) of a positive definite Hermitian form is the same as that of a scalar product, from now on we shall use only the term scalar product.] Positive Hermitian forms, which are not necessarily scalar products, satisfy the *Cauchy–Schwarz–Bunyakovski inequality*:

$$|h(\phi, \psi)|^2 \leq h(\phi, \phi) h(\psi, \psi). \tag{2.7}$$

If $h$ is positive definite, equality holds iff $\varphi = a\psi$ for some $a \in \mathbb{C}$.

Keeping $\psi$ fixed, $(\phi, \psi) = \psi(\phi)$ is a function or *functional* of $\phi \in R$, and as this functional fulfills the conditions (2.2c), (2.2d), it is called an *antilinear functional*. In general every function $F(\phi)$ that fulfills

$$F(a\phi) = \bar{a}F(\phi), \tag{2.8a}$$

$$F(\phi_1 + \phi_2) = F(\phi_1) + F(\phi_2) \qquad (\phi_1, \phi_2 \in R) \tag{2.8b}$$

is called an antilinear functional on the linear space $R$.

An antilinear functional is called *continuous* iff from the convergence of the infinite sequence of elements $\phi_n \in R$ $(n = 1, 2, 3, \ldots)$ to the element $\phi \in R$, i.e., from

$$\phi_n \to \phi \quad \text{as } n \to \infty, \tag{2.9}$$

it follows that

$$F(\phi_n) \to F(\phi) \quad \text{as } n \to \infty. \tag{2.10}$$

This definition of a continuous functional is of course in complete analogy to the definition of continuous functions of real and complex variables, except that here $\phi_1, \phi_2, \ldots, \phi$ are not numbers but elements of a linear space $R$. If $\phi_1, \phi_2, \ldots, \phi$ are numbers, then the meaning of (2.9) is clear; it means $|\phi_n - \phi| \to 0$ as $n \to \infty$. For elements of a linear space, (2.9) has to be defined; i.e., one has to define what one means by the convergence of infinite sequences of elements of a linear space. A linear space in which the convergence of infinite sequences is defined is called a *linear topological space*. One can give various kinds of definition of convergence in a linear space, i.e., introduce various topologies. For example, in a scalar-product space one can define (2.9) by

$$\|\phi_n - \phi\| \to 0 \quad \text{as } n \to \infty. \tag{2.11}$$

This definition of topology (which is, however, too narrow for the simplest mathematical formulation of quantum mechanics) leads to the well-known Hilbert space.

In a finite-dimensional scalar-product space the scalar product defines an antilinear continuous functional and vice versa; for every antilinear continuous functional $F(\phi)$, there exists an element in $R$, which we shall also call $F$, such that $F(\phi) = (\phi, F)$, i.e., $F(\phi)$ can be written as a scalar product. This is also true for the infinite-dimensional space in which (2.9) is defined by (2.11), i.e., in the Hilbert space. In general, the set of antilinear continuous functionals $F$ on $R$ [i.e., functions $F(\phi)$ that fulfill (2.8)] is larger than the linear topological space $R$. In analogy to the scalar product we shall denote these functions by

$$F(\phi) = \langle \phi | F \rangle \qquad \bar{F}(\phi) = \langle F | \phi \rangle. \tag{2.12}$$

(When we are mathematically lax we shall often call them scalar products.)

A vector $\phi$ is called *normalized* if $\|\phi\| = 1$. Two vectors $\phi, \psi$ are said to be *orthogonal* to each other if

$$(\phi, \psi) = 0.$$

A set of vectors $e_i \in R$ $(i = 1, 2, \ldots)$ is called an *orthonormal system* if

$$(e_i, e_j) = \delta_{ij}. \tag{2.13}$$

Every orthonormal system is linearly independent.

Let $\phi \in R$; then

$$a_i = \langle e_i | \phi \rangle = (e_i, \phi) \tag{2.14}$$

is called the *component* or *coordinate of $\phi$ relative to $e_i$*.

The system $e_1, e_2, \ldots$ is called a *complete orthonormal system* or *basis* of $R$ iff one of the following equivalent statements holds:

From $(e_i, \phi) = 0$ for all $i$, it follows that $\phi = 0$; \hfill (2.15a)

$$\phi = \sum_i e_i(e_i, \phi) \quad \text{for any } \phi \in R \tag{2.15b}$$

or, in a different notation, $\phi = \sum_i |e_i\rangle\langle e_i|\phi\rangle$;

$$(\phi, \psi) = \sum_i (\phi, e_i)(e_i, \psi); \qquad (2.15c)$$

$$\|\phi\|^2 = \sum_i |(\phi, e_i)|^2. \qquad (2.15d)$$

In quantum mechanics one often uses *generalized basis systems*. The elements $e_x$ ($x$ in general any real number) of the generalized basis systems for a space $R$ are not vectors of the space $R$, but are antilinear continuous functions on $R$. For the generalized basis $e_x$, where $x$ is an element of a continuous set of real numbers, the following statements equivalent to (2.15) hold:

From $\quad \langle e_x|\phi\rangle = 0$ for all $x$, it follows that $\quad \phi = 0$; $\quad$ (2.16a)

$$\phi = \int dx\, e_x\langle e_x|\phi\rangle = \int dx|x\rangle\langle x|\phi\rangle; \qquad (2.16b)$$

$$(\phi, \psi) = \int dx\, \langle\phi|e_x\rangle\langle e_x|\psi\rangle = \int dx\, \langle\phi|x\rangle\langle x|\psi\rangle; \quad (2.16c)$$

$$\|\phi\|^2 = \int dx\, \langle\phi|e_x\rangle\langle e_x|\phi\rangle = \int dx\, \langle\phi|x\rangle\langle x|\phi\rangle. \quad (2.16d)$$

## I.3 Linear Operators, Algebras

Let $R$ be a linear space. A function $A$ is called an operator in $R$ if it maps each vector $\phi$ in a linear subspace of $R$ into a vector $\psi = A(\phi)$ of the same space $R$. Often, when there can be no ambiguity, $A\phi$ is written instead of $A(\phi)$. An operator $A$ in $R$ is called *linear* if

$$A(\phi + \psi) = A\phi + A\psi, \qquad A(a\phi) = aA\phi \qquad (3.1)$$

for all $\phi, \psi \in R$ and all numbers $a$. Linear operators in a finite-dimensional space $R$ can be represented by means of matrices. For this purpose we choose a fixed basis $e_1, \ldots, e_n$ for $R$. Every vector $\phi \in R$ (and, in particular, the vectors $Ae_k$) may be expressed as linear combinations of the elements of the basis, according to (2.15):

$$Ae_k = \sum_{j=1}^{n} e_j(e_j, Ae_k) = \sum_{j=1}^{n} e_j a_{jk}. \qquad (3.2)$$

The numbers

$$a_{jk} = (e_j, Ae_k) = \langle e_j|A|e_k\rangle \qquad (3.3)$$

form a matrix of order $n$, which is called the *matrix of the operator $A$ relative to the basis $e_1, \ldots, e_n$*. For $\phi, \chi \in R$ one can form the scalar product of $A\phi$ and $\chi$: $(\chi, A\phi)$. This is called the *matrix element of $A$ between the vectors $\phi$ and $\chi$*.

If $R$ is infinite-dimensional, i.e., $n = \infty$, the matrix of the operator $A$ is an infinite matrix. However, in the infinite-dimensional case it may happen that $A\phi$ for $\phi \in R$ does not have any meaning [because, e.g., $\|A\phi\|^2$, which according to (2.15) and (3.2) would have to be $\sum_{j=1}^{\infty} (A\phi, e_j)(e_j, A\phi)$, may not exist, as the infinite sum may not converge]. We shall always assume that this never happens, i.e., in the cases we consider $A$ is defined for every $\phi \in R$ and $A\phi$ is again an element of $R$. For quantum mechanics only operators with this property seem to be important (however, not if $R$ is supposed to be the Hilbert space).

It is possible to define the operations of addition of operators, multiplication of an operator by a number, and multiplication of operators in $R$. In fact, by $A + B$, $aB$, $AB$ we shall understand the operators defined by the formulae

$$(A + B)\phi = A\phi + B\phi, \qquad (aA)\phi = a(A\phi), \qquad (AB)\phi = A(B\phi), \quad (3.4)$$

for all $\phi \in R$. It is easily verified that $A + B$, $aA$, $AB$ are linear operators if $A, B$ are linear operators. In addition, it is easy to verify that the operations of addition, multiplication by a number, and multiplication together of operators correspond to the addition, multiplication by a number, and multiplication of their matrices relative to a fixed basis.

Particular operators are the *zero operator*, which we again denote 0 and which maps every $\phi \in R$ into the zero vector $0$, $0\phi = 0$; and the *unit operator* $I$, which is defined by $I\phi = \phi$ for every $\phi \in R$.

For every linear operator $A$ defined for all $\phi \in R$, one can define an operator $A^\dagger$ by $(A^\dagger\phi, \psi) = (\phi, A\psi)$ for every $\phi, \psi \in R$. The operator $A^\dagger$ is called the *adjoint operator* of $A$. An operator for which $A^\dagger = A$ is called *self-adjoint* or Hermitian.[1]

If there exists a nonzero vector $\phi$ such that $A\phi = \lambda\phi$ ($\lambda \in \mathbb{C}$), then $\phi$ is called an *eigenvector* of $A$. The number $\lambda$ is called an *eigenvalue* of $A$.

A set $\mathscr{A}$ is an (*associative*) *algebra with unit element* iff

(a)  $\mathscr{A}$ is a linear space.
(b)  For every pair $A, B \in \mathscr{A}$, there is defined a product $AB \in \mathscr{A}$ such that

$$(AB)C = A(BC),$$
$$A(B + C) = AB + AC,$$
$$(A + B)C = AC + BC,$$
$$(aA)B = A(aB) = aAB.$$

(c)  There exists an element $I \in \mathscr{A}$ such that

$$IA = AI = A$$

for all $A \in \mathscr{A}$.

A subset $\mathscr{A}_1$ of an algebra is called a *subalgebra of $\mathscr{A}$* if $\mathscr{A}_1$ is an algebra with the same definitions of the operations of addition, multiplication by a

---

[1] We will usually use the term Hermitian if we do not want to distinguish between the mathematically precisely defined notions *self-adjoint*, *essentially self-adjoint*, and *symmetric*.

number, and multiplication as given for $\mathscr{A}$; i.e., if from $A, B \in \mathscr{A}_1$, it follows that $A + B \in \mathscr{A}_1, aA \in \mathscr{A}_1$, and $A \cdot B \in \mathscr{A}_1$.

An algebra $\mathscr{A}$ is called a *-algebra if we have on the algebra a *-operation (involution) $A \mapsto A^\dagger$ that has the following defining properties:

(d)
$$(aA + bB)^\dagger = \bar{a}A^\dagger + \bar{b}B^\dagger,$$
$$(AB)^\dagger = B^\dagger A^\dagger,$$
$$(A^\dagger)^\dagger = A,$$
$$I^\dagger = I,$$

where $A, B \in \mathscr{A}$ and $a, b \in \mathbb{C}$. From the definitions of the sum and product of operators with a number—given in (3.4)—and from the definition of the formal adjoint operator, one can see that the set of linear operators fulfill all the axioms (a), (b), (c), and (d) of a *-algebra. Thus the set of linear operators in a linear space forms a *-algebra. A *-subalgebra of this algebra is called an *operator *-algebra*. It can be shown that in a certain sense every *-algebra can be realized as an operator *-algebra in a scalar-product space (generalization of the Gelfand–Naimark–Segal reconstruction theorem). In quantum mechanics physical systems are assumed to be described by operator *-algebras.

A set $X_1, X_2, \ldots, X_n$ of elements of $\mathscr{A}$ is called a *set of generators*, and $\mathscr{A}$ is said to be *generated* by the $X_i$ ($i = 1, 2, \ldots, n$) iff each element of $\mathscr{A}$ can be written

$$A = c + \sum_{i=1}^{n} c^i X_i + \sum_{i,j}^{n} c^{ij} X_i X_j + \cdots, \tag{3.5}$$

where $c, c^i, c^{ij}, \ldots \in \mathbb{C}$.

*Defining algebraic relations* are relations among the generators

$$P(X_i) = 0, \tag{3.6}$$

where $P(X_i)$ is a polynomial with complex coefficients of the $n$ variables $X_i$. An element $B \in \mathscr{A}$,

$$B = b + \sum b^i X_i + \sum b^{ij} X_i X_j + \cdots, \tag{3.7}$$

where $b, b^i, \ldots \in \mathbb{C}$, is *equal* to the element $A$ iff (3.7) can be brought into the same form (3.5) with the same coefficients $c, c^i, c^{ij}, \ldots$ by the use of the defining relations (3.6).

# CHAPTER II

# Foundations of Quantum Mechanics—The Harmonic Oscillator

This chapter, the longest in the book, introduces three of the basic assumptions of quantum mechanics and then illustrates them, using mainly the example of the harmonic oscillator. Though some historic remarks are included, neither the historic development nor any other heuristic way towards quantum mechanics is followed. The basic assumptions are formulated, explained, and applied. In Sections II.2, II.4, II.5, the basic assumption are introduced; in Sections II.3, II.8, II.9, the harmonic oscillator is used to illustrate them. Section II.7 contains the derivations of some general consequences and might be omitted in first reading. The discussion for the continuous spectra, important for the description of the scattering and decay phenomena in the second part of the book, is given in Section II.10. Several remarks throughout this chapter emphasize the particular problems connected with generalized eigenvalues and eigenvectors and our unified treatment of continuous and discrete spectra. In Section II.11 we are ready to explain the physical meaning of the quantum-mechanical constant of nature, $\hbar$.

## II.1 Introduction

Physicists believe that there is something in nature, or in each restricted domain of it, that may be "understood"; that there is a structure in nature. To "understand" means to bring this structure into congruence with some structure in our mind, with a structure of thought objects, with a structure that has been created by our minds. For physics this structure of thought objects is a mathematical structure. So to understand part of physical nature means to map its structure on a mathematical structure. To obtain a physical theory, then, means to obtain a mathematical image of a physical system, by which we mean any suitably restricted domain of physical nature (e.g., just an atom, or even just a set of substates of it, or all atoms and molecules).

For the domains of quantum physics the mathematical structures are algebras of linear operators in linear spaces. The discovery of this, the fundamental properties of the algebra, and the other basic assumptions of quantum mechanics was a very difficult process, the history of which we do not want to describe here. It is much easier to start with the basic assumptions and to show that conclusions derived from them really describe some parts of nature. This is the path we want to follow here.

## II.2  The First Basic Assumption of Quantum Mechanics

The first basic assumption ("axiom") of Quantum Mechanics is:

**I.**  A physical observable, defined by the prescription for its measurement, is represented by a linear operator[1] in a linear space. The mathematical image of a physical system is an operator *-algebra in a linear scalar-product space $\mathscr{H}$. The algebra is generated by some basic physical quantities, and the multiplication is defined by algebraic relations.

To illustrate the axiom I, we use one of the simplest examples, the one-dimensional harmonic oscillator. The one-dimensional harmonic oscillator can be defined as the physical system whose mathematical image is the algebra of operators that is generated by the following Hermitian operators:

$H$, representing the observable energy.
$P$, representing the observable momentum.
$Q$, representing the observable position.

The defining algebraic relations that are fulfilled by these operators are

$$[P, Q] \equiv PQ - QP = \frac{\hbar}{i} I, \tag{2.1a}$$

$$H = \frac{P^2}{2\mu} + \frac{\mu\omega^2}{2} Q^2, \tag{2.1b}$$

where $\hbar$, $\mu$, and $\omega$ are numbers. $I$ is the unit operator, and $i = \sqrt{-1}$. The physical meaning of the numbers $\mu$ and $\omega$ will be found by correspondence to the classical system to be the mass and frequency of the oscillator, which are system constants. The physical meaning of the universal constant[2] $\hbar$ will be discussed in Section II.11.

---

[1] In Chapter XIX we shall also have to admit semilinear operators.
[2] $2\pi\hbar = h$ is called Planck's constant.

Figure 2.1    Model of a one-dimensional oscillator.

The justification for the above statements is that there exist in nature physical systems on which one can perform measurements that lead to numbers that can be calculated from the above assumptions. Therefore it is reasonable to say that these physical systems, which are as always "idealized physical systems," have as their mathematical image the above mathematical structure. The name "harmonic oscillator" for this physical system arises from the correspondence to a classical system that has the same name.

The classical-mechanical system called a one-dimensional harmonic oscillator has the following structure (Figure 2.1). Two mass points, with masses $m_1$ and $m_2$, are acted upon by a force $F$ along the line joining the two points. The force $F$ is proportional to the difference between the distance $r$ and its equilibrium length $r_0$:

$$F = -k(r - r_0) = -kx, \tag{2.2}$$

where $x = r - r_0$. The potential energy for this system is then

$$V = +\tfrac{1}{2}kx^2, \tag{2.3}$$

and the kinetic energy is

$$E_{\text{kin}} = \frac{1}{2\mu} p^2, \tag{2.4}$$

where we have introduced the reduced mass

$$\mu \equiv \frac{m_1 m_2}{m_1 + m_2} \tag{2.5}$$

and the momentum

$$p = \mu \frac{dx}{dt}. \tag{2.6}$$

We can therefore define a classical harmonic oscillator as a system whose potential energy is proportional to the square of the distance from its equilibrium position, or as a system whose total energy is

$$E = \frac{1}{2\mu} p^2 + \tfrac{1}{2}kx^2. \tag{2.7}$$

As is well known, such a physical system is an idealization of the mechanical system in nature that consists of two masses connected by a spring (Figure 2.2). Such a mechanical system performs oscillations with a frequency

$$\omega = \sqrt{\frac{k}{\mu}}. \tag{2.8}$$

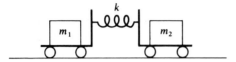

Figure 2.2   Classical one-dimensional oscillator.

If we write Equation (2.7) in terms of $\omega$ by solving Equation (2.8) for $k$, we have

$$E = \frac{1}{2\mu} p^2 + \frac{\mu\omega^2}{2} x^2. \tag{2.9}$$

This expression looks exactly like one of the relations defining the quantum-mechanical harmonic oscillator. The important difference is that in the classical case the energy $E$, the momentum $p$, and the position $x$ are real numbers, whereas in the quantum-mechanical case these quantities are represented by the operators $H$, $P$, and $Q$, respectively.

For real numbers $a$ and $b$ one always has the relation $ab - ba = 0$; thus

$$px - xp = 0. \tag{2.10}$$

Generally, for operators $A$ and $B$, $AB - BA \neq 0$, i.e., operators do not, in general, commute. For the particular case of the momentum and position operators one has the relation (2.1a).

Fundamental relations like (2.1a) and (2.1b) cannot be derived; they can only be conjectured. This is true not only for quantum mechanics but also for classical physics. Thus the relation (2.9) is essentially the result of conjectures that go back to Newton.

The process of conjecture in physical theory can be described simply as follows: First one considers the experimental data in a certain domain of physics, e.g., the atomic domain. Then one tries to construct a mathematical structure from which one can calculate numbers agreeing with the experimental data. The algebra of operators defined by the relations (2.1) is such a mathematical structure for the quantum-mechanical harmonic oscillator. The conjecture (2.1b) was simple for us, since it was obtained by correspondence with the classical case. The physical meaning of the system constants $\mu$ and $\omega$ follows from the classical analogue. The conjecture (2.1a), however, was not so simple. The replacement of the classical relation (2.10) by the quantum-mechanical relation (2.1a), and thus the realization that the quantum-mechanical observables are not represented by real numbers but by operators, was one of the greatest achievements of science. The relation (2.1a) is called the Heisenberg (or canonical) commutation relation. The physical meaning of this relation will become clear when we study its consequences in the following sections.

Since we do not want to develop quantum mechanics along its historical path we shall give only a very brief and partial sketch of the motivation for (2.1a). It was known from experiment that light was emitted from atoms at discrete frequencies. At the time the developments leading to 2.1a took place, the atom was pictured as consisting of a positively charged nucleus with a very small radius and negatively charged electrons revolving around it at a distance of about $10^{-8}$ cm. Describing this picture in terms of classical physics lead to a dilemma. The revolving and therefore accelerating electrons would emit radiation, leading to a continuous loss of its energy and a continuous change in the frequency of the emitted radiation until the electron would finally, and very rapidly, fall into the nucleus. This contradicted the experimental situation: the atoms are quite stable, and the frequency of the emitted radiation can only take certain definite values (spectral lines), which are characteristic of the atom. Hence classical theory could not decribe the observations in the atomic domain, if the above picture of the atom was kept. One did not wish to abandon the picture of a miniature planetary system for the atoms, since replacing the Hamiltonian by another classical Hamiltonian that would take only discrete values would be virtually impossible. To obtain only discrete values for the Hamiltonian, some extra conditions were imposed on the classical theory that were not only supplementary to but also in contradiction with the classical theory. This was the so-called "old quantum theory" (Bohr).

Quantum mechanics originated in the idea of Werner Heisenberg and Max Born (1925) and Erwin Schrödinger (1926) to interpret the Hamiltonian as a mathematical quantity that would take only discrete values. From an analysis of the harmonic oscillator, Heisenberg and Born and Jordan concluded that in order for the energy to be a system of discrete values, the momentum and position should be given by the matrices[3]

$$P = \frac{\hbar}{i\sqrt{2}} \begin{bmatrix} 0 & \sqrt{1} & 0 & 0 & \cdots \\ -\sqrt{1} & 0 & \sqrt{2} & 0 & \cdots \\ 0 & -\sqrt{2} & 0 & \sqrt{3} & \cdots \\ 0 & 0 & -\sqrt{3} & 0 & \cdots \\ \vdots & \vdots & \vdots & \vdots & \ddots \end{bmatrix}, \quad (2.11)$$

$$Q = \frac{1}{\sqrt{2}} \begin{bmatrix} 0 & \sqrt{1} & 0 & 0 & \cdots \\ \sqrt{1} & 0 & \sqrt{2} & 0 & \cdots \\ 0 & \sqrt{2} & 0 & \sqrt{3} & \cdots \\ 0 & 0 & \sqrt{3} & 0 & \cdots \\ \vdots & \vdots & \vdots & \vdots & \ddots \end{bmatrix}, \quad (2.12)$$

[3] This particular representation of the $P$ and $Q$ as matrices of operators with respect to a certain basis of vectors in an inner-product space will be derived in the next section (Section II.3).

where $2\pi h = h$ is Planck's constant. Schrödinger built on a partial theory due to de Broglie, according to which free particles should diffract like light and should therefore be described by a wave function that satisfies a wave equation. From this he concluded that the momentum should be the differential operator

$$P = \frac{\hbar}{i} \frac{\partial}{\partial x} \qquad (2.13)$$

applied to the wave function. Both the Heisenberg and the Schrödinger *Ansatz* lead to the relation (2.1a); and soon after the two formulations were made known, it was shown that they were two different realizations of the same mathematical entities, namely linear operators in a linear space. We shall give a brief description of subsequent historical developments in Section II.6.[4]

The quantum-mechanical harmonic oscillator is an idealization of micro-physical systems in nature; e.g., diatomic molecules are, under certain conditions, quantum-mechanical harmonic oscillators. The simplest possible geometrical (classical) picture of the diatomic molecule is two atoms, of masses $m_1$ and $m_2$, bound together by an elastic force.

There is an essential difference between the descriptions, in everyday language, of the physical systems whose idealizations are the classical harmonic oscillator and the quantum-mechanical harmonic oscillator. Everyday language gives quite an adequate description of classical systems, e.g., two carts with masses connected by a spring. Everyday language is, however, insufficient to describe quantum-mechanical systems. The quantum-mechanical harmonic oscillator is best described by the mathematical structure given by (2.1a, b). This is the justification for the above definition and the conjectures that led to it.

The procedure of conjecturing the mathematical relations outlined above for the harmonic oscillator works for many microphysical systems, so we summarize it again:

1. Obtain a classical picture of the microphysical system, and express the observables as functions of position and momentum coordinates.
2. Replace the real numbers corresponding to momentum $p$ and position $x$ in the expressions for other classical observables by the Hermitian operators $P$ and $Q$, respectively. The operators $P$ and $Q$ satisfy the Heisenberg commutation relation, and from them we obtain the algebra of observables for a quantum-mechanical system.

Not all classical observables can be expressed in terms of momentum and position: the intrinsic angular momentum or spin of a top, for example, cannot be expressed as a function of the momentum and position of the top. For these observables the above procedure of correspondence between the

---

[4] A few historical remarks will appear throughout the book. For a description of the historical developments see, e.g., Jammer (1966).

classical and the quantum-mechanical observables has to be generalized. We can give a vague formulation of this conjecturing procedure: in the transition from the mathematical description of the classical system to the mathematical description of the corresponding quantum system, we replace the numbers representing classical observables by operators that represent the corresponding quantum observables. The mathematical property of these operators is specified by the mathematical (in particular algebraic) relations between them. These relations between the observables are obtained from the relations between the classical observables and some additional relations that specify the property of noncommutativity. These relations can often be obtained from analogy to known relations between similar quantities.

There exist, (in elementary-particle physics for example) many quantum-mechanical observables without known classical counterparts. The defining relations between these observables can then be obtained only phenomeno-logically from a comparison between the consequences of these relations and the experimental data.

We shall discuss many examples of this conjecturing procedure. Rarely is the choice of the relations unambiguous, and the ultimate justification of the defining relations, like that of any fundamental theoretical assumption, is their success in providing agreement between theoretical consequences and experimental data.

The representation of physical observables by operators poses the problem of relating these mathematical quantities to experimental data, which are numbers. The solution of this problem is given by a number of further basic assumptions or axioms, which provide the actual *interpretation* of quantum mechanics. We shall introduce these axioms at suitable places when they are needed.

## II.3  Algebra of the Harmonic Oscillator

We now give a mathematical description of the quantum-mechanical harmonic oscillator. To find the mathematical properties of the algebra of the harmonic oscillator means to find how the operators of this algebra act on the vectors of the space $\mathcal{H}$. In order to do this we introduce a new operator $a$ and the operator $a^{\dagger}$ (which is its adjoint):

$$a = \frac{1}{\sqrt{2}}\left(\sqrt{\frac{\mu\omega}{\hbar}}\,Q + \frac{i}{\sqrt{\mu\omega\hbar}}\,P\right),$$

$$a^{\dagger} = \frac{1}{\sqrt{2}}\left(\sqrt{\frac{\mu\omega}{\hbar}}\,Q - \frac{i}{\sqrt{\mu\omega\hbar}}\,P\right).$$

(3.1)

Then we introduce the new operator

$$N = a^{\dagger}a,$$

(3.2)

which is Hermitian. Inserting (3.1) into (3.2) we get

$$N = \frac{\mu\omega}{2\hbar} Q^2 + \frac{1}{2\mu\omega\hbar} P^2 - \frac{i}{2\hbar} (PQ - QP). \tag{3.3}$$

Using the relations (2.1) we obtain

$$N = \frac{I}{\omega\hbar} H - \tfrac{1}{2}I. \tag{3.4}$$

Similarly we calculate

$$aa^\dagger = \frac{I}{\omega\hbar} H + \tfrac{1}{2}I. \tag{3.5}$$

Therefore as a consequence of the canonical commutation relation (2.1a) we obtain for $a$ and $a^\dagger$ the commutation relation

$$[a, a^\dagger] \equiv aa^\dagger - a^\dagger a = I. \tag{3.6}$$

Furthermore, we can express the energy operator $H$ in terms of $a$ and $a^\dagger$ to obtain

$$H = \hbar\omega a^\dagger a + \frac{\hbar\omega}{2} I = \hbar\omega(N + \tfrac{1}{2}I). \tag{3.7}$$

Now we want to find the eigenvalues and eigenvectors of $H$ and $N$. As the eigenvectors of $N$ are eigenvectors of $H$, it is sufficient to find all the eigenvectors of $N$.

Let us assume that there is at least one eigenvector of $N$ in $\mathscr{H}$, and call its eigenvalue $\lambda$. (This is not a trivial assumption, since there are many operators that do not have eigenvectors in $\mathscr{H}$. In fact it is an additional assumption without which equation 2.1a is ambiguous.) Let $\varphi_\lambda$ be this eigenvector; thus

$$N\varphi_\lambda = \lambda\varphi_\lambda. \tag{3.8}$$

Then calculate $Na\varphi_\lambda$ using Equation (3.6),

$$Na\varphi_\lambda = (a^\dagger a)a\varphi_\lambda = (aa^\dagger - I)a\varphi = a(a^\dagger a - I)\varphi_\lambda$$
$$= a(N - I)\varphi_\lambda = a(\lambda - 1)\varphi_\lambda = (\lambda - 1)a\varphi_\lambda.$$

Thus

$$N(a\varphi_\lambda) = (\lambda - 1)(a\varphi_\lambda), \tag{3.9}$$

so that either $a\varphi_\lambda = 0$ or $a\varphi_\lambda$ is an eigenvector of $N$ with eigenvalue $\lambda - 1$. Suppose the latter is the case. Then we can define

$$\varphi_{\lambda-1} \equiv a\varphi_\lambda. \tag{3.10}$$

Repeating the above calculation on $\varphi_{\lambda-1}$, we find that

$$\varphi_{\lambda-2} \equiv aa\varphi_\lambda \tag{3.11}$$

is either zero or an eigenvector with eigenvalue $\lambda - 2$. We continue in this way and obtain a sequence of vectors

$$\varphi_{\lambda-m} \equiv a^m\varphi_\lambda \quad (m = 0, 1, 2, \ldots), \tag{3.12}$$

which are eigenvectors of $N$ with eigenvalue $(\lambda - m)$, as long as $\varphi_{\lambda-m} \neq 0$.

In a similar way one calculates $N(a^\dagger \varphi_\lambda)$ to get

$$N(a^\dagger \varphi_\lambda) = (\lambda + 1)(a^\dagger \varphi_\lambda). \tag{3.13}$$

Hence $\varphi_{\lambda+1} \equiv a^\dagger \varphi_\lambda$ is either the zero vector or an eigenvector of $N$ with eigenvalue $\lambda + 1$. It is easily shown that always $\varphi_{\lambda+1} \neq 0$. Suppose $a^\dagger \varphi_\lambda = 0$. That means

$$\|a^\dagger \varphi_\lambda\| = 0. \tag{3.14}$$

But

$$\|a^\dagger \varphi_\lambda\|^2 = (a^\dagger \varphi_\lambda, a^\dagger \varphi_\lambda), \tag{3.15}$$

and by definition,

$$(a^\dagger \varphi_\lambda, a^\dagger \varphi_\lambda) = (\varphi_\lambda, a a^\dagger \varphi_\lambda). \tag{3.16}$$

By the commutation relation for $a$ and $a^\dagger$ we have

$$\begin{aligned}(\varphi_\lambda, a a^\dagger \varphi_\lambda) &= (\varphi_\lambda, a^\dagger a \varphi_\lambda) + (\varphi_\lambda, 1\varphi_\lambda) \\ &= \|a\varphi_\lambda\|^2 + \|\varphi_\lambda\|^2 \neq 0,\end{aligned} \tag{3.17}$$

since $\varphi_\lambda \neq 0$. Thus $\varphi_{\lambda+1}$ is always an eigenvector of $N$ with eigenvalue $\lambda + 1$. Again by repeating the above process one obtains a sequence of vectors $\varphi_{\lambda+n}$ ($m = 0, 1, 2, \ldots$), which are eigenvectors of $N$ with eigenvalue $\lambda + n$. These are always nonzero, as follows from the above proof.

We now determine when $\varphi_{\lambda-m}$ can be zero. We calculate

$$(\varphi_{\lambda-m}, N\varphi_{\lambda-m}) = (\lambda - m)(\varphi_{\lambda-m}, \varphi_{\lambda-m}) = (\lambda - m)\|\varphi_{\lambda-m}\|^2$$

and

$$(\varphi_{\lambda-m}, N\varphi_{\lambda-m}) = (\varphi_{\lambda-m}, a^\dagger a \varphi_{\lambda-m}) = \|a\varphi_{\lambda-m}\|^2.$$

Since the norm is always nonnegative, it follows that

$$\lambda - m = \frac{\|a\varphi_{\lambda-m}\|^2}{\|\varphi_{\lambda-m}\|^2} \geq 0. \tag{3.18}$$

Therefore the sequence of eigenvectors $\varphi_{\lambda-m}$ must terminate after a finite number of steps, and there must exist one vector $\varphi_0$ such that

$$a\varphi_0 = 0.$$

The vector $\varphi_0$ is an eigenvector of $N$ with eigenvalue zero, since $N\varphi_0 = a^\dagger a \varphi_0 = a^\dagger 0 = 0$. Now we define the normalized vectors

$$\phi_0 = \frac{\varphi_0}{\|\varphi_0\|},$$

$$\phi_1 = C_1 a^\dagger \phi_0, \tag{3.19}$$

$$\vdots$$

$$\phi_n = C_n (a^\dagger)^n \phi_0$$

$$\vdots$$

From the considerations above we know that the vectors $\phi_n$ are eigenvectors of $N$, with eigenvalue $n$, and the $C_n$ are chosen such that the $\phi_n$ are normalized. Thus we have

$$N\phi_n = n\phi_n, \qquad \|\phi_n\| = 1. \tag{3.20}$$

The $C_n$ are calculated as follows: From Equations (3.20) and (3.19),

$$1 = \|\phi_n\|^2 = (a^{\dagger n}\phi_0, a^{\dagger n}\phi_0)|C_n|^2,$$

but

$$(a^{\dagger n}\phi_0, a^{\dagger n}\phi_0) = \frac{1}{|C_{n-1}|^2}(a^{\dagger}\phi_{n-1}, a^{\dagger}\phi_{n-1}),$$

so

$$1 = \frac{|C_n|^2}{|C_{n-1}|^2}(\phi_{n-1}, aa^{\dagger}\phi_{n-1}).$$

Using Equation (3.6),

$$1 = \frac{|C_n|^2}{|C_{n-1}|^2}(\phi_{n-1}, (a^{\dagger}a + 1)\phi_{n-1})$$

$$= \frac{|C_n|^2}{|C_{n-1}|^2}(\phi_{n-1}, (N + 1)\phi_{n-1})$$

$$= n\frac{|C_n|^2}{|C_{n-1}|^2}(\phi_{n-1}, \phi_{n-1})$$

$$= n\frac{|C_n|^2}{|C_{n-1}|^2};$$

hence $C_n$ must be chosen so that

$$n|C_n|^2 = |C_{n-1}|^2. \tag{3.21}$$

Since $\phi_0$ is normalized, $C_0 = 1$, and one solution of Equation (3.21) is then

$$C_n = \sqrt{\frac{1}{n!}}. \tag{3.22}$$

There are other solutions of (3.21) which differ from (3.22) by a factor of modulus 1; we choose (3.22).

We may summarize the procedure as follows: Start with a normalized vector $\phi_0 \in \mathcal{H}$ that has the following property:

$$a\phi_0 = 0. \tag{3.23}$$

Then apply to $\phi_0$ the operator $a^{\dagger}$ to obtain a system of eigenvectors of $N$,

$$\phi_n = \frac{1}{\sqrt{n!}}a^{\dagger n}\phi_0, \tag{3.24}$$

with eigenvalues $n = 0, 1, 2, \ldots$. These eigenvectors are normalized by the construction above, and they are orthogonal to each other since they are eigenvectors of a Hermitian operator with different eigenvalue (cf. Problem). Thus we have the orthonormality relations

$$(\phi_n, \phi_{n'}) = \delta_{nn'}, \tag{3.25}$$

and the $\phi_n$ form an orthonormal system in $\mathcal{H}$. The operators $a$ and $a^\dagger$ are defined on this orthonormal system by

$$\begin{aligned} a\phi_n &= \sqrt{n}\,\phi_{n-1}, \\ a^\dagger\phi_n &= \sqrt{n+1}\,\phi_{n+1}. \end{aligned} \tag{3.26}$$

Because of this property, $a$ is called the annihilation operator and $a^\dagger$ creation operator. Both are often refered to as Ladder operators. All the elements of the algebra of the harmonic oscillator that are functions of $a$ and $a^\dagger$ are defined on the elements of this orthonormal system. Consider the set of all vectors

$$\psi = \sum_n \alpha_n \phi_n, \tag{3.27}$$

where the $\alpha_n$ are complex numbers and the sum runs through an arbitrarily large but finite number $n$. This set of vectors forms a linear space called the space spanned by $\phi_n$. We will denote this space by $\mathcal{H}$ or $\Phi$ when we ignore details.

⟦More precisely, the space of all the $\psi = \sum_{n=0}^{\infty} \alpha_n \phi_n$ for which $\sum_{n=0}^{\infty} |\alpha_n|^2 < \infty$ will be denoted by $\mathcal{H}$ (for Hilbert space), and the space of all $\psi$ for which $\sum_{n=0}^{\infty} |\alpha_n|^2 (n+1)^p < \infty$ for all $p = 0, 1, 2, 3, \ldots$ will be denoted by $\Phi$. Obviously

$$\Phi \subset \mathcal{H}. \tag{3.28}$$

The advantage of $\Phi$ over $\mathcal{H}$ is that all operators representing observables can be defined on the whole space $\Phi$, but not on all of $\mathcal{H}$. Though both spaces $\Phi$ and $\mathcal{H}$ are mathematical generalizations, $\Phi$ is more convenient for physics. ($\Phi$ is called *Schwartz space*.⟧

As all observables are functions of $a$ and $a^\dagger$, and $a$ and $a^\dagger$ are known by (3.26) and (3.27) for all $\psi$, all observables for the quantum-mechanical harmonic oscillator are in principle known on $\Phi$. So the task of determining the mathematical properties of the algebra of the harmonic oscillator is in principle completed.

Let us calculate the diagonal matrix element of the energy operator $H$ between the vector $\phi_n$ for a fixed value of $n$. Using Equations (3.7) and (3.20), we have

$$H\phi_n = \hbar\omega(N + \tfrac{1}{2})\phi_n = \hbar\omega(n + \tfrac{1}{2})\phi_n \tag{3.29}$$

and

$$(\phi_n, H\phi_n) = \hbar\omega(\phi_n, N\phi_n) + \hbar\omega(\phi_n, \tfrac{1}{2}\phi_n)$$
$$= \hbar\omega(n + \tfrac{1}{2})|\phi_n|^2 = \hbar\omega(n + \tfrac{1}{2}).$$

We call this number $E_n$:

$$E_n = \hbar\omega(n + \tfrac{1}{2}). \tag{3.30}$$

## II.4 The Relation between Experimental Data and Quantum-Mechanical Observables

We now wish to formulate the basic axiom of quantum mechanics that gives the connection between the operator representing an observable and the experimentally measured values for this observable. Before doing this, we observe some fundamental differences in the experimental properties of the classical harmonic oscillator and he quantum-mechanical oscillator.

Consider the classical picture in Fig. 2.1. Assume that the system has no energy—i.e., does not perform any vibrations, or vibrates with zero amplitude. The system can be excited into vibration if one shoots a structureless, massive projectile into one of the masses. The oscillator can perform vibrations of any amplitude, depending on the amount of energy transferred to it, which in turn depends on the momentum of the projectile. Thus for the classical oscillator one can prepare a projectile that excites the oscillator to vibrate with any desired energy $E$. The energy is given by

$$E = \tfrac{1}{2}ka^2, \tag{4.1}$$

where $a$ is the amplitude of the vibrations

$$x(t) = a\sin(\omega t + \theta_0).$$

In the realization of the classical harmonic oscillator as two massive carts (see Figure 2.2), the system may be excited, for example, by hitting one of the carts with a third cart of suitable momentum.

The situation is quite different for the quantum-mechanical harmonic oscillator. To see this we will consider an analogous experiment with microphysical systems. As is typical for experiments with microphysical systems, one does not have a single physical system but an *ensemble* of systems, i.e., a collection of systems that are in a certain sense identical. The ensemble of microphysical harmonic oscillators that is used in this experiment is a gas of CO molecules, and an electron beam serves as an ensemble of structureless particles.

Remark: It should be remembered that the statement that a physical system is structureless has no universal meaning. Any system can be considered structureless under certain conditions but not under others. For instance, in the domain of physics for which the kinetic

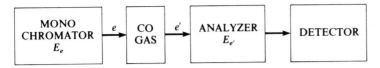

Figure 4.1   Schematic diagram of an energy-loss experiment.

theory of gases is valid, the molecules can be considered as structure-less systems, whereas when higher energies are considered, they are seen to have structure. The electron is a system that can be considered to be without "energy structure" for all currently available energies.

Experiments of this type are called *energy-loss experiments*, and are based on the original experiment performed by Franck and Hertz in 1914.

Figure 4.1 shows a schematic diagram of such experiments. A beam of electrons leaves a monochromator[5] with a very narrow energy band of energy $E_e$, and enters a collision chamber filled with an ensemble of CO molecules (or any other physical system whose structure is to be investigated.) After the electrons have passed through the CO gas, they go through an analyzer, which focuses those electrons of a similar narrow energy band $E_{e'}$ onto the detector. (The energy resolution is typically 0.005 to 0.05 eV; in particular experiment described in Figure 4.2a, it is 0.06 eV.) The energy $E_{e'}$ that is selected by the analyzer can be varied, so that one can measure the intensity $I$ (the electron current at the detector) of the electrons as a function of the energy loss $E = E_e - E_{e'}$.

The result of an actual experiment, performed with the apparatus schematically depicted in Figure 4.2a on the diatomic molecule CO, is shown in Figure 4.2b.[6] What we see is that for an analyzer setting of $E_0 = E_e - E_{e'} = 0$, a major fraction of the electron current $e$ does not lose any energy (elastically scattered electrons), as is indicated by the high value on the graph at $v = 0$. Then there is a relative maximum of intensity for the electrons that have lost the energy $E_1$ ($\approx 0.26$ eV), which means that a large portion of the electrons of energy $E_e$ lose energy inelastically to the CO molecules. A second bump in electron intensity occurs at the energy loss $E_2 = 2E_1$, and so on. As a whole, Figure 4.2 shows us that the scattered electron current $e'$ is a mixture of electrons that have lost one of the eight discrete amounts of energy $E_0, E_1, E_2, \ldots, E_7$ at $v = 0$, $v = 1, \ldots, v = 7$, respectively, to the CO molecules.

We conclude from this experimental situation: The vibrating molecule cannot be excited to any arbitrary energy value. Only a discrete number of

---

[5] A monochromator is an apparatus that prepares a beam with very well-defined momentum and, therefore, with very well-defined kinetic energy. An analyzer is the same kind of apparatus used for a different purpose.

[6] From G. J. Schulz, *Phys. Rev.* **135**, A988 (1964), with permission.

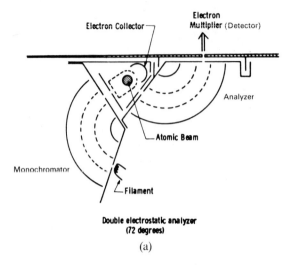

Double electrostatic analyzer
(72 degrees)

(a)

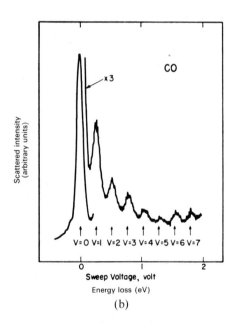

(b)

Figure 4.2   (a) Schematic diagram of double electrostatic analyzer. Electrons are emitted from the thoria-coated iridium filament. They pass between the cylindrical grids at an energy about 2.05 eV, and are accelerated into the collision chamber, where they are crossed with a molecular beam. Those electrons scattered into the acceptance angle of the second electrostatic analyzer pass between the cylindrical grids, again at an energy from 0 to $\approx 2$ eV. The electrons pass the exit slit into the second chamber and impinge onto an electron multiplier. (b) Energy spectrum of scattered electrons in CO at an incident electron energy of 2.05 eV.

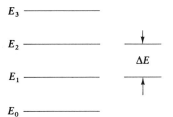

Figure 4.3   Energy-level diagram of the harmonic oscillator.

energy values are possible; i.e., the physical system has a discrete number of energy levels. This situation is represented by an energy-level diagram (Figure 4.3). In the particular case of the vibrating molecule, the energy-level diagram consists of a series of equidistant levels, as can be seen from the experimental results in Figure 4.2b.

> From Figure 4.2b, one sees that this is only approximately true and that the distance between energy levels actually decreases with increasing energy. This discrepancy indicates the limitation of the harmonic oscillator as the model for the vibrating molecule. We will discuss the corrections to this model later.

Comparing this experimental energy spectrum with the result in Equation (3.29), we see that they agree if we take $\Delta E = \hbar\omega$ for the energy difference between two neighboring energy levels. Thus the diagonal matrix elements $(\phi_n, H\phi_n)$ give the possible energy values of the quantum-mechanical oscillator (e.g., vibrating molecule). We interpret the situation as follows: The projectile (electron) excites the harmonic oscillator (vibrating molecule) into any one of a discrete number of *states* described by $\phi_n$. If no excitation takes place, the harmonic oscillator is in the ground state, described by $\phi_0$. If all the electrons that passed through the gas had lost the same amount of energy, i.e., if there were only one bump in the experimental curve at $v = n_0$, the ensemble of molecules in the collision chamber would be described by one of the $\phi_n$, say $\phi_{n_0}$. In this case the quantum-mechanical system would be said to be in a *pure state* $\phi_{n_0}$. This is often expressed by saying that "all molecules have the energy $E_{n_0}$." The results in Figure 4.2b show that the ensemble of molecules considered (i.e., those molecules that have collided with the electrons that are scattered into the acceptance angle of the analyzer, which in the experiment of Figure 4.2a is about $72°$ from the electron beam axis) is not in a pure state, since the electrons have not lost one particular value $E_{n_0}$, but rather a variety of energies $E_0, E_1, E_2, \ldots, E_7$. Thus an ensemble of molecules is said to be in a *mixture of states*, or the state of the physical system is said to be a *mixture*. This mixture can be described by the set of vectors $\phi_0, \phi_1, \phi_2, \ldots, \phi_n, \ldots$ and a set of numbers (relative probabilities) $w_0, w_1, \ldots, w_n, \ldots$ where $w_n$ is chosen to be proportional to the height $h_n$

of the bump corresponding to the energy $E_n$, and they are normalized so that

$$\sum_n w_n = 1. \tag{4.2}$$

Thus one may form a mental picture of the CO gas as a collection of $N$ molecules, where $N_n = w_n N$ is the number of molecules with energy $E_n$ in the ensemble. If it were possible to pick one molecule out of the ensemble, $w_n$ would be the probability that this molecule has energy $E_n$. However, the picking of a single specimen from a microphysical system is not possible, and should not be done even in a *Gedanken* experiment. The mere physical meaning of $w_n$ is that it represents the relative intensity of the electron current that has lost energy $E_n$.

Before we introduce a concise description of the state of a mixture, we shall give an alternate description of a pure state. To do this we shall need some more mathematics, which we now introduce.

⟦An operator $W$ is called *positive definite* if $(\phi, W\phi) \geq 0$ for all $\phi \in \mathcal{H}$. A Hermitian operator $P$ is called a *projection operator* or *projector* iff $P^2 = P$.

It is easily seen that the set $\hbar = \{\phi \in \mathcal{H}: P\phi = \phi\}$ is a subspace $\hbar \subseteq \mathcal{H}$; for if $\phi, \psi \in \hbar$ and $\alpha, \beta \in \mathbb{C}$, then $P(\alpha\phi + \beta\psi) = \alpha P\phi + \beta P\psi = \alpha\phi + \beta\psi$, i.e., $\hbar$ is closed under multiplication by a number and under vector addition. Alternatively, $\hbar = \{P\phi: \phi \in \mathcal{H}\}$, and hence we write $\hbar = P\mathcal{H}$. As an example, let $\mathcal{H} = \mathbb{R}^3$, the three-dimensional Euclidean space, and let $\hbar$ be a plane passing through the origin. Then for any $\mathbf{x} \in \mathbb{R}^3$ let $P\mathbf{x}$ be the ordinary projection of $\mathbf{x}$ onto the plane (see Figure 4.4).

Two subspaces $\hbar_1, \hbar_2 \subseteq \mathcal{H}$ are said to be *mutually orthogonal* if for each $\phi \in \hbar_1$ and for each $\psi \in \hbar_2$, we have $(\phi, \psi) = 0$. The set $\hbar = \{\phi + \psi: \phi \in \hbar_1 \text{ and } \psi \in \hbar_2\}$ is called the *direct sum* of $\hbar_1$ and $\hbar_2$, and is denoted $\hbar = \hbar_1 \oplus \hbar_2$. Similarly, given a collection $\hbar_1, \hbar_2, \hbar_3, \ldots,$ of mutually orthogonal subspaces, one can form their direct sum, $\hbar = \sum_i \oplus \hbar_i = \{\sum_i \varphi_i: \varphi_i \in \hbar_i\}$.

Given a projector $P$ that projects onto the subspace $\hbar$, the operator $I - P$ is also a projector. [*Proof*: Since $P^2 = P$, it follows that

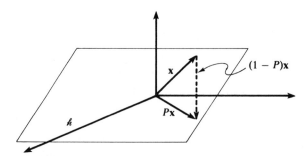

Figure 4.4    Projection operator onto a plane.

$(I - P)^2 = I - IP - PI + P^2 = I - P - (P - P^2) = I - P.]$
$I - P$ thus determines another subspace, which we will denote by $\ell^\perp$. All vectors in $\ell^\perp$ are orthogonal to those in $\ell$; for if $\phi \in \ell$ and $\psi \in \ell^\perp$, then $\phi = P\phi$ and $\psi = (I - P)\psi$; using the definition ($P^\dagger = P$ and $P^2 = P$) of a projector, we have $(\phi, \psi) = (P\phi, (I - P)\psi) = (\phi, P(I - P)\psi) = (\phi, (P - P^2)\psi) = (\phi, 0\psi) = 0$. Since for any $\psi \in \mathscr{H}$ we may write $\psi = P\psi + (I - P)\psi$, any projector $P$ gives a *decomposition* of $\mathscr{H}$ into orthogonal subspaces, $\mathscr{H} = \ell \oplus \ell^\perp$.

For any $\phi \in \mathscr{H}$, the set $\{\alpha\phi : \alpha \in \mathbb{C}\}$ is a one-dimensional subspace of $\mathscr{H}$, called the *space spanned by* $\phi$.

Finally, given a basis $\{\phi_\gamma\}$ of $\mathscr{H}$, we define the *trace of the operator A* by

$$\operatorname{Tr} A \equiv \sum_\gamma (\phi_\gamma, A\phi_\gamma). \qquad (4.3)$$

It can be shown (Problem II.21) that $\operatorname{Tr} A$ is independent of the basis used for an arbitrary operator $A$. $\operatorname{Tr} A$ need not be finite. We will in general assume that this does not occur for those operators whose trace we take. The trace has the following properties:

$$\operatorname{Tr}(AB) = \operatorname{Tr}(BA). \qquad (4.4a)$$

In general, $\operatorname{Tr}(A_1 A_2 \ldots A_n) = \operatorname{Tr}(A_2 A_3 \ldots A_n A_1)$.

$$\operatorname{Tr}((A + B)C) = \operatorname{Tr}(AC) + \operatorname{Tr}(BC).] \qquad (4.4b)$$

We now use these new mathematical concepts to describe the harmonic oscillator. Let

$\ell_0$ be the space spanned by $\phi_0$,
$\ell_1$ be the space spanned by $\phi_1$,
$\ell_2$ be the space spanned by $\phi_2$,

$\vdots$

$\ell_n$ be the space spanned by $\phi_n$.

$\vdots$

Then the space $\mathscr{H}$ is the orthogonal sum

$$\sum_{n=0}^{\infty} \oplus \ell_n = \mathscr{H}. \qquad (4.5)$$

Let $\Lambda_n$ ($n = 0, 1, 2, \ldots$) denote the projection operators that project on $\ell_n$. Then the operator form of (4.5) is written

$$\sum_{n=0}^{\infty} \Lambda_n = I. \qquad (4.6)$$

It says that every $f \in \mathscr{H}$ can be written as the sum of its components $\Lambda_n f$ in the subspaces $\ell_n$:

$$f = \sum_{n=0}^{\infty} \Lambda_n f. \qquad (4.7)$$

This is equivalent to writing

$$f = \sum_{n=0}^{\infty} \phi_n(\phi_n, f) = \sum_{n=0}^{\infty} |\phi_n\rangle\langle\phi_n|f\rangle, \tag{4.8}$$

which is, according to (I.2.15b) the statement that $\phi_n$ is a basis. Thus $\Lambda_n$ may also be written $\Lambda_n = |\phi_n\rangle\langle\phi_n|$, and (4.6) may be written

$$I = \sum_{n=0}^{\infty} |\phi_n\rangle\langle\phi_n|. \tag{4.9}$$

We will often denote the projection operator $\Lambda$ on the one-dimensional subspace spanned by a unit vector $\psi$ by

$$\Lambda = |\psi\rangle\langle\psi|. \tag{4.10}$$

If we apply the operator $H$ to the arbitrary vector $f$ and use (4.9), we have, because of (3.29) and (3.30),

$$Hf = \sum_{n=0}^{\infty} H|\phi_n\rangle\langle\phi_n|f\rangle = \sum_{n=0}^{\infty} E_n|\phi_n\rangle\langle\phi_n|f\rangle.$$

As $f$ was arbitrary, we may as well omit it and write

$$H = \sum_{n=0}^{\infty} E_n|\phi_n\rangle\langle\phi_n| = \sum_{n=0}^{\infty} E_n\Lambda_n. \tag{4.11}$$

This is called the *spectral representation* of the operator $H$. With the notion of the trace we can express the energy values $E_n = (\phi_n, H\phi_n)$ in the form

$$E_n = \text{Tr}(H\Lambda_n). \tag{4.12}$$

PROOF.   Since

$$\Lambda_n\phi_\gamma = \begin{cases} \phi_n & \text{if } n = \gamma, \\ 0 & \text{if } n \neq \gamma, \end{cases}$$

one obtains

$$\begin{aligned} \text{Tr}(H\Lambda_n) &= \sum_\gamma (\phi_\gamma, H\Lambda_n\phi_\gamma) \\ &= \sum_\gamma (\phi_\gamma, H\phi_n)\delta_{\gamma n} = E_n. \end{aligned} \tag{4.13}$$

Thus the state of the harmonic oscillator in which the energy is $E_n$ can be described by the projection operator $\Lambda_n$ in place of $\phi_n$. Clearly $\Lambda_n$ and $\phi_n$ do not represent the same mathematical objects. $\Lambda_n$ represents the subspace $\mathfrak{h}_n = \Lambda_n\mathcal{H} = \{\Lambda_n f \mid f \in \mathcal{H}\}$, which is the one-dimensional subspace spanned by $\phi_n$ (a one-dimensional subspace is often called a *ray*). The space $\mathfrak{h}_n$ contains many normalized vectors, namely, all those that can be obtained from $\phi_n$ by multiplication with a phase factor $e^{i\alpha}$, where $0 \leq \alpha \leq 2\pi$. (The set of all $e^{i\alpha}\phi$ with $\|\phi\| = 1$ is called a *unit ray*.)

   It is generally assumed that a pure state of a physical system is described by a projection operator on a one-dimensional subspace (or by a unit ray) and not by a vector. However, it is very common to call not only $\Lambda$ but also any vector $\phi \in \Lambda \mathcal{H}$ a "state" or "state vector." For the restricted case of pure states, we formulate the generalization of our above conclusions as the restricted version of the second basic axiom of quantum mechanics.

**II′.** A pure state of a physical, quantum-mechanical system is characterized by a projection operator $\Lambda$ on a one-dimensional subspace (or by the one-dimensional subspace $\Lambda \mathcal{H}$ itself). The possible values of an observable $A$ that are measured by an experiment performed on the physical system in the state $\Lambda$ are given by

$$\langle A \rangle = \mathrm{Tr}(A\Lambda). \tag{4.14}$$

$\langle A \rangle$ is called the *expectation value* of the observable $A$ in the pure state $\Lambda$.

The statement contained in II′ is incomplete as long as we do not define precisely what "expectation value" means, i.e., until we say exactly to which experimentally measured quantity $\langle A \rangle$ should correspond. $\mathrm{Tr}(A\Lambda)$ is something calculated from the mathematical image of the physical system; thus, given the mathematical image (i.e., the algebra), the observable $A$, and the state $\Lambda$, $\mathrm{Tr}\,(A\Lambda)$ is a precisely defined mathematical quantity. Equation (4.14) gives the relation between the mathematically obtained quantity $\mathrm{Tr}(A\Lambda)$ and the quantity $\langle A \rangle$, which is measured in the experimental setup that defines the observable represented by the operator $A$. $\langle A \rangle$ is obtained by the following prescription: One performs a series of $N$ measurements of the observable $A$; i.e., one either has $N$ identical systems on which one performs the measurement, or one performs the same measurement $N$ times on a physical system, making sure that the system is in the same state for each measurement. Let the numbers measured for the observable $A$ in this experiment be denoted $a_0, a_1, \ldots, a_n, \ldots$, and let $N_n$ be the number of measurements that gave the result $a_n$. Then $\langle A \rangle$ is the average of these measurements,

$$\langle A \rangle = \sum_n \frac{N_n}{N} a_n,$$

for sufficiently large $N$. The ratio $w_n = N_n/N$ is the probability of obtaining the value $a_n$ in one measurement.

If the system is in a pure state, say $\Lambda_1$ (the projection operator on the subspace spanned by $\phi_1$) then the intensity will have one bump at the energy $E_e - E_{e'} = E_1$. If $N$ measurements are performed on this system, the result is allways $E_1$. Consequently

$$\langle H \rangle = \sum_n \frac{N_n}{N} E_n = \frac{N}{N} E_1. \tag{4.15}$$

Hence, in the pure state $\Lambda_1$, the probability $w_1$ for the value $E_1$ is $w_1 = 1$ and all other $w_i = 0$.

A pure state $\Lambda$ is called an *eigenstate of the observable A* if $\Lambda \mathscr{H}$ is the one-dimensional space of all eigenvectors of $A$ with the same eigenvalue.[7] Thus the system in the above described experiment is in an eigenstate of the observable $H$ if one bump appears at energy $E_n$. In general, as we shall see later, a physical system can be in a pure state described by a one-dimensional space, and the measurement of some observable $A$, of which this state is not an eigenstate, will still yield a set of different numbers $a_1, a_2, \ldots, a_n, \ldots$. This is the point at which quantum mechanics deviates drastically from classical theories. We shall discuss this matter in detail later.

We now discuss the description of a mixture rather than a pure state. consider the gas of CO molecules in the collision chamber. If we measure the energy of the molecules $N$ times, we expect

$$N_0 = w_0 N \text{ measurements will give the value } E_0,$$
$$N_1 = w_1 N \text{ measurements will give the value } E_1,$$
$$\vdots$$
$$N_n = w_n N \text{ measurements will give the value } E_n,$$

because $w_i$ is the ratio of molecules in the ensemble, with energy $E_i$. Thus the average value $\langle H \rangle$ of the energy operator $H$ is

$$\langle H \rangle = \sum_n w_n E_n = \sum_n w_n(\phi_n, H\phi_n). \tag{4.16}$$

which is to be equal to the expectation value of $H$ in the mixed state. We want to write this expectation value as a trace of some operator; to do this we introduce the positive-definite Hermitian operator

$$W \equiv \sum_n w_n \Lambda_n, \tag{4.17}$$

where each $\Lambda_n$ is the projector on the eigenspace $h_n$ of $\mathscr{H}$ spanned by $\phi_n$. Now we calculate

$$\mathrm{Tr}(HW) = \sum_\gamma (\phi_\gamma, HW\phi_\gamma)$$

$$= \sum_\gamma \sum_n w_n(\phi_\gamma, H\Lambda_n \phi_\gamma).$$

Since

$$\Lambda_n \phi_\gamma = \delta_{n\gamma} \phi_n$$

$$= \sum_\gamma \sum_n w_n(\phi_\gamma, H\phi_n)\delta_{n\gamma},$$

we have

$$\mathrm{Tr}(HW) = \sum_n w_n(\phi_n, H\phi_n). \tag{4.18}$$

---

[7] As will be discussed shortly, projection operators on multidimensional (including infinite-dimensional) subspaces may also represent states of a physical system, which are then, however, not called pure states. Also, in that case we shall call the state $\Lambda$ an "eigenstate of the observable $A$" if $\Lambda \mathscr{H}$ is the space of all eigenvectors of $A$ with the same eigenvalue.

Thus if the state of the mixture is described by the operator $W$ defined in Equation (4.17), then the expectation value of the energy is given by

$$\langle H \rangle = \text{Tr}(HW). \tag{4.19}$$

It is important to note that in Equation (4.19), $W$ is not a projection operator as it is for a pure state.

The statement contained in Equation (4.19) is conjectured to be the general expression for quantum mechanics, and we formulate it as the second basic assumption.

**II.**   The state of a quantum-mechanical system is characterized by a positive-definite Hermitian operator $W$. The values measured in an experiment, i.e., the expectation value $\langle A \rangle$ of an observable $A$, are given by

$$\langle A \rangle = \text{Tr}(AW) \quad \text{if Tr } W = 1$$

or by

$$\langle A \rangle = \frac{\text{Tr}(AW)}{\text{Tr } W}.$$

The operator $W$ is called the *statistical operator*, and its matrix is called the *density matrix*. The state of the system is called a pure state iff $W$ is the projection operator on a one-dimensional subspace.

It is customary to normalize the probabilities $w_n$ so that

$$\sum_{n=0}^{\infty} w_n = 1, \tag{4.20}$$

i.e., to normalize the statistical operator so that

$$\text{Tr } W = 1. \tag{4.21}$$

Note, however, that Equation (4.20) as well as Equation (4.21) is only a convention. Instead of $w_n$ we could have used the height of the bumps $h_n$ to characterize the state. In that case we would get, in place of Equation (4.17),

$$\tilde{W} = \sum_n h_n \Lambda_n \tag{4.22}$$

and

$$\text{Tr } \tilde{W} = \sum_n h_n.$$

It is easy to see that

$$W = \frac{\tilde{W}}{\text{Tr } \tilde{W}}.$$

$W$ and $\tilde{W}$ describe the same state of the harmonic oscillator. However, if one uses $\tilde{W}$ to characterize the state, then the expectation value of the observable $H$ is given by

$$\langle H \rangle = \frac{\text{Tr}(H\tilde{W})}{\text{Tr } \tilde{W}} \quad \text{or in general} \quad \langle A \rangle = \frac{\text{Tr}(A\tilde{W})}{\text{Tr } \tilde{W}} \tag{4.23}$$

Thus, whenever possible, it is more convenient to use the normalized statistical operator for the description of the state.

The second basic assumption of quantum mechanics cannot be derived; it can only be conjectured. It has been conjectured, in the long process of discovering quantum mechanics, from a variety of experimental results. We have in our above considerations made this axiom plausible for the particular case of the energy measurement of the harmonic oscillator.

Consider the state of the harmonic oscillator given by

$$W = \sum_n w_n \Lambda_n.$$

The projection operator $\Lambda_p$ on the eigenspace of $H$ with eigenvalue $E_p$ is also an observable. Thus we can calculate the expectation value of $\Lambda_p$ in the state $W$:

$$\langle \Lambda_p \rangle = \text{Tr}(\Lambda_p W) = \sum_n \text{Tr}(\Lambda_p \Lambda_n) w_n. \tag{4.24}$$

From the properties of projection operators we know that

$$\begin{aligned} \Lambda_p \Lambda_n &= 0 \quad \text{for } p \neq n \\ \Lambda_n \Lambda_n &= \Lambda_n; \end{aligned} \tag{4.25}$$

consequently

$$\langle \Lambda_p \rangle = \sum_n \delta_{np} \text{Tr}(\Lambda_n) w_n \tag{4.26}$$

$$= \text{Tr}(\Lambda_p) w_p, \tag{4.27}$$

so

$$\langle \Lambda_p \rangle = w_p, \tag{4.28}$$

because $\text{Tr } \Lambda_p = 1$, since $\Lambda_p$ is the projector on a one-dimensional subspace. Therefore $\Lambda_p$ is the observable whose expectation value is the probability of obtaining the value $E_p$ in a measurement of the energy, i.e., the probability of finding the system in the state $\Lambda_p$. In particular, for the pure state $W = \Lambda_n$ we have

$$\langle \Lambda_p \rangle = \text{Tr}(\Lambda_p \Lambda_n) = \begin{cases} 0 & \text{for } p \neq n, \\ 1 & \text{for } p = n. \end{cases} \tag{4.29}$$

Generally, projection operators represent a particular kind of observable called a *property* or *proposition*, whose expectation value tells us something about the state.

The basic assumption II states how the expectation value is calculated if the statistical operator $W$ is known. We now discuss how the statistical operator has to be chosen in any particular case. To do this we use our belief, which is supported by all experimental evidence, that *a repetition of a measurement always leads to the same result*, or, in other words, that all experiments performed under the same conditions give the same experimental numbers. It is assumed that generally a certain physical system may be in a number of states $1, 2, \ldots$, which in quantum mechanics are described by statistical operators $W_1, W_2, \ldots$, and that these states can be determined by the measurements of observables $A, B, \ldots$. Thus every state of the physical system is the result of a *preparation* of the system. A preparation is a series of manipulations with physical equipment that lead to the measurement of an observable (or several observables). These measurements may or may not affect the state of the system. In classical physics the effect of a measurement upon the state of the physical system is always neglected; the essential difference in quantum physics is that here this effect is taken into account. Thus in principle, and also in practice, not all measurements can be performed simultaneously. Therefore one must always be able to reproduce a certain physical state, so that it can be determined by a series of successive measurements. After the measurement of each observable, the system will be in a state that has been prepared by this measurement. This state must be described by a statistical operator for which the calculated expectation value of the observable agree with the values just measured. That is, the $W$ must be chosen so that the calculated values agree with the measured values that one will obtain in an "immediate" repetition of the measurement, i.e., in a measurement that is performed before the state has a chance to change (develop in time). In fact, this principle was used when we wrote down the state of the CO molecules. In the example of the ensemble of CO molecules, the preparation consists of setting up the apparatus and bombarding the gas with monochromatic electrons. This prepares the CO gas in a state that is a mixture of energy eigenstates $\Lambda_n$, with the mixing ratios $w_n$. The statistical operator $W$ has to express what has been measured on the physical system, or what preparations have been performed. Therefore we choose the statistical operator to be

$$W = \sum_n w_n \Lambda_n. \tag{4.30}$$

With this $W$, one calculates for the outcome of the measurement of the energy $H$ the numbers

$$E_0, E_1, E_2, \ldots,$$
$$w_0, w_1, w_2, \ldots.$$

Thus the original values obtained in the preparation of the state are predicted for an immediate repetition of the measurement (i.e., a measurement before the state $W$ has changed), which in the present case may be any finite amount of time later, since the situation in this experiment is stationary, but before another measurement is made or before the setup is altered (e.g., by switching off the electron beam).

For the general case we formulate the above described procedure for the determination of the statistical operator as a basic assumption:

**IIIa.**   Given that an observable $A$ has been measured with the following results

$$a_1, a_2, a_3, \ldots ,$$

$$w_{a_1}, w_{a_2}, w_{a_3}, \ldots ,$$

where $w_{a_i}$ gives the relative frequency with which the value $a_i$ was measured (i.e., after $N$ measurements the value $a_i$ occurred $w_{a_i} N$ times), then the state of the system immediately following the measurement is represented by the statistical operator $W$ given by

$$W = \sum_{a_i} w_{a_i} \Lambda_{a_i} \frac{1}{\dim(\Lambda_{a_i} \mathscr{H})}, \tag{4.31}$$

where $\Lambda_{a_i}$ are the projection operators on the eigenspaces of $A$ with eigenvalues $a_i$, and where $\dim(\Lambda_{a_i} \mathscr{H}) = \operatorname{Tr} \Lambda_{a_i}$ denotes the dimension of the space $\Lambda_{a_i}$.

Before we discuss the general case, let us consider the particular situation that $a_n$ has been measured with the probability $w_n = 1$ and all other $w_i (i \neq n) = 0$. Then the statistical operator is chosen as

$$W = \Lambda_{a_n} \frac{1}{\dim(\Lambda_{a_n} \mathscr{H})}. \tag{4.32}$$

[*Note.* The $\Lambda_{a_i}$ are not necessarily projection operators on one-dimensional spaces. If $\Lambda_{a_n}$ is the projector on a one-dimensional space, the state (4.32) is said to be a pure state.] As we shall prove below, one obtains for the expectation value

$$\langle A \rangle = \operatorname{Tr}(AW) = \frac{1}{\dim(\Lambda_{a_n} \mathscr{H})} \operatorname{Tr}(A\Lambda_{a_n}) = a_n, \tag{4.33}$$

and for the probability

$$w_{a_n} = \langle \Lambda_{a_n} \rangle = \operatorname{Tr}(\Lambda_{a_n}) = \frac{1}{\dim(\Lambda_{a_n} \mathscr{H})} \operatorname{Tr} \Lambda_{a_n} = 1. \tag{4.34}$$

In other words, this statistical operator (4.32) predicts, for an immediate repetition of the measurement of $A$, the value $a_n$ with certainty. Before we prove (4.33) and (4.34), it is convenient to present some more mathematics.

[If the space $\Lambda_a \mathscr{H}$ of eigenvectors of $A$ with eigenvalue $a$ has the dimension $\dim(\Lambda_a \mathscr{H})$, then there exist $\dim(\Lambda_a \mathscr{H})$ linearly independent eigenvectors. One may then choose an arbitrary basis of

$\dim(\Lambda_a \mathcal{H})$ mutually orthonormal eigenvectors of $A$ with eigenvalue $a$, i.e., a set of vectors

$$\phi_a^1, \phi_a^2, \ldots, \phi_a^\kappa, \ldots, \phi_a^{\dim(\Lambda_a \mathcal{H})}$$

that fulfill

$$A\phi_a^\kappa = a\phi_a^\kappa \qquad (4.35)$$

and

$$(\phi_a^\kappa, \phi_a^{\kappa'}) = \delta_{\kappa\kappa'}. \qquad (4.36)$$

The basis for the whole space is given by

$$\phi_{a_1}^1, \phi_{a_1}^2, \ldots, \phi_{a_1}^{\dim(\Lambda(a_1)\mathcal{H})}$$
$$\vdots$$
$$\phi_{a_j}^1, \phi_{a_j}^2, \ldots, \phi_{a_j}^{\dim(\Lambda(a_j)\mathcal{H})} \qquad (4.37)$$
$$\vdots$$

In general, $\dim(\Lambda_{a_{(i)}} \mathcal{H}) \neq \dim \Lambda_{a_{(j)}} \mathcal{H}$. This statement (spectral theorem), given here without proof, is only true for operators with a "discrete spectrum," to which we want to restrict ourselves for the moment. Any of the $\dim(\Lambda_{a_j} \mathcal{H})$ may or may not be infinite. An arbitrary vector $f \in \mathcal{H}$ is then written, according to (I.2.15b),

$$f = \sum_{\kappa, a_i} |\phi_{a_i}^\kappa\rangle\langle\phi_{a_i}^\kappa | f\rangle. \qquad (4.38)$$

If we apply $A$ to the arbitrary vector $f$ written in this form we obtain

$$Af = \sum_{\kappa a_i} a_i |\phi_{a_i}^\kappa\rangle\langle\phi_{a_i}^\kappa | f\rangle, \qquad (4.39)$$

or, if we omit the arbitrary $f$,

$$A = \sum_{\kappa a_i} a_i |\phi_{a_i}^\kappa\rangle\langle\phi_{a_i}^\kappa| = \sum_{a_i} a_i \sum_\kappa |\phi_{a_i}^\kappa\rangle\langle\phi_{a_i}^\kappa|. \qquad (4.40)$$

The operator

$$\sum_\kappa |\phi_{a_i}^\kappa\rangle\langle\phi_{a_i}^\kappa| = \Lambda_{a_i} \qquad (4.41)$$

is the projection operator on the subspace $\Lambda_{a_i} \mathcal{H}$ of eigenvectors of $A$ with eigenvalue $a_i$, since

$$\Lambda_{a_i}\Lambda_{a_j} = \sum_\kappa \sum_{\kappa'} |\phi_{a_i}^\kappa\rangle\langle\phi_{a_i}^\kappa|\phi_{a_j}^{\kappa'}\rangle\langle\phi_{a_j}^{\kappa'}| = \sum_\kappa |\phi_{a_i}^\kappa\rangle\langle\phi_{a_j}^\kappa|\delta_{a_i a_j} = \Lambda_{a_i}\delta_{a_i a_j}$$

$$A(\Lambda_{a_i} f) = \sum_\kappa a_i |\phi_{a_i}^\kappa\rangle\langle\phi_{a_i}^\kappa | f\rangle = a_i(\Lambda_{a_i} f).$$

Therewith (4.40) can be written

$$A = \sum_{a_i} a_i \Lambda_{a_i}. \tag{4.42}$$

This is called the *spectral representation of the operator A* (with discrete spectrum). The set if $\Lambda_{a_i}$ is called the *spectral family*.

If one omits the arbitrary $f$ in (4.38), one obtains the resolution of the identity operator $I$ with respect to the spectral family of the operator $A$:

$$I = \sum_{\kappa a_i} |\phi_{a_i}^\kappa\rangle\langle\phi_{a_i}^\kappa| = \sum_{a_i} \Lambda_{a_i}.$$

We now prove (4.33). We use the basis (4.37) and calculate

$$\mathrm{Tr}(A\Lambda_{a_n}) = \sum_{a_j\kappa} (\phi_{a_j}^\kappa, A\Lambda_{a_n}\phi_{a_j}^\kappa) = \sum_{\kappa} (\phi_{a_n}^\kappa, A\phi_{a_n}^\kappa)$$

$$= \sum_{\kappa} a_n(\phi_{a_n}^\kappa, \phi_{a_n}^\kappa) = a_n \dim(\Lambda_{a_n}\mathscr{H}). \tag{4.43}$$

Thus

$$\mathrm{Tr}(AW) = \mathrm{Tr}(A\Lambda_{a_n}) \frac{1}{\dim(\Lambda_{a_n}\mathscr{H})} = a_n.$$

Equation (4.34) is established in the same way.

We now proceed with the introduction of some further mathematical notions: two Hermitian operators $A$, $B$ are called *commuting operators* iff

$$AB = BA \qquad \text{or } [A, B] = 0. \tag{4.44}$$

Let $\Lambda_{a_i}$ be the spectral family of $A$, and $P_{b_i}$ the spectral family of $B$. Then the statement that $A$ and $B$ commute is equivalent[8] to

$$\Lambda_{a_i}P_{b_j} = P_{b_j}\Lambda_{a_i} \quad \text{for every } a_i, b_j. \tag{4.45}$$

Thus, using (4.42), an operator $A$ commutes with $B$ if

$$AP_{b_j} = P_{b_j}A \quad \text{for every } b_j. \qquad ] \tag{4.46}$$

A pure state (e.g., the energy eigenstate of the one-dimensional harmonic oscillator discussed above) is a special case of (4.32).

We turn now to the general expression (4.31), and calculate the expectation value

$$\mathrm{Tr}(AW) = \sum_{a_i} w_{a_i} \frac{1}{\dim(\Lambda_{a_i}\mathscr{H})} \mathrm{Tr}(A\Lambda_{a_i}). \tag{4.47}$$

---

[8] In fact (4.45) is the precise statement of commutativity; in order to make (4.44) precise one has to add the requirement that $A^2 + B^2$ be essentially self-adjoint (its Hilbert space closure is selfadjoint).

Inserting (4.43) into this, we obtain

$$\langle A \rangle = \text{Tr}(AW) = \sum_{a_i} w_{a_i} a_i, \tag{4.48}$$

as it ought to be for the expectation value, i.e., the average value in a series of measurements of the observable $A$ in the state of the system described by $W$.

The projection operator $\Lambda_a$ on the eigenspace of $A$ with eigenvalue $a$ represents the observable "probability of $a$," as it should. To see this we calculate

$$\text{Tr}(\Lambda_a W) = \sum_{a_i} w_{a_i} \text{Tr}(\Lambda_a \Lambda_{a_i}) \frac{1}{\dim(\Lambda_{a_i} \mathcal{H})}$$

$$= \sum_{a_i} w_{a_i} \delta_{a_i a} \text{Tr}(\Lambda_a) \frac{1}{\dim(\Lambda_{a_i} \mathcal{H})} = w_a.$$

Here we have used

$$\Lambda_a \Lambda_{a_i} = 0 \quad \text{for } a_i \neq a,$$
$$\Lambda_a \Lambda_a = \Lambda_a.$$

Thus the statistical operator $W$ of (4.31) leads to the prediction we wanted, namely that an immediate repetition of the measurement of an observable gives the same value as the original measurement that constituted the preparation of the state.

If no measurement has been performed that constitutes the preparation of a state, and no information concerning the state has been obtained, then all values $a_i$ will be equally probable in the outcome of an experiment: $w_{a_i} = $ const. The statistical operator for such a state must, therefore, be taken to be proportional to the unit operator

$$\tilde{W} = I. \tag{4.49}$$

If the system is known to be in thermodynamic equilibrium, then the statistical operator $W$ is chosen to be the Gibbs distribution

$$\tilde{W} = e^{-H/kT}, \tag{4.50}$$

where $H$ is the energy operator, and $k$ and $T$ are the Boltzmann constant and absolute temperature respectively. A justification of (4.50) can be found in books on statistical physics.

One of the consequences of the basic assumption IIIa is that the possible values of an observable that can be measured in any experiment are the eigenvalues[9] of that observable's operator. This means that *in the process of*

---

[9] This applies to direct measurements, i.e., to measurements in which a state has been prepared as described in IIIa. From these directly measured values one can often calculate the matrix elements of other operators representing physical quantities. Such observables have then been indirectly measured, by an experiment that prepared a mixture of eigenstates of the directly measured observables.

*conjecturing the theory, the operator that represents an observable must be chosen in such a way that all the values obtained in the experiment that measures the observable in question are eigenvalues of the operator.*

## II.5 The Effect of a Measurement on the State of a Quantum-Mechanical System

The measurement of an observable also constitutes a preparation of the system. Generally, this means that the measurement of an observable causes alterations in the state of the system. These statements are now formulated in the second part of the basic assumption III.

**IIIb.**  Let a system be in the state $W$, and let an observable $B$ be measured on the system. Let the eigenvalues of $B$ be $b_1, b_2, b_3, \ldots$, and let $\Lambda_{b_1}, \Lambda_{b_2}, \ldots$ be the projectors on the corresponding eigenspaces. Then the state of the system after the measurement is given by

$$W' = \sum_{b_i} \Lambda_{b_i} W \Lambda_{b_i}. \tag{5.1}$$

If this measurement is used to select a certain subensemble with a definite value $b$ of $B$, then the state is given by

$$W' = \frac{1}{\text{Tr}(\Lambda_b W \Lambda_b)} \Lambda_b W \Lambda_b. \tag{5.2}$$

There is a special case in which the measurement of an observable does not alter the state $W$; this is the case if and only if the operator $B$ that represents the observable commutes with $W$. Because only then Eq. (5.1) becomes, according to (4.46),

$$W' = \sum_i \Lambda_{b_i} W \Lambda_{b_i} = \sum_i \Lambda_{b_i} \Lambda_{b_i} W$$

$$= \sum_i \Lambda_{b_i} W;$$

and, since

$$\sum_i \Lambda_{b_i} = I,$$

we get

$$W' = \sum_i \Lambda_{b_i} W = W.$$

Two observables are called *compatible* if the measurement of one does not alter the state that has been prepared by the measurement of the other. The preceding consideration then shows that compatible observables are represented by commuting operators. Observables represented by noncommuting operators are *incompatible*.

We give a brief justification of the basic assumption IIIb. The probability of obtaining the value $b_\alpha$ for an observable of the system in the state $W$ is

$$\langle \Lambda_{b_\alpha} \rangle = \mathrm{Tr}(\Lambda_{b_\alpha} W). \tag{5.3}$$

We know from experience that the immediate repetition of an experiment leads to the same results. Thus $W'$ must be chosen in such a way that the eigenvalues of $B$ and their probabilities are the same as in the state $W$. The eigenvalues of an operator are a property of that operator; thus in order to fulfill this requirement we just have to demand that the probability $\langle \Lambda_{b_i} \rangle'$ of obtaining $b_i$ in the second measurement be the same as the probability $\langle \Lambda_{b_i} \rangle$ in the first measurement:

$$\langle \Lambda_{b_i} \rangle' = \langle \Lambda_{b_i} \rangle \quad \text{for all } b_i, \tag{5.4}$$

or

$$\mathrm{Tr}(\Lambda_{b_i} W') = \mathrm{Tr}(\Lambda_{b_i} W) \quad \text{for all } b_i. \tag{5.5}$$

It is easily seen that $W'$ given by the postulate IIIb satisfies relation (5.5).

$$\begin{aligned} \mathrm{Tr}(\Lambda_{b_i} W') &= \sum_j \mathrm{Tr}(\Lambda_{b_i} \Lambda_{b_j} W \Lambda_{b_j}) \\ &= \mathrm{Tr}(\Lambda_{b_i} W \Lambda_{b_i}) \\ &= \mathrm{Tr}(\Lambda_{b_i} W). \end{aligned}$$

Thus the basic assumption IIIb leads to the relation (5.4), i.e., the repetition of the measurement gives the same result.

We now show the converse, namely, that (5.1) follows from the assumption that an immediate repetition of the measurement leads to the same result—as expressed by Equation (5.4) or (5.5)—and the additional assumption that after the measurement of $B$, $W'$ must be given according to IIIa by a linear combination of projection operators,

$$W' = \sum_{b_i} w'_{b_i} \Lambda_{b_i} \frac{1}{\dim(\Lambda_{b_i} \mathcal{H})}. \tag{5.6}$$

Generally $w'_{b_i} = \mathrm{Tr}(W'\Lambda_{b_i})$ and by (5.5) $w'_{b_i} = \mathrm{Tr}(W\Lambda_{b_i})$. Consequently under the assumption that an "immediate" repetition of the measurement leads to the same result [i.e., (5.5)], $W'$ is given by

$$W' = \sum_{b_i} \mathrm{Tr}(W\Lambda_{b_i})\Lambda_{b_i} \frac{1}{\dim(\Lambda_{b_i} \mathcal{H})}. \tag{5.7}$$

Let us consider the Hermitian operator $\Lambda_{b_j} W \Lambda_{b_j}$; this operator has the properties

$$\Lambda_{b_j} W \Lambda_{b_j} f = 0 \quad \text{for every } f \perp \Lambda_{b_j} \mathcal{H} \tag{5.8}$$

$$\Lambda_{b_i} W \Lambda_{b_j} g = \Lambda_{b_j} W g \in \Lambda_{b_j} \mathcal{H} \quad \text{for every } g \in \Lambda_{b_j} \mathcal{H}. \tag{5.9}$$

Consequently it must be proportional to the projection operator $\Lambda_{b_j}$:

$$\Lambda_{b_j} W \Lambda_{b_j} = v_j \Lambda_{b_j}. \tag{5.10}$$

From the trace of this equation we obtain the value of $v_j$:

$$\mathrm{Tr}(v_j \Lambda_{b_j}) = v_j \dim(\Lambda_{b_j} \mathcal{H}) = \mathrm{Tr}(\Lambda_{b_j} W \Lambda_{b_j}) = \mathrm{Tr}(\Lambda_{b_j} W);$$

i.e.,

$$v_j = \frac{1}{\dim(\Lambda_{b_j} \mathcal{H})} \mathrm{Tr}(\Lambda_{b_j} W).$$

Thus equation (5.10) becomes

$$\Lambda_{b_j} W \Lambda_{b_j} = \mathrm{Tr}(W \Lambda_{b_j}) \Lambda_{b_j} \frac{1}{\dim(\Lambda_{b_j} \mathcal{H})}. \tag{5.11}$$

Inserting this into Eq. (5.7), we get

$$W' = \sum_{b_j} \Lambda_{b_j} W \Lambda_{b_j},$$

which is what we wanted to derive.


## II.6   The Basic Assumptions Applied to the Harmonic Oscillator, and Some Historical Remarks

We now return to the discussion of the harmonic oscillator. Let us assume that its state has been found to be

$$W = \Lambda_n,$$

where $\Lambda_n$ is the projection operator on the space spanned by $\phi_n$. This is the case if a measurement of the energy always gives the result $E_n$. We now calculate the value we can expect to obtain if we measure the momentum $P$. According to the basic assumption II,

$$\langle P \rangle = \mathrm{Tr}(PW) = \mathrm{Tr}(P \Lambda_n)$$
$$= \sum_m (\phi_m, P \Lambda_n \phi_m)$$
$$= (\phi_n, P \phi_n), \tag{6.1}$$

so

$$\langle P \rangle = \langle n | P | n \rangle. \tag{6.2}$$

We leave it as an exercise for the reader to show that as a consequence of (3.26) and (3.1),

$$\langle P \rangle = 0. \tag{6.3}$$

Similarly the expectation value of the position $Q$ in the pure state $W = \Lambda_n$ is

$$\langle Q \rangle = \langle n | Q | n \rangle = 0. \tag{6.4}$$

We now calculate the expectation value of $P^2$ (i.e., the kinetic energy $P^2/2\mu$) and the expectation value of $Q^2$ (or the potential energy $\omega^2 \mu Q^2/2$):

$$\langle P^2 \rangle = (\phi_n, P^2 \phi_n)$$

Using the equations (3.1), we have

$$P^2 = \frac{\mu \omega \hbar}{2} (a^\dagger - a)^2$$

so

$$\langle P^2 \rangle = \frac{\mu \omega \hbar}{2} (\phi_n, (a^\dagger - a)^2 \phi_n).$$

Inserting for $a$ and $a^\dagger$ from Equation (3.26), we get

$$\langle P^2 \rangle = \mu \omega h (n + \tfrac{1}{2}) \tag{6.5}$$

In a similar fashion we calculate

$$\langle Q^2 \rangle = (\phi_n, Q^2 \phi_n).$$

Again from (3.1),

$$Q^2 = \frac{\hbar}{2\mu\omega} (a^\dagger + a)^2;$$

so

$$(\phi_n, Q^2 \phi_n) = \frac{\hbar}{2\mu\omega} (\phi_n, (a^\dagger + a)^2 \phi_n),$$

and finally

$$\langle Q^2 \rangle = w \frac{\hbar}{\mu\omega} (n + \tfrac{1}{2}). \tag{6.6}$$

Although the expectation values of the observables $P$, $Q$, $P^2$, and $Q^2$ are uniquely determined, it does not follow that one should expect to obtain the same value $\langle A \rangle$ for any measurement of an observable $A$ (although such is the case in the measurement[10] of the observable $H$ in the pure state $\Lambda_n$). What does follow—according to the definition of $\langle A \rangle$—is that the average of the values obtained in a series of many repeated measurements of $A$ on the state $\Lambda_n$ is $\langle A \rangle$, which of course is uniquely determined. In general it is possible that the values obtained in each measurement ($a_1$ for the first measurement, $a_2$ for the second, $a_3$ for the third, etc.) are entirely different from each other and widely dispersed.

---

[10] We remark that by measurement we always mean an idealized measurement in which the measuring apparatus does not lead to uncertainties; i.e., the experimental error is zero. For the above measurement on CO this would mean that the bumps are considered to be "infinitesimally" narrow.

In order to describe the dispersion of the measured values quantitatively, let us define, for any observable $A$ and state $W$, a new observable given by

$$(A - \langle A \rangle I)^2.$$

To see that this observable is suitable for the description of the dispersion of the measured values $a_1, a_2, a_3, \ldots$, we consider the expectation value of $(A - \langle A \rangle I)^2$ in the state $W$, which is called the *dispersion of A in the state W*:

$$\begin{aligned}
\text{disp}_{(W)} A &\equiv \langle (A - \langle A \rangle_W)^2 \rangle_W \\
&= \text{Tr}(A^2 W) - 2\,\text{Tr}(A \langle A \rangle_W W) + \text{Tr}(\langle A \rangle_W^2 W) \\
&= \text{Tr}(A^2 W) - (\text{Tr}(AW))^2 = \langle A^2 \rangle_W - \langle A \rangle_W^2 \qquad (6.7)
\end{aligned}$$

(unless otherwise stated we always assume that $\text{Tr}\,W = 1$); thus

$$\text{disp}_{(W)} A = \langle A^2 \rangle_W - \langle A \rangle_W^2. \qquad (6.8)$$

For the special case of the observable $H$ for the harmonic oscillator in the state $\Lambda_n$ we obtain

$$\begin{aligned}
\text{disp}_{(\Lambda_n)} H &= \text{Tr}(H^2 \Lambda_n) - (\text{Tr}(H\Lambda_n))^2 \\
&= (\phi_n, H^2 \phi_n) - E_n^2. \qquad (6.9)
\end{aligned}$$

However, since $\phi_n$ is the eigenvector of $H$ with eigenvalue $E_n$, we have

$$\begin{aligned}
H^2 \phi_n &= HE_n \phi_n = E_n H \phi_n \\
&= E_n^2 \phi_n;
\end{aligned}$$

thus

$$\text{disp}_{(\Lambda_n)} H = E_n^2 - E_n^2 = 0. \qquad (6.10)$$

A system is said to be in an *eigenstate of the observable A* if its statistical operator is the projector on a space that consists of eigenvectors of $A$ with the same eigenvalue. The above results show then that the dispersion of $H$ in an eigenstate $\Lambda_n$ of $H$ is zero. This is true in general—the dispersion of an observable $A$ in an eigenstate of $A$ is zero. Conversely, the dispersion of an observable $A$ is zero only if the system is in an eigenstate of $A$.

The energy eigenstate $\Lambda_n$ for the harmonic oscillator is not an eigenstate of either of the observables $P$ and $Q$, so its dispersion is not zero. The dispersion is easily calculated:

$$\text{disp}_{(\Lambda_n)} P = \langle P^2 \rangle - \langle P \rangle^2.$$

From Equations (6.3) and (6.5),

$$\langle P^2 \rangle - \langle P \rangle^2 = \hbar \mu \omega (n + \tfrac{1}{2}). \qquad (6.11)$$

Similarly

$$\text{disp}_{(\Lambda_n)} Q = \langle Q^2 \rangle - \langle Q \rangle^2 = \frac{\hbar}{\mu \omega}(n + \tfrac{1}{2}). \qquad (6.12)$$

The square root of the dispersion is called the *uncertainty of A in the state W*, and is denoted by $\Delta A$:

$$\Delta A = \sqrt{\text{disp } A}. \tag{6.13}$$

From Equations (6.11) and (6.12) it follows that

$$\Delta P \, \Delta Q = \hbar(n + \tfrac{1}{2}). \tag{6.14}$$

Thus for any pure energy eigenstate of the harmonic oscillator,

$$\Delta P \, \Delta Q \geq \hbar\tfrac{1}{2}. \tag{6.15}$$

This relation is true in general, as we shall see later, and is a consequence of the Heisenberg communication relations $[P, Q] = (\hbar/i)I$. It is called the *Heisenberg uncertainty relation*.

We have seen above that in an eigenstate of an observable the dispersion of that observable is zero. An eigenstate is a state in which a measurement of the observable gives with certainty one value—the eigenvalue. In particular, we saw that in the eigenstate $\Lambda_n$ of $H$ the measurement of $H$ gave the value $E_n$ with certainty, i.e., with $w_n = 1$.

Now consider the mixture $W$ of the energy-loss experiment described by (4.17). By definition

$$\text{disp}_{(W)} H = \langle H^2 \rangle - \langle H \rangle^2. \tag{6.16}$$

Using (4.18) and the corresponding expression for $\langle H \rangle^2$, we have

$$\langle H^2 \rangle - \langle H \rangle^2 = \sum_\gamma E_\gamma^2 w_\gamma - \left( \sum_\gamma w_\gamma E_\gamma \right)^2$$

$$= \sum_\gamma w_\gamma \sum_\gamma E_\gamma^2 w_\gamma - \left( \sum_\gamma w_\gamma E_\gamma \right)^2 \geq 0. \tag{6.17}$$

For the last inequality we have used the Cauchy inequality for sequences:

$$\sum_{n=0}^{\infty} x_n^2 \sum_{n=0}^{\infty} y_n^2 \geq \left( \sum_{n=0}^{\infty} x_n y_n \right)^2 \quad \text{with } x_n = \sqrt{w_n}, y_n = E_n \sqrt{w_n}, \tag{6.18}$$

which is a consequence of (I.2.7). Thus $\text{disp}_{(W)} H$ in a mixture is always greater than or equal to zero. It is equal to zero if

$$w_\gamma = \delta_{n\gamma} \quad \text{i.e., } W = \Lambda_n,$$

and it deviates from zero according to the number of and values of $w_\gamma$ that differ from zero. Thus disp $H$, and therewith $\Delta H$, is indeed a quantitative expression for the uncertainty in the measurement of the observable $H$.

It is no surprise that disp $H$ or $\Delta H$ was greater than zero for the mixed state $W$ of CO molecules in the collision chamber, since that was a mixture of CO molecules of different energies. The reason that the outcome of the measurement of $H$ in $W$ is not fully determined is that not enough preparations had been performed to separate different systems, or more precisely, to separate systems in different states, from each other.

The harmonic oscillators in the collision chamber can be separated, at least in a *Gedanken* experiment, into eight sets, each one being a set of $N_n$ molecules having the same energy $E_n$, i.e., such that a measurement of $H$ on the $n$th such set gives with certainty the value $E_n$. The state of each set is then $\Lambda_n$. It is impossible to further separate (even in a *Gedanken* experiment) any of these sets into two or more subsets of different systems or systems in different states. This is the reason for calling $\Lambda_n$ a pure state. If we now perform the measurement[11] of an observable on the system in a pure state, one would expect from classical considerations that this should lead with certainty to a well-defined number, as a set in a pure state consists of only one species. If we choose for this observable $H$ or $N$, this is indeed the case. However, if we choose $P$ or $Q$, we see from (6.11), (6.12) that it is not. Even in pure states the outcome of the measurement of an observable is in general not uniquely determined. *Thus statistics cannot in principle be eliminated from quantum mechanics*; systems in the "same pure state" do not give identical values in a measurement. All that can be said is with what probability $w_n$ a certain value $a_n$ can be expected in the measurement of an observable.

Let us illustrate this with a *Gedanken* experiment. We consider the measurement of the observable $Q$ in the state $\Lambda_n$, and ask for the probability with which we can expect the value $x$ in a measurement of $Q$. By definition, the expectation value of an observable $A$ is

$$\langle A \rangle = \sum_i w_i a_i, \tag{6.19}$$

where $a_i$ are the eigenvalues of $A$, and $w_i$ are the probabilities for finding the values $a_i$ in a measurement of $A$. We calculate $\langle Q \rangle$ as follows:

$$\langle Q \rangle = \mathrm{Tr}(Q\Lambda_n) = \sum_i \langle i | Q\Lambda_n | i \rangle, \tag{6.20}$$

where $\psi_i = |i\rangle$ is any complete basis system and we have used the following notation:

$$(\phi, \psi) = \langle \phi | \psi \rangle \quad \text{or} \quad (\psi_i, \psi_j) = \langle i | j \rangle,$$
$$(\phi, A\psi) = \langle \phi | A | \psi \rangle. \tag{6.21}$$

For $|i\rangle$ we may take any complete basis system; let us therefore choose the system of eigenvectors of the operator $Q$, which we denote by $|x\rangle$, where $x$ is the eigenvalue of $Q$:

$$Q|x\rangle = x|x\rangle. \tag{6.22}$$

Then

$$\langle Q \rangle = \sum_x \langle x | Q\Lambda_n | x \rangle$$

$$= \sum_x x \langle x | \Lambda_n | x \rangle$$

$$= \sum_x x \langle x | n \rangle \langle n | x \rangle.$$

---

[11] An idealized measurement; see footnote 10.

The last step follows from writing $\Lambda_n = |n\rangle\langle n|$.
We introduce new notation, defining

$$\phi_n(x) \equiv \langle x|n\rangle. \tag{6.23}$$

Then

$$\langle Q \rangle = \sum_x x \phi_n(x)\overline{\phi_n(x)}$$

$$= \sum_x x|\phi_n(x)|^2. \tag{6.24}$$

If we now compare this with the general expression (6.19) for the expectation value of an operator, we conclude that

$$|\langle x|n\rangle|^2 = |\phi_n(x)|^2 = w_x \tag{6.25}$$

is the probability for obtaining the value $x$ in the measurement of the observable $Q$ for an oscillator in the state $\Lambda_n$. $\phi_n(x)$ is called the *wave function of the state* with energy $E_n$. Thus we have found that the probability for finding the value $x$ for the position operator $Q$ is the absolute value squared of the wave function. It was in this form that the probability interpretation was originally introduced in quantum mechanics by M. Born [influenced by an earlier suggestion of Albert Einstein (1916)].

As we shall discuss later, the $\sum_x$ is in fact an integral, and $w_x = |\langle x|n\rangle|^2$ is a quantity of dimension $cm^{-1}$, i.e., a *probability density*. The probability is the integral of $w_x$ over a certain interval.

We have derived the fact that $|\phi(x)|^2$ describes the probability of finding the physical system at $x$ from the basic assumption II. Historically the events took place in reverse order, and II is a generalization of the discovery by Max Born in June 1926 that the wave function $\phi(x)$, which—as we shall see in Section II.8—is a solution of the Schrödinger equation, is connected with probability.

Born was at that time investigating the collision process between free particles and an atom, using the formalism of Schrödinger's wave mechanics, which for this kind of process he found superior to the matrix mechanics of Heisenberg, Born, and Jordan. However, he could not accept Schrödinger's interpretation that $\phi(x)$ represents matter waves. Recalling these events, Born said in his Nobel Prize Lecture in 1954: "On this point [Schrödinger's matter-wave interpretation] I could not follow him. This was connected with the fact that my Institute and that of James Franck were housed in the same building in Göttingen University. Every experiment by Franck and his assistants on electron collision appeared to me as a new proof of the corpuscular nature of the electron." He calculated the formula for the scattered wave in the Schrödinger formalism (Born approximation[12]) and concluded that if this formula is to allow for a corpuscular interpretation, then the most natural interpretation of $\phi(x)$, is that $|\phi(x)|^2$ describes the

[12] See Section XIV.5 below.

probability density for detecting the corpuscle. He summarized his conclusion in 1926: "The motion of particles conforms to the law of probability but the probability itself is propagated in accordance with causality."

Born's interpretation was soon generalized to include other observables in addition to the position. And the time was ripe for the general formulation of quantum mechanics. The starting point for this was the discovery that the canonical transformations can be carried over to matrix mechanics and wave mechanics (F. London, May 1926; P. Jordan, July 1926). In this way the canonical transformations were recognized as transformations of an abstract space. Transformations of objects with continuous indices were introduced by Dirac and Jordan in December 1926. The mathematical elaboration of the London–Jordan–Dirac transformation theory was undertaken in Göttingen by David Hilbert together with L. Nordheim and John von Neumann (1926–1927). Notions like state and observable evolved. Subsequently, von Neumann developed an axiomatic theory of Hilbert spaces and established the association between physical states and Hilbert-space vectors as well as the correspondence between observables and linear operators. He also introduced the statistical operator.

Parallel to the mathematical formulation went the development of the physical interpretation, which was discussed principally in Copenhagen (1926–1927). Whereas Bohr wanted to make the wave-particle duality the starting point of the physical interpretation, Heisenberg relied upon the transformation theory. His basic approach was to regard something as defined if it is measureable. In his famous *Gedanken* experiment in which he discussed the position measurement of the electron with a microscope, he explained the uncertainty of position and momentum (6.15), and therewith the probabilistic nature of quantum-mechanical predictions, as the effect of the measurement apparatus. Object and observer have no independent reality; the measurement affects the state of the object (basic assumption III).

While Heisenberg derived his uncertainty principle from the transformation theory, Bohr developed his conception of complementarity from the wave-particle duality (see Section II.11): There exist complementary properties—like position and momentum—and the exact measurement of one precludes the possibility of obtaining information of the other. Properties are not actualities; they are only possibilities for the physical system. These developments formed the basis of the so-called Copenhagen interpretation of quantum mechanics.[13] Its physical content is formulated in the basic assumptions of this chapter.

---

[13] The interpretation of quantum mechanics suggested by Born and established by Heisenberg and the Copenhagen school represented such a drastic departure from the well-established ideas of their time that decades had to pass before it could gain general acceptance, and even today new attempts are being undertaken to return to a deterministic interpretation. Born's statistical interpretation of quantum mechanics was one of the most far-reaching contributions of physics to contemporary thinking.

Von Neumann's Hilbert-space formulation could not accommodate continuous eigenvalues. Since that time, stimulated by the Dirac formalism, a new branch of mathematics—the theory of distributions—was developed and for the first time systematically presented by L. Schwartz (1950–1951). On this basis I. M. Gelfand and collaborators introduced around 1960 the rigged Hilbert space. Its use for quantum mechanics was suggested around 1965.

## II.7 Some General Consequences of the Basic Assumptions of Quantum Mechanics

After this detailed discussion of the case of the harmonic oscillator, let us investigate some general consequences of the basic assumptions of quantum mechanics. Let us consider an arbitrary state $W$ of an arbitrary microphysical system and see what we can predict for the measurement of the values of some arbitrary Hermitian observables $A, B, \ldots$ of this system. These predictions are contained in the following two statements, which we will prove below.

1.  The dispersion of any observable $A$ in an arbitrary state $W$ fulfills

$$\text{disp}_{(W)} A \geq 0 \tag{7.1}$$

   with

$$\text{disp}_{(W)} A = 0$$

   if and only if $W$ is an eigenstate of $A$.
2.  The uncertainties of two observables $A$, $B$ in an arbitrary state $W$ fulfill the uncertainty relation

$$\Delta A \, \Delta B \geq \tfrac{1}{2} |\langle [A, B] \rangle_W|. \tag{7.2}$$

A special case of this uncertainty relation is the Heisenberg uncertainty relation; from

$$[P, Q] = \frac{\hbar}{i} I \quad \text{and} \quad \left\langle \frac{\hbar}{i} I \right\rangle = \frac{\hbar}{i} \, \text{Tr} \, WI = \frac{\hbar}{i}$$

it follows that

$$\Delta P \, \Delta Q \geq \frac{\hbar}{2} \tag{7.3}$$

for any state $W$. The equality sign in the Heisenberg uncertainty relation holds only if $W = \Lambda_0$, the ground state of the one-dimensional harmonic oscillator.

The uncertainty relation is a very general property of two operators, and just depends upon the assumption that the observables are represented by linear noncommuting operators in a linear space. It says that unless $A$ and $B$ are compatible observables, there are always states for which at least one of the observables cannot be measured with exact accuracy (since for every operator $C = i[A, B]$ there is always a $W$ such that

$$|\langle C \rangle_W| = |\mathrm{Tr}(WC)| \neq 0).$$

In the case that $C$ (or $-C$) is positive definite (as, e.g., for $A = P$ and $B = Q$), $|\mathrm{Tr}(WC)| > 0$ for every state $W$, and therefore the two observables $A$ and $B$ can never be measured simultaneously with unrestricted accuracy. If the system has been prepared in an eigenstate of one of the observables $A$, $\Delta A = 0$, (i.e., $A$ has been measured to give accurately one value), then according to (7.2) the values for $B$ are completely uncertain. This is a consequence of the previously discussed fact that in quantum physics a measurement changes the state, and the measurement of one observable interferes with the measurement of another observable unless these two observables are compatible, i.e., are represented by operators that commute with each other. Thus only compatible observables can be simultaneously measured, and only for compatible observables do there exist common eigenstates.

PROOF OF STATEMENT 1.   For every linear operator $B$ and every positive definite $W$ (for which the following operations are defined),

$$\mathrm{Tr}(WB^\dagger B) = \sum_\gamma (\varphi_\gamma, WB^\dagger B\varphi_\gamma) = \sum_\gamma (B^\dagger \varphi_\gamma, WB^\dagger \varphi_\gamma) \geq 0, \qquad (7.4)$$

because $((B^\dagger \varphi_\gamma), W(B^\dagger \varphi_\gamma)) \geq 0$ for every $(B^\dagger \varphi_\gamma)$, as $W$ is positive definite. For

$$B = A - \alpha I = B^\dagger \quad \text{with } \alpha = \langle A \rangle = \mathrm{Tr}\, WA \qquad (7.5)$$

it follows from the definition (6.7) of disp $A$ that

$$\mathrm{disp}_{(W)} A = \mathrm{Tr}(W(A - \alpha I)^2) \geq 0. \qquad (7.1')$$

Assume that

$$\mathrm{disp}_{(W)} A = \mathrm{Tr}(WBB) = 0 \quad \text{for } B = B^\dagger; \qquad (7.6)$$

then

$$\sum_{\kappa, b} (\varphi_b^\kappa, WBB\varphi_b^\kappa) = \sum_{b, \kappa} b^2(\varphi_b^\kappa, W\varphi_b^\kappa) = 0, \qquad (7.7)$$

where the basis $\varphi_b^\kappa$ is chosen as in (4.35)–(4.37). As $W$ is positive definite, all terms in the sum are $\geq 0$, and it follows that

$$b^2(\varphi_b^\kappa, W\varphi_b^\kappa) = 0 \quad \text{for every } b, \kappa.$$

As $W \neq 0$, there must be at least one $\varphi_b^\kappa$ for which

$$(\varphi_{b'}^\kappa, W\varphi_{b'}^\kappa) \geq 0. \qquad (7.8)$$

For this $\varphi_{b'}^\kappa$, $b'^2 = 0$, i.e., $b' = 0$.
    Let us assume that there is only one such $b'$.
If we choose (7.5) for $B$, then $\varphi_b^\kappa$ is also an eigenvector of $A$ and

$$(A - \alpha I)\varphi_b^\kappa = b\varphi_b^\kappa = (a - \alpha)\varphi_b^\kappa. \qquad (7.9)$$

The eigenvalue $a'$ that corresponds to the eigenvalue $b'$ of (7.8) must therefore fulfill

$$a' = \alpha = \mathrm{Tr}(WA). \qquad (7.10)$$

From this it follows that

$$W = \Lambda_{a'} \frac{1}{\mathrm{Tr}\,\Lambda_{a'}},$$

the eigenstate of $A$ with eigenvalue $a'$. Let us now assume that there is more than one value $b'$ that fulfills (7.8), say $b'_1$ and $b'_2 \neq b'_1$. Then there must be two corresponding value $a'_1 \neq a'_2$ that fulfill

$$a'_1 = \mathrm{Tr}(WA) \quad \text{and} \quad a'_2 = \mathrm{Tr}(WA), \tag{7.11}$$

which is impossible.

Thus if (7.6) is fulfilled, then

$$W = \Lambda_{a'} \frac{1}{\mathrm{Tr}\,\Lambda_{a'}}, \tag{7.12}$$

where $\Lambda_{a'}$ is any of the eigenstates of $A$. That disp $A = 0$ for $W = \Lambda_a$ an eigenstate of $A$ follows immediately from (7.7). Therefore statement 1 is proven.    □

**PROOF OF STATEMENT 2.**    For two operators $A$ and $B$ we define

$$h(A, B) \equiv \mathrm{Tr}\,WA^{\dagger}B = \mathrm{Tr}(WI(A^{\dagger}B)) = h(I, A^{\dagger}B) = \langle A^{\dagger}B \rangle. \tag{7.13}$$

One easily shows from (4.3) and (4.4) that

(a)  $h(A, A) \geq 0$ (which is done in (7.4)),
(b)  $h(A, B) = \overline{h(B, A)}$,
(c)  $h(aA, B) = \bar{a}h(A, B)$,
(d)  $h(A + C, B) = h(A, B) + h(C, B)$,

i.e., $h(A, B)$ defined by (7.13) fulfills all the conditions of (I.2.5) that define a positive Hermitian form. Since $h(A, B)$ is a positive Hermitian form, the Schwarz inequality (I.2.7) holds:

$$|h(A, B)|^2 \leq h(A, A)h(B, B). \tag{7.14}$$

From this, statement 2 is easily proven (we assume $A, B$ to be Hermitian):

$$|\langle [A, B] \rangle|^2 = |h(I, [A, B])|^2 = |h(A, B) - h(B, A)|^2 \leq 4h(A, A)h(B, B)$$

so if we replace

$$A \to A - \alpha I, \qquad \alpha = \mathrm{Tr}(WA)$$

$$B \to B - \beta I, \qquad \beta = \mathrm{Tr}(WB),$$

we obtain

$$h(A - \alpha I, A - \alpha I)h(B - \beta I, B - \beta I) \geq \tfrac{1}{4}|h(I, [A - \alpha I, B - \beta I])|^2$$

of (for $A$ and $B$ Hermitian)

$$\mathrm{disp}\,A \cdot \mathrm{disp}\,B \geq \tfrac{1}{4}|h(1, [A, B])|^2 = \tfrac{1}{4}|\langle [A, B] \rangle|^2. \tag{7.15}$$

The square root of this gives (7.2), and therefore statement 2 has been proven.    □

## II.8 Eigenvectors of Position and Momentum Operators; the Wave Functions of the Harmonic Oscillator

In this section we shall investigate the properties of the wave functions. We will begin by examining the properties of the operators $P$ and $Q$; in particular, we wish to know what are the possible eigenvalues $x$ of $Q$. That is, for which values $x$ can

$$Q|x) = x|x) \tag{8.1}$$

be fulfilled? Then we wish to find the transition coefficient $(x|n)$ which transforms between a vector $|n) = \phi_n$ (an eigenstate of $H$ and $N$) and the vectors $|x)$ (eigenstates of $Q$):

$$|n) = \sum_x |x)(x|n). \tag{8.2}$$

From (3.1) and (3.26) it follows that

$$Q|n) = \sqrt{\frac{\hbar}{2\mu\omega}}\,(a + a^\dagger)|n)$$

$$= \sqrt{\frac{\hbar}{2\mu\omega}}\,(\sqrt{n}|n-1) + \sqrt{n+1}|n+1)). \tag{8.3}$$

Taking the scalar product of this equation with $|x)$, we obtain

$$(n|Q|x) = \sqrt{\frac{\hbar}{2\mu\omega}}\,(\sqrt{n}(n-1|x) + \sqrt{n+1}(n+1|x)). \tag{8.4}$$

On the other hand, taking the scalar product of (8.1) with $|n)$, we obtain

$$(n|Q|x) = x(n|x). \tag{8.5}$$

Comparing (8.4) with (8.5) yields

$$x(n|x) = \sqrt{\frac{\hbar}{2\mu\omega}}\,(\sqrt{n}(n-1|x) + \sqrt{n+1}(n+1|x)), \tag{8.6}$$

or, denoting $n + 1 = m$,

$$\sqrt{m}(m|x) = \sqrt{\frac{2\mu\omega}{\hbar}}\,x(m-1|x) - \sqrt{m-1}(m-2|x). \tag{8.7}$$

Since (8.3) is valid for $n = 1, 2, \ldots$, (8.7) is valid for $m = 2, 3, \ldots$. For $n = 0$ (i.e., $m = 1$), we obtain instead of (8.3)

$$Q|0) = \sqrt{\frac{\hbar}{2\mu\omega}}\,\sqrt{0+1}|0+1) = \sqrt{\frac{\hbar}{2\mu\omega}}\,|1); \tag{8.3a}$$

and instead of (8.7),

$$\sqrt{1}(1|x) = \sqrt{\frac{2\mu\omega}{\hbar}}\,x(0|x). \tag{8.7a}$$

Thus we see that (8.7) is a recurrence relation for $(m|x)$; if $(0|x)$ is known, we can determine $(1|x)$ by (8.7a) and then determine $(2|x)$ by (8.7). With $(1|x)$ and $(2|x)$ we can determine $(3|x)$ by (8.7), and so on.

To find out what the transition coefficients $(m|x)$ are, we introduce

$$y \equiv \sqrt{\frac{\mu\omega}{\hbar}}\, x \qquad (8.8)$$

and

$$f_n(y) \equiv \sqrt{2^n n!}\, \frac{(n|x)}{(0|x)}, \qquad (8.9)$$

which is defined for all $x$ with $(0|x) \neq 0$ (if $(0|x) = 0$, then by (8.7) and (8.7a) $(n|x) = 0$ for all $n$). Then from (8.7) it follows that

$$\sqrt{\frac{m}{2^m m!}}\, f_m(y) = \sqrt{\frac{2}{2^{m-1}(m-1)!}}\, y f_{m-1}(y) - \sqrt{\frac{m-1}{2^{m-2}(m-2)!}}\, f_{m-2}(y), \qquad (8.10)$$

or

$$f_m(y) = 2y f_{m-1}(y) - 2(m-1) f_{m-2}(y). \qquad (8.11)$$

From (8.7a) we have

$$f_1(y) = 2y f_0(y), \qquad (8.12)$$

and from (8.9)

$$f_0(y) = 1. \qquad (8.13)$$

Equations (8.11), (8.12), and (8.13) are the recurrence relations for the Hermite functions and have solutions for any complex number $y$.[14] Thus for any complex value $x$ there is a solution $(n|x)$ of the recurrence relation (8.7). For physical reasons we need to consider only real values of $x$, because $x$ is the value that is obtained in the position measurement and must be real. [Mathematically, nonreal values of $x$ can be excluded by the requirement that $(n|x)$ be square-integrable.] For real values of $y$, the solutions $f_m(y)$ of (8.11), (8.12), (8.13) are the Hermite polynomials:

$$f_n(y) = H_n(y) = (-1)^n e^{y^2} \frac{d^n(e^{-y^2})}{dy^n}. \qquad (8.14)$$

Thus from (8.9) we may obtain the transition coefficient $(n|x)$ for every real value of $x$ for which $(0|x)$ is defined. We restrict ourselves to those solutions of (8.7) for which $(0|x)$ is finite, because $|(0|x)|^2$, the probability for obtaining the value $x$ in a measurement of $Q$ in the ground state $\Lambda_0$, is assumed to be finite.

---

[14] N. N. Lebedev, *Special Functions and Their Applications*, Prentice Hall (1965) Section 10.4.

Combining (8.8), (8.9), and (8.14), we have

$$(n|x) = \frac{1}{\sqrt{2^n n!}} (0|x) H_n \left( \sqrt{\frac{\mu\omega}{\hbar}} x \right) \qquad (8.15)$$

for $-\infty < x < +\infty$.

〚The set $(\psi|x) = \sum_n \bar{a}_n(\phi_n|x)$, considered for a fixed value $x$ as a function of $\psi$, is an antilinear functional over the space $\Phi$ of the $\psi$ in Equation (3.27) [i.e., it fulfills Equations (I.2.8)]. Thus $|x)$ is a functional, and one can show that for real values of $x$, $|x)$ is even a continuous functional [Equation (I.2.9)]. As was mentioned in Chapter I, the set of continuous linear functionals on a space $\Phi$, which we denote by $\Phi^\times$, is in general larger than the space $\Phi$, except for the Hilbert space $\mathscr{H}$, where $\mathscr{H}^\times = \mathscr{H}$ (Frechet–Riesz theorem.) Taking into account (3.28), we conclude $\Phi \subset \mathscr{H} \subset \Phi^\times$. This triplet of spaces is called the *Gelfand triplet* or *rigged Hilbert space*. The $|x)$ that obey (8.1) are elements of $\Phi^\times$ and are called *generalized eigenvectors* of the operator $Q$; the set

$$|x) \qquad -\infty < x < +\infty$$

is called a *generalized basis*. The set of real eigenvalues of a Hermitian operator we call its *spectrum*.* The spectrum of the operator $N$, (2.31), or $H$, (2.37a), is called *discrete*, because it is a discrete set; the spectrum of the operator $Q$ is called *continuous*, because it is a continuous set. As $x$ can be any real value, the sum in (8.2) should actually by an integral. A famous theorem, the *nuclear spectral theorem*, states that every $\varphi \in \Phi$ can be written as the continuous form of (8.2), i.e.,

$$\varphi = \int dx \, |x\rangle\langle x|\varphi\rangle = \int dx \, |x\rangle\varphi(x). \qquad (8.16)$$

Here we denote the generalized eigenvectors of $Q$,

$$Q|x\rangle = x|x\rangle, \qquad (8.17)$$

by $|x\rangle$ instead of $|x)$ to emphasize the difference between them and the eigenvectors with discrete spectrum, $|n)$. If we take the "scalar product" of $\varphi$ in (8.16) with $|x'\rangle$, or more precisely the functional $|x'\rangle$ at the element $\varphi$, we have

$$\langle x'|\varphi\rangle = \int dx \, \langle x'|x\rangle\langle x|\varphi\rangle. \qquad (8.18)$$

$\langle x'|x\rangle$ is therefore that mathematical object which by integration (8.18) maps the "well-behaved" function $\varphi(x) = \langle x|\varphi\rangle$ into its value at the point $x'$: $\varphi(x') = \langle x'|\varphi\rangle$. One can prove that such an

---

* This is not always identical with the precise definition of the spectrum of a selfadjoint operator in the Hilbert space.

object cannot be a locally integrable function; it is called a *generalized functions* or *distribution*. The particular distribution $\langle x'|x\rangle$ is called the *Dirac δ-function* and is often denoted

$$\langle x'|x\rangle = \delta(x' - x), \tag{8.19}$$

so (8.18) is written

$$\varphi(x') = \langle x'|\varphi\rangle = \int dx\,\delta(x' - x)\langle x|\varphi\rangle$$

$$= \int dx\,\delta(x' - x)\varphi(x). \tag{8.20}$$

In the transition from the symbolic (8.2) to (8.16),

$$\sum_x |x)(x| \quad \text{was replaced by} \quad \int dx\,|x\rangle\langle x|.$$

Consequently, the states $|x)$ and the generalized states $|x\rangle$ must have different dimensions in order that

$$|x)(x| \quad \text{and} \quad dx\,|x\rangle\langle x|$$

may have the same dimension.

The eigenvectors are always chosen so that they can be normalized to 1, i.e., they are chosen to be dimensionless. This is expressed by the normalization of a system of eigenvectors of an operator with a discrete spectrum:

$$(n|n') = \delta_{nn'}, \quad \text{where } \delta_{nn'} = \begin{cases} 1 & \text{for } n = n', \\ 0 & \text{for } n \neq n'. \end{cases} \tag{8.21}$$

The generalized eigenvectors $|x\rangle$, on the other hand, do not have dimension 1 but rather the inverse of $\sqrt{\dim dx}$. For example, if $dx$ has the dimension cm, then $\langle x'|x\rangle$ has dimension cm$^{-1}$. The generalized eigenvectors are normalized in the sense of (8.19) and the δ-function has the same dimension as $x^{-1}$. Instead of generalized vectors with the δ-function normalization (8.19), we could also choose generalized eigenvectors $|x\}_\mu$ with a different normalization. Instead of (8.16) we could write

$$\varphi = \int d\mu(x)\,|x\}_\mu\,{}_\mu\{x|\varphi\} \tag{8.22}$$

with the "normalization"

$${}_\mu\{x'|x\}_\mu = g(x)\delta(x' - x), \tag{8.23}$$

where

$$d\mu(x)\,g(x) = dx \tag{8.24}$$

The most convenient choice of $g$ depends upon the property of the operator. For the operator $Q$, $g(x) = 1$ is a suitable choice; we will therefore use the generalized eigenvectors $|x\rangle$.

Equation (8.17) is the analogue, for the operator with continuous spectrum, of (3.20) for the operator $N$ (or $H$) with discrete spectrum. As $\varphi$ in (8.16) is an arbitrary vector, we can omit it and obtain the analogue of (4.9):

$$I = \int dx \, |x\rangle\langle x|. \qquad (8.25)$$

If we apply the operator $Q$ to the arbitrary vector in (8.16) we obtain

$$Q\varphi = \int dx \, x|x\rangle\langle x|\varphi\rangle,$$

or, if we omit the arbitrary $\varphi$, we obtain the continuous analogue of (4.42):

$$Q = \int dx \, x|x\rangle\langle x|. \qquad (8.26)$$

This is called the spectral representation of the operator $Q$. It is the analogue of (4.42), which was the form for operators with discrete spectrum. The $|x\rangle\langle x|$ are the analogues of the projection operators; however, they do not project on any vector of the Hilbert space $\mathcal{H}$ (or $\Phi$), because the $|x\rangle$ are not elements of these spaces. Projection operators are $\int_{\Delta x} dx \, |x\rangle\langle x|$ where the integration extends over a finite interval $\Delta x$. ⟧

We now return to the evaluation of the harmonic-oscillator function and calculate $\langle 0|x\rangle$. We write $\phi_n = |n\rangle$ in the form (8.16),

$$|n\rangle = \int dx \, |x\rangle\langle x|n\rangle, \qquad (8.16')$$

and take the scalar product of this with $\phi_m$,

$$\delta_{mn} = (\phi_m, \phi_n) = (m|n) = \int dx \, (m|x\rangle\langle x|n). \qquad (8.27)$$

We shall make use of

$$\langle x|n\rangle = \overline{(n|x\rangle}. \qquad (8.28)$$

⟦If $|n\rangle = \phi_n$ and $|x\rangle = \chi$ both were vectors in the Hilbert space then (8.28) would be the condition (I.2.2b) of the definition of the scalar product. However $(n|x\rangle$ is the functional $|x\rangle \in \Phi^\times$ at the element $|n) \in \Phi$ and $\langle x|n\rangle$ is a functional $|n) \in \Phi^{\times\times} = \Phi$ at the element $|x\rangle \in \Phi^\times$, so (8.28) is not obvious. But it can be shown that (8.28) is generally true for transition coefficients between a discrete basis and a continuous basis. Hereafter we shall drop the typographical

distinction $|\,)$ as opposed to $|\,\rangle$, and shall use generally $\rangle$, unless we
we want to stress the distinction between generalized and ordinary
vectors.$]\!]$

We insert (8.15) into (8.27) and obtain

$$\sqrt{\frac{\hbar}{2^n 2^m n!\, m!\, \mu\omega}}\int dy\, \left|\left\langle 0\left|\sqrt{\frac{\hbar}{\mu\omega}}\, y\right\rangle\right|^2 H_m(y)H_n(y) = \delta_{mn}. \qquad (8.29)$$

If we compare this with the orthogonality relations for the Hermite polynomials,[15]

$$\int dy\, e^{-y^2} H_m(y)H_n(y) = \begin{cases} 0 & \text{for } m \neq n, \\ n!\, 2^n \sqrt{\pi} & \text{for } m = n, \end{cases} \qquad (8.30)$$

we find

$$\left|\left\langle 0\left|\sqrt{\frac{\hbar}{\mu\omega}}\, y\right\rangle\right|^2 = \sqrt{\frac{\mu\omega}{\pi\hbar}}\, e^{-y^2}. \qquad (8.31)$$

Thus, up to an arbitrary phase factor (which we choose to be real),

$$\langle 0|x\rangle = \left(\frac{\mu\omega}{\pi\hbar}\right)^{1/4} e^{-(\mu\omega/2\hbar)x^2}. \qquad (8.32)$$

With this and (8.15) we obtain the transition coefficients $\langle n|x\rangle$ between
the $x$- and $n$-bases, i.e., the harmonic-oscillator wave functions $\phi_n(x)$:

$$\phi_n(x) = \langle n|x\rangle = \left(\frac{\mu\omega}{\pi\hbar}\right)^{1/4} \frac{1}{\sqrt{2^n n!}}\, H_n\left(\sqrt{\frac{\mu\omega}{\hbar}}\, x\right) e^{-(\mu\omega/2\hbar)x^2}. \qquad (8.33)$$

A few of the wave functions are plotted in Figure 8.1.
    We now repeat the procedure for the operator $P$ that we have already
gone through for the operator $Q$. The eigenvectors of $P$ will be denoted
by $|p\rangle$:

$$P|p\rangle = p|p\rangle \qquad (8.34)$$

The action of $P$ on $|n\rangle$ (calculated in the problems for Chapter IV) is

$$P|n\rangle = -i\sqrt{\frac{\hbar\mu\omega}{2}}\, (a - a^\dagger)|n\rangle$$

$$= -i\sqrt{\frac{\hbar\mu\omega}{2}}\, (\sqrt{n}\,|n-1\rangle - \sqrt{n+1}\,|n+1\rangle). \qquad (8.35)$$

---

[15] Lebedev, ibid., Section 4.13.

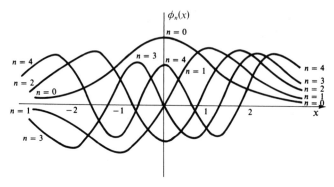

Figure 8.1   Wave functions of the harmonic oscillator [from Ludwig (1954), with permission].

If we take the "scalar product" with $|p\rangle$ and use (8.34), we obtain

$$p\langle p|n\rangle = -i\sqrt{\frac{\hbar\mu\omega}{2}}\,(\sqrt{n}\langle p|n-1\rangle - \sqrt{n+1}\langle p|n+1\rangle), \quad (8.36)$$

or

$$p\langle n|p\rangle = i\sqrt{\frac{\hbar\mu\omega}{2}}\,(\sqrt{n}\langle n-1|p\rangle - \sqrt{n+1}\langle n+1|p\rangle). \quad (8.37)$$

If we introduce the new quantities $(n|p)$ defined by

$$\langle n|p\rangle = i^n(n|p), \quad (8.38)$$

then

$$\begin{aligned} i\langle n-1|p\rangle &= ii^{n-1}(n-1|p), \\ -i\langle n+1|p\rangle &= -ii^{n+1}(n+1|p), \end{aligned} \quad (8.39)$$

so (8.37) may be written

$$p(n|p) = \sqrt{\frac{\hbar\mu\omega}{2}}\,(\sqrt{n}(n-1|p) + \sqrt{n+1}(n+1|p)). \quad (8.40)$$

We see that this is exactly the same recurrence relation as in (8.6), with $x\sqrt{\mu\omega/\hbar}$ replaced by $p/\sqrt{\mu\omega\hbar}$. Thus by the same argument as for $\langle n|x\rangle$, we find [using (8.38)] that

$$\langle n|p\rangle = i^n\left(\frac{1}{\pi\mu\omega\hbar}\right)^{1/4}\frac{1}{\sqrt{2^n n!}}\,H_n\left(\frac{1}{\sqrt{\hbar\mu\omega}}\,p\right)e^{-p^2/2\mu\omega\hbar}. \quad (8.41)$$

We have seen that the eigenvectors $|n\rangle$ of the energy operator $H$ for the harmonic oscillator have the very particular property that the transition coefficients (8.33) between these vectors and the $x$-basis are the same as the transition coefficients (8.41) between these vectors and the $p$-basis except for a phase factor.

By the same argument as above for the operator $Q$, we conclude that the spectrum of $P$ is continuous,

$$\text{spectrum } P = \{p \mid -\infty < p < \infty\}, \tag{8.42}$$

and the $|p\rangle$ are generalized eigenvectors.

The transition coefficients $\langle p|n\rangle$ in

$$|n\rangle = \int dp \, |p\rangle\langle p|n\rangle \tag{8.43}$$

are called the *wave functions in the momentum representation* and are denoted

$$\phi_n(p) \equiv \langle p|n\rangle. \tag{8.44}$$

The probability of finding the value $p$ in a measurement of the momentum $P$ in the energy eigenstate $\Lambda_n$ of the oscillator is $w_n(p) = |\langle p|n\rangle|^2$. Also, for any arbitrary vector $\varphi$ the transition coefficient

$$\varphi(p) = \langle p|\varphi\rangle \tag{8.45}$$

in

$$\varphi = \int dp \, |p\rangle\langle p|\varphi\rangle \tag{8.46}$$

is called the *momentum-space wave function* or the *wave function in the p-representation* for the state $\varphi$.

So far, we have obtained the matrix element of $Q$ in the $x$-representation,

$$\langle n|Q|x\rangle = x\langle n|x\rangle; \tag{8.47}$$

or more generally, for every well-behaved vector $\varphi$,

$$\langle \varphi|Q|x\rangle = x\langle \varphi|x\rangle \qquad \langle x|Q|\varphi\rangle = x\langle x|\varphi\rangle. \tag{8.48}$$

Likewise, we have obtained the matrix element of $P$ in the $p$-representation,

$$\langle n|P|p\rangle = p\langle n|p\rangle; \tag{8.49}$$

or more generally,

$$\langle \varphi|P|p\rangle = p\langle \varphi|p\rangle \qquad \langle p|P|\varphi\rangle = p\langle p|\varphi\rangle. \tag{8.50}$$

We now want to calculate the matrix elements of $P$ in the basis of generalized eigenvectors of $Q$ and the matrix elements of $Q$ in the basis of generalized eigenvectors of $P$. We do this in two steps: First we find out what the symbol $\langle x|p\rangle$, the transition matrix element between the $x$-basis and the $p$-basis, represents, and then we determine $\langle x|P|n\rangle$.

We begin by taking the scalar product of (8.16′) with $|p\rangle$ [or, more precisely, we will consider $\phi_n$ as a functional at the generalized eigenvector $|p\rangle \in \Phi^\times$, $p \in$ spectrum $P$, and use (8.16′)]:

$$\langle p|n\rangle = \int dx \, \langle p|x\rangle\langle x|n\rangle; \tag{8.51}$$

and then we take the scalar product of (8.43) with $\langle x|$ [or more precisely, consider $\phi_n$ as a functional at the generalized eigenvector

$$|x\rangle \in \Phi^\times, \, x \in \text{spectrum } Q,$$

and use (8.43)]:

$$\langle x|n\rangle = \int dp \, \langle x|p\rangle\langle p|n\rangle. \tag{8.52}$$

In (8.51) and (8.52), $\langle x|n\rangle$ and $\langle p|n\rangle$ are given by (8.33) and (8.41), respectively. The Hermite polynomials have the following property[16]:

$$i^n e^{-\eta^2/2} H_n(\eta) = \int_{-\infty}^{+\infty} d\xi \, \frac{e^{i\xi\eta}}{\sqrt{2\pi}} \, e^{-\xi^2/2} H_n(\xi). \tag{8.53}$$

Inserting (8.41) and (8.33) into this relation, it follows that

$$\langle n|p\rangle = \int_{-\infty}^{+\infty} dx \, \frac{e^{ixp/\hbar}}{\sqrt{2\pi\hbar}} \, \langle n|x\rangle, \tag{8.54}$$

or, taking the complex conjugate,

$$\langle p|n\rangle = \int dx \, \frac{e^{-ixp/\hbar}}{\sqrt{2\pi\hbar}} \, \langle x|n\rangle. \tag{8.55}$$

Comparing (8.55) with (8.51), we find that the $\langle p|x\rangle$ are given by

$$\langle p|x\rangle = \frac{1}{\sqrt{2\pi\hbar}} \, e^{-ixp/\hbar}. \tag{8.56}$$

In the same way one obtains from (8.52) and (8.53)

$$\langle x|p\rangle = \frac{1}{\sqrt{2\pi\hbar}} \, e^{ixp/\hbar}. \tag{8.57}$$

(8.56) and (8.57) together give

$$\langle x|p\rangle = \overline{\langle p|x\rangle}. \tag{8.58}$$

⟦We emphasize that (8.58) is not trivial, because $\langle x|p\rangle$ is not a scalar product and thus does not follow from the "Hermiticity property of the 'scalar product' $\langle x|p\rangle$." It is a very particular property of the operators $P$ and $Q$, and is not true for the general case of transition coefficients between two arbitrary systems of generalized eigenvectors.⟧

---

[16] Lebedev, ibid., Section 4.12.

It is now simple to calculate the matrix element of $P$ in the basis of generalized eigenvectors of $Q$, using (8.57):

$$\langle x|P|\varphi\rangle = \int dp\, p\langle x|p\rangle\langle p|\varphi\rangle = \int dp\, p\,\frac{e^{ixp/\hbar}}{\sqrt{2\pi\hbar}}\,\langle p|\varphi\rangle$$

$$= \int dp\, \frac{\hbar}{i}\frac{d}{dx}\langle x|p\rangle\langle p|\varphi\rangle = \frac{\hbar}{i}\frac{d}{dx}\int dp\langle x|p\rangle\langle p|\varphi\rangle.$$

Thus

$$\langle x|P|\varphi\rangle = \frac{\hbar}{i}\frac{d}{dx}\langle x|\varphi\rangle. \tag{8.59}$$

In the same way, using (8.56), one obtains

$$\langle p|Q|\varphi\rangle = -\frac{\hbar}{i}\frac{d}{dp}\langle p|\varphi\rangle. \tag{8.60}$$

Each vector in the linear space can be fully characterized by its components with respect to a basis system. In the same way as one represents the vector $\mathbf{x}$ in the 3-dimensional space by its component $x_i$ with respect to the three basis vectors $\mathbf{e}_i$, according to $\mathbf{x} = \sum_{i=1}^{3} \mathbf{e}_i x_i$, one can represent the vector $\varphi$ by its component $\langle n|\varphi\rangle$ with respect to the discrete basis $\phi_n = |n\rangle$ of eigenvectors of $H$ according to $\varphi = \sum_{n=0}^{\infty}|n\rangle\langle n|\varphi\rangle$ [Equation (4.8)]. This is also true if the basis is a continuous basis, according to

$$\varphi = \int dx\, |x\rangle\langle x|\varphi\rangle. \tag{8.16}$$

Thus to the vector $\varphi$ corresponds the function $\langle x|\varphi\rangle = \varphi(x)$ and to the vector $P\varphi$ corresponds the function $\langle x|P|\varphi\rangle$. Equation (8.59) then states that in the realization of the space of vectors $\varphi$ by the space of wave functions $\langle x|\varphi\rangle = \varphi(x)$, the momentum operator is realized by the differential operator times $\hbar/i$, and (8.48) states that the position operator is realized by the operator of multiplication with the number $x$. This *realization* is called the *Schrödinger representation*.

⟦It should be noted that the Schrödinger representation could not be derived from the commutation relation (2.1a) alone, but required an additional assumption: that the operator $H$ in (2.1b) has at least one proper eigenvector. This assumption is one of several equivalent forms of the additional requirement. Other forms are that $P^2 + Q^2$ be essentially self adjoint or that the representation of the Heisenberg algebra be integrable to the representation of the Weyl group (Weyl's form of the canonical commutation relation).⟧

The Schrödinger representation of the energy eigenvalue equation

$$H\phi_n = E_n\,\phi_n \tag{8.61}$$

is the time-independent Schrödinger equation. The energy operator for the oscillator was given by (2.1b):

$$H = \frac{1}{2\mu} P^2 + \frac{\mu\omega^2}{2} Q^2. \tag{8.62}$$

Let us take the matrix element of $H$ between $\langle x|$ and $|n\rangle$:

$$\langle x|H|n\rangle = \frac{1}{2\mu} \langle x|P^2|n\rangle + \frac{\mu\omega^2}{2} \langle x|Q^2|n\rangle. \tag{8.63}$$

We compute

$$\langle x|P^2|n\rangle = \sum_{n'} \langle x|P|n'\rangle\langle n'|P|n\rangle$$

$$= \sum_{n'} \frac{\hbar}{i} \frac{d}{dx} \langle x|n'\rangle\langle n'|P|n\rangle$$

$$= \frac{\hbar}{i} \frac{d}{dx} \sum_{n'} \langle x|n'\rangle\langle n'|P|n\rangle$$

$$= \frac{\hbar}{i} \frac{d}{dx} \langle x|P|n\rangle$$

$$= \left(\frac{\hbar}{i}\right)^2 \frac{d^2}{dx^2} \langle x|n\rangle. \tag{8.64}$$

Moreover, from (8.1),

$$\langle x|Q^2|n\rangle = x^2\langle x|n\rangle. \tag{8.65}$$

Inserting these in (8.63), we have for the matrix element of the energy operator

$$\langle x|H|n\rangle = -\frac{\hbar^2}{2\mu} \frac{d^2}{dx^2} \langle x|n\rangle + \frac{\mu\omega^2}{2} x^2\langle x|n\rangle$$

$$= \left(-\frac{\hbar^2}{2\mu} \frac{d^2}{dx^2} + \frac{\mu\omega^2}{2} x^2\right)\langle x|n\rangle. \tag{8.66}$$

Summarizing, we have seen that in the $x$-representation, the operator $P$ is given by $-i\hbar\, d/dx$,

$$P \leftrightarrow -i\hbar \frac{d}{dx}, \tag{8.67}$$

and the energy operator by

$$H \leftrightarrow \left(-\frac{\hbar}{2\mu} \frac{d^2}{dx^2} + \frac{\mu\omega^2}{2} x^2\right)$$

## II.9 Comparison between Quantum and Classical Harmonic Oscillators

As stated above, the wave functions $\langle x|n\rangle$ and $\langle p|n\rangle$ give the probability densities $w_n(x) = |\langle x|n\rangle|^2$ and $w_n(p) = |\langle p|n\rangle|^2$ for finding the value $x$ in a position measurement and the value $p$ in a momentum measurement after the energy has been measured with the result $E_n = \hbar\omega(n + \frac{1}{2})$. For physical reasons it is impossible to measure the position exactly at one point $x$. In order to do this, we would have to build an apparatus (counter) that registers when the particle is exactly at the point $x$. But this is impossible, because such an apparatus must have finite size. Thus the physical question to ask is, what is the probability $w_n(x - \epsilon, x + \epsilon)$ of finding the position in the interval between $x - \epsilon$ and $x + \epsilon$ after an energy measurement had given the result $E_n$? This probability is given by

$$w_n(x - \epsilon, x + \epsilon) = \int_{x-\epsilon}^{x+\epsilon} dx'\, w_n(x')$$

$$= \int_{x-\epsilon}^{x+\epsilon} dx'\, |\langle x|n\rangle|^2. \tag{9.1}$$

Let us consider the specific cases $n = 0$ and $n = 1$:

$$\langle x|n = 0\rangle = \frac{1}{\sqrt{x_0}\sqrt{\pi}}\, e^{-x^2/2x_0^2}, \tag{9.2}$$

$$\langle x|n = 1\rangle = \frac{1}{\sqrt{x_0}\sqrt{\pi}}\, e^{-x^2/2x_0^2}\, 2\frac{x}{x_0}, \tag{9.3}$$

where $x_0 = \sqrt{\hbar/\mu\omega}$. Then the probability of measuring the position in an interval $\Delta x$ around the value $x$ is given for small $\Delta x$ by

$$w_0\left(x - \frac{\Delta x}{2}, x + \frac{\Delta x}{2}\right) = w_0(x)\,\Delta x = \frac{1}{x_0\sqrt{\pi}}\, e^{-x^2/x_0^2}\,\Delta x \tag{9.4}$$

and

$$w_1\left(x - \frac{\Delta x}{2}, x + \frac{\Delta x}{2}\right) = w_1(x)\,\Delta x = \frac{2}{\sqrt{\pi}}\frac{x^2}{x_0^3}\,\Delta x\, e^{-x^2/x_0^2}.$$

Figure 9.1 shows $w_0(x) = w_{qm}$, and Figure 9.2 shows $w_1(x) = w_{qm}$. In order to compare the quantum-mechanical results with the classical ones, $w_{cl}(x)$ is also shown in the figures, where $w_{cl}(x)\,\Delta x$ is the probability for finding the particle of the classical oscillator in the interval $[x, x + \Delta x]$ when its energy is $E_0$ or $E_1$.

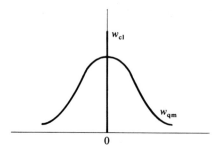

Figure 9.1   Classical and quantum probability densities for the state of an oscillator with the lowest energy $E_0$.

A simple comparison with the classical situation may be seen in the frequently used diagrams that display both the energy levels and the potential-energy function. For the potential energy

$$U(x) = \tfrac{1}{2}\mu\omega^2 x^2 \tag{9.5}$$

of the oscillator this is shown in Figure 9.3. Let us consider, for example, the level $E_1$. According to classical mechanics a particle with energy $E_1$ could be found only in the range $AB$, since $A$ and $B$ are the turning points, i.e., the points at which the potential energy is equal to the total energy. At these points the kinetic energy $T$ is zero, since

$$E = T + U \qquad T = E - U. \tag{9.6}$$

Evidently $OA = OB$ is the amplitude of oscillation of a particle having energy $E_1$.

The classical probability distribution is found in the following way: The period of oscillation is $\tau = 2\pi/\omega$. The probability of finding the particle

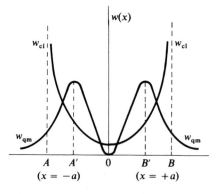

Figure 9.2   Comparison of the quantum position probability density $w_{qm}$ of a particle (for $n = 1$) with the classical density $w_{cl}$. $A$, $B$: turning points; $A'$, $B'$: points of maximum $w_{qm}$.

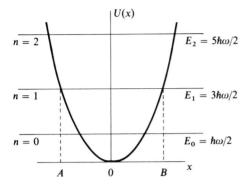

Figure 9.3    Diagram of quantum levels $E_n$ and potential energy $U(x) = \frac{1}{2}\mu\omega^2 x^2$ for a harmonic oscillator.

inside $\Delta x$ is given by the fraction of time $\Delta t/\tau$ that the particle stays inside the interval $\Delta x$:

$$w_{cl}(x)\,\Delta x = \frac{\Delta t}{\tau} = \frac{\omega}{2\pi}\frac{\Delta x}{v(x)}, \tag{9.7}$$

where $v$ is the velocity. Since

$$x = a \sin \omega t,$$

where $a = \sqrt{2E/\mu\omega^2}$ is the amplitude of oscillation, it follows that

$$v = a\omega \cos \omega t.$$

If we express the velocity as a function of $x$, we obtain

$$v(x) = a\omega\left(1 - \frac{x^2}{a^2}\right)^{1/2},$$

and inserting this in the above gives

$$w_{cl}(x)\,\Delta x = \frac{1}{2\pi a}\frac{1}{(1 - x^2/a^2)^{1/2}}\,\Delta x \tag{9.8}$$

for $-a \le x \le a$, and zero otherwise.

$w_{cl}(x)$ is shown in Figure 9.2; as we would expect, the greatest probability occurs at the turning points $A$ and $B$. Comparing this with the quantum probability density $w_{qm}$ in the same figure, we see that the latter has maxima closer to the equilibrium point (to be precise, for $E_1 = 3\hbar\omega/2$, $OA = OB = \sqrt{3\hbar/\mu\omega}$ and $OA' = OB' = \sqrt{\hbar/\mu\omega}$), but unlike the classical case, the probability of finding the particle is not zero even beyond the turning points. If $w_{cl}(x)$ were different from zero for $x > a = \sqrt{2E/\mu\omega^2}$, this would mean that the potential energy $U = (\mu\omega^2/2)x^2 > E$, the total energy. However in quantum physics $Q$, and therefore $Q^2$ and $U$, are not simultaneously measurable with $H$: after $Q$ has been measured the system is no longer in a state with energy value $E$. Thus, no contradiction arises from $w_{qm}(x) > 0$ for $x > a$.

Let us consider the lowest energy state of the oscillator, where the difference between the quantum and classical cases is particularly striking. In the classical theory the minimum energy of the oscillator is $E = 0$, corresponding to a particle at rest at the equilibrium position. The position probability density $w_{cl}(x)$ associated with this state has the form depicted in Figure 9.1, being everywhere equal to zero except at the point $x = 0$. In quantum theory the minimum energy of the oscillator is

$$E_0 = \tfrac{1}{2}\hbar\omega, \tag{9.9}$$

which is known as the zero-point energy. The position probability density $w_{qm}(x)$ associated with the state of minimum energy has its maximum at the point $x = 0$, but is not equal to zero everywhere else.

We shall now show that these nonclassical phenomena are direct consequences of the fact that momentum and position in quantum mechanics are not simultaneously measurable.

If we take the expectation value of the energy relation (2.1b) for the oscillator, we obtain

$$\langle H \rangle = \frac{1}{2\mu} \langle P^2 \rangle + \frac{\mu\omega^2}{2} \langle Q^2 \rangle. \tag{9.10}$$

We have the uncertainty relation

$$\langle P^2 \rangle \langle Q^2 \rangle = \mathrm{disp}(P^2)\,\mathrm{disp}(Q^2) \geq \frac{\hbar^2}{4} \tag{9.11}$$

because, according to (6.3) and (6.4), $\langle P \rangle = \langle Q \rangle = 0$ for every state of the oscillator. Thus, because of the uncertainty relation (9.11), the expectation value of the energy $\langle H \rangle$ cannot be zero. If we insert (9.11) into (9.10) we obtain

$$\langle H \rangle \geq \frac{1}{2\mu} \langle P^2 \rangle + \frac{\mu\omega^2\hbar^2}{8} \frac{1}{\langle P^2 \rangle}. \tag{9.12}$$

To obtain the smallest expectation value for the energy, we find the minimum of the equality part of (9.12). The condition $d\langle H \rangle/d\langle P^2 \rangle = 0$ gives

$$\langle H \rangle_{min} = \tfrac{1}{2}\,\hbar\omega. \tag{9.13}$$

Thus the ground state of the oscillator is in fact the state with lowest expectation value of the energy, and the eigenvalue $E_0$ in this state is the lowest energy compatible with the uncertainty relation.

## II.10 Basic Assumptions II and III for Observables with Continuous Spectra

The description of the basic assumptions II and III in Sections II.4 and II.5 was given in terms of operators with discrete spectra. In this section we shall give the description for the case that the observables have continuous spectra. With the help of the formalism introduced here we shall be able to discuss in more detail the problems connected with the measurement of position and momentum, the subject of the next section.

The statistical operator of a pure state,

$$W = \Lambda_{b_0} = |b_0\rangle\langle b_0|,$$

where $\Lambda_{b_0}$ is the projection operator onto the one-dimensional subspace spanned by the (proper) eigenvector $|b_0\rangle$ of an observable $B$, is written in an arbitrary basis system $\{\phi_n\}$ as

$$W = \sum_{mn} c_m c_n^* |\phi_m\rangle\langle\phi_n|, \tag{10.1a}$$

where

$$c_n = \langle\phi_n|b_0\rangle. \tag{10.1b}$$

If $\{\phi_n\}$ is the basis system $\{\phi_b = |b\rangle\}$ in which $B$ and therefore $W$ is diagonal, then

$$c_b = \langle\phi_b|b_0\rangle = \langle b|b_0\rangle = \delta_{bb_0}. \tag{10.1b'}$$

For a mixture the statistical operator is

$$W = \sum_{mn} w_{mn} |\phi_m\rangle\langle\phi_n|, \tag{10.2}$$

where $w_{mn}$ are the matrix elements of $W$ in the basis $\{\phi_n\}$; the collection $\{w_{mn}\}$ is the density matrix.

In the case of a continuous spectrum Equation (10.1a) has to be replaced by

$$W = \int dx' \, dx'' \, f(x')f(x'')^* |x'\rangle\langle x''|, \tag{10.3a}$$

where $\{|x\rangle\}$ is the basis system of generalized eigenvectors of an observable, say $Q$, with a continuous spectrum. The (well-behaved test) function $f(x)$ of the continuous variable $x$ for the pure state $W = \Lambda_{b_0}$ is given by

$$f(x) = \langle x|b_0\rangle = \langle x|b_0). \tag{10.3b}$$

Here we have used the second notation to emphasize that $|x\rangle$ is a generalized eigenvector and $|b_0)$ is a proper eigenvector. For a mixture the statistical operator is in analogy to (10.2):

$$W = \int dx' \, dx'' \, F(x', x'')|x'\rangle\langle x''|, \tag{10.4}$$

where $F(x', x'')$ is a function of the two continuous variables $x'$ and $x''$, the generalized eigenvalues of $Q$.

Let us now assume that the observable $B$ that was used to prepare the state $W$ is $Q$, and that we want to prepare and describe a pure state. Naively one would want to choose

$$f(x) = \langle x|x_0\rangle = \delta(x - x_0),$$

in analogy to the discrete case (10.1b'), so that

$$W = \Lambda_{x_0} = |x_0\rangle\langle x_0|.$$

However, it does not make sense to ask for a measurement at a particular point of an observable with a continuous spectrum. Any measurement apparatus, a macroscopic system, is of finite size, not infinitesimally small. Counters have finite extension and need a finite amount of energy; slits—as narrow as they may be—have finite extension. Therefore one cannot ask for a state of a physical system in which the measurement of an observable $Q$ with a continuous spectrum [say spectrum $Q = (-\infty, +\infty)$] gives exactly one particular value, because such a state can be neither measured nor prepared. It only makes sense to ask for a state of a physical system in which the measurement of the observable $Q$ gives values $x$ lying in some particular interval $x_0 - \epsilon < x < x_0 + \epsilon$ around a mathematically defined point $x_0$. (Physically neither an exact point nor an exact interval, i.e., an interval with precisely determined endpoints, can be defined.) Thus for physical reasons, the analogue to (10.1a) for an observable $Q$ with continuous spectrum is

$$W = \int dx'\, dx''\ f_\epsilon(x', x_0)\bar{f}_\epsilon(x'', x_0)|x'\rangle\langle x''|, \tag{10.5}$$

where $f_\epsilon(x, x_0)$ is a function that has its main support in the neighborhood $(x_0 - \epsilon, x_0 + \epsilon)$ of the point $x_0$ and describes in a more or less direct way the resolution of the measurement apparatus by which the state was prepared.

For the state $W$ given by (10.1a), the *probability* that a measurement of $B$ will yield the value $b$ is

$$\mathrm{Tr}(\Lambda_b W) = (b|W|b) = (b|b_0)(b_0|b) = \delta_{bb_0}\delta_{b_0b} = \delta_{bb_0}. \tag{10.6}$$

For the state $W$ given by (10.5), the likelihood that a measurement of $Q$ will yield the value $x$ is described not by a probability but by the probability distribution (probability density)

$$\langle x|W|x\rangle = \int dx'\, dy'\ f_\epsilon(x'; x_0)f_\epsilon^*(y'; x_0)\langle x|x'\rangle\langle x''|x\rangle$$

$$= \int dx'\, dy'\ f_\epsilon(x'; x_0)f_\epsilon^*(y'; x_0)\delta(x - x')\delta(y' - x)$$

$$= f_\epsilon(x; x_0)f_\epsilon^*(x; x_0) = |f_\epsilon(x; x_0)|^2 = F_\epsilon(x; x_0). \tag{10.7}$$

Here we have used (8.19) [or (8.23) if $dx$ is symbolic for $d\mu(x)$].

Symbolically we may write, in analogy to (10.6), $\langle x|W|x\rangle = \text{Tr}(\Lambda_x W)$ with $\Lambda_x = |x\rangle\langle x|$; however, one has to keep in mind that $\Lambda_x$ is *not* a projection operator onto a subspace of the space of physical states. The physical question, i.e., the question that has an answer that can be compared with an experimental determination, is: What is the probability of obtaining a value $x$ in the interval between $x_1$ and $x_2$ when a measurement of the observable $Q$ is made on the state $W$? Here the interval $(x_1, x_2)$ is determined by the resolution of the apparatus with which the measurement is made, and again cannot be infinitesimally small and exactly determined.

The operator which represents finding the position in the interval $x_1 < x < x_2$ is the projection operator

$$\Lambda(x_1, x_2) = \int_{x_1}^{x_2} dx\, |x\rangle\langle x|; \tag{10.8}$$

the probability of finding the position anywhere in $x_1 < x < x_2$ when the system is in the state $W$ of (10.5) is then

$$\text{Tr}(\Lambda(x_1, x_2)W) = \int dx'\, \langle x'|\Lambda(x_1, x_2)W|x'\rangle = \int_{x_1}^{x_2} dx\, \langle x|W|x\rangle$$

$$= \int_{x_1}^{x_2} dx\, |f_\epsilon(x; x_0)|^2 = \int_{x_1}^{x_2} dx\, F_\epsilon(x; x_0). \tag{10.9}$$

The function $F_\epsilon(x; x_0) = |f_\epsilon(x; x_0)|^2$ is determined by the experimental situation. It describes the resolution of the apparatus by which the state $W$ is prepared. The better the resolution, the smaller is $\epsilon$, and in the idealized but unphysical limit $\epsilon \to 0$,

$$|f_\epsilon(x; x_0)|^2 = F_\epsilon(x; x_0) \to \delta(x - x_0). \tag{10.10}$$

The limit (10.10) would correspond to an apparatus with infinite resolution, which cannot exist in reality. Thus the state $W$ in (10.5) can be only as "pure" as the experimental resolution allows, and that can never be "ideally pure" as described by a $\delta$-function.

In order that the state $W$ given in (10.5) be normalized, $f_\epsilon(x, x_0)$ must fulfill

$$\text{Tr}\, W = \int_{-\infty}^{+\infty} dx\, \langle x|W|x\rangle = \int_{-\infty}^{+\infty} dx\, |f_\epsilon(x, x_0)|^2 = \int_{-\infty}^{+\infty} dx\, F_\epsilon(x, x_0) = 1.$$

$$\tag{10.11}$$

Examples for $F_\epsilon(x) = F_\epsilon(x, 0)$ are (function sequences of $\delta$-type):

$$F_\epsilon(x) = \frac{1}{\pi} \frac{\epsilon}{x^2 + \epsilon^2} \qquad \text{(Figure 10.1)}, \tag{10.12}$$

$$F_{\epsilon}(x) = \frac{1}{\sqrt{2\pi}} \frac{1}{\epsilon} \exp\left(-\frac{1}{2}\frac{x^2}{\epsilon^2}\right) \qquad \text{(Figure 10.2)}, \qquad (10.13)$$

$$F_{\epsilon}(x) = \begin{cases} 0 & \text{if } x < -\dfrac{\epsilon}{2} \\[2mm] \dfrac{1}{\epsilon} & \text{if } -\dfrac{\epsilon}{2} < x < +\dfrac{\epsilon}{2} \\[2mm] 0 & \text{if } +\dfrac{\epsilon}{2} < x \end{cases} \qquad \text{(Figure 10.3)}, \qquad (10.14)$$

$$F_{\epsilon}(x) = \begin{cases} 0 & \text{if } x \le -\epsilon \\[2mm] \dfrac{x+\epsilon}{\epsilon^2} & \text{if } -\epsilon \le x \le 0 \\[2mm] \dfrac{-x+\epsilon}{\epsilon^2} & \text{if } 0 \le x \le \epsilon \\[2mm] 0 & \text{if } +\epsilon \le x \end{cases} \qquad \text{(Figure 10.4)}. \qquad (10.15)$$

All these functions have their main support in the interval $-\epsilon < x < +\epsilon$ and have the property

$$F_{\epsilon}(x) \to \delta(x) \quad \text{as } \epsilon \to 0.$$

Thus we have arrived at the *basic assumption IIIa for continuous spectrum*: The statistical operator $W$ given by (10.5) with a distribution function $F_{\epsilon}(x)$ as narrow as experimentally possible is the continuous-spectrum analog of a pure eigenstate for an observable with a discrete spectrum.

We shall call such a state an "almost pure" state and an "almost eigenstate" of the observable with continuous spectrum.

An arbitrary state, in whose preparation no attempt at "purity" may have been made, is described by (10.4). $F(x', x'')$ is connected with the probability distribution function of the observable $Q$ by

$$\text{Tr}\Lambda_x W = \langle x|W|x\rangle = \int dx'\, dx''\, F(x', x'')\langle x|x'\rangle\langle x''|x\rangle = F(x, x).$$

$$(10.16)$$

This is the appropriate place to mention the interrelationship between the physical measurability and preparability of a state and the mathematical description of quantum physics in a rigged Hilbert space $\Phi \subset \mathscr{H} \subset \Phi^{\times}$, though we cannot give a detailed description here.[17]

---

[17] An elementary discussion of this subject is given in A. Bohm, *The Rigged Hilbert Space and Quantum Mechanics*, Springer Lecture Notes in Physics, vol. 78 (1978).

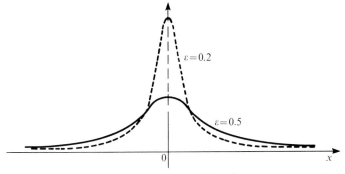

Figure 10.1   Graph of $F_\epsilon(x) = (1/\pi)\, \epsilon/(x^2 + \epsilon^2)$ for two values of $\epsilon$.

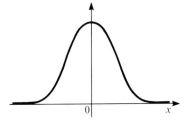

Figure 10.2   Graph of $F_\epsilon(x) = (1/\sqrt{2\pi})(1/\epsilon)\exp(-\frac{1}{2}x^2/\epsilon^2)$.

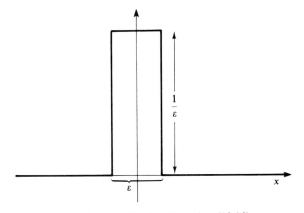

Figure 10.3   Graph of $F_\epsilon(x)$ in Equation (10.14).

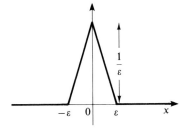

Figure 10.4   Graph of $F_\epsilon(x)$ in Equation (10.15).

The resolution function $F_\epsilon(x; x_0)$ or $f_\epsilon(x; x_0)$ that corresponds to a realistic experimental apparatus must be a well-behaved function; e.g., it can be something like that given in (10.13)–(10.16) with a small but finite $\epsilon$ for the description of a state as "pure" as the preparation allows. One can make this precise by requiring that $f_\epsilon(x; x_0)$ be an element of the Schwartz space, which is a realization of $\Phi$. More appropriate still may be to require that $F_\epsilon(x; x_0)$ have compact support. Those $f_\epsilon(x; x_0)$'s that are either distributions or elements of the Hilbert space (i.e., Lebesque square-integrable functions that are determined only up to their values on a set of measure zero) cannot describe the resolution of a realistic apparatus (e.g., the detector efficiency).

If $f_\epsilon(x, x_0)$ is an element of the Schwartz space, then the state one obtains by (10.5) or (10.4) is in the space $\Phi$ and not in either $\mathcal{H}$ or $\Phi^\times$. More precisely, $W$ is a positive definite Hilbert-space-bounded trace-class operator with $W\mathcal{H} \subset \Phi$. If $W$ is a pure state, $W = |f\rangle\langle f|$, then $f(x) = \langle x|f\rangle$ given by (10.3) is an element of the Schwartz space, and $|f\rangle$ is an element of $\Phi$. Thus the requirement that the space of physical states consist of only those states that can be prepared in a realistic experiment means that in a precise formulation of the preceding basic assumptions the space of physical states should not be the Hilbert space $\mathcal{H}$, but the nuclear topological space $\Phi$ of the rigged Hilbert space $\Phi \subset \mathcal{H} \subset \Phi^\times$. Though the elements of $\Phi^\times$ which are not in $\Phi$ do not represent physical states, $\Phi^\times$ is of immense value, as it contains the generalized eigenvectors of self-adjoint operators with continuous spectra, which are not elements of $\mathcal{H}$; cf. the mathematical note $[\![\ :\ ]\!]$ in Section II.8.

The Hilbert space $\mathcal{H}$ is infinite-dimensional, and it is infinite-dimensional in a very particular sense; namely, it is complete with respect to a particular topology, that is, with respect to a particular meaning of convergence of infinite sequences. Since an infinite number of states can never be prepared, physical measurements cannot tell us anything about infinite sequences, but can at most give us information about arbitrarily large but finite sequences. Therefore, physics cannot give us sufficient information to tell how to take the limit to infinity, i.e., how to choose the topology. The choice of the topology will be a mathematical generalization, but then one should choose the topology such that it is most convenient.

With the choice of $\Phi$ as the space of physical states, one does not only obtain a description which is closer to reality for the above-mentioned reasons connected with measurement and preparation of physical states. One also obtains a tremendous simplification of the mathematical description: Even the simplest algebra of operators that appears in quantum mechanics, the one fulfilling the relation (2.1a), cannot be represented by $\mathcal{H}$-continuous operators. Con-

sequently there are vectors that lie outside the domain of the definition of an observable that is represented by an unbounded operator
in $\mathscr{H}$, e.g., the vector that would correspond to a state of infinite
energy. In contrast to this, it seems possible that for all physical
systems the observables can be represented by $\Phi$-continuous
operators. Thus in the rigged-Hilbert-space description the mathematical image of a physical system is an algebra of continuous
operators, and all the mathematical manipulations that physicists
perform can be rigorously justified.

After having considered the continuous-spectrum analogue (10.5) of
(4.31) in the axiom IIIa we shall now consider the basic assumption IIIb for
continuous spectrum. Let the physical system be in a state $W$; this state
may have been prepared by measurements of an observable with a continuous spectrum or with a discrete spectrum, or may even be completely
unspecified. If a measurement of an observable $Q$ with continuous spectrum
$-\infty < x < +\infty$ is made on the state $W$ and is used to select out a subensemble with values $x$ of $Q$ lying in the range $x_1 < x < x_2$, then the (unnormalized) state $W'$ that describes this subensemble is

$$W' = \Lambda(x_1, x_2)W\Lambda(x_1, x_2)$$

$$= \int_{x_1}^{x_2} dx' \int_{x_1}^{x_2} dx'' \, |x'\rangle\langle x'|W|x''\rangle\langle x''|. \tag{10.17}$$

This is the continuous spectrum analog of (5.2). Note that for $x_1 \to -\infty$,
$x_2 \to +\infty$ we have $W' = W$, which says that no measurement has been
performed. Again, $x_1 < x < x_2$ must be a finite-size interval; it cannot be
infinitesimally small, due to the finite resolution and finite extension of the
measuring apparatus.

To show that (10.18) represents a state where measurement of $Q$ must
yield a value $x$ in the interval $x_1 < x < x_2$, we compute the probability
density for getting the value $x$, which according to (10.7) is given by

$$\langle \Lambda_x \rangle' = \frac{1}{\text{Tr } W'}\text{Tr } \Lambda_x W' = \frac{1}{\text{Tr } W'}\langle x|W'|x\rangle$$

$$= \frac{1}{\text{Tr } W'}\int_{x_1}^{x_2} dx' \int_{x_1}^{x_2} dx'' \, \langle x|x'\rangle\langle x'|W|x''\rangle\langle x''|x\rangle$$

$$= \begin{cases} 0 & \text{if } x < x_1 \quad \text{or} \quad x < x_2 \\ (\text{Tr } W')^{-1}\langle x|W|x\rangle & \text{if } x_1 < x < x_2. \end{cases} \tag{10.18}$$

An actual position measurement is over some range of values in a small
interval $x_0 - \epsilon < x < x_0 + \epsilon$. The probability for measuring $Q$ to be in

such an interval is

$$\langle \Lambda(x_0 - \epsilon, x_0 + \epsilon) \rangle'$$

$$= \int_{x_0-\epsilon}^{x_0+\epsilon} dx \, \langle \Lambda_x \rangle'$$

$$= \begin{cases} 0 & \text{if } (x_0 - \epsilon, x_0 + \epsilon) \cap (x_1, x_2) = \varnothing \\ \dfrac{1}{\text{Tr } W'} \int_{x_0-\epsilon}^{x_0+\epsilon} dx \, \langle x | W | x \rangle & \text{if } (x_0 - \epsilon, x_0 + \epsilon) \subseteq (x_1, x_2) \end{cases}$$

$$(10.19)$$

Thus $W'$ represents a state where the position is in the interval $(x_1, x_2)$.


## II.11 Position and Momentum Measurements—Particles and Waves

The material of Section II.10 has provided us with a formalism in which we can discuss problems connected with the measurement of position and momentum in more detail. Let us return to the classical-mechanical model of the harmonic oscillator, which consisted of two mass points vibrating around an equilibrium point. For simplicity we assume that $m_2 \gg m_1$, so we can consider $m_2$ to be at rest. The position $x$ and momentum $p$ are then essentially the position and momentum of mass point 1.

How can a measurement of the position be made on the quantum-mechanical harmonic oscillator? We have considered the diatomic molecule as a realization of this system. The question is, then, how do we measure the position, i.e., the separation between the two atoms of the CO molecules in the collision chamber? There seems to be no way one can think of to set up a measurement apparatus that would do so. The same argument also applies to a momentum measurement on this ensemble. Thus the above discussion in Section 8 of the probabilities $w(x)$ and $w(p)$ is not of much practical value if one restricts oneself to the quantum-mechanical harmonic-oscillator system. With the formalism developed in Section II.10 we can give a theoretical explanation of this situation.

According to (10.17), after a measurement of the position operator $Q$, the physical system is in an almost eigenstate of the position operator; i.e., all molecules in the ensemble have a position localized in the interval $(x - \epsilon, x + \epsilon)$. For our classical model of the CO molecule as two oscillating mass points (with $m_2 \gg m_1$ for simplicity), the above statement means that $m_1$ is located at a distance $\bar{x}$ from $m_2$ that lies in the interval $(x - \epsilon, x + \epsilon)$. However, a classical system for which $m_1$ is located in a restricted domain of space is not an oscillator. It is a localized mass point, which we call a particle. A measurement of the position operator $Q$ on the harmonic oscillator would destroy the oscillator and result in a physical system that has as its classical image a *particle*. Thus the fact that it is impossible in practice to set up an experiment that would measure the observable $Q$ on the ensemble of CO molecules in the collision chamber has a theoretical description in the basic assumption IIIb as described by (10.18).

The above arguments also exemplify the fact that for a particular physical system, such as the vibrating CO molecule, there are different kinds of observables. Of one kind (e.g., the energy operator of the vibrating CO molecule) one can easily prepare eigenstates of the particular physical system. Of the other kind (e.g., the relative-position operator of the vibrating CO molecule) the preparation of eigenstates will destroy the system.

The measurement of this second kind of observable can only be described if one enlarges the physical system. Thus in our example we need not restrict ourselves to CO molecules, but can consider the larger physical system that consists of CO molecules, O atoms, and C atoms. Then the CO molecules are just particular states of this larger physical system. Other states are the states in which the C and O atoms are not bound. In this larger system, the measurement of the position (the distance between C and O atoms) is always possible at least in principle.

For the almost pure eigenstate of the position operator $Q$, we have

$$W(x) = \int \int dx' \, dx'' \, \overline{f_\xi}(x' - x) f_\xi(x'' - x) |x'\rangle\langle x''|, \qquad (11.1x)$$

where $Q|x\rangle = x|x\rangle$ and $|f_\xi(x' - x)|^2$ has the property of the functions (10.12)–(10.15). One can define the almost pure eigenstate of the momentum operator $P$,

$$W(p) = \int_{-\infty}^{\infty} dp' \int_{-\infty}^{\infty} dp'' \, f_\epsilon(p' - p) \overline{f_\epsilon}(p'' - p) |p'\rangle\langle p''|, \qquad (11.1p)$$

where $P|p\rangle = p|p\rangle$ and $|f_\epsilon(p' - p)|^2$ is again a function of the kind given in (10.12)–(10.15). $Q$ and $P$ here are related by (2.1a). In such a state the measurement of the momentum $P$ gives a value $\bar{p}$ in the interval $p - \epsilon < \bar{p} < p + \epsilon$. By the same arguments as above, such a state cannot be realized by the oscillator; therefore we do not want to require (2.1b), and shall just require[18] that $Q$ and $P$ be related by (2.1a) and that as a consequence[18] of this $\langle p|x\rangle$ is given by (8.56). A possible relation between the energy operator $H$ and the $P$ and $Q$ that facilitates the preparation of almost momentum eigenstates is $H = P^2/2m$, where $m$ is a system constant, the mass. A quantum physical system obeying this relation is called a free nonrelativistic elementary particle.

A quantum-mechanical system with well-defined momentum has as its classical image a *wave*.

Let us discuss this last point a bit more. The characteristic or defining property of a *particle* is that it is localized in space, i.e., it can be assigned a

---

[18] As remarked in [ ] at the end of section 8, (2.1a) is not sufficient to derive (8.56). One has to make an additional assumption, e.g., that $P^2 + Q^2$ is essentially selfadjoint.

definite position. In the same way, the characteristic or defining property of a *wave*, in particular a plane wave, is that it has infinite periodic spatial extension. The mathematical description of a standing plane wave is

$$\phi(x, t) = Ae^{ikx}e^{-i\omega t}, \tag{11.2}$$

so the periodic spatial extension is described by

$$u(x) = Ae^{ikx}, \tag{11.3}$$

where $A$ is the amplitude and $k$ is the wave number. The fundamental period or wavelength is $\lambda = 2\pi/k$. If the wave is not a plane wave, i.e., does not have a well-defined wave number or wavelength but has wave numbers $k'$ inside a certain interval $(k - \Delta k, k + \Delta k)$, then one has a wave packet described by

$$\phi(x) = \int_{-\infty}^{+\infty} dk'\, A(k')e^{ik'x}, \tag{11.4}$$

where $A(k')$ is nonzero only in the interval $2\,\Delta k$ about the value $k$.

To see that a quantum-mechanical system with well-defined momentum has as its classical image a wave, let us calculate the spatial distribution of a state with well-defined momentum. Let us first do this simple-mindedly, working with "states" having exact momentum $|p\rangle$ and "states" having exact position $|x\rangle$ (which are "described" by $W(p) = dp|p\rangle\langle p|$ and $W(x) = dx|x\rangle\langle x|$, respectively), though we should keep in mind that such "states" are not physical states.

For a pure state $\Lambda_\psi$ with "well-behaved" state vector $\psi$, the spatial distribution is given by the wave function $\psi(x) = \langle x|\psi\rangle$; according to (6.25) the probability density for measuring the value $x$ in a position measurement of the state $\psi$ is $w_\psi(x) = |\langle x|\psi\rangle|^2$. Now let $\psi$ be the "not well-behaved" and therefore unphysical "state" vector $|p\rangle$. The spatial distribution is then given by $\langle x|p\rangle$, which is, according to (8.57),

$$\langle x|p\rangle = \frac{1}{\sqrt{2\pi\hbar}}\, e^{ixp/\hbar}. \tag{11.5}$$

If we compare this with the spatial distribution of the plane wave given by (11.3), we see that an "exact momentum eigenstate" of a quantum-mechanical system has the same spatial distribution as the classical system called a plane wave, with the wave number and wavelength given by the de Broglie relation:

$$k = \frac{1}{\hbar}\, p \quad \text{and} \quad \lambda = \frac{2\pi\hbar}{p} = \frac{h}{p}. \tag{11.6a}$$

This is the justification for calling an ensemble of quantum-mechanical systems with definite momentum $p$ a wave with wavelength $\lambda = h/p$.

Let us now return to the physical case where the state is given by (11.1p) and is an almost momentum eigenstate with as narrow a momentum distribution $|f_\epsilon(p' - p)|^2$ as possible. The probability density for obtaining the value $x$ in a measurement of the observable $Q$ in the state $W(p)$ is given according to (10.17) by

$$\langle x|W(p)|x\rangle = \int_{-\infty}^{+\infty} dp' \int_{-\infty}^{+\infty} dp'' \, f_\epsilon(p' - p)\bar{f}_\epsilon(p'' - p)\langle x|p'\rangle\langle p''|x\rangle$$

$$= \int_{-\infty}^{+\infty} dp' \, f_\epsilon(p' - p) \frac{e^{ixp'/\hbar}}{\sqrt{2\pi\hbar}} \int_{-\infty}^{+\infty} dp'' \, \bar{f}_\epsilon(p'' - p) \frac{e^{-ixp''/\hbar}}{\sqrt{2\pi\hbar}} \quad (11.7)$$

In order to compare this with the wave function for a (classical) wave packet (11.4), we define

$$\phi(x) \equiv \int_{-\infty}^{+\infty} dp' \, f_\epsilon(p' - p) \frac{e^{ixp'/\hbar}}{\sqrt{2\pi\hbar}}$$

$$= \int_{-\infty}^{+\infty} dk' \, \hbar \frac{f_\epsilon(k' - k)}{\sqrt{2\pi\hbar}} e^{ixk'}, \quad (11.8)$$

where $p' = \hbar k'$. With

$$A(k') = \sqrt{\frac{\hbar}{2\pi}} \, f_\epsilon(k' - k),$$

$\phi(x)$ indeed describes a wave packet (11.4), and the intensity of this wave gives, according to

$$\langle x|W(p)|x\rangle = |\phi(x)|^2, \quad (11.9)$$

the *probability density* for the value $x$.

Thus a quantum-mechanical system in a state with well-defined momentum (in a narrow momentum interval) has as its classical image a wave packet with wavelength $\lambda$ in a narrow interval:

$$\frac{2\pi\hbar}{p - \epsilon} > \lambda > \frac{2\pi\hbar}{p + \epsilon}. \quad (11.10)$$

The intensity of this wave corresponds to the probability density for the measurement of the position. $\phi(x)$ describes the probability and is called the *probability amplitude*.

We now understand why $\psi(x) = \langle x|\psi\rangle$ has been called a wave function: Let $W_\psi$ be a pure state, i.e., $W_\psi$ be a projector onto the space spanned by $\psi$. Then the probability density for obtaining the value $x$ in a measurement of the position is

$$W_\psi(x) = \langle x|W_\psi|x\rangle = \langle x|\psi\rangle\langle\psi|x\rangle$$

$$= \psi(x)\bar{\psi}(x) = |\psi(x)|^2. \quad (11.11)$$

On the other hand, we can calculate $\langle x|W_\psi|x\rangle$ by inserting a complete system of generalized states

$$I = \int dp' \, |p'\rangle\langle p'|,$$

to obtain

$$\langle x|W_\psi|x\rangle = \int dp' \int dp'' \langle x|p'\rangle\langle p'|\psi\rangle\langle\psi|p''\rangle\langle p''|x\rangle$$

$$= \int dp' \frac{e^{ixp'/\hbar}}{\sqrt{2\pi\hbar}} \psi(p') \int dp'' \frac{e^{-ixp''/\hbar}}{\sqrt{2\pi\hbar}} \psi(p''). \qquad (11.12)$$

If we compare (11.11) and (11.12), we see that the wave function, whose physical interpretation was that its absolute value squared is the probability distribution, is given by

$$\psi(x) = \int dp' \frac{e^{ixp'/\hbar}}{\sqrt{2\pi\hbar}} \psi(p') \qquad (11.13)$$

with $\psi(p') = \langle p'|\psi\rangle$. Thus the wave function $\psi(x)$ may be considered as a superposition of plane waves, $\psi(x)$ comes closer to being a plane wave as the transition coefficient $\langle p'|\psi\rangle$ between the pure state $\psi$ and the generalized momentum eigenvector $|p'\rangle$ comes closer to being a $\delta$-function $\delta(p' - p)$.

The wave character of the quantum system can be experimentally displayed by diffraction experiments. Let us consider a system that has been prepared in an almost eigenstate of momentum, i.e., a monochromatic beam of quantum particles. Such a monochromatic beam is easy to prepare if the particles are charged, but harder to prepare if they are neutral (like atoms or molecules). Let us therefore consider first a monochromatic beam of electrons, such as that prepared for the electron-loss experiment that we considered before. The electron beam is produced by a filament, then accelerated by an electric potential, and then passed through crossed electric and magnetic fields of very well-defined field strengths. Only electrons with very well-defined momenta are not deflected. If $f_e(p' - p)$ describes the momentum spread of this beam, then the spatial distribution is given by (11.8):

$$\phi(x) = \int dk' \, \hbar \frac{f_e(k' - k)}{\sqrt{2\pi\hbar}} e^{ixk'}, \qquad (11.14)$$

which becomes closer to being a plane wave,

$$\phi(x) \approx Ae^{ixk},$$

as the momentum spread about the value $\hbar k$ becomes narrower. This approximate plane wave may be diffracted by a grid, and one obtains a typical diffraction pattern.

Related to Equation (11.6a) by special relativity is another equation:

$$\omega = \frac{1}{\hbar} E, \qquad (11.6b)$$

where $\omega$ is the circular frequency, $\omega = 2\pi\nu$. At this point, where we do not yet want to consider quantum-mechanical time development, the relation (11.6b) is best obtained from the dispersion law for our waves, $\omega = f(k)$, and the relation between energy and momentum, $E = f(p)$. Thus, for the system "electromagnetic wave–photon" the dispersion law is $\omega = ck$ ($\nu = c/\lambda$), and the relation between energy and momentum is $E = cp$ [$E^2 - (cp)^2 = (mc^2)^2$ with rest mass $m = 0$]. From these classical relations (the first for a wave, the second for a massless relativistic particle) it follows immediately that (11.6b) is a consequence of (11.6a) and v..e versa. For the system "electron de Broglie wave–nonrelativistic electron" the dispersion law is $\omega = k^2/2(m/\hbar)$ and the relation between energy and momentum is $E = p^2/2m$, which again shows that (11.6a) and (11.6b) are consequences of each other.

The two physical systems "electromagnetic wave–photon" and "electron de Broglie wave–electron" are two examples of physical systems called elementary particles.[19] The historical development of the ideas about these two systems was, however, quite distinct: To begin with, all known experimental facts about light could be explained by a particle theory (Newton, 1663) as well as a wave theory (C. Huygens, 1678). The same domain of physical knowledge could be described by two different pictures. Then, diffraction of light was discovered (Young, 1801), which could only be explained by the wave picture, not the particle picture. The cathode rays, on the other hand, though considered by some to be, like light, due to a process in the ether, were soon proven to behave like negatively charged particles in an electric and magnetic field (J. J. Thomson, 1897; J. Perrin 1897). Thus light was a wave and electrons were particles. Had the photoeffect (P. Lenard, 1902) been discovered before the diffraction phenomena of light, and had the electron diffraction (Davisson and Germer, G. P. Thomson, 1927) been the first effect observed with the cathode rays, the situation might have been just the reverse.

As light was considered a wave, one had the dispersion law $\omega = ck$, or $\omega = c/\lambda$, and Einstein (1905) showed that in order to explain the photoeffect one has to associate with light, besides the energy given by Planck's (1900) formula $E = \hbar\omega$, the momentum $p = E/c$.

Electrons, on the other hand, were considered particles, and one had the energy-momentum relation $E = p^2/2m$. But de Broglie (1924) postulated that they are waves, with a wave number given by

---

[19] It would have been better to coin a new name for these physical systems (e.g., "quanta") to avoid any reference to Newtonian particles.

(11.6a). Then he derived the dispersion law for these waves from the Planck formula (11.6b) to be $\omega = k^2/2(m/h)$.

The two relations

$$E = \hbar\omega, \qquad \mathbf{p} = \hbar\mathbf{k} \qquad\qquad (11.6)$$

are justly considered the gate to quantum theory. We derived at least the second equation (11.6a) from the defining relation (2.1a). But historically they were the starting point of the long conjecturing process that led to (2.1a) and ultimately to quantum mechanics in general. The relations (11.6) connect two different classical pictures (often called dual).

The one picture is a particle, i.e., a localizable mass point. Classically, such a particle can be given definite values of energy and momentum (e.g., the momentum and the energy measured by an apparatus that moves with a velocity $-\mathbf{v}$ relative to such a localized mass point are $\mathbf{p} = m\mathbf{v}$ and $E = p^2/2m$).

The other picture is a wave, i.e., a more or less periodic disturbance of infinite spatial extension. A wave has a frequency $\omega$ and a wave vector $\mathbf{k}$.

In quantum mechanics the physical system (quantum, elementary particle), of which we should always imagine an ensemble, can occur in different states. For instance, it can be in an almost momentum eigenstate described by (11.1p). We may then ask about the position of such a state, e.g., we may want to know the probability of finding the position anywhere in an interval $x - \Delta/2 < x' < x + \Delta/2$. This is according to the basic assumption II, expressed with the projection operator of (10.8) as:

$$\mathrm{Tr}\left(\Lambda\left(x - \frac{\Delta}{2}, x + \frac{\Delta}{2}\right)W(p)\right) = \int_{x-\Delta/2}^{x+\Delta/2} dx' \, \langle x'|W(p)|x'\rangle.$$

For the sake of definiteness we may assume that the almost momentum eigenstate is given by (11.1p) with $F_\epsilon(p' - p)$ given by (10.14) or Figure 10.3, so that $f_\epsilon(p' - p) = 1/\sqrt{\epsilon}$ in the interval between $p - \epsilon/2$ and $p + \epsilon/2$. The probability density of the position operator is then obtained using (11.7) as

$$\langle x|W(p)|x\rangle = \int_{p-\epsilon/2}^{p+\epsilon/2} dp' \, \frac{1}{\sqrt{\epsilon}} \frac{e^{ixp'/\hbar}}{\sqrt{2\pi\hbar}} \int_{p-\epsilon/2}^{p+\epsilon/2} dp'' \, \frac{1}{\sqrt{\epsilon}} \frac{e^{-ixp''/\hbar}}{\sqrt{2\pi\hbar}}$$

$$= \frac{1}{2\pi\hbar} \epsilon \text{ for } \epsilon x/\hbar \ll 1.$$

This means that the probability density of the position operator is independent of the position. Thus the probability of finding for the physical system a value in the interval of length $\Delta$ around $x$ is

$$\int_{x-\Delta/2}^{x+\Delta/2} dx' \, \langle x'|W(p)|x'\rangle = \frac{1}{2\pi\hbar} \epsilon\Delta,$$

i.e., the system is located with the same small probability anywhere in space.

In the limiting case of exact momentum when the momentum spread $\epsilon \to 0$, the probability of finding the physical system in an interval $\Delta$ anywhere would be zero. However, this limiting case is unphysical, because there is no apparatus that could prepare such a state. For example, an almost momentum eigenstate of electrons, a monochromatic electron beam, can be prepared by the apparatus used in the energy-loss experiment of Section II.4. The electrons produced by a filament are accelerated by an electrical potential. Even if this potential is very well defined, the electrons will have still the same momentum spread with which they were emitted by the filament. This electron beam then passes through a monochromator (an electric or magnetic field, or in some cases crossed electric and magnetic fields), and only the electrons with a definite velocity or momentum will arrive at the position of the slit. But again, even if the field is very well defined, this slit has a finite opening, so that electrons with slightly different momenta can pass through. In general, because of the necessarily finite size of the macroscopic measurement apparatus, the preparation of a monochromatic beam with zero spread is impossible.

As we have seen above, for an almost momentum eigenstate every position has the same probability, i.e., the position is undetermined. The physical system is in a wave state with a rather well-defined wave number and wavelength; thus it appears as a wave.

Thus, according to the formulation of quantum mechanics based upon the basic assumptions I, II, and III, a quantum physical system in a momentum eigenstate behaves like a wave—it "is" a wave—until a measurement is performed that alters the state (e.g., a position measurement). *As long as such a measurement is not performed, the theoretical predictions are exactly the same as in a wave theory.*

We can go through the same arguments starting with an almost position eigenstate $W(x)$ described by (11.1x). The probability density of the momentum operator is then obtained as

$$\langle p \mid W(x) \mid p \rangle = \frac{1}{2\pi h} \xi,$$

and the probability of finding the momentum in an interval between $p - \epsilon/2$ and $p + \epsilon/2$ is

$$\int_{p-\epsilon/2}^{p+\epsilon/2} dp' \, \langle p' \mid W(x) \mid p' \rangle = \frac{1}{2\pi h} \xi\epsilon.$$

As this probability does not depend upon the central value $p$, any value of $p$ has the same probability. For an almost position eigenstate the momentum is undetermined. Such an almost position eigenstate we have called a particle, but we see now that, in distinction to a classical (Newtonian) particle, the quantum physical system in a particle state cannot have a definite value of momentum.

The constant $h$ was introduced in the defining relation (2.1a) without much discussion. With the consequences of (2.1a) derived here, we can now attempt an explanation of its physical meaning.

One can prepare an almost momentum eigenstate, e.g., by the macroscopic apparatus of Section II.4. From the electric and magnetic fields of the apparatus one obtains the velocity $v$ and the momentum $p = mv$. According to (11.6a) such a state appears as a wave with inverse wavelength $k = (1/\hbar)mv$. The wave character can be displayed experimentally, and the wavelength can be measured in a diffraction experiment. For electrons

$$(m = 9 \times 10^{-28} \text{ g}),$$

it turns out that grids with the right grid constant are crystals, whose typical separation of adjacent atoms is of the order of 1 Å $= 10^{-10}$ m. In the diffraction experiment $\lambda = 1/k$ is measured in the units of length (cm). The traditional units of momentum are erg sec/cm; if the energy is measured in eV,[20] then the units of momentum are eV sec/cm. The de Broglie relation (11.6a) states that the *inverse wavelength k of the probability waves and the momentum p are the same quantity measured in different units.*

The Planck constant $\hbar$ may then be obtained as the ratio of the value of a momentum in erg sec/cm or eV sec/cm to its value in cm$^{-1}$. The result is

$$h = 6.63 \times 10^{-27} \text{ erg sec},$$

$$\hbar = 1.0546 \times 10^{-28} \text{ erg sec},$$

or

$$\hbar = 6.58 \times 10^{-16} \text{ eV sec}.$$

It could, of course, be that $\hbar$ has a different value for different physical systems, as was the case for the system constants $\mu$ and $\omega$ in Section II.2 and for $m$ in the present section. It turns out that in fact $\hbar$ in the relation (2.1a) has the same value for all physical systems for which (2.1a) holds. Therefore one can change the units of either $P$ or $Q$ in (2.1a) so that in the new units one has $[P, Q] = -i\mathbf{1}$. For example, one can decide to keep cm as the unit of position; then $P$ is measured in cm$^{-1}$. Or, as one does in relativistic particle physics, one can choose for the unit of momentum 1 eV/$c$, where $c$ is the velocity of light, another constant of nature. Then the unit of position is $(1 \text{ eV}/c)^{-1}$. $\hbar$, *like any constant of nature, is just a conversion factor for the values of one and the same physical quantity in different units.* Before this constant of nature is discovered, the values of this same quantity in different units (e.g., the momentum $\mathbf{p}$ and the wave vector $\mathbf{k}$) are thought to be different physical quantities. After the constant has been discovered and the different physical quantities have been related by a new theory, one can measure these quantities in the same units and eliminate in this way one of the previously independent units.

---

[20] 1 eV $= 1.6 \times 10^{-12}$ erg $= 1.6 \times 10^{-12}$ dyn cm; 1 dyn $= 1$g cm/sec.

## Problems

1.  Show that the operators $a$ and $a^\dagger$, defined by the equations (3.1), are the adjoints of each other.

2.  Prove the following statements:
(a)  All eigenvalues of a Hermitian operator are real.
(b)  Let $f$ be an eigenvector of $A$ corresponding to the eigenvalue $\lambda$. Show that for each nonzero $\alpha \in \mathbb{C}$ the vector $\alpha f$ is also an eigenvector of $A$ corresponding to the eigenvalue $\alpha f$.
(c)  Let $f_1$ and $f_2$ be eigenvectors of $A$, with respective eigenvalues $\lambda_1$ and $\lambda_2$. Show that if $\lambda_1 \neq \lambda_2$, then $f_1$ and $f_2$ are orthogonal.

3.  Show that the collection of all linear operators on a (complex) linear space $R$ constitutes an associative algebra with unit element under the operations of scalar multiplication of an operator, addition of operators, and multiplication (composition) of operators.

4.  Show that the Cauchy–Schwarz–Bunyakovski inequality [Equation (I.2.7)] follows from the definition (I.2.5) of a positive Hermitian form.

5.  For the quantum-mechanical harmonic oscillator, calculate the diagonal matrix elements of the energy operator $H$ between the states

$$\tilde{\phi}_n = e^{i\alpha n}\phi_n,$$

where $\phi_n$ is the eigenvector belonging to the eigenvalue $n$ of the operator $N = a^\dagger a$, according to Equation (3.20).

6.  Let $\phi_n$ be the vectors of Equation (3.19). Define for each complex number $z$ the *coherent state vector*

$$|z\rangle \equiv e^{-|z|^2/2} \sum_{n=0}^{\infty} \frac{z^n}{\sqrt{n!}} \phi_n.$$

(a)  Show that the coherent state $|z\rangle$ is an eigenstate of the operator $a$ with eigenvalue $z$.
(b)  Show that the scalar product of $\phi_n$ with $|z\rangle$ has the property

$$|\langle \phi_n | z \rangle|^2 = \frac{|z|^2}{n!} e^{-|z|^2}.$$

(c)  Show that the scalar product $\langle z | z' \rangle$ of two coherent states $|z\rangle$ and $|z'\rangle$ is given by

$$\langle z | z' \rangle = \exp(-\tfrac{1}{2}|z|^2 + \bar{z}\,z' - \tfrac{1}{2}|z'|^2).$$

(d)  Calculate the expectation value of the operators $H$, $Q$, and $P$ in the coherent state $|z\rangle$, i.e., calculate $\langle z|H|z\rangle$, $\langle z|Q|z\rangle$, and $\langle z|P|z\rangle$.

7.  Figure PS.1a shows a schematic diagram of a slightly different experimental setup than the one of Figure 4.2a. Results of an energy-loss experiment performed on $N_2$ with this apparatus are shown in Figure PS.1b. The bump at $v = 0$ in Figure PS.1b is approximately 30 times larger than the bumps at $v = 1, 2, 3, \ldots$; this is in contrast to the situation depicted in Figure 4.2b, where the $v = 0$ bump is approximately 3 times larger than the succeeding bumps. Explain the difference. Determine from Figure PS.1b the statistical operator $W$ that describes the state of the ensemble of $N_2$ molecules of the experiment of Figure PS.1a.

8.  Suppose we have a quantum-mechanical oscillator.
(a)  Let the state of the system be given by (4.17). Show that (4.21) follows from (4.20).

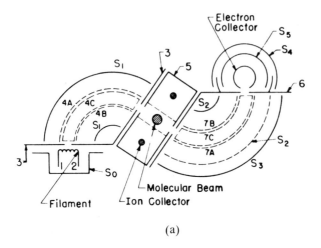

(a)

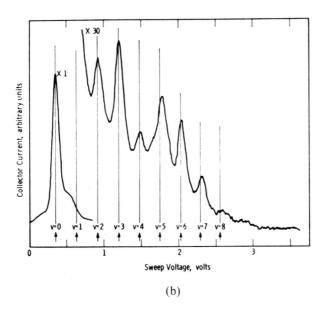

(b)

Figure PS.1   (a) Schematic diagram of the double electrostatic analyzer. Electrons are emitted by the filament, deflected by the cylindrical grids (4A and 4B) with radii of 1.0 and 1.5 cm, respectively, injected into the collision chamber 5, and analyzed by sweeping the voltage between electrodes 6 and 3. $S_1, S_2, S_3, S_4, S_5$ are shields to collect unwanted electrons; 4C and 7C are top and bottom grids. Typical operating voltages between the electrodes indicated: $(4A–4B) = 1.2$ V; $(7A–B) = 1.2$ V; $(Fil–3) = 1.4$ V; $(F–S_1) = 20$ V; $(F–S_2) = (F–S_3) = 20$ V. The electron collector is grounded. All slits are $0.5 \times 4$ mm. (b) Energy spectrum of forward-scattered electrons at an incident electron energy of 2.6 eV. [From Schulz, G. J: Physical Review *125*, 229 (1962), with permission.]

(b)  Show that $W = \tilde{W}/(\mathrm{Tr}\ \tilde{W})$ is the normalized statistical operator, where $\tilde{W}$ is given by (4.22).

(c)  Let $\Lambda$ be a projection operator. Show that $\mathrm{Tr}\ \Lambda = \dim(\Lambda\mathcal{H})$.

9.  Calculate the expectation values of the operators $Q$, $Q^2$, $P$, and $P^2$ when the quantum-mechanical oscillator is in the state $W = \Lambda_n$, where $\Lambda_n$ is the projection operator onto the energy eigenspace corresponding to the eigenvalue $E_n = \hbar\omega(n + \frac{1}{2})$.

10.  Let $\Lambda$ be the projector onto the one-dimensional subspace spanned by $\psi$ ($\|\psi\| = 1$). Show that the dispersion $\mathrm{disp}_{(\Lambda)}\ H$ of $H$ when the system is in the state $W = \Lambda$ satisfies

$$\mathrm{disp}_{(\Lambda)}\ H \geq 0.$$

11.  Calculate $\mathrm{disp}_{(W)}H$ for the state of the CO molecules in the collision chamber of the energy-loss experiment described in Section II.4. Show that $\mathrm{disp}_{(W)}\ H \geq 0$.

12.  Let $\Lambda$ be the projection operator onto an $n$-dimensional eigenspace of the operator $A$. Show that $\mathrm{disp}_{(\Lambda)}\ A = 0$.

13.  Let $W$ be given by (4.31). Calculate $\langle A \rangle$ and show that this is the result we would expect if $\langle A \rangle$ is to represent the average value of $A$. Is $W'$ as given by (5.1) normalized, i.e., does $\mathrm{Tr}\ W' = 1$?

14.  Let $H$ and $B$ be two observables with spectra $E_n$ and $b_i$, respectively. Let $\Lambda_n$ and $\Lambda_{b_i}$ be the projection operators on the eigenspaces of $H$ with eigenvalue $E_n$ and of $B$ with eigenvalue $b_i$, respectively. Let the quantum-mechanical system be in a state in which an energy measurement has always resulted in the value $E_n$.

(a)  What is the statistical operator $W$ for this state?

(b)  In this state $W$ a measurement of the observable $B$ is performed. What is the statistical operator $W'$ after the $B$ measurement, without selection of subsystems?

(c)  What is the probability of finding the value $b_j$ in a measurement of $B$ on the physical system in the state $W$? in the state $W'$? Compare these two probabilities.

(d)  The observable $H$ is now measured again. What is the probability of obtaining $E_n$ in this measurement of $H$? Compare this probability with the probability of obtaining $E_n$ in the original state $W$.

(e)  Let $A$ be a third observable with eigenvalues $a_i$ and projectors $\Lambda_{a_i}$ onto the eigenspaces. Calculate the probabilities of obtaining the value $a_k$ in a measurement of $A$ when the system is in the state $W$ and when the system is in the state $W'$. When are these probabilities equal?

15.  Calculate the matrix elements $\langle n|P|m \rangle$ and $\langle n|Q|m \rangle$ of the operators $P$ and $Q$ between the energy eigenvectors $\phi_n = |n \rangle$ of the harmonic oscillator.

16.  The Hermite polynomials are defined by

$$H_n(y) \equiv (-1)^n e^{y^2}\ \frac{d^n}{dy^n}\ (e^{-y^2}).$$

(a)  Show that the $H_n(y)$ fulfill the recurrence relations

$$H_n(y) = 2yH_{n-1}(y) - 2(n-1)H_{n-2}(y),$$

$$\frac{dH_n(y)}{dy} = 2nH_{n-1}(y).$$

(b)  Show that $H_n(y)$ satisfies the differential equation

$$H_n''(y) - 2yH_n'(y) + 2nH_n(y) = 0.$$

(c)   Give explicit expressions for the first six Hermite polynomials

$$H_n(y) \ (n = 0, 1, 2, 3, 4, 5).$$

(d)   Show that the Hermite polynomials fulfill the following orthogonality relation:

$$\int_{-\infty}^{+\infty} dy \, e^{-y^2} H_m(y) H_n(y) = 2^n n! \, \sqrt{\pi} \delta_{mn}.$$

17.   In the energy eigenstate $W = \Lambda_1$ corresponding to the energy eigenvalue $E_1 = \frac{3}{2}\hbar\omega$, the probability $w_1(x) \, \Delta x$ of measuring the position to be within an interval $(x - \Delta x/2, \, x + \Delta x/2)$ about the value $x$ is nonzero for $x > a = \sqrt{2E_1/(\mu\omega^2)}$ (cf. Figure 9.2). From this it follows that the potential energy $U$ may be larger than the total energy $E_1$ because

$$U = \tfrac{1}{2}\mu\omega^2 x^2 > \tfrac{1}{2}\mu\omega^2 a^2 = E_1.$$

Point out what is wrong with this argument, and give a correct interpretation of the situation.

18.   Show that the operators $P^{(S)} = (\hbar/i) \, d/dx$ and $Q^{(S)} = x$, considered as operators on the wave function, satisfy the Heisenberg commutation relation.

19.   It was derived in Section II.8 that, on the wave function, the momentum operator $P$ and the position operator $Q$ are represented by the operators $P^{(S)} = (\hbar/i) \, d/dx$ and $Q^{(S)} = x$. Which assumption besides the Heisenberg commutation relation has been used in this derivation?

20.   Let a quantum-mechanical particle with mass $m$ be confined by impenetrable walls to the region $-a < x < +a$ (a one-dimensional rectangular well with infinitely high walls). This means that the expectation value of the potential-energy operator $V$ between the generalized position eigenstates $|x\rangle$ is

$$\langle V \rangle_x = \frac{\langle x|V|x\rangle}{\langle x|x\rangle} = -c,$$

where $c$ is a constant; and that the probability that a measurement of the position $Q$ will give the value $x$ is nonzero only for $-a < x < +a$. Denote by $|n\rangle$ the eigenvectors of the energy operator

$$H = \frac{p^2}{2m} + V,$$

and let $\Lambda_n$ denote the projector onto the space spanned by $|n\rangle$. Calculate the eigenvalues of $H$, and compute the expectation values of $Q$, $Q^2$, $P$, and $P^2$ when the system is in an energy eigenstate $\Lambda_n$.

21.   Show that for any linear operators $A$, $B$ and for $\lambda \in C$, the trace as defined by Equation (4.3) has the following properties:
(a)   It is independent of the choice of basis.
(b)   $\text{Tr}(\lambda A) = \lambda \, \text{Tr}(A)$.
(c)   $\text{Tr}(A + B) = \text{Tr}(A) + \text{Tr}(B)$.
(d)   $\text{Tr}(AB) = \text{Tr}(BA)$.

22.   Let $W(p)$ be an almost momentum eigenstate described by the statistical operator

$$W(p) = \int_{-\infty}^{+\infty} dp' \int_{-\infty}^{+\infty} dp'' \, f_\varepsilon(p' - p) f_\varepsilon^*(p'' - p) |p'\rangle\langle p''|$$

with

$$f_\epsilon(p' - p) = \frac{1}{\sqrt{\pi}} \frac{\sqrt{\epsilon}}{p' - p - i\epsilon}$$

(Lorentzian momentum distribution). Calculate the probability density of the position operator and the probability of finding the position in the interval $x_0 - \Delta x/2 < x < x_0 + \Delta x/2$. Discuss the result.

23.   Davisson and Germer scattered low-energy electrons from metal targets. For 45-eV electrons incident normally on a crystal face, compute the angle between the incident beam and the scattering maximum if the metal is assumed to be of simple cubic structure with a lattice constant of 3.52 Å.

24.   Has it been derived that electrons are waves with wavelength $\lambda = h/p$? If so, then explain how this has been done. If not, explain what *has* been derived.

# CHAPTER III

# Energy Spectra of Some Molecules

Section III.1 discusses how the energy levels of a quantum-mechanical system emitting dipole radiation are observed. A derivation of the transition probability is not given. In Section III.2 the defining relations of angular momentum are established. Section III.3 derives the representations of the algebra of angular momentum. In Section III.4 the energy spectrum of a rotator is derived and compared with the experimental spectrum of diatomic molecules. Section III.5 contains the basic assumption about the physical combination of two nonidentical quantum-mechanical systems and the application of this assumption to the description of vibrating and rotating diatomic molecules.

## III.1 Transitions between Energy Levels of Vibrating Molecules—The Limitations of the Oscillator Model

A quantum-mechanical system in a certain stationary state [e.g., the energy eigenstate $\Lambda_n$ of the diatomic molecule (oscillator)] will remain in that state so long as it is not acted upon by outside forces. In practice, any quantum-mechanical system is acted upon by weak external forces, such as external electromagnetic fields or internal electromagnetic fields that arise from the motion of charges within the system. Under the influence of such forces, the state is liable to change. If the system has a discrete set of states (e.g., the energy eigenstates of the oscillator), then a weak external disturbance does not change these states (or, more precisely, it changes the energy levels by a negligibly small amount), but the system may jump from one state to another.

The theory of such transitions, which can be developed as a consequence of the basic assumptions of quantum mechanics, will be presented later in Chapters XIV and XXI. For the moment we shall just give some semi-classical arguments and state the result, which we shall use to obtain transition frequencies and selection rules.

We accept here as an empirical fact that under the influence of ever-present external disturbances, the quantum system may perform transitions from one energy eigenstate with energy $E_n$ to another with energy $E_m$, and emit or absorb the energy difference

$$E_n - E_m$$

as electromagnetic radiation in the form of a light quantum or photon of frequency

$$\nu_{nm} = \frac{E_n - E_m}{h} = \frac{E_n - E_m}{2\pi\hbar}. \tag{1.1}$$

If the electromagnetic field has the frequency $\nu_{nm}$, then the quantum-mechanical system can absorb a photon of this frequency and jump from a state of energy $E_n$ to a state of higher energy $E_m$. On the other hand, if a quantum system is in an excited state $E_n$ (a state of higher energy than the ground state), it can emit a photon of frequency $\nu_{nm}$ and drop to a state of lower energy $E_m$.

Transitions between two states cannot occur under the influence of electromagnetic radiation if the matrix element of the total electric displacement operator $D$ of the system vanishes between these two states. Also, the probability for such a transition, and thus the intensity of the emitted (or absorbed) electromagnetic radiation, is proportional to the square of the modulus of this matrix element.

To illustrate, let us return to our classical picture of the diatomic molecule. If the molecule consists of unlike atoms (e.g., CO) then it has an electric dipole moment, since the centers of the positive and negative charges do not coincide. The dipole moment is the vector directed from the center of negative charges to the center of positive charges and is given by

$$\mathbf{D} = q\mathbf{d},$$

where $q$ is the charge and $d$ is the distance between the centers of the charges. The permanent dipole moment $\mathbf{D}_0$ of the molecule lies along the internuclear axis.

If the interatomic (or internuclear) distance changes, the dipole moment will change, and to a good approximation it may be assumed that the dipole moment is a linear function of the deviation from the equilibrium position of the interatomic distance:

$$\mathbf{D} = \mathbf{D}_0 + q\mathbf{x}. \tag{1.2}$$

Therefore the dipole moment changes with the frequency of the mechanical vibration. Oscillating charges radiate an electromagnetic field, and on the basis of classical electrodynamics the emitted light should have a frequency equal to the frequency of the oscillator, i.e.,

$$v = \frac{\omega}{2\pi},$$  (1.3)

where $\omega = \sqrt{k/m}$ is the angular frequency of the classical oscillator [see (II.2.8)].

If the molecule consists of two like atoms (e.g., $O_2$, $N_2$), then the dipole moment is zero, because the centers of positive and negative charge coincide and oscillations of the molecule about its equilibrium position do not lead to emission or absorption of electromagnetic radiation.

> *Note*: The above considerations apply only to dipole radiation; there may be quadrupole or higher multipole radiation even for molecules for which the dipole moment is zero. However, the magnitude of the higher multipole radiation is exceedingly small.

Let us now turn to the quantum-mechanical molecule. Quantum-theoretically, the emission of radiation takes place as a result of a transition of the oscillator from a higher to a lower state, and absorption takes place by the reverse process. The frequency of the emitted light is given by

$$v = \frac{E_n - E_m}{h}.$$  (1.1)

The intensity of the emission, classically proportional to the time-averaged value (over one period) of the square of the dipole moment $\mathbf{D}$, is in quantum theory proportional to the absolute value squared of the transition matrix elements

$$\langle m|\mathbf{D}|n\rangle = \mathbf{D}_{mn},$$  (1.4)

where $\mathbf{D}$ is the dipole operator, obtained from (1.2) by the usual procedure of replacing the number $x$ with the operator $\mathbf{Q}$:

$$\mathbf{D} = \mathbf{D}_0 + q\mathbf{Q}.$$  (1.5)

The transition probability (intensity) $A_{nm}$ for spontaneous dipole emission[1] in the transition from an energy state with energy value $E_n$ to a state with energy $E_m$ is given by

$$A_{nm} = \frac{4}{3\hbar c^3} \omega_{nm}^3 |\mathbf{D}_{mn}|^2,$$  (1.6)

where $\omega_{nm} = (E_n - E_m)/\hbar$, $c$ is the velocity of light, and

$$|\mathbf{D}_{mn}|^2 = \frac{1}{\dim(\Lambda_n \mathcal{H})} \mathrm{Tr}(\Lambda_n \mathbf{D} \cdot \Lambda_m \mathbf{D}) = \frac{1}{\dim(1_n \mathcal{H})} \sum_{i=1}^{3} \sum_{v,\mu} |\langle \mu, m|D_i|n, v\rangle|^2.$$  (1.7)

---

[1] The probabilities for induced emission and absorption are also proportional to $|\mathbf{D}_{mn}|^2$ and to the intensity of the incident radiation.

**D** is given by (1.5), $\Lambda_n$ is the projection operator on the energy eigenspace with eigenvalue $E_n$, and $\dim(\Lambda_n \mathscr{H})$ is the dimension of this energy eigenspace.

Since we are now not interested so much in knowing the intensity as we are in knowing when this intensity is zero, the precise form of the transition probability is not of primary interest to us at the moment. Equation (1.6) can be derived using the general formalism developed in Chapter XXI. For the special case of the one-dimensional oscillator we replace the dipole and transition vectors by the one-dimensional quantities $D$ and $Q$ and (1.7) goes over into $|\langle m|D|n\rangle|^2$.

For many quantum-mechanical systems, a great majority of the matrix elements of $D$ vanishes, so there is a severe limitation on the possibilities for transitions. The rules that express this limitation are called *selection rules*. In order to determine which particular transitions can actually occur for the harmonic oscillator, we have to calculate the matrix elements

$$\langle m|D|n\rangle = q\langle m|Q|n\rangle. \tag{1.8}$$

The matrix element of the position operator in the energy eigenstates has already been calculated [see Equation (II.8.3)] and is given by

$$\langle m|Q|n\rangle = \sqrt{\frac{\hbar}{2\mu\omega}}(\sqrt{n}\langle m|n-1\rangle + \sqrt{n+1}\langle m|n+1\rangle). \tag{1.9}$$

Thus we see that the transition probability and hence the emission and absorption intensity of light are zero except when the quantum numbers $n$ and $m$ differ by unity. Thus the selection rule for the harmonic oscillator is

$$n - m = \pm 1. \tag{1.10}$$

Transitions in the harmonic oscillator are possible only between neighboring energy levels. The frequency of light that is emitted (for $E_n > E_m$) or absorbed (for $E_n < E_m$) is given according to (1.1) and (1.10) by

$$\nu_{nm} = \frac{E_n - E_m}{h} = \frac{\hbar\omega}{h}[(n+\tfrac{1}{2})-(m+\tfrac{1}{2})] = \frac{\omega}{2\pi}. \tag{1.11}$$

Thus quantum-theoretically the frequency of the radiated light is equal to the frequency $\omega/2\pi$ of the oscillator, and is independent of the energy level $n$. Similar arguments apply for absorption. Thus we have seen that for the particular case of the quantum-mechanical harmonic oscillator, the frequency of emitted and absorbed light is the same as it would be for the classical oscillator.

If we recall the energy-level diagram for the harmonic oscillator (Figure II.4.3) we can indicate the allowed transitions by vertical lines (see Figure 1.1). All these transitions give rise to the same frequency. This is a consequence of the equal spacing between energy levels.

For a diatomic molecule consisting of two like atoms (e.g., $O_2$), the dipole moment operator (1.5) is the zero operator, and therefore no transitions between different energy levels occur.

Let us now turn to the comparison of our theoretical results with the experimental situation.

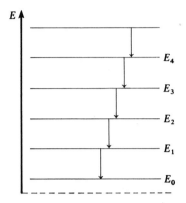

Figure 1.1   Dipole transitions between the energy levels of the harmonic oscillator.

In order to find out what frequency we should expect, we turn first to the energy-loss spectrum of the CO molecules (Figure II.4.2). From the distance between the bumps in the energy-loss spectrum we find that the difference between the various energy levels of the vibrating CO molecule is

$$\Delta E = 0.265 \text{ eV}. \tag{1.12a}$$

If we calculate the frequency from this according to (1.11) we find

$$v = \frac{\Delta E}{2\pi \hbar} = \frac{0.265 \text{ eV}}{2\pi \times 6.58 \times 10^{-16} \text{ eV sec}} = 6.4 \times 10^{13} \text{ sec}^{-1} \tag{1.12b}$$

and

$$\lambda = \frac{c}{v} = 0.466 \times 10^{-3} \text{ cm} = 4.66 \, \mu \tag{1.12c}$$

$(1 \, \mu = 10^{-4} \text{ cm}; 1 \text{ Å} = 10^{-8} \text{ cm} = 10^{-4} \, \mu)$.

In spectroscopy it is customary to give the frequency not in $\text{sec}^{-1}$ but in $\text{cm}^{-1}$, i.e., give instead of the frequency $v$ the wave number $v/c = 1/\lambda$, which indicates the number of waves per cm. We shall not introduce a new symbol for it but also call it $v$, and the unit next to the number will then tell us what is meant. The frequency in $\text{cm}^{-1}$, or wave number, of the radiation emitted by the transition between the vibrational levels of CO is then

$$v = 2140 \text{ cm}^{-1}. \tag{1.12d}$$

Thus we expect, from the energy-loss spectrum of CO, that the vibrating CO molecules emit or absorb electromagnetic radiation only with the frequency given by (1.12), i.e., we expect one spectral line in the near infrared region.[2] If we compare this with the absorption or emission spectrum, we find that

[2] In order to give a feeling for the orders of magnitude of various quantities involved in molecular spectroscopy, we show a table of the regions of the electromagnetic spectrum in Figure 1.2.

THE ELECTROMAGNETIC SPECTRUM

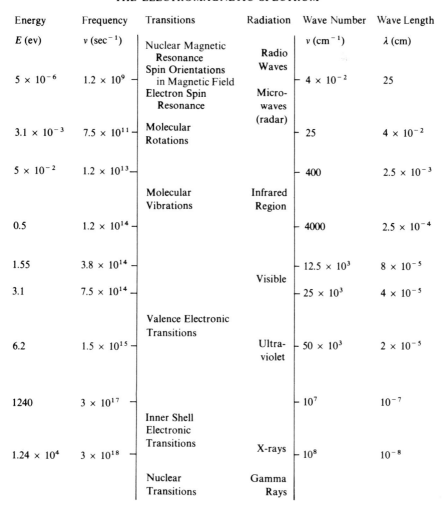

| Energy | Frequency | Transitions | Radiation | Wave Number | Wave Length |
|---|---|---|---|---|---|
| $E$ (ev) | $\nu$ (sec$^{-1}$) | | | $\nu$ (cm$^{-1}$) | $\lambda$ (cm) |
| | | Nuclear Magnetic Resonance | Radio Waves | | |
| $5 \times 10^{-6}$ | $1.2 \times 10^9$ | Spin Orientations in Magnetic Field | | $4 \times 10^{-2}$ | 25 |
| | | Electron Spin Resonance | Micro-waves (radar) | | |
| $3.1 \times 10^{-3}$ | $7.5 \times 10^{11}$ | Molecular Rotations | | 25 | $4 \times 10^{-2}$ |
| $5 \times 10^{-2}$ | $1.2 \times 10^{13}$ | | | 400 | $2.5 \times 10^{-3}$ |
| | | Molecular Vibrations | Infrared Region | | |
| 0.5 | $1.2 \times 10^{14}$ | | | 4000 | $2.5 \times 10^{-4}$ |
| 1.55 | $3.8 \times 10^{14}$ | | | $12.5 \times 10^3$ | $8 \times 10^{-5}$ |
| | | | Visible | | |
| 3.1 | $7.5 \times 10^{14}$ | | | $25 \times 10^3$ | $4 \times 10^{-5}$ |
| | | Valence Electronic Transitions | | | |
| 6.2 | $1.5 \times 10^{15}$ | | Ultra-violet | $50 \times 10^3$ | $2 \times 10^{-5}$ |
| 1240 | $3 \times 10^{17}$ | | | $10^7$ | $10^{-7}$ |
| | | Inner Shell Electronic Transitions | X-rays | | |
| $1.24 \times 10^4$ | $3 \times 10^{18}$ | | | $10^8$ | $10^{-8}$ |
| | | Nuclear Transitions | Gamma Rays | | |

Figure 1.2   Schematic diagram of the electromagnetic spectrum. Note that the scale is nonlinear. Boundaries between regions are generally quite arbitrary.

it is indeed correct. If the absorption spectrum is obtained with a thin layer of absorbing gas, one finds only a single, broad, intense absorption line (or band) in the near infrared region, with a wavelength around $\lambda = 4.66\ \mu$. For other diatomic molecules consisting of unlike atoms, one finds the same situation; e.g., for HCl this band lies at $\lambda = 2.46\ \mu$. One also finds that such bands do not appear for molecules consisting of like atoms, such as $O_2$, $N_2$, $H_2$.

If the absorption is observed with thicker layers of gas, the intensity of absorption of the fundamental band naturally increases, and in addition a

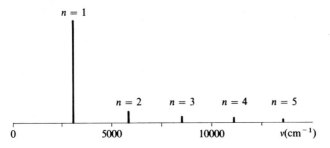

Figure 1.3  Coarse structure of the infrared spectrum of HCl (schematic).
The intensity actually falls off five times faster than indicated by the
height of the vertical lines. Herzberg [1966] vol. 1.

second band of similar form appears quite weakly, at approximately half the
wavelength or double the frequency (wave number). If the thickness of the
layer is still further increased (up to several meters at atmospheric pressure),
a third and possibly even a fourth and a fifth band appear whose wavelengths
are approximately a third, a fourth, and a fifth, respectively, of that of the
first band; that is to say, their frequencies are three, four, and five times as
great. Figure 1.3 gives schematically the complete infrared spectrum of HCl.
In this figure the lengths of the vertical lines that represent the bands give an
indication of their intensity. However, the actual decrease in intensity is five
times as fast as is indicated in the drawing.

   The explanation of these additional bands with lower intensity is that the
diatomic molecule is not quite a harmonic oscillator. In a harmonic oscillator
the restoring force increases indefinitely with increasing distance from the
equilibrium point. However, it is clear that in an actual molecule, when the
atoms are at a great distance from one another, the attractive force is zero.
Thus the quantum-mechanical harmonic oscillator is only a simplified model
of the vibrating molecule, and if one wants to describe the finer details of

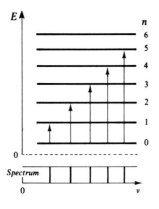

Figure 1.4  Energy levels and infrared transitions of the anharmonic
oscillator. The absorption spectrum is given schematically beneath.

vibrating molecules, then the anharmonic forces also have to be taken into account. The energy levels of the anharmonic oscillator are not equidistant like those of the harmonic oscillator, but rather their separation decreases slowly with increasing $n$.

The energy levels and absorption spectrum for an anharmonic but almost harmonic oscillator are shown in Figure 1.4. (For the sake of clarity a faster decrease of $\Delta E$ is drawn than is actually found in most observed cases.) The selection rule (1.10),

$$n - m = \pm 1,$$

holds only approximately for the anharmonic oscillator and applies only to the most intense transitions. But now transitions with $n - m = \pm 2, \pm 3, \ldots$ can also appear—though with rapidly decreasing intensity. All these results can be calculated using perturbation theory, which we shall introduce in Chapter VIII. We describe these facts here to demonstrate that the simple soluble quantum-mechanical models like the harmonic oscillator describe only the principal structures of a microphysical system in nature, and cannot be expected to describe all the details. This is not a deficiency of the harmonic oscillator model but a general property of physical theories. Models are only idealizations and cannot be expected to reproduce the experimental results up to the last digit. The explanation of a new decimal place in an experimental number often requires a new model, and sometimes a completely new theory.

We shall see this presently when we consider the transition frequencies in the near infrared region in more detail, as obtained with a spectrometer of sufficiently high resolution. The broad spectral lines for the CO molecule around $v = 2140 \text{ cm}^{-1}$ is then resolved into a number of individual narrow lines, as shown in Figure 1.5. That is, around $v = 2140 \text{ cm}^{-1}$ one does not have a single line but a band, called the vibration-rotation band. As one sees from this figure, this band consists of a series of almost equidistant lines, with one line missing in the center of the band. Going out from the gap, there are two branches, which are called the $P$ branch (towards longer wavelengths) and the $R$ branch (towards smaller wavelengths). Figure 1.6 shows the same effect for the $n = 1$ line of Figure 1.3 for HCl.

One would expect such fine structure in the absorption or emission spectrum of electromagnetic radiation for the CO molecule if the energy levels of the vibrating molecule of Figure 1.1 were split into a series of sub-levels as shown in Figure 1.7, which shows only any two neighboring energy levels of the energy spectrum of the vibrating molecule as given in Figure 1.1.

The description of such a splitting lies outside the capability of an oscillator model. It can only mean that a state characterized by the quantum number $n$ is not a pure state but is in fact a mixture of states with different energies. In the oscillator, however, the state characterized by $n$ was a pure state described by a projection operator $\Lambda_n$ on a one-dimensional subspace spanned by $\phi_n$, namely $\Lambda_n \mathscr{H}$. The state of the diatomic molecule characterized by the quantum number $n$ must have at least as many dimensions as energy levels (when number of energy levels equals the dimension then to any energy value

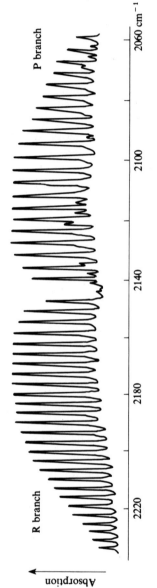

Figure 1.5    The vibration-rotation band of carbon monoxide.

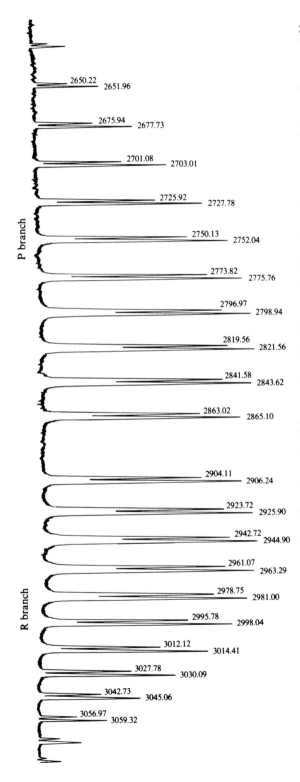

Figure 1.6   The fundamental absorption band for HCl under high resolution. (The lines are doubled due to the presence of the two isotopes Cl$^{35}$ and Cl$^{37}$ in the ratio 3:1; we do not discuss this effect here.) [From Alpert Keiser and Szymanski (1970), with permission.]

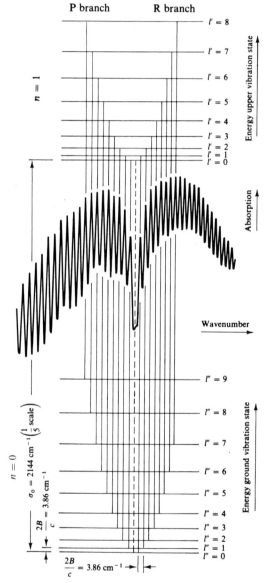

Figure 1.7   Origin and appearance of rotational structure. P and R branches are shown to the left and right, respectively, on the spectrometer tracing of the CO fundamental absorption band at 2144 cm$^{-1}$. The Q branch (dashed line) is missing. Energy levels are shown to scale, except that the distance between upper and lower vibrational states (2144 cm$^{-1}$) should be about five times as great as in the figure. [From Bauman (1962), with permission.]

there belongs a one-dimensional subspace or a projector on a one-dimensional subspace). Consequently the oscillator model alone describes only part of the properties of a diatomic molecule, and in order to describe the finer details of the spectrum we have to combine the oscillator model with another model, which describes these finer details and which reflects additional features of the diatomic molecule that we have so far not taken into account. This new model is the *rotator model*.

If we return to our classical picture of the CO molecule, considering it as two atoms of mass $m_1$ and $m_2$ at a distance $x$, then we see that this classical object not only can perform vibrations along the $x$-axis, but also can perform rotations in three-dimensional space around its center of mass. As long as it is in the vibrational ground state, that is, as long as the energy involved is less than 0.26 eV for CO, it is a rigid rotator, i.e., it can be considered as two pointlike masses $m_1$, $m_2$ fastened to the ends of a weightless rigid rod of length $x$. We shall therefore first study the rigid-rotator model by itself. This will provide us with a description of the CO states that are characterized by the quantum number $n = 0$, and will also approximately describe each set of states with a given vibrational quantum number $n$. Then we shall see how these two models are combined to form the vibrating rotator or the rotating vibrator.

## III.2  The Rigid Rotator

To conjecture the mathematical image (the algebra of operators) for the rotator, we proceed in the same way as for the oscillator: We consider the classical rotator, and we replace the three coordinates of momentum $p_i$ and the three coordinates of position $x_i$ in all expressions for the observables with operators $P_i$ and $Q_i$ that fulfill the canonical commutation relations

$$[P_i, Q_j] = \frac{\hbar}{i}\delta_{ij}I \qquad [Q_i, Q_j] = 0 \qquad [P_i, P_j] = 0. \tag{2.1}$$

—the obvious generalization of (II.2.1a) to the three-dimensional case.

In classical mechanics the energy $E$ of rotation of a rigid body is given by

$$E = \tfrac{1}{2}I\omega^2. \tag{2.2}$$

Here $\omega$ is the angular velocity of the rotation,[3] and $I$ is the moment of inertia of the system about the axis of rotation. The angular velocity is related to the number of rotations per second, $\nu_{\text{rot}}$ (the rotational frequency) by

$$\omega = 2\pi\nu_{\text{rot}}. \tag{2.3}$$

The angular momentum of the system is given by $\mathbf{I} = I\vec{\omega}$. Introducing this into (2.2), the energy may also be expressed by

$$E = \frac{\mathbf{I}^2}{2I}. \tag{2.4}$$

---

[3] We use the same symbol $\omega$ for the rotational angular velocity that we used for the vibrational angular velocity in Chapter II.

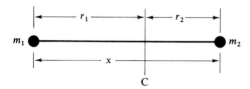

Figure 2.1  Dumbbell model of a diatomic molecule.

The moment of inertia about the axis of rotation for the rotator model is given by

$$I = m_1 r_1^2 + m_2 r_2^2,$$

where

$$r_1 = \frac{m_2}{m_1 + m_2} x \quad \text{and} \quad r_2 = \frac{m_1}{m_1 + m_2} x \tag{2.5}$$

are the distances of the masses $m_1$ and $m_2$ from the center of mass $C$, and $x$ is the distance between the two mass points (see Figure 2.1). Substitution gives

$$I = \frac{m_1 m_2}{m_1 + m_2} x^2 \tag{2.6}$$

that is, the moment of inertia about the axis of rotation is the same as that of a mass point of mass

$$\mu = \frac{m_1 m_2}{m_1 + m_2} \tag{2.7}$$

at a distance $x$ from the axis; $\mu$ is called the *reduced mass* of the molecule.

Thus, instead of considering the rotation of the rigid rotator, we can equally well consider the rotation of a single mass point of mass $\mu$ with coordinates $x_i$, where the vector $\mathbf{x} = (x_1, x_2, x_3)$ is the position vector for the mass point $\mu$ with the center of mass $C$ as the origin. If we denote the momentum of the mass point $\mu$ in this coordinate system by $\mathbf{p} = (p_1, p_2, p_3)$, then the angular momentum about the point $C$ is given by

$$\mathbf{l} = \mathbf{x} \times \mathbf{p}, \tag{2.8}$$

so its components are given by

$$l_i = \sum_{j, k} \epsilon_{ijk} x_j p_k \equiv \epsilon_{ijk} x_j p_k. \tag{2.9}$$

In this equation $\epsilon_{ijk} = +1$ for $(ijk) = (123)$ and every even permutation thereof, $\epsilon_{ijk} = -1$ for $(ijk)$ an odd permutation of $(123)$, and $\epsilon_{ijk} = 0$ otherwise. It is understood from now on that two identical indices in a term indicate a summation from 1 to 3 on those indices unless stated otherwise.

The operators that represent the corresponding quantum-mechanical observables are then obtained by our general principle of replacing $x_i$, $p_j$ by $Q_i$, $P_j$ fulfilling (2.1). Thus the angular-momentum operator $\mathbf{L}$ is given by

$$\mathbf{L} = \mathbf{Q} \times \mathbf{P},$$

or

$$L_i = \epsilon_{ijk} \, Q_j P_k, \tag{2.10}$$

and the energy operator is given in correspondence to (2.4) by

$$H = \frac{1}{2I} \mathbf{L}^2 = \frac{1}{2I} \sum_{i=1}^{3} L_i L_i = \frac{1}{2I} L_i L_i. \tag{2.11}$$

As $Q_j$ and $P_k$ are Hermitian operators, it follows that $L_i$ and $H$ are Hermitian:

$$L_i^\dagger = \epsilon_{ijk} P_k^\dagger Q_j^\dagger = \epsilon_{ijk} P_k Q_j = \epsilon_{ijk} Q_j P_k = L_i. \tag{2.11a}$$

From the Heisenberg commutation relations (2.1) one obtains the commutation relations of the operators $L_i$ that represent the components of angular momentum. The calculation goes as follows:

$$[L_i, L_l] = \epsilon_{ijk} \epsilon_{lmn} [Q_j P_k, Q_m P_n]$$

$$= \epsilon_{ijk} \epsilon_{lmn} (Q_j [P_k, Q_m P_n] + [Q_j, Q_m P_n] P_k)$$

$$= \epsilon_{ijk} \epsilon_{lmn} (Q_j \{[P_k, Q_m] P_n + Q_m [P_k, P_n]\}$$
$$+ \{[Q_j, Q_m] P_n + Q_m [Q_j, P_n]\} P_k)$$

$$= \epsilon_{ijk} \epsilon_{lmn} \left( Q_j P_n \frac{\hbar}{i} \delta_{km} - Q_m P_k \frac{\hbar}{i} \delta_{jn} \right)$$

$$= \frac{\hbar}{i} (\epsilon_{ijk} \epsilon_{lkn} Q_j P_n - \epsilon_{ijk} \epsilon_{lmj} Q_m P_k).$$

By changing the summation indices we can write this as

$$\frac{\hbar}{i} (\epsilon_{imk} \epsilon_{lkn} Q_m P_n - \epsilon_{ikn} \epsilon_{lmk} Q_m P_n) = \frac{\hbar}{i} (\epsilon_{imk} \epsilon_{knl} + \epsilon_{ink} \epsilon_{klm}) Q_m P_n. \tag{2.12}$$

We now make use of the following property of the $\epsilon_{imk}$, which is easily verified:

$$\epsilon_{imk} \epsilon_{knl} = \delta_{in} \delta_{ml} - \delta_{mn} \delta_{il}. \tag{2.13}$$

Writing the second term in (2.12) in the form (2.13):

$$\epsilon_{ink} \epsilon_{klm} = \delta_{il} \delta_{nm} - \delta_{im} \delta_{nl},$$

and adding it to (2.13) gives

$$\epsilon_{imk} \epsilon_{knl} + \epsilon_{ink} \epsilon_{klm} = \delta_{in} \delta_{ml} - \delta_{im} \delta_{nl} = -\epsilon_{ilk} \epsilon_{kmn}.$$

The last equality again follows from (2.13). Substitution of this expression into (2.12) gives

$$[L_i, L_l] = i\hbar \epsilon_{ilk} \epsilon_{kmn} Q_m P_n.$$

If we insert the definition (2.10) of $L_i$ on the right-hand side, we obtain the commutation relation of the angular-momentum operators $L_i$:

$$[L_i, L_l] = i\hbar\epsilon_{ilk}L_k. \tag{2.14}$$

We have obtained the commutation relations (2.14) of the angular-momentum operators (2.10) from the Heisenberg commutation relations (2.1). The expression of the energy operator (2.11) does not contain the $P_i$ and $Q$ explicitly; this is true for all physical observables of the rotator. In fact the operators $P_i$ and $Q_i$ are unphysical observables for the quantum-mechanical rotator system in the following sense: It is not possible to prepare the quantum-mechanical rotator so that it is in a generalized eigenstate of momentum or position. Therefore, for the rotator, the $L_i$ that obey the commutation relations (2.14) are the fundamental physical observables, and the fact that they may be obtained from the momentum and position operators can be ignored. In fact, the operators $L_i$ defined by (2.10) or the classical quantities $l_i$ defined by (2.9) are a particular case of an observable that is connected with new degrees of freedom of physical systems in the three-dimensional space.

A physical object in the three-dimensional physical space has six degrees of freedom: three translational degrees of freedom described by the three coordinates $x_i$, and three rotational degrees of freedom, described by a rotation $R(\alpha, \beta, \gamma)$ that depends upon three angles $\alpha$, $\beta$, $\gamma$ (e.g., the three Euler angles, or the three angles of rotation around three fixed coordinate axes). The momentum $p_i$ is the canonical variable conjugate to the coordinate $x_i$, and the canonical variable conjugate to the angular coordinates $\alpha_i$ is the angular momentum $l_i$.

If we forget about the constituents of the diatomic molecule and consider the dumbbell as an entire physical system whose center of mass is fixed at a particular point in space, then the remaining degrees of freedom are the rotations by the three angles, and the dynamical variables are these angles and their conjugates, the angular momenta $l_i$. In our particular case above the $l_i$ are obtained from the momenta and coordinates of the constituents, which are thought of as mass points, i.e., not possessing rotational degrees of freedom.

In general, a classical extended particle, in the three-dimensional physical space, has as variable the momenta $p_i$ and the angular momenta $s_i$ (spin) associated with it, which are the conjugates of the linear coordinates $x_i$ and the angular coordinates $\alpha_i$ respectively. For the case that the extended physical object is a dumbbell the spin $s_i$ is the orbital angular momentum $l_i = \epsilon_{ijk}x_jp_k$ of its two constituents.

For the quantum physical extended particle, the momentum is represented by the operator $P_i$, and the spin by the operator $S_i$. It is now easy to conjecture that the defining commutation relations of the spin operators $S_i$ are

$$[S_i, S_j] = i\hbar\epsilon_{ijk}S_k. \tag{2.15}$$

For the special case of the spin of the dumbbell, these relations were derived in (2.14) from the relations (2.1), which had already proven to be successful. Equation (2.15) can also be derived from the properties of the group of rotations, if one assumes that the rotation $R(\alpha, \beta, \gamma)$ of the physical object is represented (continuously) by a (unitary) operator $U(\alpha, \beta, \gamma)$ in the space of physical states of this quantum-physical object. This latter assumption can be derived from the fact that the rotations are symmetry transformations and that the rotation group is a symmetry group of the physical system (Wigner Theorem). We will not discuss this connection here, but we take the algebra of observables defined by (2.15) as the starting point.

We will now investigate the properties of the algebra of operators generated by the $J_i$, which fulfill the commutation relations

$$[J_i, J_k] = i\hbar\epsilon_{ikl}J_l \qquad (i, k = 1, 2, 3), \qquad (2.16)$$

where $J_i$ may be either an $L_i$ as given by (2.10) or an $S_i$. In particular we shall find the properties of all the $J_i$, which are linear Hermitian operators in a linear space. It will turn out that the set of all $J_i$ is richer than the set of the $L_i$ given by (2.10). The algebra generated by the $J_i$ is called [because of its connection to the rotation group denoted SO(3)] $\mathscr{E}(SO(3))$ or $\mathscr{E}(SU(2))$.

## III.3 The Algebra of Angular Momentum

We shall now find all possible solutions of the commutation relation (2.16) that fulfill $J_i^\dagger = J_i$. (We remark that to be mathematically rigorous, the condition $J_i^\dagger = J_i$ would have to be given a mathematically precise formulation, which we replace with the additional assumption that there exists at least one eigenvector of $J_3$ in the Hilbert space.)

Instead of working with the $J_i$ ($i = 1, 2, 3$), we introduce the following linear combinations:

$$H_3 = \hbar^{-1}J_3, \qquad H_+ = \hbar^{-1}(J_1 + iJ_2) \qquad H_- = \hbar^{-1}(J_1 - iJ_2). \quad (3.1)$$

The Hermiticity condition $J_i^\dagger = J_i$ is expressed by

$$H_3^\dagger = H_3, \qquad H_+^\dagger = H_-, \qquad H_-^\dagger = H_+. \qquad (3.2)$$

From (2.16) it then follows that

$$[H_3, H_\pm] = \pm H_\pm,$$
$$[H_+, H_-] = 2H_3. \qquad (3.3)$$

The operator $\mathbf{J}^2$ can be written

$$\mathbf{J}^2 = \hbar^2\mathbf{H}^2 \qquad (3.4)$$

with

$$\mathbf{H}^2 = H_+H_- + H_3^2 - H_3 = H_-H_+ + H_3^2 + H_3. \qquad (3.5)$$

$H^2$ and therefore $J^2$ have the property that

$$[H^2, H_3] = 0, \qquad [H^2, H_\pm] = 0, \qquad (3.6a)$$

or in general

$$[H^2, A] = 0, \qquad (3.6b)$$

where $A$ is any element of the algebra of operators $\mathscr{E}(SU(2))^*$ generated by $J_i$ $(i = 1, 2, 3)$, i.e., $A$ is any element of the form

$$A = a + a^i J_i + a^{ij} J_i J_j + a^{ijk} J_i J_j J_k + \cdots, \qquad (3.7)$$

where $a, a^i, a^{ij}, a^{ijk}, \ldots,$ are complex numbers. Because of the property (3.6), $H^2$ and $J^2$ are called *invariant operators (Casimir operators)* of the algebra $\mathscr{E}(SU(2))$.

We now find all *ladder representations* of $\mathscr{E}(SU(2))$, i.e., all solutions of the commutation relations (2.16) by linear operators in linear spaces that are obtained by applying to one eigenvector of $H_3$ the whole algebra $\mathscr{E}(SU(2))$. (This one eigenvector is assumed to exist.)

Let $f_m$ be normalized and be an eigenvector of $H_3$ with eigenvalue $m$, i.e.,

$$(f_m, f_m) = 1, \qquad H_3 f_m = m f_m. \qquad (3.8')$$

($f_m$ are called *weight vectors*; $m$ is called a *weight*.) Then we define

$$f_{m\pm} = H_\pm f_m$$

and calculate using (3.3)

$$
\begin{aligned}
H_3 f_{m\pm} &= H_3 H_\pm f_m = (H_\pm H_3 \pm H_\pm) f_m \\
&= (H_\pm m \pm H_\pm) f_m = (m \pm 1) H_\pm f_m \qquad (3.8) \\
&= (m \pm 1) f_{m\pm}.
\end{aligned}
$$

Thus $f_{m\pm}$ is, if it is different from 0, an eigenvector of $H_3$ with eigenvalue $m \pm 1$.

$f_m$ can be chosen so that it is not only an eigenvector of $H_3$ but also an eigenvector of $H^2$, because $H_3$ and $H^2$ commute.

PROOF.

$$H_3 H^2 f_m = H^2 H_3 f_m = m H^2 f_m.$$

Let us denote the eigenvalue of $H^2$ belonging to $f_m$ by $c$:

$$H^2 f_m = c f_m. \qquad (3.9)$$

Then the above relation is an identity. Note that if two operators do not commute then in general they have no common eigenvectors. $\qquad \square$

We note some properties of (3.9):

1. The number $c$ is nonnegative.

---

* The symbol $\mathscr{E}$ stands for enveloping algebra.

PROOF.

$$c = (f_m, \mathbf{H}^2 f_m) = \hbar^{-2}(f_m, (J_1^2 + J_2^2 + J_3^2)f_m)$$
$$= \hbar^{-2}(\|J_1 f_m\|^2 + \|J_2 f_m\|^2 + \|J_3 f_m\|^2) \geq 0. \qquad \square$$

2.  Any vector $f$ obtained from $f_m$ by applying any element $A$ of the form (3.7) on it: $f^{(c)} = Af_m^{(c)}$ is again an eigenvector of $\mathbf{H}^2$ and has the same eigenvalue $c$.

PROOF.

$$\mathbf{H}^2 f = \mathbf{H}^2 A f_m = A\mathbf{H}^2 f_m = cAf_m = cf, \text{ where we have used (3.6).} \qquad \square$$

3.  Any eigenvalue $m$ of $H_3$ fulfills

$$m^2 \leq c. \tag{3.10}$$

PROOF.

$$0 \leq \hbar^{-2}\{(J_1 f_m, J_1 f_m) + (J_2 f_m, J_2 f_m)\}$$
$$= \hbar^{-2}\{(f_m, J^2 f_m) - (f_m, J_3^2 f_m)\} = \hbar^{-2}\hbar^2(c - m^2)\|f_m\|^2,$$

from which follows $(c - m^2) \geq 0$.

Note that in the proofs the Hermiticity property of $J_i$ has been used.

If we start with an arbitrary eigenvector $f_{m_0}^c$ of $H_3$ and $\mathbf{H}^2$ and apply $H_+$ successively, then we obtain according to (3.8) new eigenvectors $f_m^c$ of $H_3$ with ever-increasing eigenvalues. Because of (3.10), after a finite number of steps we must reach the eigenvector $f_l^{(c)}$ with the largest eigenvalue $l$ of $H_3$, i.e., with

$$f_{l+}^c = H_+ f_l^c = 0. \tag{3.11}$$

From (3.5) it then follows that

$$\mathbf{H}^2 f_l^c = (H_3^2 + H_3)f_l^c = l(l + 1)f_l^c. \tag{3.12}$$

Consequently the eigenvalue $c$ of $\mathbf{H}^2$ and the largest eigenvalue $l$ of $H_3$ (highest weight) are connected by

$$c = l(l + 1).$$

Instead of characterizing the eigenvectors of $\mathbf{H}^2$ and $H_3$ by $c$ and $m$, we can characterize them by $l$ and $m$.

If we apply $H_-$ successively to $f_m^l$, then after a finite number of steps, because of (3.10), we must reach the vector $f_\mu^l$ with the lowest eigenvalue $\mu$ of $H_3$, i.e., with

$$f_{\mu-}^l = H_- f_\mu^l = 0. \tag{3.13}$$

From (3.5) it then follows

$$\mathbf{H}^2 f_\mu^l = (H_3^2 - H_3)f_\mu^l = \mu(\mu - 1)f_\mu^l. \tag{3.14}$$

Comparing (3.14) with (3.12) we find

$$l(l + 1) = \mu(\mu - 1),$$

and the only solution of this equation for $\mu$ that fulfills (3.10) is

$$\mu = -l. \tag{3.15}$$

Thus if we start with the vector $f_l^l$, apply $H_-$ successively to it and normalize, we obtain the sequence of vectors

$$f_{l-1}^l = (\alpha_l)^{-1} H_- f_l^l,$$
$$f_{l-2}^l = (\alpha_{l-1})^{-1} H_- f_{l-1}^l, \tag{3.16}$$
$$\vdots$$
$$f_{m-1}^l = (\alpha_m)^{-1} H_- f_m^l,$$

where the $\alpha_m$ are

$$\alpha_m = \sqrt{(H_- f_m^l, H_- f_m^l)}.$$

We shall finally reach the vector with the smallest weight $\mu$:

$$f_\mu^l = (\alpha_{\mu+1})^{-1} H_- f_{\mu+1}^l.$$

As $\mu = -l$, we have $2l + 1$ vectors in the sequence (3.16).

$$f_m^l \quad \text{with } m = l, l-1, l-2, \ldots, -l+1, -l, \tag{3.17}$$

which fulfill

$$(f_m^l, f_{m'}^l) = \delta_{mm'}. \tag{3.18}$$

As $2l + 1$ is a number of vectors, it must be an integer; consequently $l$ can be only one of the following numbers:

$$l = 0, \tfrac{1}{2}, 1, \tfrac{3}{2}, \ldots \tag{3.19}$$

Thus for a given number $l$, which can be one of the numbers in (3.19), we have $2l + 1$ vectors $f_m^l$ that are orthogonal to each other and span a space, which we shall call $\mathcal{R}^l$:

$$\mathcal{R}^l = \left\{ f : f = \sum_{m=-l}^{+l} a^m f_m^l \right\}. \tag{3.20}$$

We shall now determine the value of the normalization constant $\alpha_m$:

$$\alpha_m \overline{\alpha_m} = (H_- f_m, H_- f_m) = (f_m, H_+ H_- f_m)$$
$$= (f_m, (\mathbf{H}^2 - H_3^2 + H_3) f_m) = l(l+1) - m^2 + m.$$

Thus, except for a phase factor (which remains undetermined),

$$\alpha_m = \sqrt{l(l+1) - m^2 + m} = \sqrt{(l+m)(l-m+1)}, \tag{3.21}$$

and we have

$$H_- f_m^l = \sqrt{(l+m)(l-m+1)} \, f_{m-1}^l = \alpha_m f_{m-1}^l. \tag{3.22}$$

It remains to determine $H_+ f_m$. We know already that $H_+ f_m^l \propto f_{m+1}^l$. We set $H_+ f_m^l = \beta_m f_{m+1}^l$ and calculate

$$\overline{\beta_m}(f_{m+1}^l, f_{m+1}^l) = (H_+ f_m^l, f_{m+1}^l) = (f_m^l, H_- f_{m+1}^l) = \alpha_{m+1}(f_m^l, f_m^l).$$

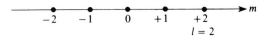

Figure 3.1  Example of a weight diagram of an irreducible representation of SU(2).

Therefore

$$\overline{\beta}_m = \alpha_{m+1} = \sqrt{(l + m - 1)(l - m)} = \beta_m,$$

and we have

$$H_+ f^l_m = \sqrt{(l - m)(l + m + 1)}\, f^l_{m+1} = \alpha_{m+1} f^l_{m+1}. \qquad (3.23)$$

Summarizing, we have found the following: For every integer or half-integer value $l$ there is a space $\mathscr{R}^l$ spanned by $2l + 1$ orthogonal vectors $f^l_m$ $(m = -l, \ldots, l)$. In this space the operators $H_+$, $H_-$, $H_3$ are given by (3.8′), (3.22), (3.23); and thus the action of any element $A \in \mathscr{E}(SU(2))$ given by (3.7) on any vector $f \in \mathscr{R}^l$ given by (3.20) is determined. To indicate that for every $l$ one obtains a different operator, one could also write $H^{(l)}_+$, $H^{(l)}_-$, $H^{(l)}_3$ for the operators in (3.8′), (3.22), (3.23). The space $\mathscr{R}^l$ is called an *irreducible representation space* of $\mathscr{E}(SU(2))$. In this space the elements $H_+$, $H_-$, $H_3$ defined by (3.3) are represented by the operators given in (3.8′), (3.22), (3.23). These operators are called the $(2l + 1)$-*dimensional irreducible representation* of the $H_+$, $H_-$, $H_3$. As can be seen, they depend upon $l$; for each $l = 0, \frac{1}{2}, 1, \ldots$, there is one "different" set of operators. All of the vectors in $\mathscr{R}^l$ are eigenvectors of $\mathbf{H}^2$ with the same eigenvalue $l(l + 1)$. There is no element $A \in \mathscr{E}(SU(2))$ that can transform from a vector $f^l \in \mathscr{R}^l$ to a vector $f^{l'} \in \mathscr{R}^{l'}$ with $l \neq l'$. This fact is expressed by the statement that $\mathscr{R}^l$ is "left invariant by all $A \in \mathscr{E}(SU(2))$." In particular, it is left invariant by the $J_i$ $(i = 1, 2, 3)$ and the $H_\pm$, $H_3$.

If we plot the possible values of $m$ in an irreducible representation along a line, then we obtain the weight diagram of the representation characterised by $l$, of SU(2). This is the simplest example of a weight diagram for Lie groups. For $l = 2$ this is shown in Figure 3.1. To each · there corresponds a basis vector $f^l_m$ in the representation space $\mathscr{R}^l$, or equivalently there corresponds the one-dimensional subspace spanned by $f^l_m$. Each such subspace (or basis vector) represents a (pure) physical state. Thus to each point in the weight diagram there corresponds a pure physical state.

The smallest space is $\mathscr{R}^0$, which is one-dimensional, as shown in Figure 3.2.

In general, we have a weight diagram for each representation, and the action of the operators $H_+$, $H_-$ can be represented in this diagram, as shown

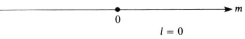

Figure 3.2  Weight diagram of the one-dimensional representation of SU(2).

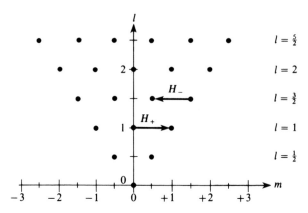

Figure 3.3   Collections of weight diagrams of irreducible representations of SU(2).

in Figure 3.3. For each weight diagram, there is a space $\mathscr{R}^l$, and for each space $\mathscr{R}^l$ there is a possible state (or set of states) of the quantum-mechanical system characterized by the value $l$. Because of the correspondence between the operator $\mathbf{L}$ (2.10) and the classical angular momentum $\mathbf{l}$ (2.8), the number $l$ is called the *angular momentum quantum number*:

$$\mathbf{L}^2 f^l = \hbar^2 l(l + 1) f^l.$$

Thus the quantum-mechanical angular momentum can take only a discrete number of values. The physical state that corresponds to a given space $\mathscr{R}^l$, and which is described by the statistical operator $W = (\dim \mathscr{R}^l)^{-1}\Lambda^l = (2l + 1)^{-1}\Lambda^l$ where $\Lambda^l$ is the projector onto $\mathscr{R}^l$, has a definite angular momentum $l$. In general (except for $l = 0$) this is not a pure state; it can be a mixture of states with different values of $m: m = -l, -l + 1, \ldots, l$.

The space $\mathscr{R}^l$ is the direct sum of one-dimensional spaces $\mathscr{R}^l_m$,

$$\mathscr{R}^l = \sum_{m=-l}^{+l} \oplus \mathscr{R}^l_m, \tag{3.24}$$

and each $\mathscr{R}^l_m$ is spanned by the vector $f^l_m$.

It has already been mentioned that not all sets of linear operators $J_i$ ($i = 1, 2, 3$) that fulfill (2.16) can be given by (2.10). It can be proven (Problem III.1) that for $L_i$ given by (2.10) the number $l$ can only take the values $l = 0, 1, 2, \ldots$; in other words, the operators $L_i$ given by (2.10) can be represented by operators in the spaces $\mathscr{R}^l$ with $l = 0, 1, 2, \ldots$ only. Thus for the operators given by (2.9) there are a countable number of representatives $L_i^{(l)}$ in spaces $\mathscr{R}^l$ that are left invariant by the $L_i^{(l)}$. These spaces $\mathscr{R}^l$ are not, however, left invariant by the $Q_j$ and $P_j$.

The operators $L_i$ in $\mathscr{R}^l$ with $l = \frac{1}{2}, \frac{3}{2}, \frac{5}{2}, \ldots$ or any direct sum of them,

$$\sum_{\substack{l=\text{half} \\ \text{integer}}} \oplus \mathscr{R}^l$$

cannot be expressed as functions of the operators $Q_i$ and $P_i$. For $l = \frac{1}{2}$ the operators $L_i^{(l=1/2)}$ are called spin operators. The $2 \times 2$ matrices $\sigma_i$ with the matrix elements $2(f_{m'}^{l=1/2}, L_i f_m^{l=1/2})$ are called Pauli matrices (Problem III.7).

## III.4  Rotation Spectra

As mentioned before, the algebra $\mathscr{E}(\mathrm{SU}(2))$ does not contain operators that transform out of a given $\mathscr{R}^l$. The algebra of observables of the quantum-mechanical rotator is, however, larger than $\mathscr{E}(\mathrm{SU}(2))$; additional elements can be formed, e.g., algebraic functions of $J_i$ and $P_i$ or of $J_i$ and $Q_i$. The observables $Q_i$, for instance, have the property that they transform from a given $\mathscr{R}^l$ into the neighboring $\mathscr{R}^{l+1}$ and $\mathscr{R}^{l-1}$:

$$Q_i : \mathscr{R}^l \to \mathscr{R}^{l-1} \oplus \mathscr{R}^{l+1}, \tag{4.1}$$

but not into $\mathscr{R}^{l\pm n}$ for $n > 1$. We shall prove this in the section on parity.

The space of physical states $\mathscr{R}$ of the quantum-mechanical rotator is the direct sum of the spaces $\mathscr{R}^l$:

$$\mathscr{R} = \sum_{l=0}^{\infty} \oplus \mathscr{R}^l. \tag{4.2}$$

Before we justify this statement, we want to give a brief description of the properties of $\mathscr{R}$.

$\mathscr{R}$ is not an irreducible representation space of $\mathscr{E}(\mathrm{SO}(3))$, the algebra of angular momenta. It is called a *reducible representation space*, and it reduces as given in (4.2) into a direct sum of irreducible representation spaces $\mathscr{R}^l$ ($l = 0, 1, 2, \ldots$). The operators $H_+$, $H_-$, $H_3$, now considered as operators in the big space $\mathscr{R}$, transform every element of a given space $\mathscr{R}^l$ into an element that is again in the same $\mathscr{R}^l$. Thus the subspaces $\mathscr{R}^l$ of $\mathscr{R}$ are left "invariant" by the $H_+$, $H_-$, $H_3$, and consequently by any $A \in \mathscr{E}(\mathrm{SU}(2))$. The operator $\mathbf{H}^2$—which is a number in the space $\mathscr{R}^l$, namely $l(l+1)$—has in $\mathscr{R}$ a nontrivial spectrum, namely

$$\text{spectrum } \mathbf{H}^2 = \{l(l+1), l = 0, 1, 2, 3, \ldots\} \tag{4.3}$$

The weight diagram for the representation in $\mathscr{R}$ is shown in Figure 4.1.

We shall now justify the statement (4.2). We assume that the $J_i$ are the angular momenta $L_i = \epsilon_{ijk} Q_j P_k$; then, according to the result stated in Section III.3 and proved in Problem III.1, only integer values of $l$ are allowed, i.e., $\mathscr{R}$ contains only $\mathscr{R}^l$ with $l = 0, 1, 2, \ldots$. That $\mathscr{R}$ contains all the $\mathscr{R}^l$ ($l = 0, 1, 2, \ldots$) follows from the fact that there are observables for the rotator (e.g., the operators $Q_i$) that transform from a given $\mathscr{R}^l$ to the neighboring $\mathscr{R}^{l-1}$ and $\mathscr{R}^{l+1}$ according to (4.1). That each $\mathscr{R}^l$ appears only once follows from the fact that for the rotator no additional quantum number is necessary; if one $\mathscr{R}^{l_0}$ were to appear twice or more, then there would be two or more vectors $f_m^{l_0}(1), f_m^{l_0}(2), \ldots$ with the same quantum numbers $l_0, m$, and a new

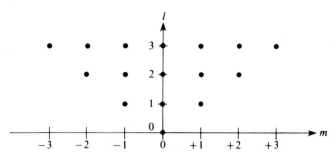

Figure 4.1   Weight diagram of E(3) or of SO(3, 1).

quantum number would be necessary to distinguish between these two or more vectors. But the rotator is just that model for which there is no other diagonal observable besides angular momentum ($L^2$ and $L_3$). (In other words, a real physical system can be a rotator only to the extent that no other quantum numbers are necessary for the description of its properties; e.g., polyatomic (symmetric top) molecules cannot in general be described by the rotator model, and even for the diatomic molecule, the rotator model is only an approximate description that neglects all but the rotational properties of the dumbbell.) Thus, as always, the justification of (4.2) is that in nature there are physical systems whose physical states are (up to a certain limitation) described by $\mathscr{R}$.

> [Mathematically this can be stated more briefly: "The spectrum-generating algebra of the rotator is $\mathscr{E}(E_3)$." $\mathscr{E}(E_3)$ is generated by $L_i$, $Q_i$ that fulfill the commutation relations
>
> $$[L_i, L_j] = i\hbar\epsilon_{ijk}L_k, \qquad [L_i, Q_j] = i\hbar\epsilon_{ijk}Q_k, \qquad [Q_i, Q_j] = 0,$$
>
> and $\mathscr{R}$ is a particular irreducible representation space of $\mathscr{E}(E_3)$.]

Each dot on the weight diagram of $\mathscr{R}$ represents the pure state that is described by the one-dimensional subspace $\mathscr{R}_m^l$ ($l$, $m$ fixed) spanned by $f_m^l$. The (normalized) statistical operator for the pure state $W = \Lambda_m^l$, where $\Lambda_m^l$ is the projector on $\mathscr{R}_m^l$, represents a quantum-mechanical system for which the angular momentum has the definite value $l$ and the 3-component of angular momentum, $H_3$, has a definite value $m$. As there is no distinguished direction in space and the coordinate system has been chosen arbitrarily, $H_3$ represents the angular momentum around an arbitrarily choosable direction; it is also called the helicity.

The values of the energy operator in $\mathscr{R}$, i.e., the energy spectrum of the rotator is obtained from (2.11) as

$$\text{spectrum } H = E_l = \frac{1}{2I}\hbar^2 l(l + 1). \tag{4.4}$$

Thus we see that the energy levels depend upon $l$, as represented in the diagram of Figure 4.2. If we compare this with Figure 1.7, we see that the

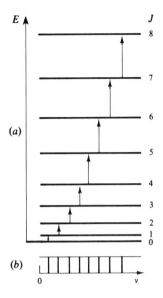

Figure 4.2    Energy levels and infrared transitions of a rigid rotator:
(a) The energy-level diagram, (b) the resulting spectrum (schematic).
[From Herzberg (1966), with permission.]

rotator has indeed the energy spectrum required to interpret the infrared
spectrum of diatomic molecules.

In contrast to the oscillator, the energy eigenspaces (i.e., the spaces of
vectors with the same energy eigenvalue) for the rotator are not one-
dimensional, except for $l = 0$. Therefore the state of a rotator with a definite
energy value $E_{l_0}$ ($l_0 = 0$) is not a pure state. If energy measurements have
been made with the result $E_{l_0}$, the statistical operator is given by

$$W = \Lambda^{l_0} \quad \text{(unnormalized)}, \qquad (4.5)$$

or

$$W = (\text{Tr } \Lambda^{l_0})^{-1} \Lambda^{l_0} = (2l_0 + 1)^{-1} \Lambda^{l_0} \quad \text{(normalized)}, \qquad (4.5')$$

where $\Lambda^{l_0}$ is the projector on the $(2l_0 + 1)$-dimensional space $\mathcal{R}^{l_0}$. By an
energy measurement alone it is not possible to prepare a pure state of the
rotator. Only under certain additional conditions—if a direction in space is
distinguished (e.g., by an external magnetic field)—can one prepare a state
with definite helicity, i.e., a pure state $\Lambda^{l_0}_{m_0}$. If only the energy of the rotator
has been measured but not the helicity, then the states with different helicities
are assumed to appear with equal weight, which is why one chooses

$$W = \Lambda^{l_0} = \Lambda^{l_0}_{-l_0} + \Lambda^{l_0}_{-l_0+1} + \cdots + \Lambda^{l_0}_{m} + \cdots + \Lambda^{l_0}_{l_0} \qquad (4.6)$$

for the (unnormalized) statistical operator.

In order to calculate the frequencies that can be emitted and absorbed by a rotator, we have to know the selection rules. In our classical picture of the rotator we can consider it as a rotating dipole moment $\mathbf{D}$, with

$$\mathbf{D} = \text{const } \mathbf{Q}, \tag{4.7}$$

where $\mathbf{Q}$ is the vector between the centers of positive and negative charge. Classically the radiation is then a consequence of the rotation of this electrical dipole moment. Quantum-mechanically, the intensity of the absorbed or emitted radiation is proportional to the absolute value squared of the matrix element of the operator $\mathbf{D}$, i.e., proportional to

$$|\langle f_{m'}^{l'}|\mathbf{Q}|f_m^l\rangle|^2. \tag{4.8}$$

Thus dipole radiation will only be obtained in transitions from states $f_{m'}^{l'}$ to $f_m^l$ between which the matrix element (4.8) is nonzero (quadrupole and higher-order radiation is negligibly small). It will be shown in the section on vector operators that [as was stated in (4.1)]

$$\langle f_{m'}^{l'}|Q_i|f_m^l\rangle = 0 \quad \text{unless } l = l' = \pm 1. \tag{4.9}$$

That is, the selection rule for dipole radiation of the rotator is

$$\Delta l = l - l' = \pm 1. \tag{4.10}$$

If we compare this result with the experimental situation for CO, depicted in Figure 1.7, we observe complete agreement. Figure 1.7 shows the transitions between states not only with different values of angular momentum $l$ but also with different values of the vibrational quantum number $n$.

We also expect radiation from transitions between different rotator states of the diatomic molecule that belong to the same oscillator state $n = 0$. These transitions (absorption) are indicated by the $\uparrow$ in Figure 4.2. The wave number of this radiation is given by Equation (1.1):

$$|\nu_{l'l}| = \frac{E_{l'} - E_l}{2\pi\hbar c}.$$

With (4.4) and (4.10) we calculate

$$|\nu_{l+1,l}| = \frac{\hbar^2}{2I}\frac{(l+1)(l+2) - (l+1)l}{2\pi\hbar c}$$

$$= \frac{\hbar}{4\pi c I}2(l+1) = B2(l+1), \tag{4.11}$$

where

$$B = \frac{\hbar}{8\pi^2 c I}. \tag{4.11'}$$

Thus the spectrum of a simple rigid rotator consists of a series of equidistant lines, as schematically drawn at the bottom of Figure 4.2.

We expect the frequency for pure rotational transitions to be much

**Table 4.1** Absorption Spectrum of HCl in the Far Infrared [From Herzberg (1966), with permission.]

| $l + 1$ | $v_{obs}$ | $\Delta v_{obs}$ | $v_{calc} =$ $20.68(l + 1)$ | $v_{calc} =$ $20.79(l + 1)$ $-0.0016(l + 1)^3$ |
|---|---|---|---|---|
| 1 | | | 20.68 | 20.79 |
| 2 | | | 41.36 | 41.57 |
| 3 | | | 62.04 | 62.33 |
| 4 | 83.03 | | 82.72 | 83.06 |
| 5 | 104.1 | 21.1 | 103.40 | 103.75 |
| 6 | 124.30 | 20.2 | 124.08 | 124.39 |
| 7 | 145.03 | 20.73 | 144.76 | 144.98 |
| 8 | 165.51 | 20.48 | 165.44 | 165.50 |
| 9 | 185.86 | 20.35 | 186.12 | 185.94 |
| 10 | 206.38 | 20.52 | 206.80 | 206.30 |
| 11 | 226.50 | 20.12 | 227.48 | 226.55 |

smaller than the vibrational frequency, because the spacing between the rotational energy levels is much smaller than between the vibrational energy levels as seen in Figure 1.7 (note the scale factor of $\frac{1}{5}$ there).

The pure rotation spectrum lies in the far infrared. The absorption spectrum of HCl in the far infrared has been measured, and the experimental results are given in the second column of Table 4.1. From (4.11) we expect that the frequencies will have an equidistant spacing. Therefore in the third column of the table the differences between the neighboring levels are given. This difference must be—according to (4.11):

$$\Delta v = 2B_{HCl}. \tag{4.12}$$

From the $\Delta v_{obs}$ we obtain

$$B_{HCl} = \frac{h}{8\pi^2 c I_{HCl}} \approx 10.34 \text{ cm}^{-1}. \tag{4.13}$$

The fourth column of the table gives the values calculated from (4.11) with the value (4.13). We observe fairly good agreement between the calculated and observed values if we compare column 2 with column 4. (The first observed value is for $l + 1 = 4$ because the frequencies for $l$'s lower than 3 lie in the far infrared and outside the region investigated.) The last column of the table gives a fit to the expansion

$$v_{l+1,l} = 2b(l + 1) - 4d(l + 1)^3 \tag{4.14}$$

($b, d$ are constants). Comparing this last column with the observed values in the second column, we see that the agreement of (4.14) with the experimental data is far better than that of (4.11). The energy spectrum that corresponds to (4.14) is given by

$$E_l = [bl(l + 1) - dl^2(l + 1)^2]2\pi\hbar c \tag{4.15}$$

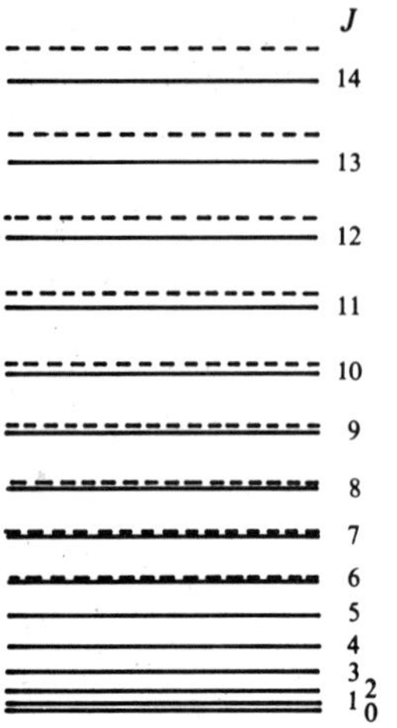

Figure 4.3   Energy levels of the nonrigid rotator. For comparison, the energy levels of the corresponding rigid rotator are indicated by broken lines (for $J < 6$, they cannot be drawn separately). [From Herzberg (1966), with permission.]

($b, d$ are constants). The energy levels (4.15) have been drawn in Figure 4.3 with an exaggerated value of $d$.

The explanation for the better fit of (4.15) to the experimental values is that the diatomic molecule HCl is not exactly a rigid rotator. The bonds between atoms are not rigid, and the interatomic distance varies with the speed of rotation, giving rise to a centrifugal distortion. Equation (4.15) can be obtained if we return to the classical picture in which the molecule is considered as two hard spheres (atoms) joined, not as in Figure 2.1 by a rigid rod, but by a spring. If the molecule rotates about an axis perpendicular to this spring, then at equilibrium the centrifugal force $l^2/(\mu x^3)$ equals the centripetal force $k(x - x_e)$, where $k$ is the spring constant and $x_e$ the interatomic distance of the stationary molecule. Thus

$$k(x - x_e) = \frac{l^2}{\mu x^3}. \tag{4.16}$$

The energy of this system is [cf. (II.2.3) and (2.4)]

$$E \doteq \frac{l^2}{2\mu x^2} + \tfrac{1}{2}k(x - x_e)^2. \tag{4.17}$$

Making use of the expansion

$$x^2 = x_e^2 \left( 1 + 2\frac{x - x_e}{x_e} + \cdots \right) \tag{4.18}$$

and (4.16), one obtains for $E$

$$E = \frac{1}{2\mu x_e^2}l^2 + \frac{1}{2\mu^2 k x_e^6}(l^2)^2 + O((l^2)^3). \tag{4.19}$$

The first term is the energy of the rigid rotator, and the second term is the contribution due to the centrifugal forces. Going to the quantum system operator by replacing the number $l^2$ with the operator $\mathbf{L}^2$, one obtains the energy operator

$$H = \frac{1}{2\mu x_e^2}\mathbf{L}^2 - \frac{1}{2\mu^2 k x_e^6}(\mathbf{L}^2)^2, \tag{4.20}$$

from which the spectrum (4.14) follows. The better fit of (4.14) to the experimental values confirms the above classical consideration. But we also observe that the empirical value of $d$ ($d = 0.0004$ cm$^{-1}$ for HCl) is orders of magnitude smaller than that of $b$, which is obtained from the above fit:

$$b_{HCl} = \frac{h}{8\pi^2 c I_{HCl}} = 10.395 \text{ cm}^{-1}. \tag{4.21}$$

This shows that the rigid rotator is a remarkably good model of the rotating diatomic molecule.

As we shall see below, the spacings between the levels of the rotating CO molecule are considerably smaller than for the HCl molecule. Therefore the pure rotation spectrum of CO lies at a considerably longer wavelength, where experimental investigation is very difficult.

We now want to obtain some quantitative features of the classical picture for the diatomic molecule. From the value (4.13) we calculate the moment of inertia of HCl:

$$I_{HCl} = 2.69 \times 10^{-40} \text{ g cm}^2.$$

With

$$m_{Cl} = \frac{36}{N_A} = 6.0 \times 10^{-23} \text{ g}$$

and

$$m_H = \frac{1}{N_A} = 0.17 \times 10^{-23} \text{ g},$$

one calculates (2.7):

$$\mu_{HCl} = \frac{m_{Cl} m_H}{m_{Cl} + m_H} = 1.63 \times 10^{-24} \text{ g}$$

From (2.6) we may calculate the internuclear distance of the HCl molecules using the values $I_{HCl}$ and $\mu_{HCl}$:

$$x_{HCl} = 1.29 \times 10^{-8} \text{ cm}$$

Thus we have calculated from the infrared absorption spectrum that the size of the molecule is of the order of $10^{-8}$ cm. This order of magnitude agrees very well with the values of atomic and molecular radii obtained from other classical considerations. We want to stress, however, that $x$ is the value for the classical picture of the quantum-mechanical system and is not the expectation value of a quantum-mechanical observable.

## III.5  Combination of Quantum Physical Systems—The Vibrating Rotator

We shall now combine the quantum-mechanical rotator model with the quantum-mechanical oscillator model to form the quantum-mechanical vibrating rotator (or rotating oscillator) model. This will provide a description of the experimental situation shown in Figure 1.5, 1.6, and 1.7. We first discuss the general case of a combination of two quantum physical systems. For this we require a new mathematical notion: the *direct product* or *tensor product* of linear spaces.

⟦Let $R_1$ and $R_2$ be two linear spaces, let $u_i \in R_1$ and $v_j \in R_2$, and let $a_{ij} \in \mathbb{C}$ (complex numbers). The set of all (arbitrarily large but finite) sums

$$f = \sum_{i,j} a_{ij} u_i v_j, \tag{5.1}$$

where $a_{ij}$ takes any value in $\mathbb{C}$, forms a linear space, which is called the *direct-product space* and is denoted $R_1 \otimes R_2$. $u_i v_j$ is the formal product, which is also written $u_i v_j = u_i \otimes v_j$. If $(u, u')_1$ denotes the scalar product in $R_1$, and if $(v, v')_2$ denotes the scalar product in $R_2$, then the scalar product in $R_1 \otimes R_2$ is defined by

$$\left( \sum_{ij} a_{ij} u_i v_j, \sum_{lm} b_{lm} u'_l v'_m \right) = \sum_{ijlm} \overline{a_{ij}} b_{lm} (u_i, u'_l)_1 (v_j, v'_m)_2. \tag{5.2}$$

(*Remark*: If $R_1$ and $R_2$ are Hilbert spaces, then the "completion" of $R_1 \otimes R_2$ with respect to this scalar product is the direct Hilbert-space product.)
If $\phi_\nu$ is a basis in $R_1$ and $\psi_\mu$ is a basis in $R_2$, then

$$f_{\nu\mu} = \phi_\nu \otimes \psi_\mu = \phi_\nu \psi_\mu \tag{5.3}$$

is a basis in $R_1 \otimes R_2$.
If $A_1$ is a linear operator in $R_1$ and $A_2$ is a linear operator in $R_2$, then the operators

$$C = A_1 \otimes I \qquad B = I \otimes A_2 \tag{5.4}$$

in $R_1 \otimes R_2$ are defined in the following way:

$$Cf = \sum_{ij} a_{ij} (A_1 u_i) \otimes v_j, \tag{5.4'}$$

$$Bf = \sum_{ij} a_{ij} u_i \otimes (A_2 v_j).$$

The linear operator $A = A_1 \otimes A_2$ in $R_1 \otimes R_2$ is defined by

$$A(u_j \otimes v_k) = (A_1 \otimes A_2)(u_j \otimes v_k) =_{\text{def}} (A_1 u_j) \otimes (A_2 v_k). \quad (5.5)$$

It is easily seen that if $A_1$, $B_1$ are linear operators in $R_1$ and $A_2$, $B_2$ are linear operators in $R_2$, then

$$A_1 B_1 \otimes A_2 B_2 = (A_1 \otimes A_2)(B_1 \otimes B_2). \quad (5.6)$$

Every operator $A$ in the direct-product space is a linear combination of direct products of operators, i.e.,

$$A = \sum_i A_1^{(i)} \otimes A_2^{(i)}, \quad (5.7)$$

with $A_1^{(i)}$ linear operators in $R_1$ and $A_2^{(i)}$ linear operators in $R_2$.⟧

With the notion of the direct product of spaces we can formulate the basic assumption about the physical combination of two quantum-mechanical systems:

**IVa.** Let one physical system be described by an algebra of operators, $\mathscr{A}_1$, in the space $R_1$, and the other physical system by an algebra $\mathscr{A}_2$ in $R_2$. The direct-product space $R_1 \otimes R_2$ is then the space of physical states of the physical combination of these two systems, and its observables are operators in the direct-product space [given in the form (5.7)]. The particular observables of the first system alone are given by $A_1 \otimes I$, and the observables of the second system alone by $I \otimes A_2$ ($I =$ identity operator).

We reemphasize that IVa is a basic assumption of quantum mechanics and can only be justified by the fact that such physical systems exist.

We shall now apply this basic assumption IVa to the diatomic molecule that vibrates and rotates.

We called the space of physical states of the oscillator $\mathscr{H}$. In $\mathscr{H}$ we introduced a basis of eigenvectors of the operator $N$ or $H_{\text{osc}}$:

$$\text{basis of } \mathscr{H}: \quad \phi_n = |n\rangle \quad (n = 0, 1, 2, \ldots). \quad (5.8)$$

The action of all observables (all elements of the algebra of observables of the quantum-mechanical oscillator) of the harmonic oscillator on the basis vectors $|n\rangle$ is known from Chapter II.

We called the space of physical states of the rotator $\mathscr{R}$. In $\mathscr{R}$ we introduced a basis of eigenvectors of the operators $L_3$ and $\mathbf{L}^2$ or $H_{\text{rot}}$:

$$\text{basis of } \mathscr{R}: \quad f_m^l = |lm\rangle$$

$$(l = 0, 1, 2, \ldots, \quad m = \text{integer with } -l \leq m \leq +l). \quad (5.9)$$

The space of physical states of the vibrating rotator is, according to IVa, the direct-product space

$$\mathfrak{S} = \mathscr{H} \otimes \mathscr{R}, \quad (5.10)$$

and the observables are the operators $\sum_i A_{\text{osc}}^{(i)} \otimes A_{\text{rot}}^{(i)}$, where $A_{\text{osc}}^{(i)}$ is any observable of the oscillator and $A_{\text{rot}}^{(i)}$ is any observable of the rotator. The basis system in $\mathfrak{S}$ is obtained as the direct product of the basis systems $\phi_n$ in $\mathcal{H}$ and $f_m^l$ in $\mathcal{R}$, and is denoted by $|nlm\rangle$:

$$|nlm\rangle = |n\rangle \otimes |lm\rangle = \phi_n \otimes f_m^l. \tag{5.11}$$

We have already mentioned that the rotating diatomic molecule is not a rigid rotator and the vibrating diatomic molecule is not a harmonic oscillator. Furthermore, the rotations and vibrations are not independent motions of the molecule: In our classical picture the diatomic molecule is a system of two mass points which are connected by a massless spring. Consequently there are interactions between the vibration and rotation caused, e.g., by the fact that during the vibration the internuclear distance $x = (x_i x_i)^{1/2}$ changes, and consequently the moment of inertia $I = \mu x^2$ changes.

For the moment we want to neglect all these finer details and consider the idealized system that is simultaneously a rigid rotator and a harmonic oscillator—keeping in mind, however, that his is an idealized system, which can at best be only approximately correct.

The energy operator of this idealized physical combination of the harmonic oscillator and rigid rotator is given by

$$H = H_{\text{osc}} \otimes I + I \otimes H_{\text{rot}}, \tag{5.12}$$

where

$$H_{\text{osc}} = \hbar\omega(N + \tfrac{1}{2}I) = \frac{1}{2\mu}P^2 + \frac{\mu\omega^2}{2}Q^2$$

and

$$H_{\text{rot}} = \frac{1}{2I}\mathbf{L}^2.$$

If the interactions between the two systems is neglected, all observables are given by

$$A = A_{\text{osc}} \otimes I + I \otimes A_{\text{rot}}. \tag{5.13}$$

From (5.12) we obtain the energy spectrum of the idealized vibrating rotator:

$$\text{spectrum } H = E_{nl} = \hbar\omega(n + \tfrac{1}{2}) + \frac{1}{2I}\hbar^2 l(l + 1). \tag{5.14}$$

The experiments show that the system constants $\omega$ for oscillator and $I$ for the rotator fulfill $\hbar\omega > \hbar^2/(2I)$ (the pure vibrational transitions are in the near infrared and the pure rotation transitions are in the far infrared). The energy-level diagram that we obtain under these conditions from (5.14) is shown in Figure 5.1. To obtain the transition frequencies we use the selection rules

$$\Delta n = \pm 1 \quad \text{and} \quad \Delta l = \pm 1 \tag{5.15}$$

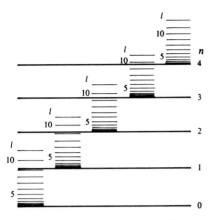

Figure 5.1   Energy levels of the vibrating rotator. For each of the first
five vibrational levels, a number of rotational levels are drawn (short
horizontal lines.) [From Herzberg (1966), with permission.]

given by (1.10) and (4.10). If we consider a particular vibrational transition
from $n$ to $n + 1$ (absorption) or $n + 1$ to $n$ (emission), we obtain from (5.14)
the frequencies (in cm$^{-1}$)

$$v_R = \frac{E_{n+1,l+1} - E_{n,l}}{2\pi\hbar c}$$

$$= v_0 + 2B + 2Bl \qquad (\Delta l = \pm 1), \qquad (5.16R)$$

$$v_P = \frac{E_{n+1,l-1} - E_{n,l}}{2\pi\hbar c}$$

$$= v_0 - 2Bl \qquad (\Delta l = -1) \qquad (5.16P)$$

for absorption, where $B$ is given by (4.11′):

$$B = \frac{h}{8\pi^2 cI} = \frac{\hbar}{4\pi cI} \qquad (4.11')$$

and

$$v_0 = \frac{\omega}{2\pi c}.$$

Thus we have two series of equidistant lines, which are called the $R$ and $P$
branches, with a gap at $v_0$ (as $\Delta l = 0$ is excluded by the selection rule). The
corresponding transitions are indicated in the energy-level diagram, Figure
5.2. The frequency spectrum calculated from (5.16) is depicted in strip (b).

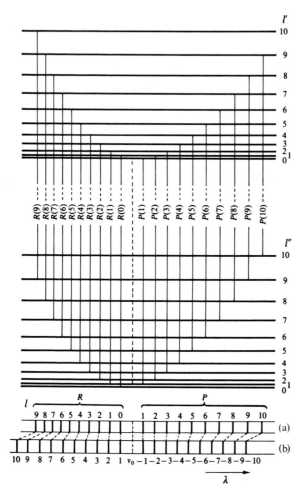

Figure 5.2   Energy-level diagram explaining the fine structure of a rotation-vibration band. In general, the separation of the two vibrational levels is considerably larger compared to the spacing of the rotational levels than shown in the figure (indicated by the broken parts of the vertical lines representing the transitions). The schematic spectrograms (a) and (b) give the resulting spectrum with and without allowance for the interaction between rotation and vibration. In these spectrograms, unlike most of the others, short wavelengths are at the left. [From Herzberg (1966), with permission.]

The observed spectrum from Figure 1.7 is depicted in strip (a). Thus the prediction of the vibrating-rotator model without interaction between vibration and rotation agrees rather well, but is not quite accurate. The observed lines in the R branch draw closer together, and those in the P branch draw farther apart, than the predicted equidistant lines. This is due to the interaction of rotation and vibration.

If one assumes that the moment of inertia is different in different vibrational states, then one obtains

$$E_{nl} = \hbar\omega(n + \tfrac{1}{2}) + \frac{1}{2I_n} \hbar^2 l(l + 1) \tag{5.17}$$

instead of (5.14), and for the wave numbers of the resulting lines

$$v = \frac{E_{n'l'} - E_{n''l''}}{2\pi\hbar c} = v_0(n' - n'') + B_{n'} l'(l' + 1) - B_{n''} l''(l'' + 1), \tag{5.18}$$

where

$$B_n = \frac{h}{8\pi^2 c I_n}.$$

From this one obtains the absorption frequencies for transitions $n''l'' \to n'l'$ between neighboring vibrational levels:

$$v_R = v_0 + 2B_{n'} + (3B_{n'} - B_{n''})l + (B_{n'} - B_{n''})l^2$$
$$(l' = l + 1, \qquad l'' = l, \qquad \Delta l = +1), \tag{5.19R}$$

$$v_P = v_0 - (B_{n'} + B_{n''})l + (B_{n'} - B_{n''})l^2$$
$$(l' = l - 1, \qquad l'' = l, \qquad \Delta l = -1). \tag{5.19P}$$

Equations (5.19) give excellent agreement with the empirical fine structure of the infrared bands.

For the HCl molecule the values of $B_n$ have been obtained for the various bands $n' \leftrightarrow n''$:

$$0 \leftrightarrow 1 \qquad 0 \leftrightarrow 2 \qquad 0 \leftrightarrow 3 \qquad 0 \leftrightarrow 4 \qquad 0 \leftrightarrow 5$$

(transitions with $\Delta n > 1$ occur as a consequence of the small anharmonicity; cf. Figure 1.3). The results are summarized in Table 5.1. The difference $\Delta B_n$ between successive values is very nearly a constant, so that $B_n$ can be fitted by the formula

$$B_n = B_e - \alpha_e(n + \tfrac{1}{2}), \tag{5.20}$$

**Table 5.1** Rotational Constants of HCl in the Different Vibrational Levels of the Electronic Ground State. [From Hertzberg (1966), p. 800, with permission.]

| $n$ | $B_n$ (cm$^{-1}$) | $\Delta B_n$ (cm$^{-1}$) |
|---|---|---|
| 0 | 10.4400 | |
| | | 0.3034 |
| 1 | 10.1366 | |
| | | 0.3037 |
| 2 | 9.8329 | |
| | | 0.2986 |
| 3 | 9.5343 | |
| | | 0.302 |
| 4 | 9.232 | |
| | | 0.299 |
| 5 | 8.933 | |

where $\alpha_e$ is a constant small compared to $B_e = 10.5909$ cm$^{-1}$, the equilibrium value of $B_n$. The value of $B_n$ given in the table for the rotation vibration spectrum agree within the accuracy of the measurements with the value $B^{HCl} = 10.395$ cm$^{-1}$ obtained from the pure rotation spectrum of HCl, Equation (4.21).

Experimental values for the vibration-rotation spectra of the CO molecule are not as numerous or as accurate. From the spectrum depicted in Figure 1.5 one obtains

$$B^{CO} = 1.96 \text{ cm}^{-1}. \tag{5.21}$$

This value, and hence the fine structure in the energy spectrum, is considerably smaller than that for HCl.

The diatomic molecule with the largest rotational constant $B_e$, and thus the largest energy difference between rotational levels, is the H$_2$ molecule, for which $B_e^{H_2} = 60.80$ cm$^{-1}$.

A qualitative theoretical explanation of (5.17) with (5.20) follows from the classical picture of the diatomic molecule as two rigid spheres connected by a spring. When this system is in a state of higher vibrational energy, it has a larger amplitude and consequently a larger moment of inertia. Consequently $I_n^{-1}$ decreases with increasing $n$, as expressed by (5.20).

For the quantum-mechanical observables, the empirical formula (5.20) means that in the presence of an interaction between vibrational and rotational degrees of freedom, the form (5.13) is not sufficient. For the energy operator one has in addition to (5.12) an interaction term, for which one may try as a first guess

$$H_{int} = gH_{osc} \otimes H_{rot}, \tag{5.22}$$

where $g$ is a coupling constant of dimension (eV)$^{-1}$. Thus the energy operator for the vibrating, rotating, interacting diatomic molecule in this approximation will be given by

$$H = H_{osc} \otimes I + I \otimes H_{rot} + gH_{osc} \otimes H_{rot}, \tag{5.23}$$

where

$$H_{osc} = \hbar\omega(N + \tfrac{1}{2}I) \tag{5.24}$$

and

$$H_{rot} = \frac{1}{2I_e}\mathbf{L}^2. \tag{5.25}$$

$I_e$ is the moment of inertia that corresponds to the equilibrium separation $x_e$: $I_e = \mu x_e^2$.

The energy values of the diatomic molecule with vibration-rotation interaction are the expectation values of $H$ in the physical states. It is, of course, not obvious that the $|nlm\rangle = \phi_n \otimes f_m^l$ of (5.11), where $\phi_n$ are eigenvectors of $N$ and $f_m^l$ are eigenvectors of $\mathbf{L}^2$ and $L_3$, represent the pure states of this physical system. However, as they happen to be also eigenstates of the energy operator $H$ of (5.23), they are the obvious choice for states in an energy measurement. Thus the energy values are the eigenvalues of $H$ in the basis $|nlm\rangle$ given by (5.11) which are calculated to be:

$$E_{nl} = \hbar\omega(n + \tfrac{1}{2}) + \frac{4}{2I_e}\hbar^2 l(l + 1) + g\hbar\omega\frac{\hbar^2}{2I_e}(n + \tfrac{1}{2})l(l + 1). \quad (5.26)$$

The wave number of the radiation quantum corresponding to the energy value, i.e.,

$$v_{nl} = \frac{E_{nl}}{hc} = \frac{E_{nl}}{2\pi\hbar c} \quad (5.27)$$

is called the *term value*[4] [cf. equation (1.12)]. The term value of the vibrating rotator are therefore, by (5.26),

$$v_{nl} = v_0(n + \tfrac{1}{2}) + (B_e - \alpha_e(n + \tfrac{1}{2}))l(l + 1), \quad (5.28)$$

where we have used the standard notation of molecular spectroscopy:

$$B_e = \frac{\hbar}{4\pi c I_e} = \frac{\hbar}{4\pi c \mu x_e^2}, \quad (5.29)$$

$$\alpha_e = -\frac{g v_0 \hbar^2}{2I_e}, \quad (5.30)$$

$$B_n = B_e - \alpha_e(n + \tfrac{1}{2}). \quad (5.31)$$

According to the above qualitative considerations, $\alpha_e$ should be larger than zero, which is experimentally always fulfilled. Equation (5.28) with (5.31) gives for the wave numbers of the transitions in the $R$ branch

$$v_R = v_{n' l+1} - v_{n''l}$$
$$= v_0(n' - n'') + 2B_{n'} + (3B_{n'} - B_{n''})l + (B_{n'} - B_{n''})l^2, \quad (5.32R)$$

and for the wave numbers in the $P$ branch

$$v_P = v_{n' l-1} - v_{n''l}$$
$$= v_0(n' - n'') - (B_{n'} + B_{n''})l + (B_{n'} - B_{n''})l^2, \quad (5.32P)$$

i.e., the empirically well-established formulas (5.19).

Two previously mentioned effects have not been taken into account in (5.23) with (5.24) and (5.25). These are the anharmonicity of the oscillator and the influence of centrifugal forces. Thus (5.28) is not the end of the story of the vibrating and rotating diatomic molecule. If these effects are also taken into account, then up to a certain degree of accuracy one obtains for the term values of a vibrating rotator

$$v_{nl} = \omega_e(n + \tfrac{1}{2}) - \omega_e x_e(n + \tfrac{1}{2})^2 + B_n l(l + 1) - D_n l^2(l + 1)^2, \quad (5.33)$$

---

[4] Note that we use the same symbol $v$ for the frequency (in sec$^{-1}$) and the wave number (in cm$^{-1}$); cf. statement following equation (1.12.)

where

$$B_n = B_e - \alpha_e(n + \tfrac{1}{2}), \tag{5.34}$$

$$D_n = D_e + \beta_e(n + \tfrac{1}{2}). \tag{5.35}$$

$\omega_e$ is the standard notation for

$$\omega_e = \frac{\omega}{2\pi c} = \frac{1}{2\pi c}\sqrt{\frac{k}{\mu}}. \tag{5.36}$$

$B_e$ is given by (5.29):

$$B_e = \frac{\hbar}{4\pi c I_e} = \frac{\hbar}{4\pi c \mu x_e^2}. \tag{5.29}$$

According to the semiclassical consideration leading to (4.20), $D_e$ may be expressed in terms of the reduced mass $\mu$, equilibrium separation $x_e$ and spring constant $k$ as

$$D_e = \frac{\hbar^3}{4\pi c k \mu^2 x_e^6}. \tag{5.37}$$

From (5.36), (5.29), and (5.37) it follows that the three system parameters $D_e, B_e, \omega_e$ are not independent but are related by

$$D_e = \frac{4B_e^3}{\omega_e^2}. \tag{5.38}$$

The parameters $x_e, \alpha_e, \beta_e$, expressing the degree of anharmonicity, are known empirically to be small:

$$x_e \ll 1, \qquad \frac{\alpha_e}{B_e} \ll 1, \qquad \frac{\beta_e}{D_e} \ll 1,$$

which must be the case, as they represent the effect of corrections to models which are very well realized by microphysical systems in nature. The system parameters $\omega_e, x_e, B_e, D_e, \alpha_e$, and $\beta_e$ have been experimentally determined for many diatomic molecules and are collected in tables (cf. Herzberg (1966]). Equation (5.33) gives a very good description of the vibration-rotation spectra of diatomic molecules, and only in exceptional cases are higher correction terms needed.

Diatomic molecules are vibrating rotators only as long as the internal energy is sufficiently low—roughly, in the region of energy of infrared radiation. For higher energies (1–20 eV) which correspond to the visible and ultraviolet regions, the molecules are no longer just vibrating rotators, since new degrees of freedom become accessible to electronic transitions. In each electronic state the molecule is, however, still a vibrator, in the same way as in each vibrational state the molecule is a rotator. This leads to energy spectra as depicted schematically in Figure 5.3 for two electronic states. We shall not describe the electronic structure of molecules here, they are of the

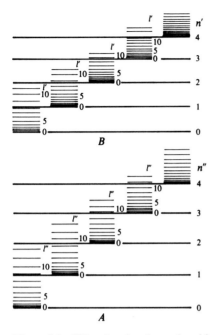

Figure 5.3    Vibrational and rotational levels of two electronic states $A$ and $B$ of a molecule (schematic). Only the first few rotational and vibrational levels are drawn in each case. [From Herzberg (1966), with permission.]

same nature as the electronic structure of atoms which will be discussed in later chapters of the book.

## Problems

1.    Let $P_i$ and $Q_j$ $(i, j = 1, 2, 3)$ satisfy the canonical communication relation $[P_i, Q_j] = (\hbar/i)\delta_{ij} I$. Define the orbital angular momentum $L_i \equiv \epsilon_{ijk} Q_j P_k$ $(\mathbf{L} = \mathbf{Q} \times \mathbf{P})$. The component operators are Hermitian [Equation (2.11a)] and satisfy the commutation relation (2.14). Define $H_3$, $H_+$, and $H_-$ as follows:

$$H_3 \equiv \hbar^{-1} L_3 \quad \text{and} \quad H_\pm \equiv \hbar^{-1}(L_1 \pm iL_2).$$

(a)    Show that

$$[H_3, H_\pm] = \pm H_\pm$$

and that

$$[H_+, H_-] = 2H_3.$$

Show that

$$\mathbf{L}^2 \equiv L_i L_i = \hbar^2 (H_+ H_- + H_3^2 - H_3)$$
$$= \hbar^2 (H_- H_+ + H_3^2 + H_3).$$

(b)   Show that $\mathbf{H}^2 \equiv \hbar^{-2}\mathbf{L}^2$ is an invariant operator of $\mathscr{E}(\mathrm{SU}(2))$; i.e., show that

$$[\mathbf{H}^2, H_3] = 0, \qquad [\mathbf{H}^2, H_+] = 0$$

and

$$[\mathbf{H}^2, H_-] = 0,$$

and consequently that

$$[\mathbf{H}^2, A] = 0,$$

where $A$ is any element of the algebra generated by the $L_i$,

$$A = aI + a^i L_i + a^{ij}L_i L_j + \cdots$$

$(a, a^i, a^{ij}, \ldots \in \mathbb{C})$.

(c)   Prove that the spectrum of $\mathbf{L}^2$ is $\hbar^2 l(l + 1)$ $(l = 0, 1, 2, 3, \ldots)$. *Hint*: Express the operator $L_3$ in terms of the annihilation operators $a_i$ and creation operators $a_i^\dagger$:

$$a_i = \frac{1}{\sqrt{2\hbar}}(Q_i + iP_i),$$

$$a_i^\dagger = \frac{1}{\sqrt{2\hbar}}(Q_i - iP_i)$$

of the three-dimensional harmonic oscillator, and show that as a consequence of the spectrum of the harmonic oscillator $L_3$ can have only integral eigenvalues.

2.   Calculate the internuclear distance for the CO molecule using the absorption spectrum given in Figure 1.7.

3.   The energy-loss spectrum of vibrating $H_2$ molecules (Figure PS.1) shows two bumps at 0 and at 0.52 eV with respective intensities 3.5 and $7.8 \times \frac{1}{30} = 0.26$ (arbi-

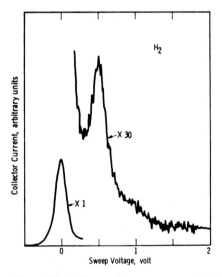

Figure PS.1   Energy-loss spectrum of $H_2$ [From G. J. Schulz, Phys. Rev. *135*, A988 (1964) with permission.]

trary units), respectively. What is the statistical operator $W$ for the ensemble of $H_2$ molecules in this energy-loss experiment? Can dipole transitions occur in this $H_2$ gas? What would be the frequency for these dipole transitions?

4.   In an infrared absorption experiment with HCl molecules in the ground state it is observed that the following frequencies $v$ $(cm^{-1})$ are absorbed:

|       |        |        |        |
|-------|--------|--------|--------|
| 20.68 | 82.72  | 144.76 | 206.80 |
| 41.36 | 103.40 | 165.44 | 227.48 |
| 62.04 | 124.08 | 186.12 |        |

Assume that in this absorption experiment only transitions between neighboring energy levels take place (dipole transitions). What are the energy levels of this (rotating) HCl molecule if the zero of the energy scale is fixed by $E_0 = 0$?

5.   Let $\{f^{j=1/2}_{-1/2}, f^{j=1/2}_{1/2}\}$, be the orthonormal basis for the space $\mathscr{R}^{1/2}$, defined in Section III.3.
(a)   Show that the matrices $\sigma_i$ with matrix elements $\sigma_i = 2(f^{j=1/2}_{m'}, J_i f^{j=1/2}_m)$ are given by

$$\sigma_1 = \begin{bmatrix} 0 & 1 \\ 1 & 0 \end{bmatrix}, \qquad \sigma_2 = \begin{bmatrix} 0 & -i \\ i & 0 \end{bmatrix}, \qquad \sigma_3 = \begin{bmatrix} 1 & 0 \\ 0 & -1 \end{bmatrix}$$

with respect to the basis $\{f^{j=1/2}_{-1/2}, f^{j=1/2}_{1/2}\}$. The $\sigma_i$ are the Pauli spin matrices.
(b)   Show that the Pauli spin matrices together with the unit matrix are a basis for the vector space of operators on $\mathscr{R}^{1/2}$.

6.   Let $\mathscr{R}_1$ and $\mathscr{R}_2$ be two finite-dimensional linear scalar-product spaces and $\mathscr{R} = \mathscr{R}_1 \oplus \mathscr{R}_2$ be their direct sum. Let $A_1$ be a linear operator in $\mathscr{R}_1$ and $A_2$ be a linear operator in $\mathscr{R}_2$. The direct sum $A_1 \oplus A_2$ is defined as the map

$$(A_1 \oplus A_2)\Psi = (A_1 \oplus A_2)(\Psi_1 \oplus \Psi_2) = (A_1\Psi_1 \oplus A_2\Psi_2)$$

for $\Psi_i \in \mathscr{R}_i$ and $\Psi = \Psi_1 \oplus \Psi_2 \in \mathscr{R}$. Let $\mathscr{H} = \mathscr{R}_1 \otimes \mathscr{R}_2$ be the direct product of $\mathscr{R}_1$ and $\mathscr{R}_2$. The direct product $A_1 \otimes A_2$ is defined as the map

$$(A_1 \otimes A_2)h = (A_1 \otimes A_2)(h_1 \otimes h_2) = A_1h_1 \otimes A_2h_2$$

for $h_i \in \mathscr{R}_i$ and $h = h_1 \otimes h_2 \in \mathscr{H}$.
(a)   Show that in an appropriate basis the matrix $a$ of the operator $A_1 \oplus A_2$ has the form

$$\begin{pmatrix} a_1 & 0 \\ 0 & a_2 \end{pmatrix},$$

where $a_1$ is the matrix of $A_1$ and $a_2$ is the matrix of $A_2$.
(b)   Show that the matrix of the operator $A_1 \otimes A_2$ when written in an appropriate basis has the property

$$a_{jk, il} = a^{(1)}_{ji} a^{(2)}_{kl},$$

i.e., may be written as a block matrix

$$\begin{pmatrix} a_{11}^{(1)}a^{(2)} & \cdots & a_{1n}^{(1)}a^{(2)} \\ a_{21}^{(1)}a^{(2)} & \cdots & a_{2n}^{(1)}a^{(2)} \\ \vdots & & \vdots \\ a_{n1}^{(1)}a^{(2)} & \cdots & a_{nn}^{(1)}a^{(2)} \end{pmatrix}$$

or

$$\begin{pmatrix} a_{11}^{(2)}a^{(1)} & \cdots & a_{1m}^{(2)}a^{(1)} \\ \vdots & & \vdots \\ a_{m1}^{(2)}a^{(1)} & \cdots & a_{mm}^{(2)}a^{(1)} \end{pmatrix}$$

where $a^{(i)}$ is the matrix of $A_i$ ($i = 1, 2$) and $a_{rs}^{(i)}$ are the matrix elements.

(c)  Show that
   (i)   $\operatorname{Tr}(A_1 \oplus A_2) = \operatorname{Tr} A_1 + \operatorname{Tr} A_2$,
   (ii)  $\operatorname{Tr}(A_1 \otimes A_2) = \operatorname{Tr} A_1 \operatorname{Tr} A_2$.

# CHAPTER IV

# Complete Systems of Commuting Observables

In this chapter it is explained that the question of what constitutes a complete system of commuting observables is not a mathematical question but can only be answered by experiment.

For the algebra of the quantum-mechanical harmonic oscillator, the eigenvectors of the operator $N$ [or of $H = \hbar\omega(N + \frac{1}{2}I)$],

$$N|n\rangle = n|n\rangle \qquad (n = 0, 1, 2, \ldots) \tag{1.1}$$

constituted a complete orthonormal system of $\mathcal{H}$ [see (II.3.25)]. For the algebra of the quantum-mechanical rotator a complete orthonormal system in $\mathcal{R}$ was given by the eigenvectors of $\mathbf{J}^2$ and $J_3$:

$$\mathbf{J}^2|jj_3\rangle = j(j + 1)\hbar^2|jj_3\rangle \qquad J_3|jj_3\rangle = j_3\hbar|jj_3\rangle. \tag{1.2}$$

(We shall use either $j_3$ or $m$ to denote the quantum number labeling eigenvalues of $J_3$, depending on typographical convenience for the topic being discussed. In this chapter we use $j_3$.) Instead of the eigenvector $|n\rangle$ one can use the generalized eigenvectors $|x\rangle$ or $|p\rangle$ as a generalized basis for $\mathcal{H}$, as expressed by Equations (II.8.16′) and (II.8.34). (Generalized basis vector of $\mathcal{R}$ also exist for the algebra of the quantum-mechanical rotator.)

The distinction between (1.1) and (1.2) is that for the one-dimensional oscillator (1.1), one operator is sufficient to define the basis. If for any vector $\phi \in \mathcal{H}$ it follows that

$$N\phi = a\phi, \tag{1.3}$$

125

then $a$ is one of the nonnegative integers, say $a = n'$, and $\phi = \alpha|n'\rangle$ where $\alpha \in \mathbb{C}$. For the rotator model, two operators are needed in general to define the basis, and two are sufficient. If for any vector $\psi \in \mathscr{R}$ it follows that

$$\mathbf{J}^2\psi = a\psi \quad \text{and} \quad J_3\psi = b\psi, \tag{1.4}$$

then (1) $a = j'(j' + 1)\hbar^2$ and $b = j_3'\hbar$, where $j'$ is an integer or half integer, and $j_3'$ is one of the numbers $-j'$, $-j' + 1, \ldots, j'$; and (2) $\psi = \alpha|j'j_3'\rangle$, where $\alpha \in \mathbb{C}$. Instead of $\mathbf{J}^2$ and $J_3$ one could of course use any two independent (algebraic) functions $f_1(\mathbf{J}^2, J_3)$ and $f_2(\mathbf{J}^2, J_3)$ of $\mathbf{J}^2$ and $J_3$, i.e., functions $f_i$ such that $j$ and $j_3$ are uniquely determined by the numbers $f_i(j, j_3)$ and vice versa. One would then obtain the same basis system.

Instead of the basis system (1.2) one could also use another basis system, e.g.,

$$\mathbf{J}^2|jj_2\} = j(j + 1)\hbar^2|jj_2\} \qquad J_2|jj_2\} = j_2\hbar|jj_2\} \tag{1.5}$$

or any two other functions of the operators $J_i$ $(i = 1, 2, 3)$ that commute with each other. To obtain a (generalized) basis system of $\mathscr{R}$ one need not restrict oneself to functions of $J_i$, but could even use two functions of the $J_i$, $Q_i$ $(i = 1, 2, 3)$ that commute with each other (e.g. $Q_i$ or $P_i$).* The eigenvalues of these functions usually specify a basis system of $\mathscr{R}$ completely.

⟦Sometimes, but only in the case that at least one of these functions has a continuous spectrum, it may happen that in addition to the two (generalized) eigenvalues of these functions a further label is needed to specify the generalized basis system completely.⟧

The system of commuting Hermitian operators that specifies the (generalized) basis system completely is called (following Dirac) a *complete system of commuting observables* (c.s.c.o.). The (generalized) eigenvalues of a c.s.c.o. are called quantum numbers.

For the quantum-mechanical one-dimensional harmonic oscillator, the c.s.c.o. consists of the one operator $N$ (or $H$), but the operator $Q$ may also serve as the c.s.c.o. For the rotator the c.s.c.o. consists of two operators; $\mathbf{J}^2$ and $J_3$ are a convenient choice.

From what was said above, and the fact that in the direct-product space one basis system is obtained as the direct product of the basis vectors in the two factor spaces, it is clear that one c.s.c.o. of the combination of two systems is given by the combination of the two c.s.c.o.'s for each subsystem.

For example, a c.s.c.o. for the vibrating rotator is given by

$$N, \quad \mathbf{J}^2, \quad J_3. \tag{1.6}$$

For different algebras there are different c.s.c.o.'s. The larger the algebra (i.e., the more complicated the physical system), the larger is the number of operators in a c.s.c.o.

* Such a basis system is given in (3.76) of the Appendix to Section V.3.

⟦The problem of determining for which operator *-algebras there exists a c.s.c.o. is unsolved; the requirement of the existence of a c.s.c.o. is certainly a restrictive condition. For certain classes of operator *-algebras $\mathscr{A}$ there does exist a c.s.c.o. (e.g., if $\mathscr{A}$ is the enveloping algebra of a nilpotent or semisimple group). This question is connected with the fulfillment of the conditions for the nuclear spectral theorem, of which we shall give a vague statement below.⟧

The physicist's problem is not to find a c.s.c.o. for a given algebra,[1] but is usually the reverse: From the experimental data he finds out how many quantum numbers are required, and what are the possible values of these quantum numbers. This gives him, according to the basic assumption IIIa, the complete commuting system $\{A_k\}$ and its spectrum. He then conjectures the total algebra $\mathscr{A}$ by adjoining to $\{A_k\}$ a minimum of other operators such that the matrix elements of elements of $\mathscr{A}$ calculated from the properties of this algebra agree with the experimental values of the corresponding observables.

Thus, the question of what is a c.s.c.o. for a particular physical system and the question of when a system of commuting operators is complete are physical questions. If an experiment gives more values than can be supplied by a given system of commuting operators, then this system is not complete; one has to introduce a new quantum number, i.e., enlarge the system of commuting operators. This usually requires a further enlargement of the algebra.

We already treated one example of this procedure when we described the transition from the oscillator model of the diatomic molecule to the vibrating-rotator model. As long as one ignored the fine structure of the order of $10^{-3}$ eV, the quantum number $n$ was sufficient to describe the infrared spectrum. However, to explain the results of higher-resolution experiments, a new quantum number $j$—and therewith a whole new algebra, the rotator algebra—had to be introduced and combined with the oscillator algebra to give the algebra of the vibrating rotator.

Because of the above described procedure for obtaining the algebra of observables $\mathscr{A}$ for a given physical system, we can assume that for the algebras in quantum mechanics one always has a c.s.c.o.,[2] i.e., there exists a set of commuting operators

$$A_1, A_2, \ldots, A_n \in \mathscr{A}$$

such that

$$A_k | \lambda_1, \ldots, \lambda_n \rangle = \lambda_k | \lambda_1, \ldots, \lambda_n \rangle, \tag{1.7}$$

---

[1] Although it would be nice to know the answer, as this restricts the class of admissible algebras.

[2] Mathematically this assumption can be stated: the conditions of the nuclear spectral theorem are fulfilled.

where the set of numbers $\Lambda = \{\lambda\} = \{(\lambda_1, \lambda_2, \ldots, \lambda_N)\}$, called the *spectrum* of $A_k$, is real and in general continuous, and the set of generalized eigenvectors $|\lambda_1 \cdots \lambda_N\rangle^3$ is a generalized basis system. That is, every physical state vector $\phi$ can be represented as

$$\phi = \int_\Lambda d\mu(\lambda) |\lambda_1 \cdots \lambda_N\rangle\langle\lambda_1 \cdots \lambda_N|\phi\rangle, \tag{1.8}$$

where $\int_\Lambda d\mu(\lambda)$ means summation and integration with certain weight factors (measures) over all values $\lambda$ in $\Lambda$.

The statement (1.7) or (1.8) is the basis of the Dirac formulation of quantum mechanics. Recent results in mathematics assure us that the class of algebras that fulfill this statement is not empty. This was not the case when Dirac made his famous conjecture; none of the mathematical structures known at that time fulfilled the conditions justifying this statement.

The statement (1.7) or (1.8) is the mathematical basis for the solution of actual problems; it does not tell us what the spectrum $\Lambda$ and what the measure $\mu(\lambda)$ (weight factor) is for the cases of physical interest. This depends upon the particular problem, and has to be solved for every particular case.

An immediate consequence of the statement (1.7) or (1.8) is the following statement: Let $A_1, \ldots, A_N$ be a c.s.c.o. with (generalized) eigenvectors $|\lambda_1 \cdots \lambda_N\rangle = |\lambda\rangle$ such that

$$A_k|\lambda\rangle = \lambda_k|\lambda\rangle \qquad (k = 1, 2, \ldots, N).$$

Then from

$$A_i|a\} = a_i|a\} \qquad (i = 1, 2, \ldots, N) \tag{1.9}$$

it follows that

$$a_i = \lambda_i \quad \text{and} \quad |a\} = \alpha|\lambda\rangle, \tag{1.10}$$

where $\alpha$ is a phase factor.

For the case that the spectrum $\Lambda$ is continuous, i.e., that one of the $\lambda_i$ can take continuous values, the formulation of (1.10) has to be given a precise meaning. In the following we shall generally use this statement only for the case of a discrete spectrum.

---

[3] [In general, $|\lambda \cdots \lambda_N\rangle \in \Phi^\times$ and the physical state vector $\phi \in \Phi$.]

# CHAPTER V

# Addition of Angular Momenta— The Wigner–Eckart Theorem

In Section V.1 the elementary rotator is defined as the system described by an irreducible representation of the algebra of angular momentum. Section V.2 discusses then the direct product of two irreducible representations of angular momentum and its reduction with respect to the total angular momentum. The Clebsch–Gordan coefficients are introduced, their recursion relations are derived, and their most frequently used values are tabulated. In Section V.3 tensor operators are introduced and the Wigner–Eckart theorem for the rotation group is stated without derivation. In Section V.4 a new observable, parity, is introduced. Parity is then applied to discuss the spectrum of diatomic symmetric-top molecules. In an Appendix to Section V.3 the irreducible representations of the algebras of SO(3.1), SO(4) and E(3) are derived.

## V.1 Introduction—The Elementary Rotator

In Chapter III we have treated the quantum-mechanical rotator. Its space of physical states is given by (III.4.2):

$$\mathcal{R} = \sum_{l=0}^{\infty} \oplus \mathcal{R}^l, \tag{1.1}$$

where $\mathcal{R}^l$ is the irreducible-representation space of the algebra of angular momentum $\mathcal{E}(SO(3)_{L_i})$, and

$$[L_i, L_j] = i\hbar\epsilon_{ijk}L_k. \tag{1.2}$$

$\mathcal{E}(SO(3)_{L_i})$ is a subalgebra of the algebra of observables for the rotator.

Let us now consider the physical system whose space of physical states is $\mathcal{R}^l$, with an arbitrary fixed value $l = 0, \frac{1}{2}, 1, \frac{3}{2}, \ldots$, and whose algebra of observables is the angular momentum algebra. Every rotator can, under certain conditions, be considered as such a physical system—namely, if the energy $E^{\text{ext}}$ that can be transferred to or from this rotator is small compared to the energy differences $E_l - E_{l-1}$ and $E_{l+1} - E_l$. In nature, this condition is rarely fulfilled, because a state of higher energy always tends to decay into a state of lower energy "spontaneously," and therefore the excited state has only a limited lifetime, unless $E_l$ is the ground-state energy value. It is, nevertheless, customary to consider these unstable states often as independent physical systems, not only in nonrelativistic quantum mechanics, but also in the relativistic case of elementary-particle physics. We call a physical system whose space of states is $\mathcal{R}^l$ an *elementary rotator*. Elementary rotators are not just substructures of the diatomic molecule; they appear in other quantum-mechanical systems in all areas of physics.

## V.2  Combination of Elementary Rotators

According to the basic assumption IV, the space of physical states of the combination of two elementary rotators with spaces $\mathcal{R}^{j_1}_{(1)}$ and $\mathcal{R}^{j_2}_{(2)}$ is the direct-product space $\mathcal{R}^{j_1}_{(1)} \otimes \mathcal{R}^{j_2}_{(2)} = \mathcal{R}$. Let $J^{(1)}_i$ and $J^{(2)}_i$ be the angular-momentum operators in $\mathcal{R}^{j_1}_{(1)}$ and $\mathcal{R}^{j_2}_{(2)}$ respectively, and let

$$|j_1 m_1\rangle \quad (m_1 = -j_1, \ldots, +j_1)$$

and

$$|j_2 m_2\rangle \quad (m_2 = -j_2, \ldots, +j_2)$$

be the basis systems in $\mathcal{R}^{j_1}_{(1)}$ and $\mathcal{R}^{j_2}_{(2)}$ respectively. Then a basis system in $\mathcal{R}^{j_1}_{(1)} \otimes \mathcal{R}^{j_2}_{(2)}$ is given by

$$|j_1 m_1\rangle \otimes |j_2 m_2\rangle = |j_1 m_1 j_2 m_2\rangle. \tag{2.1}$$

The $J^{(\alpha)}_i$ ($\alpha = 1, 2$) fulfill the commutation relations

$$[J^{(\alpha)}_i, J^{(\alpha)}_j] = i\hbar \epsilon_{ijk} J^{(\alpha)}_k. \tag{2.2}$$

We define operators in $\mathcal{R} = \mathcal{R}^{j_1}_{(1)} \otimes \mathcal{R}^{j_2}_{(2)}$ by

$$J_i = J^{(1)}_i \otimes I + I \otimes J^{(2)}_i. \tag{2.3}$$

As a consequence of (2.2) and (2.3) it follows that the $J_i$ fulfill the commutation relation

$$[J_i, J_j] = i\hbar \epsilon_{ijk} J_k, \tag{2.4}$$

Thus the $J_i$ of (2.3) fulfill the defining relations of the generators of the algebra of angular momentum, and are thus a representation of angular momentum in the space $\mathcal{R}^{j_1}_{(1)} \otimes \mathcal{R}^{j_2}_{(2)}$. The $J_i$ are called the *total-angular-momentum operators* of the combined system.

The basis system (2.1) of $\mathcal{R}$ consists of eigenvectors of the following complete set of commuting operators (c.s.c.o.):

$$\mathbf{J}^{(1)2} \otimes I, \quad J^{(1)}_3 \otimes I, \quad I \otimes \mathbf{J}^{(2)2}, \quad I \otimes J^{(2)}_3 \tag{2.5}$$

with the eigenvalues

$$
\begin{aligned}
\mathbf{J}^{(1)\,2} \otimes I |j_1 m_1 j_2 m_2\rangle &= j_1(j_1 + 1)\hbar^2 |j_1 m_1 j_2 m_2\rangle \\
J_3^{(1)} \otimes I |j_1 m_1 j_2 m_2\rangle &= m_1 \hbar |j_1 m_1 j_2 m_2\rangle \\
I \otimes \mathbf{J}^{(2)\,2} |j_1 m_1 j_2 m_2\rangle &= j_2(j_2 + 1)\hbar^2 |j_1 m_1 j_2 m_2\rangle \\
I \otimes J_3^{(2)} |j_1 m_1 j_2 m_2\rangle &= m_2 \hbar |j_1 m_1 j_2 m_2\rangle.
\end{aligned}
\tag{2.6}
$$

We now introduce the abbreviated notation for the operators in $\mathcal{R}$:

$$
J_i^{(1)} = J_i^{(1)} \otimes I, \qquad J_i^{(2)} = I \otimes J_i^{(2)}.
\tag{2.7}
$$

These $J_i^{(\alpha)}$ are operators in the space $\mathcal{R} = \mathcal{R}_{(1)}^{j_1} \otimes \mathcal{R}_{(2)}^{j_2}$ in contrast to the original $J_i^{(\alpha)}$, which were operators in $\mathcal{R}_{(\alpha)}^{j_\alpha}$ only. The $J_i^{(\alpha)}$ of (2.7) clearly fulfill the commutation relation (2.2), so we denote them by the same symbol. However, we should keep in mind that the old $J_i^{(\alpha)}$ are the restrictions of the new $J_i^{(\alpha)}$ to the space $\mathcal{R}_{(\alpha)}^{j_\alpha}$.

The basis (2.1) is in general not a physical basis for the combined system. A basis is physical if it consists of eigenstates in which the physical system can be prepared. If all physical states are eigenstates of the energy operator $H$ of the physical system, then the physical basis must consist of eigenvectors of operators that commute with the energy operator. In general not all the $J_i^{(\alpha)}$ commute with the energy operator (in particular, not if there is an interaction term between the two angular momenta).

The operators that in general commute with the energy operator are the total angular momenta $J_i$ (they commute with $H$ whenever $H$ is rotationally invariant). Therefore it is convenient to choose a basis in $\mathcal{R}$ that consists of eigenvectors of the system of commuting operators

$$
\mathbf{J}^{(1)\,2} = \mathbf{J}^{(1)\,2} \otimes I, \quad \mathbf{J}^{(2)\,2} = I \otimes \mathbf{J}^{(2)\,2}, \quad \mathbf{J}^2 = \sum_i J_i^2 \quad J_3.
\tag{2.8}
$$

We denote this basis by

$$
|j_1 j_2 j\, m\rangle.
\tag{2.9}
$$

It has the properties

$$
\begin{aligned}
\mathbf{J}^{(\alpha)\,2} |j_1 j_2 j\, m\rangle &= j_\alpha(j_\alpha + 1)\hbar^2 |j_1 j_2 j\, m\rangle, \\
J_3 |j_1 j_2 j\, m\rangle &= m\hbar |j_1 j_2 j\, m\rangle, \\
\mathbf{J}^2 |j_1 j_2 j\, m\rangle &= j(j + 1)\hbar^2 |j_1 j_2 j\, m\rangle.
\end{aligned}
\tag{2.10}
$$

Each vector, and consequently each basis vector of the basis (2.1), can be expanded according to (I.2.15) with respect to the basis system (2.9):

$$
|j_1 m_1 j_2 m_2\rangle = \sum_{j,\,m} |j_1 j_2 j\, m\rangle \langle j_1 j_2 j\, m|j_1 m_1 j_2 m_2\rangle;
\tag{2.11}
$$

and each basis vector of the basis (2.9) can be expanded with respect to the basis system (2.1),

$$
|j_1 j_2 j\, m\rangle = \sum_{m_1,\, m_2} |j_1 m_1 j_2 m_2\rangle \langle j_1 m_1 j_2 m_2|j_1 j_2 j\, m\rangle.
\tag{2.12}
$$

The transition coefficients $\langle j_1 m_1 j_2 m_2 | j_1 j_2 j m \rangle$ are called the *Clebsch–Gordan* or *Wigner coefficients*, and are denoted by

$$
\begin{aligned}
\langle j_1 m_1 j_2 m_2 | j_1 j_2 j m \rangle &= \langle j_1 m_1 j_2 m_2 | j m \rangle \\
&= C(j_1 j_2 j\, m_1 m_2 m) \\
&= \langle m_1 m_2 | j m \rangle \\
&= C^{j_1 j_2 j}_{m_1 m_2 m}
\end{aligned} \tag{2.13}
$$

for fixed values of $j_1$ and $j_2$. In (2.13) we have given various notations used in the literature.

Taking the scalar product of (2.11) with $|j_1 m_1' j_2 m_2' \rangle$ we obtain the orthogonality relation of the Clebsch–Gordan coefficients:

$$
\delta_{m_1' m_1} \delta_{m_2' m_2} = \sum_{j,m} \langle j_1 m_1' j_2 m_2' | j m \rangle \langle j m | j_1 m_1 j_2 m_2 \rangle, \tag{2.14}
$$

and taking the scalar product of (2.12) with $|j_1 j_2 j' m'\rangle$, we obtain

$$
\delta_{jj'} \delta_{mm'} = \sum_{m_1, m_2} \langle j'm' | j_1 m_1 j_2 m_2 \rangle \langle j_1 m_1 j_2 m_2 | j m \rangle. \tag{2.15}
$$

We now want to find the spectrum of $\mathbf{J}^2$ and $J_3$ in $\mathscr{R} = \mathscr{R}^{j_1}_{(1)} \otimes \mathscr{R}^{j_2}_{(2)}$, i.e., the values that $j$ and $m$ in basis system (2.9) can take. As $j_1$ and $j_2$ are fixed values, we denote

$$|j_1 j_2 j m\rangle = |j m\rangle \quad \text{and} \quad |j_1 m_1\rangle \otimes |j_2 m_2\rangle = |m_1\rangle \otimes |m_2\rangle = |m_1 m_2\}$$

For the sake of clarity we denote in the following calculation the direct product basis by $|j_1 m_1 j_2 m_2\} = |j_1 m_1\rangle \otimes |j_2 m_2\rangle$ and the total angular momentum basis by $|j_1 j_2 jm\rangle$.

(A)   The vectors $|j_1 j_2 j m\rangle$ and $|j_1 m_1 j_2 m_2\}$ are eigenvectors of the operator $J_3 = J_3^{(1)} + J_3^{(2)}$ with eigenvalue $m$ and $m_1 + m_1$ respectively. The highest possible value of $m$ is $j$, and the highest possible value of $m_1 + m_2$ is $j_1 + j_2$. The highest possible eigenvalue of $J_3$ is therefore on the one hand $j$ and on the other $j_1 + j_2$. Consequently $j = j_1 + j_2$ is the highest possible value for $m = m_1 + m_2$. Let us therefore consider $|j_1 j_2 j j\rangle$ and

$$|j_1 j_1 j_2 j_2\} = |j_1 j_1\rangle \otimes |j_2 j_2\rangle.$$

(B)   The vectors $|j_1 m_1 j_2 m_2\}$ are eigenvectors not only of the c.s.c.o. (2.5) but also of the three operators $\mathbf{J}^{(1)2}, \mathbf{J}^{(2)2}, J_3$ of the c.s.c.o. (2.8). The particular vector $|j_1 j_1 j_2 j_2\}$ is however also an eigenvector of $\mathbf{J}^2$.

PROOF.

$$
\begin{aligned}
\mathbf{J}^2 &= \mathbf{J}^{(1)2} + 2\mathbf{J}^{(1)} \cdot \mathbf{J}^{(2)} + \mathbf{J}^{(2)2} \\
&= \mathbf{J}^{(1)2} + \mathbf{J}^{(2)2} + 2J_3^{(1)}J_3^{(2)} + (J_+^{(1)}J_-^{(2)} + J_-^{(1)}J_+^{(2)}), \tag{2.16}
\end{aligned}
$$

because

$$J_1^{(1)}J_1^{(2)} + J_2^{(1)}J_2^{(2)} = \tfrac{1}{2}(J_+^{(1)}J_-^{(2)} + J_-^{(1)}J_+^{(2)}).$$

Thus

$$\mathbf{J}^2 |j_1 j_1\rangle \otimes |j_2 j_2\rangle = \{j_1(j_1 + 1) + j_2(j_2 + 1) + 2j_1 j_2)\}\, \hbar^2 |j_1 j_1\rangle \otimes |j_2 j_2\rangle, \tag{2.17}$$

because

$$\tfrac{1}{2}(J^{(2)}_- J^{(1)}_+ + J^{(1)}_- J^{(2)}_+)|j_1 j_1\rangle \otimes |j_2 j_2\rangle = 0,$$

since $m_1 = j_1$ and $m_2 = j_2$ are the highest possible values of $m_1$ and $m_2$, and thus $J^{(1)}_+|j_1 j_1\rangle = 0$ and $J^{(2)}_+|j_2 j_2\rangle = 0$. But

$$j_1(j_1 + 1) + j_2(j_2 + 1) + 2j_1 j_2 = (j_1 + j_2)(j_1 + j_2 + 1)$$
$$= j(j + 1), \tag{2.18}$$

since, according to (A), $j = j_1 + j_2$. Consequently $|j_1 j_1\rangle \otimes |j_2 j_2\rangle$ is an eigenvector of the complete commuting set (2.8) with eigenvalues

$$j_1(j_1 + 1) \qquad j_2(j_2 + 1) \qquad m = j = j_1 + j_2$$

and

$$j(j + 1) = (j_1 + j_2)(j_1 + j_2 + 1).$$

Thus $|j_1 j_1\rangle \otimes |j_2 j_2\rangle$ and $|j_1 j_2 jj\rangle$ are both eigenvectors of the complete commuting set (2.8) with the same eigenvalues, and consequently according to (IV.1.10)

$$|j_1 j_1\rangle \otimes |j_2 j_2\rangle = \alpha|j_1 j_2 jj\rangle, \qquad \alpha \in \mathbb{C}.$$

If they are both normalized, they differ only by a phase factor, which we choose to be unity; thus

$$|j_1 j_1 j_2 j_2\} = |j_1 j_1\rangle \oplus |j_2 j_2\rangle = |j_1 j_2 jj\rangle \qquad (j = j_1 + j_2) \tag{2.19}$$

or

$$\{j_1 j_1 j_2 j_2|j\, m\rangle = 1 \tag{2.20}$$

This vector (2.19) spans the one-dimensional space $\mathscr{R}_{j_1 + j_2}$ of eigenvectors of $J_3 = J^{(1)}_3 + J^{(2)}_3$ with highest eigenvalue $m = j = j_1 + j_2$.

(C)   Let us now consider the space $\mathscr{R}_{j_1 + j_2 - 1}$ of eigenvectors of $J_3$ with eigenvalues

$$m = j_1 + j_2 - 1. \tag{2.21}$$

This space is spanned by

$$|j_1 j_1 - 1\rangle \otimes |j_2 j_2\rangle \quad \text{and} \quad |j_1 j_1\rangle \otimes |j_2 j_2 - 1\rangle, \tag{2.22}$$

because both vectors (2.22) are eigenvectors of $J_3$ with eigenvalue $j_1 + j_2 - 1$. Thus $\mathscr{R}_{j_1 + j_2 - 1}$ is two-dimensional. One basis vector of the type (2.9) in $\mathscr{R}_{j_1 + j_2 - 1}$ is

$$|j_1 j_2 j = j_1 + j_2\, m = j - 1\rangle, \tag{2.23}$$

which is obtained from (2.19) by applying

$$J_- = J^{(1)}_- + J^{(2)}_-. \tag{2.24}$$

The second one, which is orthogonal to it, must therefore be

$$|j_1 j_2 = j_1 + j_2 - 1\, m = j_1 + j_2 - 1\rangle, \tag{2.25}$$

because only such a value of $j$ will have an $m = j_1 + j_2 - 1$. This vector is specified by the condition that it must be an eigenvector of the c.s.c.o. (2.8) with the indicated eigenvalues and is therefore [according to (IV.1.10)] determined only up to a phase factor. We fix this phase factor by requiring

$$\{j_1 j_1 j_2 j_2 - 1|j = j_1 + j_2 - 1\, m = j\rangle \text{ real positive.} \tag{2.26}$$

We proceed and consider next the space $\mathscr{R}_{j_1+j_2-2}$ of eigenvectors of $J_3$ with eigenvalue

$$m = j_1 + j_2 - 2.$$

This space is spanned by

$$|j_1 - 2\rangle \otimes |j_2\rangle \qquad |j_1 - 1\rangle \otimes |j_2 - 1\rangle \quad \text{and} \quad |j_1\rangle \otimes |j_2 - 2\rangle \tag{2.27}$$

(where we are now using the abbreviation $|m_1\rangle \otimes |m_2\rangle$ for $|j_1 m\rangle \otimes |j_2 m_2\rangle$) and is thus three-dimensional. One basis vector of the type (2.9) in $\mathscr{R}_{j_1+j_2-2}$ is obtained by applying $J_-$ to (2.23) and is

$$|j = j_1 + j_2\, m = j - 2\rangle. \tag{2.28}$$

The second vector of the type (2.9) is obtained by applying $J_-$ to (2.25) and is

$$|j = j_1 + j_2 - 1\, m = j_1 + j_2 - 2\rangle. \tag{2.29}$$

The third one is orthogonal to these two and must therefore be

$$|j = j_1 + j_2 - 2\, m = j_1 + j_2 - 2\rangle, \tag{2.30}$$

because only such a value of $j$ will have $m = j_1 + j_2 - 2$. To fix the arbitrary phase we require

$$\{j_1 j_1 j_2 j_2 - 2 | j = j_1 + j_2 + 2\, m = j\rangle \text{ real positive.} \tag{2.31}$$

(D)   In this way we proceed to the spaces

$$\mathscr{R}_{j_1+j_2-3},\, \mathscr{R}_{j_1+j_2-4},\ldots \tag{2.32}$$

with dimensions 4, 5,... and with new values of $m = j_1 + j_2 - 3, j_1 + j_2 - 4,\ldots$, respectively.

Without loss of generality we assume $j_1 \le j_2$ (if $j_1 > j_2$ we just rename $j_1 \to j_2, j_2 \to j_1$). Then the highest dimension of the eigenspaces of $J_3$ must be $2j_1 + 1$. This is the dimension of the space $\mathscr{R}_{j_2-j_1}$ spanned by

$$|-j_1\rangle \otimes |j_2\rangle, |-j_1 + 1\rangle \otimes |j_2 - 1\rangle, \ldots, |j_1\rangle \otimes |j_2 - 2j_1\rangle. \tag{2.33}$$

This introduces the ultimate new value for $j: j = j_2 - j_1$. The arbitrary phase of the vector

$$|j = j_2 - j_1\, m = j\rangle \tag{2.34}$$

is again chosen so as to make

$$\{j_1 j_1 j_2 j_2 - 2j_1 | j = j_2 - j_1\, m = j\rangle \text{ real positive} \tag{2.35}$$

Equations (2.20), (2.26), and (2.31)–(2.35) are summarized in the phase convention

$$\{j_1 j_2 m_2 | j j\rangle \text{ real positive.} \tag{2.36}$$

This convention is the standard phase convention for the Clebsch–Gordan coefficients. Thus the spectrum of $j$ is

$$j = j_1 + j_2, j_1 + j_2 - 1, j_1 + j_2 - 2, \ldots, j_2 - j_1,$$

or if we also admit $j_2 \le j_1$,

$$j = j_1 + j_2, j_1 + j_2 - 1, j_1 + j_2 - 2, \ldots, |j_2 - j_1|. \tag{2.37}$$

(E)   The space spanned by the $2j + 1$ vectors

$$|j_1 j_2 j m\rangle \qquad (m = -j, -j + 1, \ldots, +j)$$

we denote by $\mathscr{R}^j$. This is the space that is left invariant by the operators $J_i = J_i^{(1)} + J_i^{(2)}$. Then (2.37) can also be expressed as

$$\mathscr{R}_{(1)}^{j_1} \otimes \mathscr{R}_{(2)}^{j_2} = \mathscr{R}^{j_1 + j_2} \oplus \mathscr{R}^{j_1 + j_2 - 1} \oplus \cdots \oplus \mathscr{R}^{|j_2 - j_1|}. \tag{2.38}$$

This is called the reduction of $\mathscr{R}$ into a sum of irreducible angular-momentum spaces.

Summarizing, we have seen that the space $\mathscr{R}_{(1)}^{j_1} \otimes \mathscr{R}_{(2)}^{j_2}$ is in general not an irreducible-representation space or ladder-representation space of the algebra of total angular momentum $\mathscr{E}(SO(3)_{J_i})$. That is, not all vectors of $\mathscr{R}_{(1)}^{j_1} \otimes \mathscr{R}_{(2)}^{j_2}$ can be obtained by applying the $J_\pm$ a sufficient number of times to one of its vectors, but rather it is the direct sum of several such irreducible representation spaces $\mathscr{R}^j$ as given in (2.38). We have seen in particular that the Clebsch–Gordan coefficients, which are the transition coefficients between the two basis systems (2.1) and (2.9), are zero unless $j$ is as in (2.37) and $m = m_1 + m_2$:

$$\{m_1 m_2 | j m\rangle = 0$$

for $m \neq m_1 + m_2$,  $j \neq j_1 + j_2, j_1 + j_2 - 1, \ldots, |j_2 - j_1|$. \qquad (2.39)

(F)[1]   The Clebsch–Gordan coefficients are calculated recursively. The recursion relations are obtained if one takes the matrix element of $J_\pm = J_\pm^{(1)} + J_\pm^{(2)}$ between the states $|j\, m\rangle$ and $|m_1 m_2\rangle$:

$$
\begin{aligned}
\{m_1 m_2 | J_\pm | j\, m\rangle &= \sqrt{j(j + 1) - m(m \pm 1)}\, \hbar \{m_1 m_2 | j\, m \pm 1\rangle \\
&= \{m_1 m_2 | J_\pm^{(1)} + J_\pm^{(2)} | j\, m\rangle \\
&= ((J_\mp^{(1)} + J_\mp^{(2)}) | m_1 m_2\}, |j\, m\rangle) \\
&= \sqrt{j_1(j_1 + 1) - m_1(m_1 \mp 1)}\, \hbar \{m_1 \mp 1\, m_2 | j\, m\rangle \\
&\quad + \sqrt{j_2(j_2 + 1) - m_2(m_2 \mp 1)}\, \hbar \{m_1 m_2 \mp 1 | j\, m\rangle. \quad (2.40\pm)
\end{aligned}
$$

Here we have used (III.3.23),

$$L_\pm f_m^l = \sqrt{l(l + 1) - m(m \pm 1)}\, \hbar f_{m \pm 1}^l \tag{III.3.23}$$

for

$$L_\pm = J_\pm, J_\pm^{(1)}, J_\pm^{(2)}$$

and

$$J_\pm^{(2)\dagger} = J_\mp^{(2)}.$$

In the first step we shall describe the calculation of the Clebsch–Gordan coefficients $\{m_1 m_2 | j j\rangle$; in the second step we then turn to the calculation of Clebsch–Gordan coefficients for all values of $m$, $\{m_1 m_2 | j\, m\rangle$.

Setting $m = j$ in (2.40+), we obtain

$$-\sqrt{j_1(j_1 + 1) - m_1(m_1 + 1)}\{m_1 - 1\, m_2 | j j\rangle$$

$$= \sqrt{j_2(j_2 + 1) - m_2(m_2 - 1)}\{m_1 m_2 - 1 | j j\rangle. \quad (2.41)$$

---

[1] In the following a brief outline of the calculation of the Clebsch-Gordan coefficients is given, only the results given in the table will be needed later and this part may, therefore, be omitted at a first reading.

Because of (2.39) we can restrict our attention to the values $m_1 + (m_2 - 1) = (m_1 - 1) + m_2 = j(=m)$. Then (2.41) can be written

$$\{m_1 - 1\, j - m_1 + 1 | j j\rangle = -\left[\frac{(j_2 + j - m_1 + 1)(j_2 - j + m_1)}{(j_1 + m_1)(j_1 - m_1 + 1)}\right]^{1/2} \{m_1 j - m_1 | j j\rangle$$

(2.42)

Starting from $m_1 = j_1$ one can calculate all $\{m_1 m_2 | j j\rangle$ successively from (2.42). One obtains for an arbitrary $m_1$ and $m_2 = j - m_1 + 1$

$$\{m_1 m_2 | j j\rangle = (-1)^{j_1 - m_1}$$
$$\times \frac{(j_2 + j - j_1 + 1)(j_2 + j - j_1 + 2)\cdots(j_2 + j - m)(j_2 - j + j_1)}{2j_1(2j_1 - 1)\cdots(2j_1 - j_1 + m_1 + 1)}$$
$$\times \frac{(j_2 - j + j_1 - 1)\cdots(j_2 - j + m_1 + 1)}{1\cdot 2\cdots\cdots(j_1 - m_1)} \{j_1 j - j_1 | j j\rangle$$
$$= (-1)^{j_1 - m_1} \frac{(j_1 + m_1)!(j_2 - j + j_1)!(j_2 + j - m_1)!}{(2j_1)!(j_1 - m_1)!(j_2 - j + m)!(j_2 + j - j_1)!} \{j_1 j - j_1 | j j\rangle.$$

(2.43)

From the orthogonality relation (2.15), it follows that for $j = j' = m = m'$

$$1 = \sum_{m_1 = -j_1}^{+j_1} \{m_1 m_2 | j j\rangle|^2.$$

(2.15′)

Inserting (2.43) into (2.15′) and making use of the equality[2]

$$\sum_{m_1} \frac{(j_1 + m_1)!(j_2 + j - m_1)!}{(j_1 - m_1)!(j_2 - j + m_1)!} = \frac{(j_1 + j_2 + j + 1)!(-j_1 + j_2 + j)!(j_1 - j_2 + j)!}{(2j + 1)!(j_1 + j_2 - j)!},$$

(2.44)

one obtains

$$|\{j_1 j - j_1 | j j\rangle|^2 = \frac{(2j_1)!(2j + 1)!}{(j_1 + j_2 + j + 1)!(j_1 - j_2 + j)!}.$$

(2.45)

And with the phase convention (2.36) one obtains

$$\{j_1 j - j_1 | j j\rangle = +\sqrt{\frac{(2j_1)!(2j + 1)!}{(j_1 + j_2 + j + 1)!(j - j_2 + j)!}}.$$

(2.46)

This equation gives all coefficients that occur on the right-hand side of Equation (2.40) for $m = j$. Equation (2.40−) then gives the values of

$$\{m_1 m_2 | j\, m = j - 1\rangle$$

and successively the values of $\{m_1 m_2 | j\, m\rangle$ for all $m = j - 1, j - 2, \ldots, -j$.

The recursion relation (2.40−) for fixed values $m_1, m_2, j$ can be solved, and an expression of $\{m_1 m_2 | j\, m\rangle$ in terms of $\{m_1 m_2 | j j\rangle$ can be given. From this, using (2.46), a general formula for the Clebsch–Gordan coefficients

---

[2] A derivation of (2.44) can be found in Edmonds (1957, Appendix 1).

**Table 2.1** $\langle j_1 m_1 \tfrac{1}{2} m_2 | jm \rangle$

| | $m_2 = \tfrac{1}{2}$ | $m_2 = -\tfrac{1}{2}$ |
|---|---|---|
| $j = j_1 + \tfrac{1}{2}$ | $\sqrt{\dfrac{j_1 + m + \tfrac{1}{2}}{2j_1 + 1}}$ | $\sqrt{\dfrac{j_1 - m + \tfrac{1}{2}}{2j_1 + 1}}$ |
| $j = j_1 - \tfrac{1}{2}$ | $-\sqrt{\dfrac{j_1 - m + \tfrac{1}{2}}{2j_1 + 1}}$ | $\sqrt{\dfrac{j_1 + m + \tfrac{1}{2}}{2j_1 + 1}}$ |

can be derived. We shall not give this derivation here,[3] but give just one of the formulas for the Clebsch–Gordan coefficients:

$$\{ j_1 j_1 m_2 | j\, m \rangle = \langle j_1 j_1 j_2 m_2 | j\, m \rangle$$

$$= \delta_{m_1 + m_2,\, m} \left[ \frac{(2j + 1)(j_1 + j_2 - j)!(j_1 - j_2 + j)!(-j + j_2 + j)!}{(j_1 + j_2 + j + 1)!} \right]^{1/2}$$

$$\times \left[ (j_1 + m_1)!(j - m_1)!(j_2 + m_2)!(j_2 - m_2)!(j + m)!(j - m)! \right]^{1/2}$$

$$\times \sum_z \{ (-1)^z / [z!(j_1 + j_2 - j - z)!(j_1 - m_1 - z)!(j_2 + m_2 - z)!$$

$$\times (j - j_2 + m_1 + z)!(j - j_1 - m_2 + z)! ] \} \tag{2.47}$$

We also list some of the properties (symmetry relations) of the Clebsch–Gordan coefficients:

$$\langle j_1 m_1 j_2 m_2 | j_1 j_2 j_3 m_3 \rangle = (-1)^{j_2 + m_2} \left[ \frac{2j_3 + 1}{2j_1 + 1} \right]^{1/2} \langle j_2 - m_2 j_3 m_3 | j_2 j_3 j_1 m_1 \rangle. \tag{2.48}$$

We now give explicit expressions for the Clebsch–Gordan coefficients for the cases $j_2 = 0, \tfrac{1}{2}, 1$.
   For $j_2 = 0$;

$$\langle j_1 m_1 00 | j\, m \rangle = \delta_{j_1 j} \delta_{m_1 m}. \tag{2.49}$$

The values for $j_2 = \tfrac{1}{2}$ and $j_2 = 1$ are given in Tables 2.1 and 2.2.
   Instead of the Clebsch–Gordan coefficients it is frequently more convenient to use the Wigner 3-$j$ symbols, since these display symmetry properties more clearly. The 3-$j$ symbol is defined by

$$\begin{pmatrix} j_1 & j_2 & j_3 \\ m_1 & m_2 & m_3 \end{pmatrix} = (-1)^{j_1 - j_2 - m_3}(2j_3 + 1)^{-1/2} \langle j_1 m_1 j_2 m_2 | j_3 - m_3 \rangle \tag{2.50}$$

---

[3] A derivation using the same notation as used here is given in Edmonds (1957, pp. 44–45).

**Table 2.2** $\langle j_1 m_1 1 m_2 | jm \rangle$

| $j =$ | $m_2 = 1$ | $m_2 = 0$ | $m_2 = -1$ |
|---|---|---|---|
| $j_1 + 1$ | $\sqrt{\dfrac{(j_1 + m)(j_1 + m + 1)}{(2j_1 + 1)(2j_1 + 2)}}$ | $\sqrt{\dfrac{(j_1 - m + 1)(j_1 + m + 1)}{(2j_1 + 1)(j_1 + 1)}}$ | $\sqrt{\dfrac{(j_1 - m)(j_1 - m + 1)}{(2j_1 + 1)(2j_1 + 2)}}$ |
| $j_1$ | $-\sqrt{\dfrac{(j_1 + m)(j_1 - m + 1)}{2j_1(j_1 + 1)}}$ | $\dfrac{m}{\sqrt{j_1(j_1 + 1)}}$ | $\sqrt{\dfrac{(j_1 - m)(j_1 + m + 1)}{2j_1(j_1 + 1)}}$ |
| $j_1 - 1$ | $\sqrt{\dfrac{(j_1 - m)(j_1 - m + 1)}{2j_1(2j + 1)}}$ | $-\sqrt{\dfrac{(j_1 - m)(j_1 + m)}{j_1(2j_1 + 1)}}$ | $\sqrt{\dfrac{(j_1 + m + 1)(j_1 + m)}{2j_1(2j_1 + 1)}}$ |

Its symmetry properties are given by:

$$\begin{pmatrix} j_1 & j_2 & j_3 \\ m_1 & m_2 & m_3 \end{pmatrix} = \begin{pmatrix} j_2 & j_3 & j_1 \\ m_2 & m_3 & m_1 \end{pmatrix} = \begin{pmatrix} j_3 & j_1 & j_2 \\ m_3 & m_1 & m_2 \end{pmatrix}, \quad (2.51)$$

$$(-1)^{j_1 + j_2 + j_3} \begin{pmatrix} j_1 & j_2 & j_3 \\ m_1 & m_2 & m_3 \end{pmatrix} = \begin{pmatrix} j_2 & j_1 & j_3 \\ m_2 & m_1 & m_3 \end{pmatrix}$$

$$= \begin{pmatrix} j_1 & j_3 & j_2 \\ m_1 & m_3 & m_2 \end{pmatrix} = \begin{pmatrix} j_3 & j_2 & j_1 \\ m_3 & m_2 & m_1 \end{pmatrix}, \quad (2.52)$$

and

$$\begin{pmatrix} j_1 & j_2 & j_3 \\ m_1 & m_2 & m_3 \end{pmatrix} = (-1)^{j_1 + j_2 + j_3} \begin{pmatrix} j_1 & j_2 & j_3 \\ -m_1 & -m_2 & -m_3 \end{pmatrix}. \quad (2.53)$$

Table 2.3[4] lists several values for the 3-$j$ symbols from which many others can be calculated using the symmetry properties (2.51)–(2.53)

## V.3 Tensor Operators and the Wigner–Eckart Theorem

We have considered the angular-momentum operators in some detail in the previous section. Now we consider a more general class of operators, called *tensor operators*,[5] which are defined by their relationship with the angular-momentum operators. The simplest example of such an operator is a scalar operator, which is defined to be any operator $S$ satisfying

$$[J_i, S] = 0. \quad (3.1)$$

Another example is any set of three operators $V_1$ ($i = 1, 2, 3$) that satisfy

$$[J_i, V_j] = i\hbar\epsilon_{ijk}V_k \quad (i, j, k = 1, 2, 3). \quad (3.2)$$

---

[4] From Edmonds (1957), with permission.
[5] We shall actually only consider what are called "irreducible tensor operators."

Such a set of operators is called a *vector operator* or *regular tensor operator*. We note that the angular momentum itself is a vector operator.

Rather than using the "Cartesian" components $V_1$, it is more convenient to use the "spherical" components $V_0$, $V_{\pm 1}$ defined by

$$V_0 = V_3, \qquad V_{\pm 1} = \frac{\mp 1}{\sqrt{2}}(V_1 \pm iV_2). \tag{3.3}$$

Note that the $H_\pm$ of (III.3.1) (or the $J_\pm = \hbar H_\pm$) are *not* the spherical components of a vector operator, but differ from them by the factor $+1/\sqrt{2}$ or $-1/\sqrt{2}$; therefore we use the notation $J_{\pm 1}$ for the spherical components of the angular momentum operator, in contrast to $J_\pm$ for the raising and lowering operators. Equation (3.2) becomes

$$[J_\kappa, V_\kappa] = 0 \qquad (\kappa = -1, 0, +1), \tag{3.4a}$$

$$[J_\pm, V_{\mp 1}] = \sqrt{2}\hbar V_0 \qquad [J_\pm, V_0] = \sqrt{2}\hbar V_{\pm 1} \qquad [J_0, V_\kappa] = \kappa \hbar V_{\pm 1}. \tag{3.4b}$$

In general, we define a *tensor operator* of rank $j$ to be (in spherical components) a set of $2j + 1$ operators[6] $T_\kappa^{(j)}$ ($\kappa = -j, -j + 1, \ldots, +j$) that satisfy

$$[J_0, T_\kappa^{(j)}] = \kappa \hbar T_\kappa^{(j)},$$

$$[J_\pm, T_\kappa^{(j)}] = \sqrt{j(j + 1) - \kappa(\kappa \pm 1)}\hbar T_{\kappa \pm 1}^{(j)}. \tag{3.5}$$

This can be written in a more compact form using the Clebsch–Gordan coefficients:

$$[J_\mu, T_\kappa^{(j)}] = \sqrt{j(j + 1)}\langle j\kappa 1\mu | j\kappa + \mu\rangle T_{\kappa + \mu}^{(j)} \tag{3.5'}$$

We note that scalar and vector operators are tensor operators of rank 0 and 1, respectively.

The matrix elements of tensor operators have an important property which is expressed by the *Wigner–Eckart theorem*:

*Let $T_\kappa^{(J)}$ be a tensor operator. The matrix element of $T_\kappa^{(J)}$ between the angular momentum eigenstates may be written as*

$$\langle j'm' | T_\kappa^{(J)} | j\, m\rangle = \langle j\, m\, J\, \kappa | j'm'\rangle\langle j' \| T^{(J)} \| j\rangle, \tag{3.6}$$

*where $\langle j\, m\, J\, \kappa | j'm'\rangle$ is a Clebsch–Gordan coefficient and the symbol $\langle j' \| T_\kappa^{(J)} \| j\rangle$ [defined by (3.6)] denotes a quantity that depends on $j'$, $j$, $J$ and*

---

[6] We emphasize the use of spherical components by the use of Greek letters such as $\kappa$ in labeling the components. While such emphasis is not needed for the general tensor operator, in the case that the tensor operator is a vector operator the use of Greek letters will enable us to distinguish between the Cartesian components (Latin letters) and the spherical components (Greek letters).

# Table 2.3

$$\begin{pmatrix} j_1 & j_2 & j_3 \\ 0 & 0 & 0 \end{pmatrix} = (-1)^{1/2J} \left[ \frac{(j_1 + j_2 - j_3)!(j_1 + j_3 - j_2)!(j_2 + j_3 - j_1)!}{(j_1 + j_2 + j_3 + 1)!} \right]^{1/2} \frac{(\tfrac{1}{2}J)!}{(\tfrac{1}{2}J - j_1)!(\tfrac{1}{2}J - j_2)!(\tfrac{1}{2}J - j_3)!}$$

if $J$ is even.

$$\begin{pmatrix} j_1 & j_2 & j_3 \\ 0 & 0 & 0 \end{pmatrix} = 0 \text{ if } J \text{ is odd, where } J = j_1 + j_2 + j_3$$

$$\begin{pmatrix} J+\tfrac{1}{2} & J & \tfrac{1}{2} \\ M & -M-\tfrac{1}{2} & \tfrac{1}{2} \end{pmatrix} = (-1)^{J-M-1/2} \left[ \frac{J - M + \tfrac{1}{4}}{(2J+2)(2J+1)} \right]^{1/2} \qquad (J+\tfrac{1}{2}, J, \tfrac{1}{2})$$

$$\begin{pmatrix} J+1 & J & 1 \\ M & -M-1 & 1 \end{pmatrix} = (-1)^{J-M-1} \left[ \frac{(J - M)(J - M + 1)}{(2J+3)2J + 2)(2J+1)} \right]^{1/2} \qquad (J+1, J, 1)$$

$$\begin{pmatrix} J+1 & J & 1 \\ M & -M & 0 \end{pmatrix} = (-1)^{J-M-1} \left[ \frac{(J + M + 1)(J - M + 1) \cdot 2}{(2J+3)(2J+2)(2J+1)} \right]^{1/2}$$

$$\begin{pmatrix} J & J & 1 \\ M & -M-1 & 1 \end{pmatrix} = (-1)^{J-M} \left[ \frac{(J - M)(J + M + 1) \cdot 2}{(2J+2)(2J+1)(2J)} \right]^{1/2} \qquad (J, J, 1)$$

$$\begin{pmatrix} J & J & 1 \\ M & -M & 0 \end{pmatrix} = (-1)^{J-M} \frac{M}{[(2J+1)J(J+1)]^{1/2}}$$

$$\begin{pmatrix} J+\tfrac{3}{2} & J & \tfrac{3}{2} \\ M & -M-\tfrac{3}{2} & \tfrac{3}{2} \end{pmatrix} = (-1)^{J-M+1/2} \left[ \frac{(J - M + \tfrac{1}{2})(J - M + \tfrac{1}{2})(J + M + \tfrac{3}{2})}{(2J+4)(2J+3)(2J+2)(2J+1)} \right]^{1/2} \qquad (J+\tfrac{3}{2}, J, \tfrac{3}{2})$$

$$\begin{pmatrix} J+\tfrac{3}{2} & J & \tfrac{3}{2} \\ M & -M-\tfrac{1}{2} & \tfrac{1}{2} \end{pmatrix} = (-1)^{J-M+1/2} \left[ \frac{3(J - M + \tfrac{1}{2})(J - M + \tfrac{3}{2})(J + M + \tfrac{3}{2})}{(2J+4)(2J+3)(2J+2)(2J+1)} \right]^{1/2}$$

$$\begin{pmatrix} J+\tfrac{1}{2} & J & \tfrac{3}{2} \\ M & -M-\tfrac{3}{2} & \tfrac{3}{2} \end{pmatrix} \qquad (-1)^{J-M-1/2}\left[\frac{3(J-M-\tfrac{1}{2})(J-M+\tfrac{1}{2})(J+M+\tfrac{3}{2})}{(2J+3)(2J+2)(2J+1)2J}\right]^{1/2} \qquad (J+\tfrac{1}{2},\,J,\,\tfrac{3}{2})$$

$$\begin{pmatrix} J+\tfrac{1}{2} & J & \tfrac{3}{2} \\ M & -M-\tfrac{1}{2} & \tfrac{1}{2} \end{pmatrix} \qquad (-1)^{J-M-1/2}\left[\frac{J-M+\tfrac{1}{2}}{(2J+3)(2J+2)(2J+1)2J}\right]^{1/2}(J+3M+\tfrac{3}{2})$$

$$\begin{pmatrix} J+2 & J & 2 \\ M & -M-2 & 2 \end{pmatrix} \qquad (-1)^{J-M}\left[\frac{(J-M-1)(J-M)(J-M+1)(J-M+2)}{(2J+5)(2J+4)(2J+3)(2J+2)(2J+1)}\right]^{1/2} \qquad (J+2,\,J,\,2)$$

$$\begin{pmatrix} J+2 & J & 2 \\ M & -M-1 & 1 \end{pmatrix} \qquad 2(-1)^{J-M}\left[\frac{(J+M+2)(J-M+2)(J-M+1)(J-M)}{(2J+5)(2J+4)(2J+3)(2J+2)(2J+1)}\right]^{1/2}$$

$$\begin{pmatrix} J+2 & J & 2 \\ M & -M & 0 \end{pmatrix} \qquad (-1)^{J-M}\left[\frac{6(J+M+2)(J+M+1)(J-M+2)(J-M+1)}{(2J+5)(2J+4)(2J+3)(2J+2)(2J+1)}\right]^{1/2}$$

$$\begin{pmatrix} J+1 & J & 2 \\ M & -M-2 & 2 \end{pmatrix} \qquad 2(-1)^{J-M+1}\left[\frac{(J-M-1)(J-M)(J-M)(J+M+1)(J+M+2)}{(2J+4)(2J+3)(2J+2)(2J+1)2J}\right]^{1/2} \qquad (J+1,\,J,\,2)$$

$$\begin{pmatrix} J+1 & J & 2 \\ M & -M-1 & 1 \end{pmatrix} \qquad (-1)^{J-M+1}2(J+2M+2)\left[\frac{(J-M+1)(J-M)}{(2J+4)(2J+3)(2J+2)(2J+1)2J}\right]^{1/2}$$

$$\begin{pmatrix} J+1 & J & 2 \\ M & -M & 0 \end{pmatrix} \qquad (-1)^{J-M+1}2M\left[\frac{(6J+M+1)(J-M+1)}{(2J+4)(2J+3)(2J+2)(2J+1)2J}\right]^{1/2}$$

$$\begin{pmatrix} J & J & 2 \\ M & -M-2 & 2 \end{pmatrix} \qquad (-1)^{J-M}\left[\frac{6(J-M-1)(J-M)(J+M+1)(J+M+2)}{(2J+3)(2J+2)(2J+1)(2J)(2J-1)}\right]^{1/2} \qquad (J,\,J,\,2)$$

$$\begin{pmatrix} J & J & 2 \\ M & -M-1 & 1 \end{pmatrix} \qquad (-1)^{J-M}(1+2M)\left[\frac{6(J+M+1)(J-M)}{(2J+3)(2J+2)(2J+1)(2J)(2J-1)}\right]^{1/2}$$

$$\begin{pmatrix} J & J & 2 \\ M & -M & 0 \end{pmatrix} \qquad (-1)^{J-M}\frac{2[3M^2-J(J+1)]}{[(2J+3)(2J+2)(2J+1)(2J)(2J-1)]^{1/2}}$$

*the nature of the tensor operator $T_\kappa^{(J)}$ but does not depend on $m$, $m'$, or $\kappa$. $\langle j' \| T_\kappa^{(J)} \| j \rangle$ is called the* reduced matrix element[7] *of the tensor operator.*

This theorem will not be proven here, since the proof is pure mathematics and does not give any additional insight into the physics. If in addition to $j$ and $m$ there are other quantum numbers, say $\eta = (a_1, a_2, \ldots, a_N)$, then the reduced matrix element will in general also depend upon these quantum numbers, i.e.,

$$\langle \eta' j' m' | T_\kappa^{(J)} | \eta\, j\, m \rangle = \langle j\, m\, J\, \kappa | j' m' \rangle \langle \eta' j' \| T^{(J)} \| \eta\, j \rangle. \tag{3.6'}$$

It should be emphasized that the Wigner–Eckart theorem is both a theorem and a definition. It is a theorem in that (3.6) says that the matrix element of $T_\kappa^{(J)}$ can be factored in a certain way, and it is a definition in that (3.6) defines the reduced matrix elements.

The Wigner–Eckart theorem has become one of the most important tools for the understanding of physics. Equation (3.6) is the Wigner–Eckart theorem for the rotation group, which is connected with the algebra of angular momenta $J_i$ as an algebra of observables. Many physical systems have an (enveloping) algebra of a group as a subalgebra of the algebra of observables. And they have observables that are tensor operators with respect to this group. For those observables the Wigner–Eckart theorem, a generalization of (3.6), holds.

The Wigner–Eckart theorem expresses the matrix elements of the tensor operators, which are directly connected with the numbers measured in an experiment, in terms of Clebsch–Gordan coefficients and reduced matrix elements. The Clebsch–Gordan coefficients are known mathematical quantities and are calculated from the properties of the group. The reduced matrix elements are physical parameters whose values have to be obtained from the experimental data. [If an observable has further additional properties—e.g., Equation (VI.3.12) of the Lenz vector in Chapter VI—then the reduced matrix elements can be reduced to a still smaller number of parameters.] Thus the importance of the Wigner–Eckart theorem is that it allows one to fit the large number of experimentally observable matrix elements in terms of a much smaller number of more fundamental quantities, the reduced matrix elements. The property of being a tensor operator is often all one knows of an observable, and the Wigner–Eckart theorem then is the only tool at hand.

A particular consequence of the Wigner–Eckart theorem (3.6) is

$$\langle j' m' | T_\kappa^{(J)} | j\, m \rangle = 0$$

$$\text{if } m' \neq \kappa + m \quad \text{or } j' \neq J + j, J + j - 1, \tag{3.7}$$

---

[7] In some textbooks, a constant factor or function of $j$ appears explicitly in the Wigner–Eckart theorem. In our notation these factors have been absorbed into the reduced matrix element.

which follows immediately from the properties of the Clebsch–Gordan coefficients (2.39).

As an illustration of the Wigner–Eckart theorem, we note that for a scalar operator $S$, Equation (3.6) becomes

$$\langle j'm'|S|j\,m\rangle = \delta_{m'm}\delta_{j'j}\langle j\|S\|j\rangle. \tag{3.8}$$

This says that $S$ cannot change the angular-momentum quantum number, which is just as we would expect. For the angular-momentum operators themselves, Equation (3.6) gives

$$\langle j'm'|J_\kappa|j\,m\rangle = \langle j\,m\,1\,\kappa|j'm'\rangle\langle j'\|J\|j\rangle, \tag{3.9}$$

where the Clebsch–Gordan coefficients are listed in Table 2.2. By comparing the Clebsch–Gordan coefficient in the table with the formula (III.3.8′), (III.3.22), and (III.3.23), we see that

$$\langle j'\|J\|j\rangle = \delta_{j'j}\hbar\sqrt{j(j+1)}. \tag{3.10}$$

If $|jm\rangle$ is a basis in the space in which $V_\kappa$ is a general vector operator, then we can expand $V_\kappa|jm\rangle$ with respect to this basis:

$$V_\kappa|jm\rangle = \sum_{j'm'}|j'm'\rangle\langle j'm'|V_\kappa|j\,m\rangle$$

and obtain, according to the Wigner–Eckart theorem (3.6),

$$V_\kappa|j\,m\rangle = \sum_{j'm'}|j'm'\rangle\langle j\,m\,1\,\kappa|j'm'\rangle\langle j'\|V\|j\rangle. \tag{3.11}$$

If the c.s.c.o. contains, in addition to $\mathbf{J}^2$ and $J_3$, $N$ other operators $A_1, \dots, A_N$ with spectrum $\eta = (a_1, a_2, \dots, a_N)$, then the basis is $|\eta\,j\,m\rangle$, and instead of (3.11) one has, according to (3.6′),

$$V_\kappa|\eta\,j\,m\rangle = \sum_{\eta'j'm'}|\eta'j'm'\rangle\langle j\,m\,1\,\kappa|j'm'\rangle\langle\eta'j'\|V\|\eta\,j\rangle \tag{3.11′}$$

We shall for the remainder of this section suppress the additional quantum numbers $\eta$, but it should be understood that whenever additional quantum numbers $\eta$ are needed, the reduced matrix elements will depend upon these quantum numbers and summation over $\eta'$ is implied.

According to (3.7), the only nonvanishing terms in (3.11) are those for which $j' = j + 1, j, j - 1$ and $m' = m + \kappa$:

$$\begin{aligned}
V_\kappa|j\,m\rangle = &|j-1\,m+\kappa\rangle\langle j\,m\,1\,\kappa|j-1\,m+\kappa\rangle\langle j-1\|V\|j\rangle \\
&+ |j\,m+\kappa\rangle\langle j\,m\,1\,\kappa|j\,m+\kappa\rangle\langle j\|V\|j\rangle \\
&+ |j+1\,m+\kappa\rangle\langle j\,m\,1\,\kappa|j+1\,m+\kappa\rangle\langle j+1\|V\|j\rangle.
\end{aligned} \tag{3.12}$$

This is the most general form possible for the action of a vector operator. According to (3.12), then, any vector operator can be completely specified by three reduced matrix elements (which in general may depend upon $\eta$, $\eta'$). Actually, as we shall show below, only two quantities are needed to determine a vector operator.

By using the Clebsch–Gordan coefficients of Table 2.2, we can write (3.12) explicitly. For the 0-component,

$$V_0|j\,m\rangle = |j-1\,m\rangle\left(-\sqrt{\frac{(j-m)(j+m)}{j(2j+1)}}\right)\langle j-1\|V\|j\rangle$$

$$+ |j\,m\rangle\left(\frac{m}{\sqrt{j(j+1)}}\right)\langle j\|V\|j\rangle$$

$$+ |j+1\,m\rangle\left(\sqrt{\frac{(j-m+1)(j+m+1)}{(2j+1)(j+1)}}\right)\langle j+1\|V\|j\rangle,$$

or

$$V_0|j\,m\rangle = \sqrt{j^2 - m^2}\,c_j|j-1\,m\rangle - m a_j|j\,m\rangle - \sqrt{(j+1)^2 - m^2}\,d_j|j+1\,m\rangle,$$

(3.13)

where the $c_j$, $a_j$, and $d_j$ are defined by Equation (3.13) itself:

$$c_j = -\frac{\langle j-1\|V\|j\rangle}{\sqrt{j(2j+1)}},$$

$$a_j = -\frac{\langle j\|V\|j\rangle}{\sqrt{j(j+1)}},$$

(3.14)

$$d_j = -\frac{\langle j+1\|V\|j\rangle}{\sqrt{(2j+1)(j+1)}}.$$

Likewise,

$$V_{+1}|j\,m\rangle = -|j-1\,m+1\rangle\sqrt{(j-m-1)(j-m)/2}\,c_j$$

$$+ |j\,m+1\rangle\sqrt{(j+m+1)(j-m)/2}\,a_j$$

$$- |j+1\,m+1\rangle\sqrt{(j+m+1)(j+m+2)/2}\,d_j \quad (3.15)$$

and

$$V_{-1}|j\,m\rangle = -|j-1\,m-1\rangle\sqrt{(j+m)(j+m-1)/2}\,c_j$$

$$- |j\,m-1\rangle\sqrt{(j-m+1)(j+m)/2}\,a_j$$

$$- |j+1\,m-1\rangle\sqrt{(j-m+1)(j-m+2)/2}\,d_j. \quad (3.16)$$

Equations (3.13), (3.15), and (3.16) say that a vector operator is completely determined by the three functions $c_j$, $a_j$, and $d_j$ (which may in general depend upon $\eta'$, $\eta$) of the discrete parameter $j$. In the mathematical note below it will be shown that in fact only two such functions are needed, as it is possible to choose the basis $|j\,m\rangle$ in such a way that

$$d_j = c_{j+1}, \quad (3.17)$$

i.e.,

$$\sqrt{2j + 3}\langle j + 1\|V\|j\rangle = \sqrt{2j + 1}\langle j\|V\|j + 1\rangle. \tag{3.17'}$$

⟦We define a new basis

$$|h_m^j\rangle = \omega(j)|j\,m\rangle, \tag{3.18}$$

where $\omega(j)$ is a complex number. The angular-momentum operators have the same form (Section III.3) in the $|h_m^j\rangle$ basis as they do in the $|j\,m\rangle$ basis. In general, the $|h_m^j\rangle$ are not normalized unless

$$|\omega(j)| = 1. \tag{3.19}$$

In this basis, Equation (3.13) becomes

$$V_0|h_m^j\rangle = \sqrt{j^2 - m^2}\,c_j\,\frac{\omega(j)}{\omega(j-1)}\,|h_m^{j-1}\rangle - ma_j|h_m^j\rangle$$

$$- \sqrt{(j+1)^2 - m^2}\,d_j\,\frac{\omega(j)}{\omega(j+1)}\,|h_m^{j+1}\rangle. \tag{3.20}$$

We define $A_j = a_j$, and we wish to choose $\omega(j)$ such that

$$c_j\,\frac{\omega(j)}{\omega(j-1)} = C_j \quad\text{and}\quad d_j\,\frac{\omega(j)}{\omega(j+1)} = D_j = C_{j+1}. \tag{3.21}$$

This will be possible if

$$c_j\,\frac{\omega(j)}{\omega(j-1)} = \frac{\omega(j-1)}{\omega(j)}\,d_{j-1}, \tag{3.22}$$

i.e., if

$$\omega^2(j) = \omega^2(j-1)\,\frac{d_{j-1}}{c_j}. \tag{3.23}$$

Suppose $j_0$ is the smallest value of $j$ in the space, i.e.,

$$\langle j_0 - 1\|V\|j_0\rangle = 0. \tag{3.24}$$

Then (3.23) will be satisfied if we choose

$$\omega(j) = \sqrt{\omega^2(j_0)\,\frac{d_{j_0}}{c_{j_0+1}}\,\frac{d_{j_0+1}}{c_{j_0+2}}\,\cdots\,\frac{d_{j-1}}{c_j}} \tag{3.25}$$

so (3.20) becomes

$$V_0|h_m^j\rangle = \sqrt{j^2 - m^2}\,C_j|h_m^{j-1}\rangle - mA_j|h_m^j\rangle$$

$$\equiv \sqrt{(j+1)^2 - m^2}\,C_{j+1}|h_m^{j+1}\rangle. \tag{3.26}$$

It can be shown that the $|h_m^j\rangle$ can be normalized in the following two cases:

1. For Hermitian $V_\kappa$ $(V_\kappa^\dagger = V_{-\kappa})$ we have $C_j = -\bar{C}_j$.
2. For skew-Hermitian $V_\kappa$ $(V_\kappa^\dagger = -V_{-\kappa})$ we have $C_j = \bar{C}_j$. ⟧

Thus every Hermitian or skew-Hermitian vector operator can be written in the form (3.13), (3.15), (3.16) with (3.17), where the $c_j$ are purely imaginary or real functions of $j$. It should be noted, however, that in general we may have $d_j = c_{j+1}$ for only one vector operator $V_\kappa$ at a time. If two different vector operators are involved in one problem and the basis has been chosen so that for the reduced matrix element of one of them (3.17) holds, then in this basis, the other or any additional vector operator is expressed in terms of three independent reduced matrix elements.


## Appendix to Section V.3—Representations of the Algebras $\mathscr{E}(\mathrm{SO}(3, 1))$, $\mathscr{E}(\mathrm{SO}(4))$, and $\mathscr{E}(E_3)$

Algebras that are generated by the angular-momentum operators $J_i$ and a particular vector operator are needed for various problems of quantum physics. We have already seen an example of such an algebra in Section III.4, where $\mathscr{E}(E_3)$ was the algebra generated by the angular momentum $J_i$ and the dipole operator $Q_i$. In Chapter VI we shall encounter the algebra $\mathscr{E}(\mathrm{SO}(4))$ generated by the angular-momentum operator $L_i$ and the Lenz vector $A_i$. $\mathscr{E}(\mathrm{SO}(3,1))$ is the algebra of the Lorentz group and has various applications in quantum physics.

In this appendix we give a derivation of the representations of these three algebras. These algebras are the enveloping algebras of the groups SO(3,1) [the pseudo-orthogonal group in $(3 + 1)$ dimensions], SO(4) [the orthogonal group in 4 dimensions] and $E_3$ [the three-dimensional Euclidean group], but we shall not discuss these group-theoretic connections here.

The defining relations of this algebra are given by

$$[H_i, H_j] = i\epsilon_{ijk} H_k, \tag{3.27}$$

$$[H_i, F_j] = i\epsilon_{ijk} F_k, \tag{3.28}$$

$$[F_i, F_j] = \lambda^2 i\epsilon_{ijk} H_k, \tag{3.29$_{\lambda^2}$}$$

where $\lambda^2 = -1$ for $\mathscr{E}(\mathrm{SO}(3,1))$, $\lambda^2 = +1$ for $\mathscr{E}(\mathrm{SO}(4))$, and $\lambda^2 = 0$ for $\mathscr{E}(E_3)$. Equation (3.27) defines the algebra $\mathscr{E}(\mathrm{SO}(3))$ of angular momentum. Equation (3.28) specifies that $F_i$ is a vector operator with respect to $\mathscr{E}(\mathrm{SO}(3))$; the $F_i, i = 1, 2, 3$, are the Cartesian components. Equation (3.29) then specifies that this vector operator together with $H_k$ generates one of the above algebras. If in addition to (3.27) and (3.28), (3.29) is also fulfilled, then the $F_i$ are very particular vector operators and its reduced matrix elements have very particular values. We shall now determine these values. We shall derive the representations of (3.27)–(3.29) for $\lambda^2 = -1$ and obtain the other two cases $\lambda^2 = 1$ and $\lambda^2 = 0$ by a change of the Hermiticity property and by a limiting process, respectively.

To find all linear representations of the algebra (3.27)–(3.29) means to find all linear operators in all linear spaces that fulfill these commutation relations. We shall restrict ourselves here to the subclass of those representations whose representation space $\mathscr{R}$ contains each of the representation spaces $\mathscr{R}^l$ ($l = 0$, or $\frac{1}{2}$, or etc of $\mathscr{E}(SO(3))$ at most once.[8] In each $\mathscr{R}^l$ we have, according to Section III.3, the basis

$$f_m^l \qquad m = -l, -l + 1, \ldots, +l, \tag{3.30}$$

and the space $\mathscr{R}$ is then spanned by these basis vectors $f_m^l$, where $l$ runs through a set of values that is to be determined. In other words,

$$\mathscr{R} = \sum_l \oplus \mathscr{R}^l. \tag{3.31}$$

If one of the $\mathscr{R}^l$, say for $l = l_0$, were to appear more than once, then $f_m^l$ would not be a basis system of the space $\mathscr{R}$, and a new label (quantum number) would be needed to distinguish orthogonal vectors with the same value $l_0$ and $m$: $f_m^{l_0 \eta}$.

The action of the linear operators $H_k$ on all $f_m^l$ is already known from Section III.3, and we have now to determine the action of the $F_k$ upon the $f_m^l$. We introduce the components [in analogy to (III.3.1)]

$$F_0 = F_3, \qquad F_\pm = F_1 \pm iF_2 = \mp 2F_{\pm 1}. \tag{3.32}$$

$F_0$ and $F_{\pm 1}$ are the standard spherical components of the vector operator $\mathbf{F}$. The commutation relations of the $F_0, F_\pm$ with $H_\pm, H_0 = H_3$ and with each other are then

$$
\left.
\begin{aligned}
[F_\kappa, H_\kappa] &= 0 \qquad \kappa = 0, +, - \\
[H_0, F_\pm] &= \pm F_\pm, \qquad [H_\pm, F_0] = \mp F_\pm
\end{aligned}
\right\} \tag{3.33}
$$

$$[F_0, F_\pm] = \mp H_\pm \qquad [F_+, F_-] = -2H_0. \tag{3.34}$$

Equation (3.33) corresponds to (3.28) and expresses the vector-operator property of $\mathbf{F}$, and (3.34) corresponds to (3.29) with $\lambda^2 = -1$ and specifies that this vector operator is a generator of $\mathscr{E}(SO(3,1))$. We now utilize the vector-operator property and use the Wigner–Eckart theorem; then we obtain for $F_0$ and $F_\pm$ the expressions (3.13), (3.15), and (3.16). If we also choose the phase factor of the basis vectors appropriately, as discussed in the mathematical note just before this appendix, then (3.17) can be used,

---

[8] These include all the representations that are connected with unitary representations of the groups SO(3,1) and SO(4).

and we obtain for the action of $F_0$ and $F_\pm$ on the basis vectors the expressions

$$F_0 f^l_m = \sqrt{l^2 - m^2}\, c_l f^{l-1}_m - m a_l f^l_m - \sqrt{(l+1)^2 - m^2}\, c_{l+1} f^{l+1}_m, \qquad (3.35_0)$$

$$F_+ f^l_m = \sqrt{(l-m)(l-m-1)}\, c_l f^{l-1}_{m+1} - \sqrt{(l-m)(l+m+1)}\, a_l f^l_{m+1}$$
$$+ \sqrt{(l+m+1)(l+m+2)}\, c_{l+1} f^{l+1}_{m+1}, \qquad (3.35_+)$$

$$F_- f^l_m = -\sqrt{(l+m)(l+m-1)}\, c_l f^{l-1}_{m-1} - \sqrt{(l+m)(l-m+1)}\, a_l f^l_{m-1}$$
$$\sqrt{(l-m+1)(l-m+2)}\, c_{l+1} f^{l+1}_{m-1}. \qquad (3.35_-)$$

These have been obtained using only (3.33); and every vector operator, in a space in which $\mathcal{R}^l$ appears at most once, has this form.

The vectors $f^l_m$ have, according to the results of Section III.3, the property

$$H_3 f^l_m = m f^l_m \qquad \mathbf{H}^2 f^l_m = l(l+1) f^l_m,$$
$$H_\pm f^l_m = \sqrt{(l \mp m)(l \pm m + 1)}\, f^l_{m \pm 1}. \qquad (3.36)$$

The $f^l_m$ have in addition been chosen so that (3.17) is fulfilled, but, due to the redefinition (3.18), they are in general not normalized. Let us now see when the $f^l_m$ can be normalized without destroying (3.35). This can be done when $\| f^l_m \|$ is independent of $l$. From $(3.35_0)$ one obtains

$$(F_0 f^{l-1}_m, f^l_m) = -\sqrt{l^2 - m^2}\, c^*_l (f^l_m, f^l_m) \qquad (3.37_1)$$

and

$$(f^{l-1}_m, F_0 f^l_m) = \sqrt{l^2 - m^2}\, c_l (f^{l-1}_m, f^{l-1}_m). \qquad (3.37_2)$$

In order that $\| f^l_m \| = \| f^{l-1}_m \|$ (i.e., $\| f^l_m \|$ be independent of $l$), one must have according to (3.37)

$$(F_0 f^{l-1}_m, f^l_m) c_l = -(f^{l-1}_m, F_0 f^l_m) c^*_l. \qquad (3.38)$$

This can be fulfilled in two cases:

1. $F_0$ is Hermitian, $F^\dagger_0 = F_0$; then $c_l$ must be imaginary, $c_l = -c^*_l$; but then it follows by the same argument that $a_l = a^*_l$.
2. $F_0$ is skew-Hermitian, $F^\dagger_0 = -F_0$; then $c_l$ must be real, $c_l = c^*_l$; but then it follows by the same argument that $a_l = -a^*_l$.

If $F_0 = F_3$ is Hermitian (skew-Hermitian) then by (3.28) all other components of $\mathbf{F}$ are also Hermitian (skew-Hermitian), and the spherical components fulfill $F^\dagger_\pm = F_\mp (F^\dagger_\pm = -F_\mp)$, in which case $\mathbf{F}$ is called a Hermitian (skew-Hermitian) vector operator. Thus the basis vectors $f^l_m$ can be normalized in two cases: if $\mathbf{F}$ is a Hermitian or if $\mathbf{F}$ is a skew-Hermitian vector operator.

So far we have only exploited (3.27) and (3.28); we shall now make use of (3.29) or (3.34) to determine the unknown coefficients $a_l$ and $c_l$ [related to the reduced matrix elements by (3.14)]. It is sufficient to use just one of the three relations (3.34) because the other two are consequences of it and the relations (3.33). We choose

$$[F_+, F_0] = H_+. \tag{3.34'}$$

Applying both sides of (3.34') to the vector $f_m^l$ using (3.35) and comparing coefficients of $f_{m+1}^l$, $f_{m+1}^{l+1}$, and $f_{m+1}^{l-1}$ leads to the following equations:

$$[a_l(l+1) - (l-1)a_{l-1}]c_l = 0, \tag{3.39a}$$

$$[a_{l+1}(l+2) - la_l]c_{l+1} = 0, \tag{3.39b}$$

$$(2l-1)c_l^2 - (2l+3)c_{l+1}^2 - a_l^2 = 1. \tag{3.39c}$$

One starts with an arbitrary basis vector $f_m^l$, about which one assumes that it is contained in the particular representation space $\mathscr{R}$ that one wants to construct. Applying the operators $H_\kappa$ to $f_m^l$ a sufficient number of times, one obtains, as described in Section III.3, the whole space $\mathscr{R}^l$. Applying the vector operators $F_\kappa$ to $f_m^l$ a sufficient number of times, one reaches all the other $\mathscr{R}^j \subset \mathscr{R}$ where $j$ differs from $l$ by a positive integer, because $F_\kappa$ changes $l$ by $0$, $+1$, $-1$. As $j > 0$, there must always exist a smallest value of $j$ in any $\mathscr{R}$. This smallest $j$ we call $k_0$; it can be any of the allowed values for $j$ and will characterize the space $\mathscr{R}$. If $k_0$ is the smallest value of $\mathscr{R}$, then according to (3.35)

$$c_{k_0} = 0.$$

If $k_0$ is integer, then $\mathscr{R}$ contains only integer $l \geq k_0$, and if $k_0$ is half-integer, then $\mathscr{R}$ contains only half-integer $l \geq k_0$:

$$\mathscr{R} = \sum_{l=k_0, k_0+1, \ldots} \oplus \mathscr{R}^l. \tag{3.40}$$

For all $l$ for which $c_l \neq 0$ one obtains from (3.39a)

$$a_l(l+1) - (l-1)a_{l-1} = 0, \tag{3.41a}$$

and for all $l$ for which $c_{l+1} \neq 0$ one obtains from (3.39b)

$$a_{l+1}(l+2) - la_l = 0. \tag{3.41b}$$

Defining

$$\rho_l = l(l+1)a_l, \tag{3.42}$$

these two conditions lead to

$$\begin{aligned} \rho_l - \rho_{l-1} &= 0, \\ \rho_{l+1} - \rho_l &= 0, \end{aligned} \tag{3.43}$$

which means that $\rho_l$ is independent of $l$. We call this arbitrary complex constant $\rho_l = ik_0 c$, where $c$ is arbitrary complex, and we shall show below that the factoring out of $k_0$ is always possible. $a_l$ can then be written

$$a_l = \frac{ik_0 c}{l(l+1)} \quad (c \text{ arbitrary complex}). \tag{3.44}$$

For the case that both $c_l = 0$ and $c_{l+1} = 0$, it follows from (3.39c) that $a_l^2 = -1$, and the matrix elements (3.35) of $F_\kappa$ are completely determined. With (3.44) this means that $k_0 = l$, $c = l + 1$ and that $\mathscr{R} = \mathscr{R}^l$.

It remains to show that the factoring out of $k_0$ is always possible, i.e., that $a_l = 0$ for $k_0 = 0$. But for $k_0 = 0$, (3.41a) holds for $l = 1$ and leads to $a_1 = 0$. For $l = 2$ this in turn leads to $a_2 \cdot 3 - a_1 = a_2 \cdot 3 = 0$, and proceeding in this way we obtain $a_l = 0$ for $l = 1, 2, 3, \ldots$. For $l = 0$ we are free to choose $a_0 = 0$, as then according to (3.35) the factor in front of $a_0$ is zero. Thus (3.44) holds generally.

To determine $c_l$ we use (3.39c). Defining

$$\sigma_l = (2l - 1)(2l + 1)c_l^2, \tag{3.45}$$

we can write (3.39c) as

$$\sigma_l - \sigma_{l+1} - (2l + 1)a_l^2 = 2l + 1, \tag{3.46}$$

or using (3.44),

$$\sigma_l - \sigma_{l+1} = (2l + 1) - k_0^2 c^2 \left( \frac{1}{l^2} - \frac{1}{(l+1)^2} \right). \tag{3.47}$$

We calculate now for any value $k \geq k_0$

$$\sigma_{k_0} - \sigma_k = \sum_{l=k_0}^{k-1} (\sigma_l - \sigma_{l+1}) = \sum_{l=k_0}^{k-1} (2l + 1) - k_0^2 c^2 \sum_{l=k_0}^{k} \left( \frac{1}{l^2} - \frac{1}{(l+1)^2} \right)$$

$$= k^2 - k_0^2 - k_0^2 c^2 \left( \frac{1}{k_0^2} - \frac{1}{k^2} \right)$$

$$= \frac{(k^2 - k_0^2)(k^2 - c^2)}{k^2}.$$

As $k_0$ is the smallest value of $l$, $c_{k_0} = 0$, and consequently $\sigma_{k_0} = 0$, so that one obtains

$$\sigma_k = -\frac{(k^2 - k_0^2)(k^2 - c^2)}{k^2}, \tag{3.48}$$

and by (3.45),

$$c_l = \frac{i}{l} \sqrt{\frac{(l^2 - k_0^2)(l^2 - c^2)}{4l^2 - 1}} \quad \text{for any } l \geq k_0. \tag{3.49}$$

Therewith we have found all possible ways the operators $H_i$ and $F_i$ that fulfill (3.27)–(3.29) act upon the vectors $f_m^l$. All these possibilities are characterized by two numbers $(k_0, c)$, where $k_0$ is $0, \frac{1}{2}, 1, \frac{3}{2}, \ldots$, $c$ is any complex number, and the operators are given by (3.36) and (3.35) with (3.44) and (3.49). These are the linear representations of the algebra $\mathscr{E}(SO(3,1))$ that contain a given value of the angular momentum $l$ at most once. They will be, in general, infinite-dimensional, and only for certain values of $c$ will they be finite-dimensional. For an arbitrary complex value of $c$, $c_l$ will not fulfill conditions 1 and 2 below (3.38), the $f_m^l$ will not be normalizable, and a scalar product cannot be introduced in $\mathscr{R}$.

We shall now find those representations of $\mathscr{E}(SO(3,1))$ for which a scalar product can be defined in $\mathscr{R}$. According to conditions 1 and 2, this can be done in two cases:

1.    For Hermitian vector operators $a_l$ must be real and $c_l$ purely imaginary. This means according to (3.44) that $ic$ must be real, and according to (3.49) that $(l^2 - k_0^2)(l^2 - c^2)/(2l - 1)$ must be positive. The latter is possible only when $c^2$ is real, i.e., when $c$ is real or purely imaginary.

Thus for any integer of half-integer value of $k_0$ and any purely imaginary value of $c$, condition 1 can be fulfilled. If $c$ is real, then $k_0$ must be equal to zero in order that $a_l = a_l^*$. Then the expression under the square root in (3.49) becomes $l^2(l^2 - c^2)/(4l^2 - 1)$, which will be positive for all $l = 0, 1, 2\ldots$ only if $c^2 < 1$.

Thus we have seen that condition 1 can be fulfilled in two cases:

a.  For $k_0 = 0, \frac{1}{2}, 1, \ldots$ and $ic = $ real, $-\infty < ic < +\infty$
b.  For $k_0 = 0, 0 < c < 1$.

In neither of these two cases can $c_{l+1}$ ever become zero. Consequently, according to (3.35), one obtains from any given $f_m^l$ also an $f_m^{l+1}$. One can apply $F_\kappa$ an arbitrary number of times arriving, at ever higher values for $l$. The representation space $\mathscr{R} = \mathscr{R}(k_0, c)$ with $(k_0, c)$ fulfilling case a or b is infinite-dimensional:

$$\mathscr{R}(k_0, c) = \sum_{l = k_0, k_0 + 1, \ldots}^{\infty} \oplus \mathscr{R}^l. \tag{3.50}$$

It is the (algebraic)[9] direct sum of the irreducible representation spaces $\mathscr{R}^l$ ($l = k_0, k + 1, \ldots$) of the algebra of angular momentum, $\mathscr{E}(SO(3))$.

2.   For skew-Hermitian vector operators according to conditions 2, $a_l$ must be purely imaginary; this means that $c$ must be real. And $c_l$ must be real, which means according to (3.49) that $l^2 < c^2$. Thus the possible values of $l$ for a representation space $\mathscr{R}(k_0, c)$ with $c$ real are bounded, and consequently there must be a highest value of $l$, which we call $k_1$. According to (3.35) this means that $c_{k_1 + 1} = 0$, because only then would successive application of $F_\kappa$ not lead to higher values of $l$. This in turn means, according to (3.49), that $(k_1 + 1)^2 = c^2$ and $c = \pm(k_1 + 1)$. Thus we have seen that condition 2 can be fulfilled for

$$k_0 = 0, \tfrac{1}{2}, 1, \ldots \quad \text{and} \quad c = \pm(k_1 + 1) \quad \text{where } k_1 = k_0, k_0 + 1, \ldots$$

$$\tag{3.51}$$

---

[9] It can be completed with respect to various topologies, e.g., with respect to the Hilbert-space topology defined by the scalar product, or with respect to the topology defined by the countable number of scalar products $(\phi, \psi)_n = (\phi, (\mathbf{H}^2 + \mathbf{F}^2 + 1)^n \psi)$ leading to the nuclear space $\Phi$. Cf. Böhm [1978].

The representation space $\mathscr{R} = \mathscr{R}(k_0, \pm(k_1 + 1))$ is finite-dimensional and given by

$$\mathscr{R}(k_0), \pm(k_1 + 1)) = \sum_{l=k_0, k_0+1, \ldots}^{k_1} \oplus \mathscr{R}^l. \tag{3.52}$$

Though $\mathscr{R}(k_0, (k_1 + 1))$ and $\mathscr{R}(k_0, -(k_1 + 1))$ are the same direct sum of irreducible representation spaces of $\mathscr{E}(SO(3))$, the $F_\kappa$ act differently in these two spaces, as can be seen from (3.44) and (3.35). They are inequivalent representation spaces of $\mathscr{E}(SO(3, 1))$ with the same reduction with respect to $\mathscr{E}(SO(3))$.

Summarizing our results, we have found that the irreducible-representation[10] spaces of $\mathscr{E}(SO(3,1))$ [the algebra defined by (3.27), (3.28), (3.29$_{\lambda^2 = -1}$), for which $H_i$ and $F_i$ are Hermitian] are characterized by two numbers $(k_0, c)$ that can take any of the values in case a or b. The reduction of this representation space $\mathscr{R}(k_0, c)$ with respect to the algebra of $\mathscr{E}(SO(3))$ is given by (3.50), and the operators $H_\kappa$ and $F_\kappa$ are given by (3.36) and (3.35) with (3.44) and (3.49).

The irreducible representation spaces of $\mathscr{E}(SO(3,1))$ with skew-Hermitian $F_i$ are characterized by the two numbers $(k_0, c) = (k_0, \pm(k_1 + 1))$ which can take any of the values (3.51). The reduction of this representation space with respect to $\mathscr{E}(SO(3))$ is given by (3.52), and $H_\kappa$ and $F_\kappa$ are again given by (3.36), (3.35), (3.44), (3.49).

It is now easy to obtain the representations of $\mathscr{E}(SO(4))$ and $\mathscr{E}(E_3)$. To obtain the representations of $\mathscr{E}(SO(4))$ we define

$$A_j = -iF_j, \quad A_\kappa = -iF_\kappa \quad (\kappa = 0, +, -). \tag{3.53}$$

If $F_j$ fulfills the commutation relation (3.29$_{\lambda^2 = -1}$) then $A_i$ fulfills the commutation relation (3.29$_{\lambda^2 = +1}$), i.e., it fulfills together with $H_i$ the commutation relations of $\mathscr{E}(SO(4))$:

$$[A_i, A_j] = i\epsilon_{ijk} H_k \tag{3.29$_{\lambda^2 = +1}$}$$

If $F_j$ are skew-Hermitian ($F_j^\dagger = -F_j$), then $A_j$ are Hermitian ($A_j^\dagger = A_j$). Therefore we conclude that the Hermitian irreducible representations of $\mathscr{E}(SO(4))$ are characterized by two numbers $(k_0, \pm(k_1 + 1))$ that can take the values (3.51). The irreducible-representation spaces are all finite-dimensional and given by (3.52). The operators $A_\kappa$ are again given by (3.35) with (3.44) and (3.49), where $F_\kappa$ is given by (3.53).

The algebra $\mathscr{E}(SO(3,1))$ has two invariant operators, i.e., two operators that commute with all generators $H_i$ and with $F_i$. They are

$$C_1 = \mathbf{F}^2 - \mathbf{H}^2, \quad C_2 = \mathbf{F} \cdot \mathbf{H}. \tag{3.54}$$

---

[10] The irreducible representations are also called ladder representations because they are obtained by climbing from one value of $l$ to the next.

As they commute with all $H_\kappa$ and $F_\kappa$ their eigenvalues cannot change by applying $H_\kappa$, $F_\kappa$ to the corresponding eigenvector. Therefore their eigenvalues must be the same in the whole irreducible representation space. One can calculate these eigenvalues applying $C_1$ and $C_2$ to a suitable chosen $f_m^l$ using (3.36), (3.35) with (3.44) and (3.49). The result is

$$C_1 f_m^l(k_0, c) = (-k_0^2 - c^2 + 1) f_m^l(k_0, c), \tag{$3.55_1$}$$

$$C_2 f_m^l(k_0, c) = ik_0 c f_m^l(k_0, c). \tag{$3.55_2$}$$

The two invariant operators for $\mathscr{E}(SO(4))$ are

$$Q_1^{SO(4)} = \mathbf{A}^2 + \mathbf{H}^2, \qquad Q_2^{SO(4)} = \mathbf{A} \cdot \mathbf{H}, \tag{3.56}$$

with the eigenvalues

$$\begin{aligned}
C_1^{SO(4)} f_m^l(k_0, \pm(k_1 + 1)) &= (k_0^2 + c^2 - 1) f_m^l(k_0, \pm(k_1 + 1)) \\
&= (k_0^2 + k_1^2 + 2k_1) f_m^l(k_0, \pm(k_1 + 1)),
\end{aligned} \tag{$3.57_1$}$$

$$\begin{aligned}
C_2^{SO(4)} f_m^l(k_0, \pm(k_1 + 1)) &= k_0 c f_m^l(k_0, \pm(k_1 + 1)) \\
&= \pm k_0 (k_1 + 1) f_m^l(k_0, \pm(k_1 + 1)).
\end{aligned} \tag{$3.57_2$}$$

Thus, the eigenvalues of the invariant operators specify the irreducible representation completely; it is, however, more convenient to characterize the irreducible representations by the values of $(k_0, c)$ or by the values of $(k_0, k_1, \pm)$, where $\pm$ refers to the two inequivalent representations with the same value of $k_0$ and $k_1$.

We finally mention the special case where $k_0 = 0$ or $C_2 = 0$. This is the case one encounters most frequently in the application to simple quantum-mechanical systems. We shall meet this case for $\mathscr{E}(SO(4))$ again in Chapter VI in the derivation of the hydrogen spectrum. It is the case that does not require parity doubling, as we shall see below in Section V.4 [in particular Equation (4.25)]. We have encountered it already in Section III.4; it leads to a spin spectrum given by (III.4.2) and depicted by Figure III.4.1. For $k_0 = 0$ it follows from (3.50) for Hermitian representations of $\mathscr{E}(SO(3,1))$ that

$$\mathscr{R}(0, c) = \sum_{l=0, 1, \ldots} \oplus \mathscr{R}^l, \tag{3.58}$$

and from (3.52) for Hermitian representations of $\mathscr{E}(SO(4))$ that

$$\mathscr{R}(0, k_1 + 1 = n) = \sum_{l=0, 1, \ldots}^{n-1} \mathscr{R}^l \qquad (n = 1, 2, 3, \ldots). \tag{3.59}$$

From (3.44) follows that for this case

$$a_l = 0 \quad \text{for all } l \qquad (\text{or } \langle l \| F \| l \rangle = 0) \tag{3.60}$$

and the $F_\kappa$ and $A_\kappa$ change the value of $l$ by $\pm 1$. Therefore, the representations $(0, k_1 + 1)$ and $(0, -k_1 - 1)$ are equivalent.

We shall now derive the irreducible representations of $\mathscr{E}(E_3)$ from the irreducible representations $(k_0, c)$ of $\mathscr{E}(SO(3,1))$ by a limiting process, called Inönü–Wigner contraction. We define the operators

$$P_i = \lambda F_i, \qquad J_i = H_i. \tag{3.61}$$

If $H_i$ and $F_i$ fulfill the commutation relations (3.27, (3.28), (3.29$_{\lambda^2 = -1}$) of $\mathscr{E}(SO(3,1))$, then $P_i$ and $J_i$ fulfill the commutation relations

$$[J_i, J_k] = i\epsilon_{ikl}J_l, \tag{3.62}$$

$$[J_i, P_k] = i\epsilon_{ikl}P_l, \tag{3.63}$$

$$[P_i, P_k] = -\lambda^2\epsilon_{ikl}J_l, \tag{3.64}$$

and $P_i$ depends upon $\lambda$. The invariant operators are now

$$\lambda^2 C_1 = P_i P_i - \lambda^2 J_i J_i, \tag{3.65}$$

$$\lambda C_2 = P_i J_i. \tag{3.66}$$

In the limit $\lambda^2 \to 0$ the commutation relation (3.64) goes into the commutation relation

$$[P_i, P_k] = 0. \tag{3.67}$$

(3.62), (3.63), (3.67) are the commutation relations that define the algebra $\mathscr{E}(E_3)$. Thus in the limit $\lambda^2 \to 0$ the commutation relations of $\mathscr{E}(SO(3,1))$ go into those of $\mathscr{E}(E_3)$. However, if we simply take the limit $\lambda \to 0$, then we see from (3.61) that $P_i \to 0$. In order to obtain the algebra of operators of $\mathscr{E}(E_3)$ from the algebra of operators $\mathscr{E}(SO(3,1))$ in $\mathscr{R}(k_0, c)$ by the limit $\lambda \to 0$, we must simultaneously increase the $F_\kappa$ so that $P_\kappa$ does not go to zero. This can be done, as seen from (3.35), (3.44) and (3.49), by increasing the value of $|ic| \to \infty$ when $\lambda \to 0$. The contraction in the representation is therefore performed in the following way:

$$\lambda \to 0 \qquad |ic| \to \infty \qquad \text{such that } ic\lambda \to \epsilon \tag{3.68}$$

where $\epsilon$ is a finite real value. From (3.35), (3.44), and (3.49) we see that e.g. $P_3$ is then given by

$$P_3 f_m^l = \lim_{ic\lambda \to \epsilon} \lambda F_0 f_m^l$$

$$= \sqrt{l^2 - m^2}\, \tilde{c}_l f_m^{l-1} - m\tilde{a}_l f_m^l - \sqrt{(l+1)^2 - m^2}\, \tilde{c}_{l+1} f_m^{l-1}, \tag{3.69}$$

where

$$\tilde{c}_l = \lim \lambda c_l = \lim \frac{i}{l}\sqrt{\frac{(l^2 - k_0^2)(\lambda^2 l^2 - \lambda^2 c^2)}{4l^2 - 1}} = \epsilon\frac{i}{l}\sqrt{\frac{l^2 - k_0^2}{4l^2 - 1}} \tag{3.70}$$

and

$$\tilde{a}_l = \lim \lambda a_l = \lim \frac{\lambda i c k_0}{l(l+1)} = \frac{k_0\epsilon}{l(l+1)}. \tag{3.71}$$

The invariant operators of $\mathscr{E}(E_3)$ are $P_i P_i$ and $P_i J_i$. From (3.65) we see that in this limiting process

$$P_i P_i = \lim(\lambda^2 C_1 + \lambda^2 J_i J_i) = \lim(-k_0^2 \lambda^2 - \lambda^2 c^2 + \lambda^2 1) + \lambda^2 J_i J_i = \epsilon^2,$$

(3.72)

where we have used $(3.55_1)$ for the irreducible representation $(k_0, c)$. In the same way it follows from (3.66), using $(3.55_2)$, that

$$P_i J_i = \lim \lambda c_2 = \lim \lambda i c k_0 = \epsilon k_0.$$ (3.73)

The representation of $\mathscr{E}(E_3)$ obtained in this contraction process (3.68) is therefore characterized by the two parameters $(k_0, \epsilon)$ with $-\infty < \epsilon < +\infty$.

The $\mathscr{E}(SO(3))$ subalgebra and the reduction with respect to this subalgebra are not affected by the contraction process (3.68). Therefore the irreducible representation space $\mathscr{R}(k_0, \epsilon)$ of $\mathscr{E}(E_3)$ obtained from the representation space $\mathscr{R}(k_0, c)$ is again given by [cf. (3.50)]

$$\mathscr{R}(k_0, \epsilon) = \sum_{l=k_3, k_3+1, \ldots}^{\infty} \oplus \mathscr{R}^l.$$ (3.74)

Summarizing our results, we have found that the Hermitian irreducible-representation spaces of $\mathscr{E}(E_3)$, defined by (3.62), (3.63), (3.64), are characterized by two numbers $(k_0 \, \epsilon)$ which take the values $k_0 = 0, \frac{1}{2}, 1, \ldots, -\infty < \epsilon < +\infty$. The reduction of this space with respect to $\mathscr{E}(SO(3))$ is given by (3.74), and the operators $J_\kappa = H_\kappa$ and $P_\kappa$ are given by (3.36) and (3.35) with $F_\kappa$ replaced by $P_\kappa$ and $a_l, c_l$ replaced by $\tilde{a}_l$, and $\bar{c}_l$ given in (3.70) and (3.71).

The basis system $f_m^l$ of the representation space $\mathscr{R}(k_0, \epsilon)$ is the one in which the following complete system of commuting operators is diagonal:

$$J_3, J_i J_i; P_i P_i, P_i J_i.$$ (3.75)

As the $P_i$ commute [Equation (3.67)], one usually chooses a basis system of the representation space in which the $P_i$ are diagonal. As a complete system of commuting operators one chooses therefore

$$P_1, P_2, P_3, \quad P_i J_i$$ (3.76)

with the corresponding basis vectors $|p_i, k_0\rangle$, which have the property

$$\langle k_0' p_1' | p_i k_0 \rangle = \delta_{k_0 k_0'} \delta^3(\mathbf{p} - \mathbf{p}'),$$ (3.77)

$$P_i | p_i k_0 \rangle = p_i | p_i k_0 \rangle \qquad P_i J_i | p_i k_0 \rangle = k_0 \epsilon | p_i k_0 \rangle,$$ (3.78)

$$\epsilon^2 = p_i p_i.$$

Without a derivation,[11] we mention that the transformation matrices from the basis $f_m^l(k_0, \epsilon)$ to the basis $|p_i k_0\rangle$,

$$|p_i k_0\rangle = \sum_{l, m} f_m^l(k_0, \epsilon)\langle \epsilon k_0 \, lm | p_i k_0 \rangle,$$ (3.79)

[11] A derivation is given in Appendix I of A. Bohm, R. B. Teese, Spectrum Generating Group of the Symmetric Top Molecule, *J. Math. Phys.* 17, 94 (1976).

are given by

$$\langle \epsilon k_0\, lm | p_i k_0\rangle = \sqrt{\frac{2l+1}{4\pi}}\,\frac{1}{|\epsilon|}\,D^l_{mk_0}(\phi, \theta, -\phi). \qquad (3.80)$$

Here $(\phi, \theta)$ are the spherical coordinates of $\mathbf{p}$; $p_1 = |\epsilon| \sin \theta \cos \phi$, $p_2 = |\epsilon| \sin \theta \sin \phi$, $p_3 = |\epsilon| \cos \theta$; and $D^l_{mk_0}(\phi, \theta, -\phi)$ is the rotation matrix.[12]

In this appendix we have derived a huge class of irreducible representations of $\mathscr{E}(SO(4))$, $\mathscr{E}(SO(3,1))$, and $\mathscr{E}(E_3)$, many of which are used for applications in this book or for applications to other problems in quantum physics. We have used here the same method that was used in Section III.3 for the derivation of the representation of $\mathscr{E}(SO(3))$. Though group theory has not been introduced, we want to mention that these are all the unitary representations of the groups $SO(4)$, $SO(3,1)$ (the homogeneous Lorentz group), and $E_3$ (the Euclidean group in three dimensions).

## V.4   Parity

Parity is a very important observable in quantum physics, as all quantum-mechanical systems are usually in states that are eigenstates of parity. The *parity operation* $P$, which is also called *space inversion*, is the operation of taking the mirror image. If a physical object has the position $x_i$ and momentum $p_i$, then its mirror image has the position $x_i^R = -x_i$ and momentum $p_i^R = -p_i$. For a quantum physical system this transformation $P$ is, according to the basis assumption I, represented by a linear operator in the space of physical states. We call this operator the parity operator $U_P$.

Some properties of $U_P$ follow directly from the physical interpretation of the parity $P$; others are convention. As we shall justify below, $U_P$ is required to fulfill the following defining relations:

$$U_P Q_i U_P^{-1} = -Q_i \qquad (i = 1, 2, 3), \qquad (4.1)$$

$$U_P P_i U_P^{-1} = -P_i \qquad (i = 1, 2, 3), \qquad (4.2)$$

$$U_P J_i U_P^{-1} = +J_i \qquad (i = 1, 2, 3), \qquad (4.3)$$

where $Q_i$, $P_i$, and $J_i$ are the position, momentum, and angular-momentum operators, respectively. A further requirement on $U_P$ is

$$U_P U_P^\dagger = I \qquad (4.4)$$

and

$$U_P^2 = I. \qquad (4.5)$$

A consequence of (4.4) and (4.5) is

$$U_P^\dagger = U_P^{-1}, \qquad U_P^{-1} = U_P, \qquad U_P^\dagger = U_P. \qquad (4.6)$$

---

[12] The rotation matrices can be found in, e.g., L. C. Biedenharn and J. D. Louck [1979], Chapter 3.

Linear operators that fulfill (4.4) are called *unitary* operators;[13] thus $U_P$ is unitary and according to (4.6) also Hermitian.

Equation (4.3) follows from (4.1) and (4.2) if $J_i = L_i = \epsilon_{ijk} Q_j P_k$; for other angular momenta it is a postulate.

From (4.1), (4.2), and (4.3) we obtain the relation of $U_P$ with any observable that is a function of momentum, position, and angular momentum. When new independent observables are introduced, their relations with $U_P$ will have to be postulated.

The parity provides a classification of tensor operators. A tensor operator $T_\kappa^{(k)}$ for which

$$U_P T_\kappa^{(k)} U_P^{-1} = (-1)^k T_\kappa^{(k)} \tag{4.7}$$

is called a (*proper*) *tensor operator*; if

$$U_P T_\kappa^{(k)} U_P^{-1} = -(-1)^k T_\kappa^{(k)}, \tag{4.8}$$

then $T_\kappa^{(k)}$ is called a *pseudotensor operator*. Thus the operators $Q_i$ and $P_i$ are, according to (4.1) and (4.2), proper vector operators, whereas the operator $J_i$ is a pseudovector (also called an axial-vector) operator, according to (4.3).

All energy operators that we have met so far fulfill the condition

$$U_P H U_P^{-1} = H; \tag{4.9}$$

i.e., they are proper scalar operators. Equation (4.9) is, however, not universally fulfilled; the part of the energy operator that causes the weak decay is known not to fulfill (4.9) (parity nonconservation).

In the following we shall give some justification of the above postulates:

Let $\psi$ be a vector in the space of states $\mathcal{H}$, and denote by $\psi^R$ the vector we obtain from $\psi$ by applying $U_P$:

$$\psi^R = U_P \psi. \tag{4.10}$$

Let us consider the pure physical state of which $\psi$ is the representative vector, i.e., the state $\Lambda_\psi$, which is the projector on the one-dimensional space spanned by $\psi$. Let us denote by $\Lambda_\psi^R$ the projector on the space spanned by the $\psi^R$, and let us make the "physically reasonable" assumption that this is again a pure state (i.e., the $\psi^R$ span a one-dimensional space), $\psi^R$ is one of the vectors that represent the state obtained from $\Lambda_\psi$ by a space reflection, because $U_P$ is an operator that represents this space reflection. Thus $\Lambda_\psi^R$ is the space-reflected state of $\Lambda_\psi$, and these two states are connected by

$$\Lambda_\psi^R = |\psi^R\rangle\langle\psi^R| = U_P|\psi\rangle\langle\psi|U_P^\dagger = U_P \Lambda_\psi U_P^\dagger. \tag{4.11}$$

As a generalization of (4.11) we obtain [cf. (II.4.31)] that for any arbitrary state $W$,

$$W^R = U_P W U_P^\dagger \tag{4.12}$$

---

[13] See also Section XV.3 for the definition of a unitary operator.

will describe the state that is obtained by a space reflection of the state $W$. When $W$ is normalized ($\text{Tr } W = 1$), the space-reflected state should also be normalized ($\text{Tr } W^R = 1$), or more generally,

$$\text{Tr } W^R = \text{Tr } W, \tag{4.13}$$

i.e., space reflection should not change the probability. Thus

$$\text{Tr } W = \text{Tr } (U_P W U_P^\dagger) = \text{Tr}(W U_P^\dagger U_P),$$

which is fulfilled if one requires (4.4).

If one takes the mirror image of an object two times, then one gets the original object back. Therefore

$$W = (W^R)^R = U_P W^R U_P^\dagger = U_P U_P W U_P^\dagger U_P^\dagger. \tag{4.14}$$

This is fulfilled if one requires (4.5), which is sufficient to ensure (4.14) but not necessary for (4.14); i.e., (4.5) is also partially a convention.

Let $\bar{x}_i$ be the expectation value of the position operator of a quantum-mechanical system in the state $W$,

$$\bar{x}_i = \text{Tr}(Q_i W);$$

and let $\bar{x}_i^R$ be the expectation value in the state $W_R$ of the quantum-mechanical system obtained from the state $W$ by space reflection, $W^R = U_P W U_P^\dagger$:

$$\bar{x}_i^R = \text{Tr}(Q_i W^R).$$

Then, because of the physical meaning of space inversion,

$$\bar{x}_i^R = -\bar{x}_i; \tag{4.15}$$

i.e., space reflection changes the value of $\bar{x}_i$ into the value $-\bar{x}_i$. Thus we should have

$$\text{Tr}(Q_i W) = -\text{Tr}(Q_i W^R) = -\text{Tr}(Q_i U_P W U_P^\dagger)$$
$$= -\text{Tr}(U_P^\dagger Q_i U_P W). \tag{4.16}$$

This is fulfilled if

$$Q_i = U_P^\dagger Q_i U_P,$$

or, using (4.4), if

$$U_P Q_I U_P^{-1} = -Q_i,$$

which is the requirement (4.1). The other defining relations (4.2) and (4.3) can be justified by similar arguments.

We have therefore justified the defining relations (4.1)–(4.5) of the operator that represents the observable parity by showing that these relations lead to consequences that we expect of the physical operation of space reflection. Thus $U_P$ defined by (4.1)–(4.5) can represent the observable parity. We have not, however, shown that this $U_P$ is the only operator that can represent parity. (A few remarks concerning this question are given in Section XIX.2, in particular in the Appendix of that section.)

We shall now study the properties of the parity operator $U_P$ in angular-momentum eigenstates—for example, as they occur in the rotator space (III.4.2):

$$\mathscr{R} = \sum_{l=0}^{\infty} \oplus \mathscr{R}^l. \tag{4.17}$$

As the system of commuting observables

$$\mathbf{J}^2, J_3 \tag{4.18}$$

commutes with $U_P$, it follows that

$$\mathbf{J}^2, J_3, U_P \tag{4.19}$$

is also a system of commuting observables. We have to distinguish two cases: (a) that (4.18) is already a c.s.c.o.—as is the case for the rotator—and (b) that (4.19) is the c.s.c.o. This latter case is called *parity doubling*.

*Case (a)*: If (4.18) is the c.s.c.o., then $|jj_3\rangle$ is a basis system. Because of (4.3) it follows that

$$U_P|jj_3\rangle = \pi|jj_3\rangle, \tag{4.20}$$

$[\pi \neq 0$ because of (4.4)], i.e., $|jj_3\rangle$ are eigenvectors of $U_P$. The eigenvalue $\pi = \pi(j,j_3)$, which we call the parity of the state $|jj_3\rangle$, could in general depend upon $j, j_3$. We will now show that

$$\pi(j, j_3) = (-1)^j \eta, \tag{4.21}$$

where $\eta$, called the *intrinsic parity*, is independent of $j$ and $j_3$, and where $|\eta| = 1$.

From (III.3.22) and (III.3.23) it follows that[14]

$$U_P J_{\pm}|jj_3\rangle = U_P\sqrt{j(j+1) - j_3(j_3 \pm 1)}|jj_3 \pm 1\rangle$$
$$= \pi(j, j_3 \pm 1)\sqrt{j(j+1) - j_3(j_3 \pm 1)}|jj_3 \pm 1\rangle.$$

Because of (4.3) the left-hand side of this must be equal to

$$J_{\pm} U_P|jj_3\rangle = \pi(j, j_3)J_{\pm}|jj_3\rangle$$
$$= \pi(j, j_3)\sqrt{j(j+1) - j_3(j_3 \pm 1)}|jj_3 \pm 1\rangle.$$

Comparing the last two equations, we conclude that

$$\pi(j, j_3) = \pi(j, j_3 \pm 1), \tag{4.22}$$

---

[14] In this and succeeding sections we shall use units such that $\hbar \equiv 1$. We shall on occasion restore the $\hbar$'s to an equation, in which case we are using the usual (Gaussian) c.g.s. units.

i.e., that $\pi$ is independent of $j_3$. To find the dependence of $\pi$ upon $j$, we have to use (4.1) and the vector-operator property of $Q_i$. Writing

$$Q_{\kappa=\pm 1} = \frac{Q_1 \pm iQ_2}{\mp\sqrt{2}}, \qquad Q_{\kappa=0} = Q_3,$$

and using (3.12), we have

$$
\begin{aligned}
Q_\kappa |j\,j_3\rangle &= |j - 1\,j_3 + \kappa\rangle\langle j\,j_3 1\kappa|j - 1\,j_3 + \kappa\rangle\langle j - 1\,\|Q\|j\rangle \\
&+ |j\,j_3 + \kappa\rangle\langle j\,j_3 1\kappa|j\,j_3 + \kappa\rangle\langle j\|Q\|j\rangle \\
&+ |j + 1\,j_3 + \kappa\langle\rangle j\,j_3 1\kappa|j + 1\,j_3 + \kappa\rangle\langle j + 1\,\|Q\|j\rangle, \quad (4.23)
\end{aligned}
$$

where $\langle j'\|Q\|j\rangle$ are the reduced matrix elements of $Q$. If we apply $U_P$ to both sides of (4.23), then from (4.1) it follows that

$$
\begin{aligned}
Q_\kappa U_P|j\,j_3\rangle &= \pi(j)Q_\kappa|j\,j_3\rangle = -U_P Q_\kappa|j\,j_3\rangle \\
&= -|j - 1\,j_3 + \kappa\rangle\pi(j - 1)\langle j\,j_3 1\kappa|j - 1\,j_3 + \kappa\rangle\langle j - 1\,\|Q\|j\rangle \\
&- |j\,j_3 + \kappa\rangle\pi(j)\langle j\,j_3 1\,\kappa|j\,j_3 + \kappa\rangle\langle j\|Q\|j\rangle \\
&- |j + 1\,j_3 + \kappa\rangle\pi(j + 1)\langle j\,j_3 1\,\kappa|j + 1\,j_3 + \kappa\rangle\langle j + 1\,\|Q\|j\rangle.
\end{aligned}
$$
$$(4.24)$$

If we insert (4.23) into the second term of this equation and compare it with the last term we conclude

$$\langle j\|Q\|j\rangle = 0, \tag{4.25}$$

$$\pi(j - 1) = -\pi(j) \quad \text{if} \quad \langle j - 1\,\|Q\|j\rangle \neq 0, \tag{4.26}$$

$$\pi(j + 1) = -\pi(j) \quad \text{if} \quad \langle j + 1\,\|Q\|j\rangle \neq 0. \tag{4.27}$$

According to (3.17'),

$$\sqrt{2j + 3}\langle j + 1\|Q\|j\rangle = \sqrt{2j + 1}\langle j\|Q\|j + 1\rangle, \tag{4.28}$$

so if $\langle j\|Q\|j + 1\rangle = 0$ for every $j$, then this and (4.25) would mean that $Q_\kappa$ is the zero operator. Thus for some value(s) of $j$, $\langle j\|Q\|j + 1\rangle \neq 0$; for all such values we have by (4.26), (4.27)

$$\pi(j) = -\pi(j + 1). \tag{4.29}$$

From this it follows that

$$\pi(j) = (-1)^j \eta \qquad (\eta = \text{const}). \tag{4.21}$$

To determine what $\eta$ can be we use (4.5):

$$U_P U_P|j\,j_3\rangle = \pi(j)\pi(j)|j\,j_3\rangle = I|j\,j_3\rangle, \tag{4.30a}$$

and (4.4):

$$\langle j\,j_3|I|j\,j_3\rangle = \langle j\,j_3|U_P^\dagger U_P|j\,j_3\rangle = \bar\pi(j)\pi(j)\langle j\,j_3|j\,j_3\rangle. \tag{4.30b}$$

Thus

$$\pi(j)\pi(j) = 1, \qquad \bar\pi(j)\pi(j) = 1, \tag{4.31}$$

and consequently $\pi(j) = \bar{\pi}(j)$, so $\pi(j)$ is real and $\pi(j)^2 = 1$, i.e., $\pi(j)$ is $+1$ or is $-1$. Thus

$$\eta = +1 \quad \text{or} \quad \eta = -1. \tag{4.32}$$

We can now give a complete justification for the selection rule (III.4.10):

$$\Delta j = \pm 1. \tag{4.33}$$

It is a consequence of the fact that $Q_\kappa$ is a proper vector operator.

From the property

$$[J_k, Q_l] = i\epsilon_{klm} Q_m \quad (k, l, m = 1, 2, 3) \tag{4.34}$$

it follows that $\Delta j = \pm 1, 0$ [see Equation (3.7)]. From (4.25), i.e., the property (4.1) of the $Q_i$ together with the assumption that there is no parity doubling, it follows that $\Delta j \neq 0$, which establishes III(4.10).

*Case (b)*: If (4.19) is the c.s.c.o., then the parity $\pi$ is an additional label for the basis vectors:

$$U_P |j\, j_3 \pi\rangle = \pi |j\, j_3\, \pi\rangle. \tag{4.35}$$

The question now is what is the spectrum of $U_P$, i.e., what are the possible values of $\pi$. From (4.4) and (4.5) it follows in the same way as above that (4.31) must hold. Thus

$$\pi = +1 \quad \text{or} \quad \pi = -1. \tag{4.36}$$

Instead of (4.23) we now have

$$\begin{aligned}
Q_\kappa |j\, j_3 \pi\rangle &= |j - 1\, j_3 + \kappa, -\pi\rangle\langle j\, j_3\, 1\, \kappa | j - 1\, j_3 + \kappa\rangle\langle j - 1, -\pi\| Q \| j\, \pi\rangle \\
&+ |j\, j_3 + \kappa, -\pi\rangle\langle j\, j_3\, 1\, \kappa | j\, j_3 + \kappa\rangle\langle j, -\pi\| Q \| j\, \pi\rangle \\
&+ |j + 1\, j_3 + \kappa, -\pi\rangle\langle j\, j_3\, 1\, \kappa | j + 1\, j_3 + \kappa\rangle\langle j + 1, -\pi\| Q \| j\, \pi\rangle,
\end{aligned} \tag{4.37}$$

where in addition to the vector-operator property (4.34) of $Q_\kappa$ we have taken (4.1) into account, i.e., that $Q_\kappa$ changes the eigenvalue of $U_P$, and also (4.36). Thus unlike case (a), (4.1) with (4.36) does not lead to any conditions on the reduced matrix elements $\langle j'\, \pi' \| Q \| j\, \pi\rangle$ and parities. Consequently for every choice of $j, j_3$ there are two vectors

$$|j\, j_3\, \pi = +1\rangle \quad \text{and} \quad |j\, j_3\, \pi = -1\rangle. \tag{4.38}$$

That is the reason case (b) is called the parity-doubling case.

It is customary to obtain an analogy with case (a) by labeling the states not with $\pi$ but rather with the intrinsic parity $\eta$, which is defined by

$$U_P |j\, j_3\, \eta\rangle = \eta(-1)^j |j\, j_3\, \eta\rangle \quad (\eta = \pm 1). \tag{4.39}$$

The parity $\pi(j)$ in a state with angular momentum $j$ and intrinsic parity $\eta$ is then given by

$$\pi(j) = (-1)^j \eta. \tag{4.21}$$

The selection rule is now seen from (4.37) to be

$$\Delta j = \pm 1, 0, \tag{4.40}$$

$$\Delta \pi = -1. \tag{4.41}$$

Equation (4.41), which is a consequence of (4.1), was also fulfilled in case (a), only there it was already implicit in the selection rule (4.33) as a consequence of (4.21).

If parity is an independent observable (case (b)) then, according to (4.38), for every value of $j$ we have two angular momentum spaces

$$\mathscr{R}^{j^+} \quad \text{and} \quad \mathscr{R}^{j^-}. \tag{4.42}$$

As $Q_\kappa$ transforms from a given value of $j^\pi$ to $(j + 1)^\pi$, $j^\pi$, and $(j - 1)^\pi$, one obtains all angular-momentum spaces unless there is a $j_0$ such that

$$\langle j_0 - 1\, \eta \| Q \| j_0\, \eta \rangle = 0 \tag{4.43a}$$

or a $j_1$ such that

$$\langle j_1 + 1\, \eta \| Q \| j_1\, \eta \rangle = 0. \tag{4.43b}$$

That is, $j_0$ is the smallest value of $j$ that can be reached by repeated application of $Q_\kappa$, and $j_1$ is the largest. The space of all states may then be written

$$\mathscr{R} = \sum_{j=j_0}^{j_1} \oplus\, \mathscr{R}^{j}_{\eta=+1} \oplus \sum_{j=j_0}^{j_1} \oplus\, \mathscr{R}^{j}_{\eta=-1} = \mathscr{R}_{\eta=+1} \oplus \mathscr{R}_{\eta=-1}. \tag{4.44}$$

Are there physical systems in nature whose space of physical states is given by (4.44)?

When we considered the rotating diatomic molecule, we used as its classical picture the dumbbell, i.e., we assumed that the amount of inertia about the line joining the two atoms was zero. However, because there are a number of electrons revolving about the two nuclei, a better classical model in many cases is a dumbbell with a flywheel on its axis. Thus in many cases the diatomic molecule is not a simple quantum-mechanical rotator but a quantum-mechanical symmetric top; the total angular momentum $\mathbf{j}$ is no longer perpendicular to the direction of the figure axis $\mathbf{a} = \mathbf{x}/|\mathbf{x}|$, but instead has a constant component in the direction $\mathbf{a}$, i.e.,

$$\mathbf{j} \cdot \mathbf{a} = \text{const}, \tag{4.45}$$

due to the revolution of the electrons.

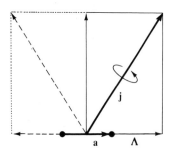

Figure 4.1   Vector diagram for the symmetric top. The curved arrow indicates the rotation of the whole diagram about $\mathbf{j}$. The dashed part of the figure gives the vector diagram when the sense of the direction of $\mathbf{a}$ is reversed.

A classical nonrelativistic symmetric top (Figure 4.1) is a physical system that has constant moments of inertia $I_A$, $I_B$, $I_B$ about its principal axes. It has angular velocity $\boldsymbol{\omega}$ and angular momentum $\mathbf{j}$, and the axis of symmetry lies along the direction of the unit vector $\mathbf{a} = \mathbf{x}/|\mathbf{x}|$. Its energy is given by

$$E = \tfrac{1}{2}\mathbf{j}\cdot\boldsymbol{\omega}. \qquad (4.46)$$

$\boldsymbol{\omega}$ may be expressed in terms of $\mathbf{j}$ and $\mathbf{a}$ by using

$$\mathbf{j} = I_B\boldsymbol{\omega} + \frac{I_A - I_B}{I_A}(\mathbf{x}\cdot\mathbf{j})\mathbf{x}\frac{1}{\mathbf{x}^2}. \qquad (4.47)$$

From Equations (4.46) and (4.47) one obtains the classical expression for the energy in terms of $\mathbf{j}$ and $\mathbf{a}$:

$$E = \frac{1}{2I_B}\left(\mathbf{j}^2 - \frac{I_A - I_B}{I_A}\frac{1}{\mathbf{x}^2}(\mathbf{x}\cdot\mathbf{j})^2\right). \qquad (4.48)$$

For the quantum-mechanical system, the numbers $j_i$ and $x_i$ are replaced by the operators $J_i$ and $Q_i$. The energy operator of the quantum-mechanical symmetrical top is therefore given by

$$H = \frac{1}{2I_B}\left(\mathbf{J}^2 - \frac{I_A - I_B}{I_A}\frac{1}{\mathbf{Q}^2}(\mathbf{Q}\cdot\mathbf{J})^2\right). \qquad (4.49)^{15}$$

The $Q_i$ and $J_i$ fulfill the commutation relations:

$$[L_i, L_j] = i\hbar\epsilon_{ijk}L_k \qquad [L_i, Q_j] = i\hbar\epsilon_{ijk}Q_k \qquad [Q_i, Q_j] = 0$$
$$\qquad (4.50)$$

It has been shown in Appendix to Section V.3 that as a consequence of these commutations relations, which are identical with (3.62) (3.63) and (3.67) the following equations (identical to (3.72) and (3.73)) are fulfilled:

$$\mathbf{Q}^2 = Q_iQ_i = \mathbf{x}^2 = \text{const},$$
$$\qquad (4.51)$$
$$\mathbf{Q}\cdot\mathbf{J} = Q_iJ_i = j_0 x,$$

where $x = \pm|\mathbf{x}|$ and $j_0$ is the smallest angular momentum in $\mathscr{R}$, defined by (4.43a). The matrix element of $Q_i$ are given by (4.37) with the reduced matrix elements (determined from (3.14) (3.17) (3.70) and (3.71)) given by:

$$\langle j\,\eta' \|Q\| j\,\eta\rangle = \frac{j_0 x}{[j(j+1)]^{1/2}}\delta_{\eta,-\eta'}, \qquad (4.52)$$

$$\langle j-1\,\eta' \|Q\| j\,\eta\rangle = -i\left[\frac{(j^2-j_0^2)x^2}{j(2j-1)}\right]^{1/2}\delta_{\eta\eta'}, \qquad (4.53)$$

$$\langle j+1\,\eta' \|Q\| j\,\eta\rangle = -i\left[\frac{((j+1)^2-j_0^2)x^2}{(j+1)(2j+3)}\right]^{1/2}\delta_{\eta\eta'}. \qquad (4.54)$$

---

[15] $1/\mathbf{Q}^2$ is the operator defined as the inverse of $\mathbf{Q}^2$: $1/\mathbf{Q}^2 = (\mathbf{Q}^2)^{-1}$; see $[\![M]\!]$ in Section VI.3.

The $\eta$-dependence follows from (4.1) and (4.3). These reduced matrix elements, together with (4.37) and the usual matrix elements of $J_\kappa$ [as given by (III.3.8′), (III.3.22), and (III.3.23)], constitute a representation in $\mathscr{R}$ of the enveloping algebra $\mathscr{E}(E_3)$ of the three-dimensional Euclidean group extended by parity. When $j_0 = 0$, then $\langle j\|Q\|j\rangle = 0$ and the space is given by (4.17). Equation (4.54) shows that $j$ can become arbitrarily large, i.e., $j_1 = \infty$.

The energy spectrum of the symmetrical top, i.e., the matrix element of the energy operator $H$ in the basis $|j j_3\, \eta\rangle$ of $\mathscr{R}$, is obtained from (4.49) using (4.51):

$$\text{spectrum } H = \frac{1}{2I_B}\left(j(j+1) - \frac{I_A - I_B}{I_A}\Lambda^2\right) \tag{4.55}$$

$$(j = \Lambda, \Lambda + 1, \Lambda + 2, \ldots), \tag{4.55a}$$

where $\Lambda = j_0$ is the conventional notation used in molecular spectroscopy. The angular momentum spectrum (4.55a) follows from (4.44) or (3.74). It is also intuitively (classically) clear, as the value of total angular momentum $j$ must always be larger than its component $\Lambda$ in the direction of $\mathbf{Q}$. The component along the direction $\mathbf{Q}$ (axis of the molecule) can take the two values $\pm\Lambda = j_0 x/|x|$ (the $-$ sign corresponding to the dashed line in Figure 4.1). As

$$Q_i J_i U_P = -U_P Q_i J_i, \tag{4.56}$$

parity states are not eigenstates of the operator $Q_i J_i$ that represents this component. Thus in parity eigenstates the molecule does not have definite values for the component of angular momentum along the molecular axis.

From (4.55) it follows that the energy levels of the symmetric top, in which the component of angular momentum along the top axis is fixed, are the same as those of the simple rotator except that there is a shift of magnitude proportional to $\Lambda^2$, and except that levels with $j < \Lambda$ are absent. Each energy level has, in addition to the usual $(2j + 1)$-fold degeneracy, a twofold degeneracy because of the two possible values of $\eta = \pm 1$. The energy diagram of this kind of symmetric top for the case $\Lambda = 1$ is shown in Figure 4.2, with that of the rotator for comparison. The $+$ or $-$ sign gives the values of the parity $\pi$. The levels with a given value of $j$ and $\pi = \pm$, which according to (4.55) should be degenerate, are drawn slightly separated. The selection rule $\Delta j = \pm 1, 0$ has already been derived. Thus for these kinds of molecules, in addition to the $R$ and $P$ branches given by (III.5.19), there also appear in the vibration-rotation bands transitions with $\Delta j = 0$, i.e., a series of lines given by

$$v_Q = \gamma_0 + (B_{n'} - B_{n''})j + (B_{n'} - B_{n''})j^2 \tag{4.57}$$

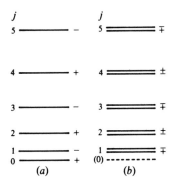

Figure 4.2   Energy diagram (a) of the rotator and (b) of the symmetric top. For the symmetric top $\Lambda = 1$ is assumed. The dashed level with $J = 0$ therefore does not occur. [From G. Herzberg Vol. 1 (1966) with permission.]

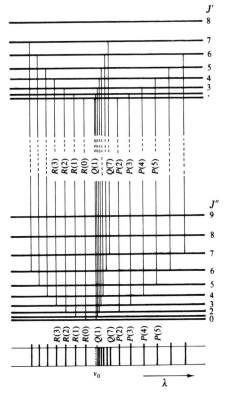

Figure 4.3   Energy level diagram for a band with $P$, $Q$, and $R$ branches. For the sake of clarity, in the spectrogram below, the lines of the $P$ and $R$ branches which form a single series are represented by longer lines than those of the $Q$ branch. The separation of the lines in the $Q$ branch has been made somewhat too large in order that the lines might be drawn separately. The convergence in the $P$ and $R$ branches is frequently much more rapid than shown. [From Herzberg (1966), with permission.]

(Here and also in the formula (III.5.19) for the $R$ and $P$ branches, $v_0$ is not the oscillator's wave number but

$$v_0^{osc} - [(I_A - I_B)/I_A](B_{n'}\Lambda_{(n')}^2 - B_{n''}\Lambda_{(n'')}^2).$$

But as $\Lambda_{(n')}$ and $\Lambda_{(n'')}$ are constant, this last term can be absorbed in a suitably redefined $\Lambda_0$.) This series of lines is called the $Q$-branch. The energy-level diagram and the transitions for the case that the upper rotational band has the value $\Lambda = 1$ and the lower band the value $\Lambda = 0$ are depicted in Figure 4.3.

$\Delta j = 0$ transitions are not observed in the infrared spectrum of diatomic molecules. Therefore $\Lambda = 0$ for the electronic ground state of the diatomic molecules, i.e., the motion of the electrons in the electronic ground states is such that no angular momentum about the interatomic axis results, so the diatomic molecule in the ground state is a dumbbell. However, all three branches have been observed in transitions between different electronic levels of diatomic molecules. Therefore in excited electronic states $\Lambda$ is in general different from zero but has a constant value as long as the electronic state does not change. Thus in an excited electronic state, the diatomic molecule is a dumbbell with a flywheel rotating around its axis. Figure 4.4a shows an example of the energy levels and transitions between an electronic

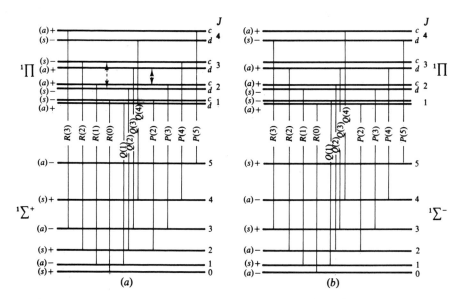

Figure 4.4   Energy-level diagram for the first lines of (a) a $^1\Pi - {}^1\Sigma^+$ transition and (b) a $^1\Pi - {}^1\Sigma$ transition. For the sake of clarity, the $\Lambda$-type doubling in the $^1\Pi$ state has been greatly exaggerated. The broken arrow to the left in (a) gives $R(2)-Q(2)$, and the one to the right gives $Q(3)-P(3)$. Their difference gives the sum of the $\Lambda$ doublings for $J = 2$ and $J = 3$ in the upper state. [From Herzberg (1966), with permission]

state with the electronic angular-momentum component $\Lambda = 0$ and intrinsic parity $\eta = +1\ (^1\Sigma^+)$, and an electronic state with $\Lambda = 1\ (^1\Pi)$. Figure 4.4b shows the same for an electronic state with $\Lambda = 0$ and $\eta = -1\ (^1\Sigma^-)$ and the $^1\Pi$ state. The splitting between the two "degenerate" levels of opposite parity, which was indicated in Figure 4.2, and which should not be there according to (4.55), does in fact occur. It is a consequence of the interaction between the electronic motion and the rotation of the molecule ($\Lambda$-type doubling), which was not taken into account in the derivation of (4.55).

The transition frequencies given by (4.57) are in the visible or ultraviolet region, as $v_0$ is in this region for transitions between different electronic states.

## Problems

1.  The wave numbers of the spectral lines corresponding to the transitions between two electronic states are given by

$$v = v_{n'l'}^{(e')} - v_{n''l''}^{(e'')}.$$

Here the $v_{nl}^{(e)}$ are the term values of the vibrating rotator in the electronic state $(e)$ and are given by

$$v_{nl}^{(e)} = T_e + \omega_e(n + \tfrac{1}{2}) - \omega_e x_e(n + \tfrac{1}{2})^2 + \omega_e y_e(n + \tfrac{1}{2})^3 + B_n l(l + 1)$$

where $T_e$ is a constant for a given electronic transition and the remainder follows as for Equation (III.5.33) the only difference being that the above formula includes a third order term in the vibration and no second order term for the rotation. The vibrational and rotational constants are different for different electronic states $(e)$. In general the selection rule $\Delta n = 1$ does not hold for vibrational transitions between electronic states.

Figure PS.1 is the spectrum of the emitted radiation obtained by exciting $N_2$ with a beam of $Ne^+$. The spectrum contains transitions due to both $N_2$ and $N_2^+$. The peaks labeled A and B represent transitions due to $N_2^+$. Neglecting the rotational structure, determine which electronic and vibrational transitions peaks A and B correspond to by using the table of data on $N_2^+$ (Table PS.1).

(a)  Locate the $R$ and $P$ branches and the null line.
(b)  What does the large peak in the rotational spectrum at about 4280 Å represent?
(c)  Why does this rotational spectrum look significantly different from the CO spectrum in Figure III.1.5?
(d)  What are the two large peaks at about 4239 Å and 4242 Å?
(e)  Calculate the spectrum for this rotation band using Table PS.1, and determine which peaks in the rotation spectrum correspond to which $l$-values.
(f)  Compare your calculated spectrum to the spectrum measured from Figure PS.2. If there are any differences, explain why.
(g)  Why do the peaks in the rotation spectrum alternate in intensity in a 2:1 ratio?

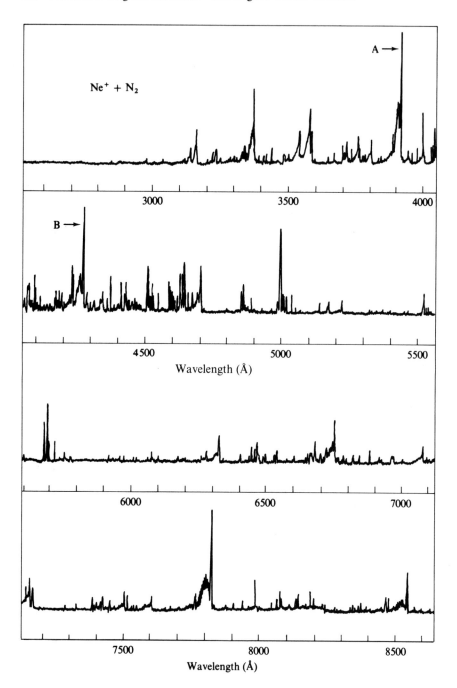

Figure PS.1   Spectrum of $N_2$ and $N_2^+$ obtained in $N_2$–$Ne^+$ collisions, from J. R. White, dissertation, University of Texas at Austin (1971). with permission. Figure PS.2 is an expansion of the spectrum at B, which clearly shows the rotational band of $N_2^+$.

**Table PS.1** From Herzberg [1966] vol. 1, p. 554. All numbers in the table are in cm$^{-1}$ units (except for $r_e$).

| State | $T_e$ | $\omega_e$ | $\omega_e x_e$ | $\omega_e y_e$ | $B_e$ | $\alpha_e$ | $r_e$ $(10^{-8}$ cm) | Observed transitions | | |
| --- | --- | --- | --- | --- | --- | --- | --- | Designation | $\nu_{00}$ | |
| $C\,^2\Sigma^+$ | (64622) | (2050) | | | 1.65 | [0.05] | 1.21 | $C \rightarrow X$ | 64547 | R |
| $B\,^2\Sigma_u^+$ | 25461.5 | 2419.84 Z | 23.19$_0$ | $-0.5375$ | 2.083 | 0.0195 | 1.075 | $B$   $X$ | 25566.0 | $V_R$ |
| $X\,^2\Sigma_g^+$ | 0 | 2207.19 Z | 16.136 | $-0.0400$ | 1.932$_2$ | 0.020$_2$ | 1.116$_2$ | | | |

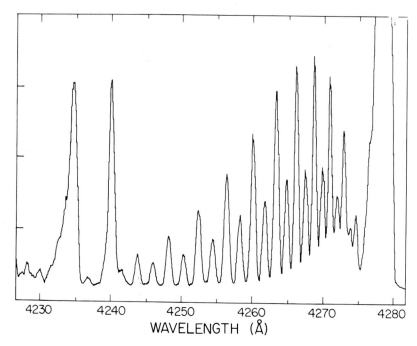

Figure PS.2   Expansion of spectrum of $N_2^+$ at point B of Figure PS.1, from J. R. White (1971) with permission.

2.   Determine the energy levels of the (isotropic) three-dimensional harmonic oscillator and the degree of degeneracy of each energy level. Determine the possible values of the orbital angular momentum in the $n$th energy level. To do this, proceed as follows:

(a)   Use the basic assumption IV to obtain the space of physical states and the product basis from the combination of three one-dimensional oscillators. In this product basis the energy operator $H = P^2/(2m) + (m\omega^2/2)Q^2$ is diagonal. Use this fact to obtain the spectrum of $H$ and the degeneracy of its eigenvalues.

(b)   Introduce the angular momentum operator $L_i = \epsilon_{ijk}Q_jP_k$ and the c.s.c.o. $H$, $\mathbf{L}^2$, and $L_3$. This c.s.c.o. is not diagonal in the product basis obtained in (a). Introduce a basis of eigenvectors of the c.s.c.o. $H$, $\mathbf{L}^2$, and $L_3$ (the angular momentum basis), and determine the transition coefficients between the product basis and the angular-momentum basis. Determine the possible values of $l$ for a given energy level $E_n$.

# CHAPTER VI

# Hydrogen Atom— The Quantum-Mechanical Kepler Problem

The classical Kepler problem is described in Section VI.2, where the Lenz vector is introduced. Section VI.3 gives the algebra of angular momentum and the Lenz vector, and Section VI.4 describes the representation of this algebra. In Section VI.5 we present the algebraic derivation of the hydrogen spectrum and discuss, at the end, fine-structure effects.

## VI.1 Introduction

An atom is described in everyday language as a nucleus with positive charge $+Ze$ and $Z$ electrons, each with negative charge $-e$, that move around the nucleus in fixed, stable orbits. The simplest atom is the hydrogen atom, which —in this language—consists of the proton as the nucleus and one electron moving around it in the field of the Coulomb electrostatic force of the proton. Thus the hydrogen atom is considered to be the realization of the quantum-mechanical analog of the classical Kepler problem.

The quantum-mechanical Kepler problem was first solved by Pauli (in 1925) using the Heisenberg commutation relations. Later it became customary to treat the hydrogen atom as a solution of the Schrödinger equation for the $1/r$ potential, which was introduced in 1926. Our presentation is based on Pauli's and Bargmann's treatment.[1]

---

[1] W. Pauli: Z. Phys. 36:336, 1926; V. Bargmann: Z. Phys. 99:576, 1936.

To conjecture the algebra of observables for this problem we shall follow our general procedure for quantum-mechanical systems that have classical analogs and obtain the relations between the quantum-mechanical observables by correspondence from the relations between the classical observables. We therefore recall some facts about the classical Kepler problem.

## VI.2  Classical Kepler Problem

Let $x_i$ be the coordinates of the electron (with charge $-e$), and let $p_i$ be its momentum components. Let the proton (with charge $+e$) be located at the origin of the coordinate system. Then the Coulomb force between electron and proton is given by

$$\mathbf{F} = -\frac{e^2}{r^2}\mathbf{n},$$

where

$$\mathbf{n} = \mathbf{x}/r \quad \text{and} \quad r = |\mathbf{x}| = \sqrt{\sum_i x_i x_i},$$

and the potential energy (Coulomb potential) is given by

$$V(r) = -\frac{e^2}{r}.$$

[Here the charge $e$, the system constant, is measured in electrostatic units: $e = 4.803 \times 10^{-10}$ esu; 1 esu $= 1$ cm dyn$^{1/2} = 1$ g$^{1/2}$ cm$^{3/2}$ sec$^{-1}$. If $e$ is measured in the practical unit of charge, $e = 1.602 \times 10^{-19}$ coulomb, then $F$ is given by $|\mathbf{F}| = (\frac{1}{4\pi\epsilon_0})(e^2/r^2)$ with $\frac{1}{4\pi\epsilon_0} = 8.99 \times 10^9$ kg m$^3$ sec$^{-2}$ C$^{-2}$]. Therefore the energy of the electron is given by

$$E = \frac{p^2}{2m} - \frac{e^2}{r}, \tag{2.1}$$

where $m$ is the mass of the electron. [The electron mass $m_e = 9.11 \times 10^{-31}$ kg is much smaller than the proton mass $m_p = 1836 m_e$; if one wants to take into account that $m_p$ is not infinite, then $m$ in (2.1) is given by the reduced mass

$$m = \frac{m_e m_p}{m_e + m_p},$$

and $\mathbf{p}$ is not the momentum of the electron but rather

$$\mathbf{p} = \frac{-m_e \mathbf{P}_p + m_p \mathbf{P}_e}{m_e + m_p}.$$

—the reduced momentum.]

Besides the energy $E$, there are two other constants of the motion: the angular momentum

$$\mathbf{l} = \mathbf{x} \times \mathbf{p}, \tag{2.2}$$

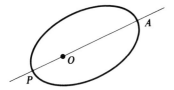

Figure 2.1

and the *Lenz vector* [or *Runge–Lenz vector*; cf. Barger and Olsson (1973)]

$$\tilde{\mathbf{a}} \equiv -\frac{\mathbf{p} \times \mathbf{l}}{m} + \frac{e^2}{r}\mathbf{x}. \tag{2.3}$$

The closed orbits of the classical Kepler problem are ellipses (Figure 2.1). $\mathbf{l}$ is an axial vector that is perpendicular to the plane of the orbit. As $\mathbf{l}$ is a constant of the motion, the orbit lies in some plane through $O$. For the Kepler problem, $V = e^2/r$, not only is the plane of the orbit fixed, but the orbit itself is also fixed, i.e., the major axis $PA$ is fixed and the ellipse does not precess. The constant of motion that is used to characterize the orientation of the major axis in the orbital plane is the Lenz vector $\tilde{\mathbf{a}}$, which is directed along the major axis from $O$ to $A$.

The constant of motion $\tilde{\mathbf{a}}$ is a special feature of the $1/r$ potential. Whereas $\mathbf{l}$ is a constant of the motion for every spherically symmetric system [i.e., for $V(\mathbf{x}) = V(|\mathbf{x}|)$], $\tilde{\mathbf{a}}$ is a constant of the motion only for $V(\mathbf{x}) \propto 1/r$; a small deviation of the potential from this form causes the axis $PA$ of the ellipse to precess slowly in the plane perpendicular to $\mathbf{l}$, so that the orbit is note closed.

For later comparison with the quantum-mechanical case; we list here some relations between the classical constants of the motion.[2] Let a particular orbit have the semimajor axis with length $a$, the semiminor axis with length $b$, and eccentricity $\epsilon = (a^2 - b^2)^{1/2}/2$. Then

$$E = -\frac{e^2}{2a} \qquad \mathbf{l}^2 = me^2 a(1 - \epsilon^2),$$

$$\tilde{\mathbf{a}}^2 = \frac{2E}{m}\mathbf{l}^2 + e^4 = (e^2\epsilon)^2. \tag{2.4}$$

Moreover,

$$\mathbf{l} \cdot \tilde{\mathbf{a}} = \mathbf{l} \cdot \mathbf{a} = 0,$$

where

$$\mathbf{a} \equiv \tilde{\mathbf{a}}/\sqrt{-2E}.$$

---

[2] See, e.g., Barger and Olsson (1973), Chapters 4–3.

## VI.3  Quantum-Mechanical Kepler Problem

The quantum-mechanical observables of the quantum-mechanical Kepler problem for the hydrogen atom are obtained by replacing the numbers $p_i$, $x_i$ in (2.1), (2.2), (2.3) with the Hermitian operators $P_i$ and $Q_i$ that fulfill the Heisenberg commutation relations (III.2.1).

For the angular momentum this has been done before; the angular momentum operators are given by

$$L_i = \epsilon_{ijk} Q_j P_k \qquad (i = 1, 2, 3). \tag{3.1}$$

The energy operator for the hydrogen atom is conjectured from (2.1) to be

$$H = \frac{P_2}{2m} - e^2 Q^{-1} \qquad Q = \sqrt{Q_1^2 + Q_2^2 + Q_3^2}, \tag{3.2}$$

where $Q^{-1}$ is the inverse of the operator $Q$.

> ⟦The mathematical meaning of $Q^{-1}$ requires some explanation. As $Q_i$ are Hermitian, $A = \sum_i Q_i^2$ is a positive-definite operator. An operator $A$ is *positive definite* if and only if $(f, Af) \geq 0$ for all $f \in \mathcal{H}$. In particular, $A$ has the property that its spectrum $\{a\}$ (all its eigenvalues) is nonnegative. For every Hermitian positive-definite operator $A$, one uniquely defines a positive-definite Hermitian operator $B$ such that $B^2 = A$. $B$ is written as $B = \sqrt{A}$. It has the same system of eigenvectors as $A$, commutes with every operator that commutes with $A$, and has the eigenvalues $\{\sqrt{a}\}$. Often this definition of the operator $\sqrt{A}$ is inadequate, and one must allow for eigenvalues $\{-\sqrt{a}\}$.
>
> The inverse operator $B^{-1}$ of an operator $B$ is the operator that (if it exists) has the property that $B^{-1}B = BB^{-1} = I$. If $B$ is a Hermitian operator with spectrum $\{b\}$, then the spectrum of $B^{-1}$ is $\{1/b\}$, and $B^{-1}$ has the same system of eigenvectors as $B$.⟧

To conjecture the quantum-mechanical Lenz vector from (2.3), we have to agree first what the quantum-mechanical correspondence of the numbers $(\mathbf{p} \times \mathbf{l})_i = \epsilon_{ijk} p_j l_k$ should be. As the operators $P_j$ and $L_k$ do not commute, the operator $\epsilon_{ijk} P_j L_k$ is not Hermitian, but instead

$$(\epsilon_{ijk} P_j L_k)^\dagger = \epsilon_{ijk} L_k P_j = \epsilon_{ijk} P_j L_k - 2i\hbar P_i.$$

As we want the quantum-mechanical observables $\tilde{A}_i$ that correspond to the components of the Lenz vector to be Hermitian operators, we replace the classical quantity by the symmetrized product,

$$\epsilon_{ijk} p_j l_k \rightarrow \epsilon_{ijk} \tfrac{1}{2}\{P_j, L_k\},$$

where the *anticommutator* $\{A, B\}$ of two operators $A$ and $B$ is defined by

$$\{A, B\} \equiv AB + BA.$$

Thus the quantum-mechanical Lenz vector is

$$\tilde{A}_i \equiv \frac{1}{2m} \epsilon_{ikl}\{P_l, L_k\} + \frac{e^2 Q_i}{Q}. \tag{3.3}$$

For the sake of convenience we shall now introduce some new units. So far the momentum $P$ has been measured in eV/$c$, where $c$ is the velocity of light, and the position $Q$ has been measured in cm. The canonical commutation relations read

$$[P_i, Q_j] = \frac{\hbar}{i} I\delta_{ij},$$

where $\hbar = 6.58 \times 10^{-18}$ eV sec $= 1.97 \times 10^{-5}$ eV cm/$c$ with the velocity of light $c = 2.998 \times 10^{10}$ cm/sec. The Planck constant $2\pi\hbar$ was the conversion factor between the measurement of energy $E$ in units of eV and the measurement of energy $v$ in units of sec$^{-1}$ (frequency units): $E = (2\pi\hbar)v = hv$ (Planck–Einstein relation); or it is the conversion factor between the momentum in units of inverse length (cm$^{-1}$)—i.e., as the reciprocal of the de Broglie wavelength, $1/\lambda$—and the momentum in standard units (erg sec/cm): $p = 2\pi\hbar/\lambda$ (the de Broglie relation). After we have convinced ourselves that frequency and energy or inverse wavelength and momentum are the same physical quantities measured in different units, it is no longer necessary to carry the conversion factor $\hbar$ between these different units.[3] Therefore we put $\hbar \equiv 1$ in all our calculations, which means that we replace the position $Q$ measured in cm with the position $Q/\hbar$ measured in (eV/$c$)$^{-1}$. We call $Q/\hbar$ simply $Q$. Thus, the canonical commutation relations now read

$$[P_i, Q_j] = -iI\delta_{ij}. \tag{3.4}$$

For the particular case of the electron in a hydrogen atom we simplify the notation further by using $m_e H$ as the energy operator, i.e., instead of $\mathbf{P}^2/2m - e^2/Q$ we use the energy operator measured in units of eV $\cdot m_e$, which we simply call $H$:

$$H = \frac{\mathbf{P}^2}{2} - \frac{a}{Q}, \quad \text{where} \quad a = \frac{m_e e^2}{\hbar} = m_e c\alpha. \tag{3.5}$$

The quantity

$$\alpha \equiv \frac{e^2}{\hbar c}$$

is the *Sommerfeld fine-structure constant*. In these units the Lenz vector is given by

$$\tilde{A}_i = \tfrac{1}{2}\epsilon_{ikl}\{P_l, L_k\} + aQ_i Q^{-1}. \tag{3.6}$$

---

[3] That energy and momentum are also the same physical quantities measured in different units follows from relativity.

The commutation relation of the angular-momentum operators is given in these units by

$$[L_i, L_j] = i\epsilon_{ijk} L_k, \tag{3.7}$$

which is immediately seen from (III.2.13).

Now that we have conjectured the quantum-mechanical observables and thus the algebra of operators for the quantum-mechanical hydrogen-atom we want to derive the properties of this algebra and compare the results with the experimental data. In particular, we would like to know the spectrum of $H$, which will give us the energy levels of the hydrogen atom. It is clear that if we knew all algebras of operators in linear spaces generated by $P_i$, $Q_j$ that fulfill (3.4), then the algebra of observables and space of states for the hydrogen atom would be one of them. For the hydrogen atom we need only one representation, i.e., one particular algebra of operators in one particular space of states. Instead of giving the precise mathematical speci-fications for this representation, we shall just go ahead and derive it. We remark that the space of physical states (subspace of the Hilbert space) for the hydrogen atom is different from the one for the three-dimensional oscillator, though the basic commutation relations (3.4) are the same.

As in the case of the oscillator, the momentum $P_i$ and the position $Q_i$ of the electron are not physical observables, in the sense that the physical system (hydrogen atom) cannot be prepared in approximate eigenstates of $P_i$ or $Q_i$; such a preparation would break the hydrogen atom apart. The hydrogen atom, like every nonrelativistic quantum-mechanical system, appears in eigenstates (pure or mixtures) of the energy operator. The energy operator (3.2), (3.5) of the hydrogen atom differs from that of the oscillator, which is of the form $H_{\text{osc}} = \mathbf{P}^2 + \mathbf{Q}^2$ [cf. (II.2.1b)]. The space of physical states is the space spanned by the physically preparable states. For the oscillator these are eigenstates of $H_{\text{osc}}$, and for the hydrogen atom they are eigenstates of $H$ in (3.5). Since the energy operators differ, the spaces of physical states differ.

In the previously considered cases of the oscillator and rotator, new observables were introduced, as functions of the $P_i$ and $Q_j$, which were more directly related to the physically preparable states (e.g., the $L_i$ for the rotator). After the relations between these new observables had been derived as consequences of (3.4), we could forget about the origin of these relations and consider them as new fundamental (defining) relations for the particular physical system under consideration. We shall follow the same procedure here for the hydrogen atom.

In order to decide what to choose for these fundamental observables in the case of the hydrogen atom, we recall that a state vector can be an eigen-vector of two different operators only if these two operators commute. Physical states for stationary nonrelativistic quantum-mechanical systems appear to be in general eigenstates (or mixtures thereof) of the energy operator. Therefore we choose operators that commute with $H$ as these new observables; these are the quantum-mechanical constants of the motion, as we shall discuss later when we consider time development (Chapter XII).

As $l_i$ and $\tilde{a}_i$ are the classical constants of the motion, we expect that the angular-momentum operator $L_i$ and the Lenz operator $\tilde{A}_i$ will be constants of the motion, i.e., commute with $H$, and are therefore to be chosen as our new fundamental observables for the hydrogen atom. A straightforward calculation (Problem VI.1) shows that as a consequence of the Heisenberg commutation relations (3.4) it follows that indeed

$$[H, L_i] = 0 \quad \text{and} \quad [H, \tilde{A}_i] = 0. \tag{3.8}$$

As $H$ commutes with $L_i$ and $\tilde{A}_i$, it will commute with the whole algebra generated by $L_i$ and $\tilde{A}_i$. Consequently there is no operator in this algebra that transforms from a given eigenvector of $H$ with eigenvalue $E$ to a vector that is not an eigenvector of $H$ with the same eigenvalue $E$. Let us denote the space of eigenvectors of $H$ with eigenvalue $E$ by $\mathscr{R}(E)$. Then all $L_i$, $\tilde{A}_i$ and all (well-defined) functions $A = A(\mathbf{L}, \tilde{\mathbf{A}})$ of them transform a vector $f \in \mathscr{R}(E)$ into a vector $Af = g$ that is again in $\mathscr{R}(E)$ with the same eigenvalue $E$. In other words, the algebra of $L_i$, $\tilde{A}_i$ leaves the eigenspaces $\mathscr{R}(E)$ *invariant*. Let us therefore first consider this algebra and investigate the structure of the eigenspaces $\mathscr{R}(E)$ of $H$.

To identify this algebra better (and to see that it is a simple, well-known mathematical structure, we define a new set of operators

$$A_i = (\sqrt{-2H})^{-1}\tilde{A}_i = \tilde{A}_i(\sqrt{-2H})^{-1}. \tag{3.9}$$

We have to ask ourselves whether the definition (3.9) makes sense. On the space $\mathscr{R}(E)$ there is no problem, because on $\mathscr{R}(E)$ $\sqrt{-2H} = \sqrt{-2E}$, i.e., $\sqrt{-2H}$ is a number, so on $\mathscr{R}(E)$ $A_i$ and $\tilde{A}_i$ differ just by a constant factor. Thus the restriction of $A_i$ to $\mathscr{R}(E)$, which we again call $A_i$, is well defined. $A_i$ is also well defined wherever $\sqrt{-2H}$ is well defined. This is the case for the space of vectors $f$ for which $(f, (-2H)f) > 0$, according to the definition of the square root and inverse of an operator. This space is called the negative-energy space, or space of bound states. We restrict ourselves to this space. [For the space where $+2H$ is positive definite, one defines $A_i$ by $A_i' = \sqrt{2H}\,\tilde{A}_i$ and find that these $A_i'$ fulfill commutation relations that differ from those of $A_i$ by a factor of $(-1)$, i.e., $A_i$ and $iA_i'$ fulfill the same commutation relations; see below.]

We denote the negative-energy space by $\mathscr{R}$. On $\mathscr{R}$, $A_i$ given by (3.9) is well defined. From (3.8) it follows immediately that

$$[H, A_i] = 0. \tag{3.10}$$

By a straightforward but lengthy calculation (Problem VI.1) one can show that as a consequence of the Heisenberg commutation relation (3.4) the $A_i$ fulfill the following commutation relations with $L_i$ and with each other:

$$[L_i, A_j] = i\epsilon_{ijk}A_k, \tag{3.11}$$

$$[A_i, A_j] = i\epsilon_{ijk}L_k \qquad A_i^\dagger = A_i. \tag{3.12}$$

With the $L_i$ and $A_i$ one defines the operators

$$C_1 \equiv \mathbf{A}^2 + \mathbf{L}^2 = A_1^2 + A_2^2 + A_3^2 + L_1^2 + L_2^2 + L_3^2, \qquad (3.13)$$

$$C_2 \equiv \mathbf{A} \cdot \mathbf{L} = \mathbf{L} \cdot \mathbf{A} = A_1 L_1 + A_2 L_2 + A_3 L_3. \qquad (3.14)$$

These operators have the property that they commute with the $A_i$ and $L_j$ (and consequently with the algebra generated by $A_i$, $L_j$) if $A_i$ and $L_j$ fulfill the commutation relations (3.7), (3.11), (3.12):

$$[C_1, L_i] = 0, \qquad [C_1, A_i] = 0, \qquad (3.15)$$

$$[C_2, L_i] = 0, \qquad [C_2, A_i] = 0. \qquad (3.16)$$

The algebra generated by the $L_i$ and $A_i$ that fullfill the commutation relations (3.7), (3.11), (3.12) is called $\mathscr{E}(\mathrm{SO}(4))$.

⟦It is the enveloping algebra of the four-dimensional rotation group SO(4). The algebra generated by $L_i$ and $A_i' = \sqrt{2H}\tilde{A}_i$, where $2H$ is positive definite, is the enveloping algebra $\mathscr{E}(\mathrm{SO}(3, 1))$ of the group SO(3, 1). cf Appendix to Section V.3.⟧

The operators $C_1$ and $C_2$ are the *invariant* or *Casimir operators* of $\mathscr{E}(\mathrm{SO}(4))$. In an irreducible representation space of $\mathscr{E}(\mathrm{SO}(4))$, i.e., a space that is obtained by applying every element of $\mathscr{E}(\mathrm{SO}(4))$ to one vector (ladder representation), the operators $C_1$ and $C_2$ are multiples of the identity operator. That is, they each have only one eigenvalue, $c_1$ and $c_2$ respectively, and these values characterize the representation space [in the same way as the value $j(j + 1)$ characterizes the representation space of $\mathscr{E}(\mathrm{SO}(3))$]. For the particular case where $L_i$ and $A_i$ are defined by (3.1), (3.6), (3.9), one can calculate (Problem VI.1) that as a consequence of (3.4),

$$C_1 = a^2(-2H)^{-1} - I, \qquad (3.17)$$

$$C_2 = 0; \qquad (3.18)$$

i.e., the operator $C_1$ is related to the energy operator (3.5), and $C_2$ is the zero operator. The relations (3.17) and (3.18) are the quantum-mechanical analogue of the classical relations (2.4) between $a_i$ and $l_i$. Thus the space $\mathscr{R}(E)$ is a particular representation space of $\mathscr{E}(\mathrm{SO}(4))$, namely the one in which $c_2 = 0$ and $c_1 = a^2(-2E)^{-1} - 1$.

If we know all possible representation spaces of $\mathscr{E}(\mathrm{SO}(4))$ that fulfill (3.18), then we know the properties of all the operators $L_i$, $A_i$ given by (3.1), (3.3), (3.9) in the spaces $\mathscr{R}(E)$, for all possible values $E$ with $E < 0$. It will turn out that $c_1$ cannot take any arbitrary real value, but rather that the spectrum of $C_1$ is discrete. [For the invariant operator $\mathbf{J}^2$ of $\mathscr{E}(\mathrm{SO}(3))$ we found the same result, namely that only the discrete set of eigenvalues $j(j + 1)$ with $j = 0, \frac{1}{2}, 1, \frac{3}{2}, \ldots$ is possible.] Consequently from (3.17) it will follow that the spectrum of the energy operator $H$ is discrete and the spectrum of $H$ is obtained by (3.17) from the spectrum of $C_1$.

Our task is therefore to find all the representation spaces of $\mathscr{E}(\mathrm{SO}(4))$, in particular those that fulfill (3.18). This can be done in essentially the same way as it was done for $\mathscr{E}(SO(3))$ in Section III.4, except that the calculations for $\mathscr{E}(SO(4))$ are more involved. In the next section we shall give a description of the properties of the representation spaces of $\mathscr{E}(SO(4))$ which should suffice for the understanding of the subsequent material. A derivation of the representations of $\mathscr{E}(SO(4))$ has been given in the Mathematical Appendix to Section V.3.

## VI.4 Properties of the Algebra of Angular Momentum and the Lenz Vector [4]

We choose as basis vectors in $\mathscr{R}(E)$ the angular-momentum vectors $|lm\rangle$. This is possible because

$$H, \mathbf{L}^2, L_3$$

is a system of commuting operators. Whether this is a c.s.c.o. for the hydrogen atom can only be decided by comparison with experiment. Therefore we make the assumption that this is a c.s.c.o.—i.e., that $l$, $m$ and the energy are the only quantum numbers of the hydrogen atom—and we shall later see that this is true to a high degree of accuracy (neglecting spin effects). Thus $|l\,m\rangle$ is a basis system in $\mathscr{R}(E)$. Since the $A_i$ are proper vector operators with respect to the angular momentum, as stated by (3.11) and $U_p A_i U_p = -A_i$, it follows from the Wigner–Eckart theorem that

$$
\begin{aligned}
A_\kappa |lm\rangle = &\; |l-1\,m+\kappa\rangle\langle l\,m\,1\,\kappa|l-1\,m+\kappa\rangle\langle l-1\|A\|l\rangle \\
&+ |l\,m+\kappa\rangle\langle l\,m\,1\,\kappa|l\,m+\kappa\rangle\langle l\|A\|l\rangle \\
&+ |l+1\,m+\kappa\rangle\langle l\,m\,1\,\kappa|l+1\,m+\kappa\rangle\langle l+1\|A\|l\rangle. \quad (4.1)
\end{aligned}
$$

Since $|l\,m\rangle$ has been assumed to be a complete basis system, there is no parity doubling, and since $A_i$ is a proper vector operator, it follows as in (V.4.25) that the reduced matrix element $\langle l\|A\|l\rangle = 0$. [This can also be seen to be a direct consequence of (3.18), which is related to the $U_p$ transformation property of $\mathbf{A} \cdot \mathbf{J}$]. If we insert into (4.1) the explicit values of the Clebsch–Gordon coefficients we obtain that as a consequence of (3.11) and (3.18), $A_\kappa$ is given by

$$
A_0|l\,m\rangle = \sqrt{l^2 - m^2}\,C_l|l-1\,m\rangle - \sqrt{(l+1)^2 - m^2}\,C_{l+1}|l+1\,m\rangle,
$$
$$(4.2_0)$$

$$
\begin{aligned}
A_{+1}|l\,m\rangle = &-\sqrt{(l-m)(l-m-1)/2}\,C_l|l-1\,m+1\rangle \\
&- \sqrt{(l+m+1)(l+m+2)/2}\,C_{l+1}|l+1\,m+1\rangle, \quad (4.2_+)
\end{aligned}
$$

$$
\begin{aligned}
A_{-1}|l\,m\rangle = &-\sqrt{(l+m)(l+m-1)/2}\,C_l|l-1\,m-1\rangle \\
&- \sqrt{(l-m+1)(l-m+2)/2}\,C_{l+1}|l+1\,m-1\rangle, \quad (4.2_-)
\end{aligned}
$$

where the $C_l$ are as yet undetermined functions of the discrete variable $l$, related to $\langle l-1\|A\|l\rangle$ by Equation (V.3.14).

[4] For a derivation see Mathematical Appendix V.3.

The $C_l$ are determined by (3.12). We state the result: For every natural number $n = 1, 2, 3, \ldots$, there is a function $C_l^{(n)}$ such that (3.12) is fulfilled. This function is

$$C_l^{(n)} = i \sqrt{\frac{n^2 - l^2}{4l^2 - 1}}, \qquad C_0^{(n)} = 0. \qquad (4.3)$$

From (4.2) and (4.3) one finds the following properties: For every number $n = 1, 2, 3, \ldots$, there is an irreducible representation space $\mathscr{R}(n)$. The lowest angular momentum in $\mathscr{R}(n)$ is $l = 0$, because starting from a given value $l$ by application of $A_k$ one can always reach the value $l - 1$, according to (4.2) and (4.3), unless $l = 0$, because then $C_0^{(n)} = 0$. The highest angular momentum is $n - 1$, because according to (4.2) one can always reach the value $l + 1$ from a given value $l$ by applying $A_\kappa$, unless $C_{l+1}^{(n)} = 0$—which, however, is the case according to (4.3) for $l = n - 1$. As the $|l\,m\rangle$ are orthogonal for different values of $l$, the space $\mathscr{R}(n)$ is the orthogonal direct sum of the irreducible representation spaces $\mathscr{R}^l$ of angular momentum:

$$\mathscr{R}(n) \underset{\mathscr{E}(SO(3))}{\Longrightarrow} \sum_{l=0}^{n-1} \oplus \,\mathscr{R}^l, \qquad (4.4)$$

where the symbol $\Rightarrow$ is to indicate that the two spaces are the same when only the algebra $\mathscr{E}(SO(3))$ acts in them.

The action of $L_3$, $L_\pm$ in each of the spaces $\mathscr{R}^l$ and therefore in $\mathscr{R}(n)$ has been derived in Section III.3. In particular,

$$L_3|l\,m\rangle = m|l\,m\rangle \qquad \mathbf{L}^2|l\,m\rangle = l(l + 1)|l\,m\rangle \qquad (4.5)$$

Thus $L_i$ and $A_i$, and consequently the whole algebra $\mathscr{E}(SO(4))$, is known in each $\mathscr{R}(n)$ $(n = 1, 2, 3, \ldots)$. To indicate that the vector $|l\,m\rangle$ is an element of $\mathscr{R}(n)$, i.e., that $A_\kappa$ is given by (4.2) with the particular $C_l^{(n)}$, we label it with this number: $|n\,l\,m\rangle$. A straightforward calculation, using (4.2) and (4.3), gives that the eigenvalue of the operator $C_1$ in the space $\mathscr{R}(n)$ is $c_1 = n^2 - 1$:

$$C_1|n\,l\,m\rangle = (n^2 - 1)|n\,l\,m\rangle. \qquad (4.6)$$

This gives the set of discrete numbers that are possible for $c_1$.

The space $\mathscr{R}(n)$ is not only an irreducible representation space of $\mathscr{E}(SO(4))$, but also of $\mathscr{E}(SO(4))$ extended by parity $U_P$:

$$U_P L_i U_P = +L_i, \qquad (4.7)$$

$$U_P A_i U_P = -A_i. \qquad (4.8)$$

Applying the considerations of Section V.4 to the space $\mathscr{R}(n)$, where there is no parity doubling, one has

$$U_P|n\,l\,m\rangle = (-1)^l \eta_n |n\,l\,m\rangle, \qquad (4.9)$$

where $\eta_n = +1$ or $\eta_n = -1$, but may be different for different spaces $\mathscr{R}(n)$. We will study the $n$-dependence of $\eta$ below.

## VI.5 The Hydrogen Spectrum

In the previous section we described all the irreducible representation spaces $\mathcal{R}(n)$ of the algebra of angular momentum and the Lenz vector [i.e., all irreducible representations of $\mathscr{E}(SO(4))$ that fulfill (3.18)]. The relation (3.17) connects these spaces with the energy eigenspaces $\mathcal{R}(E)$ of negative energy $E < 0$:

$$c_1 = (n^2 - 1) = a^2(-2E)^{-1} - 1. \tag{5.1}$$

Thus the energy in each $\mathcal{R}(n)$ is given by

$$E = E_n = -\frac{a^2}{2n^2},$$

and the $\mathcal{R}(n)$ are all possible energy eigenspaces of the hydrogen atom.

The space of physical states of the hydrogen atom $\mathcal{R}$ is the direct sum of all these $\mathcal{R}(n)$

$$\mathcal{R} = \sum_{n=1}^{\infty} \oplus \mathcal{R}(n), \tag{5.2}$$

and the spectrum of the energy operator in $\mathcal{R}$ is

$$\text{spectrum } H = E_n = -a^2/(2n^2) \qquad (n = 1, 2, 3, \ldots). \tag{5.3}$$

$n$ is called the principal quantum number.

[The meaning of infinite linear combinations of $f_n \in \mathcal{R}(n)$ (i.e., $\sum_{n=1}^{\infty} f_n$) requires a precise definition. It may mean infinite sum in the Hilbert-space sense ($\sum_{n=1}^{\infty} \|f_n\|^2 < \infty$), in the Schwartz-space sense ($\sum_{n=1}^{\infty} \|f_n\|^2 n^{2p} < \infty$), or in some other sense. Even if (5.2) is meant in the Hilbert-space sense, $\mathcal{R}$ is "different" from the Hilbert space $\mathcal{H} = \{\sum_{v=0}^{\infty} \phi_v | H_{\text{osc}} \phi_v = (P^2 + Q^2)\phi_v = (v + \frac{3}{2})\phi_v \text{ and } \sum_{v=0}^{\infty} \|\phi_v\|^2 < \infty\}$, which is the usual Hilbert space of quantum mechanics. The isometric map between $\mathcal{R}$ and $\mathcal{H}$ is physically meaningless, as it would map elements that represent different physical states into each other.

Physical states of the hydrogen atom are in any event only vectors in the dense subspace $\sum_n \mathcal{R}(n)$ of the Hilbert space, where the sum extends over an arbitrarily large but finite number of $n$. Any limit $n \to \infty$ is a mathematical idealization, and physically for $n \to \infty$ the physical system ceases to be a hydrogen atom; it ionizes and becomes a system consisting of a proton and electron. One can, of course, consider the enlarged physical system, consisting of hydrogen-atom states as well as electron-and-proton states. For such a physical system $\mathcal{R}$ is not the total space of physical states, though this space may still be a dense subspace of $\mathcal{H}$. The space of physical states of this system is the direct sum of $\mathcal{R}$ and a space $\mathcal{R}'$

that is the continuous direct sum of irreducible-representation spaces of the algebra $\mathscr{E}(SO(3, 1))$ generated by $L_i$ and $A_i' = \sqrt{2H}\,\tilde{A}_i$.

The space $\mathscr{R}$ is an irreducible-representation space of the algebra $\mathscr{E}(SO(4, 1))$ [and also of $\mathscr{E}(SO(4, 2))$ and of $\mathscr{E}(E_4)$]; the space $\mathscr{R}'$ is an irreducible-representation space of $\mathscr{E}(SO(3, 2))$. Such algebras have been called spectrum-generating algebras, the underlying groups are called spectrum generating groups.]

Transitions between different energy levels are performed by the position operator $Q_i$ (dipole transitions) and powers thereof (multipole transitions). Thus the $Q_i$ are observables that transform between different spaces $\mathscr{R}(n)$. Using the property that the $Q_\kappa$ ($\kappa = 0, \pm 1$) are the components of a proper vector operator, one obtains

$$Q_\kappa|n\,l\,m\rangle = \sum_{n'} |n'\,l-1\,m+\kappa\rangle\langle l\,m\,1\,\kappa|l-1\,m+\kappa\rangle(2l+1)^{1/2}q_l^{n'n}$$

$$+ \sum_{n'} |n'\,l+1\,m+\kappa\rangle\langle l\,m\,1\,\kappa|l+1\,m+\kappa\rangle(2l+1)^{1/2}\tilde{q}_l^{n'n},$$

(5.4)

where the reduced matrix elements

$$q_l^{n'n} = (2l+1)^{-1/2}\langle n'\,l-1\|Q\|nl\rangle,$$

$$\tilde{q}_l^{n'n} = (2l+1)^{-1/2}\langle n'\,l+1\|Q\|nl\rangle.$$

depend upon $n'$, $n$. They can be computed (Problem VII.1), but at present we only need the property that they are different from zero for all $n'$ and $n$. The physical meaning of this statement is that there exist dipole transitions between all energy levels of the hydrogen atom.

We can determine the $n$-dependence of the intrinsic parity $\eta_n$ using the $U_P$ transformation property of the $Q_\kappa$. Applying $U_p$ to (5.4) gives, according to (4.9),

$$U_P Q_\kappa|n\,l\,m\rangle = \sum_{n'} (-1)^{l-1}\eta_{n'}|n'\,l-1\,m+\kappa\rangle\langle l\,m\,1\,\kappa|l-1\,m+\kappa\rangle$$

$$\times (2l+1)^{1/2}q_l^{n'n}$$

$$+ \sum_{n'} (-1)^{l+1}\eta_{n'}|n'\,l+1\,m+\kappa\rangle\langle l\,m\,1\,\kappa|l+1\,m+\kappa\rangle$$

$$\times (2l+1)^{1/2}\tilde{q}_l^{n'n}.$$

(5.5)

But use of (V.4.1) gives

$$U_P Q_\kappa|n\,l\,m\rangle = -Q_\kappa U_P|n\,l\,m\rangle = -(-1)^l\eta_n Q_\kappa|n\,l\,m\rangle$$

(5.6)

Inserting (5.4) into the right-hand side of (5.6) and comparing the co-efficients with those of (5.5) (i.e., making use of the linear independence of the basis vectors), one finds

$$\eta_{n'}\,q_l^{n'n} = \eta_n q_l^{n'n} \qquad \eta_{n'}\,\tilde{q}_l^{n'n} = \eta_n\tilde{q}_l^{n'n}.$$

(5.7)

As the reduced matrix elements are different from zero, $q_l^{n'n} \neq 0$, and $\tilde{q}_l^{n'n} \neq 0$, it follows that

$$\eta_{n'} = \eta_n = \eta, \tag{5.8}$$

i.e., $\eta_n$ is independent of $n$, and the parity of the energy eigenstate is

$$U_P|n\,l\,m\rangle = (-1)^l\eta|n\,l\,m\rangle, \tag{5.9}$$

where $\eta$ is either $+1$ or $-1$. Equation (5.9) means that the parity of a state with a definite angular-momentum value $l$ is the same in all energy spaces $\mathscr{R}(n)$. By convention one chooses $\eta = +1$; only the relative parity has a physical meaning.

Each energy eigenspace $\mathscr{R}(n)$ of the hydrogen atom is, according to (4.4), the direct sum of angular-momentum eigenspaces $\mathscr{R}^l$; thus each energy level $E_n$ contains the $n$ angular momenta

$$l = 0, 1, 2, \ldots, n - 1.$$

Each angular-momentum eigenspace $\mathscr{R}^l$ in $\mathscr{R}(n)$ is, according to (III.3.24), the direct sum of the $2l + 1$ one-dimensional spaces $\mathscr{R}_m^l$:

$$\mathscr{R}^l \xrightarrow[\mathscr{E}(SO(2))]{} \sum_{m=-l}^{+l} \oplus \mathscr{R}_m^l, \tag{5.10}$$

which are spanned by each vector $|n\,l\,m\rangle$. $\mathscr{E}(SO(2))$ denotes the subalgebra of $\mathscr{E}(SO(3))$ that is generated by $L_3$. The dimension of $\mathscr{R}(n)$ is therefore

$$\sum_{l=0}^{n-1} (2l + 1) = n^2, \tag{5.11}$$

i.e., there are $n^2$ linearly independent eigenstates of $H$ with eigenvalue $E_n$, or the eigenvalue $E_n$ is $n^2$-fold degenerate.

In order to compare the calculated energy spectrum with the experimental data measured in eV, we insert $a = m_e e^2/\hbar$ from (3.5) into (5.3):

$$E_n' = \frac{E_n}{m_e} = -\frac{m_e e^4}{\hbar^2}\frac{1}{2n^2} = -R''\frac{1}{n^2} \tag{5.12}$$

(recall that $E_n$ is the energy in eV $\cdot$ $m_e$, $E_n'$ is the energy in eV). The number

$$R'' = \frac{m_e e^4}{2\hbar^2} = \frac{c^2 m_e}{2}\left(\frac{e^2}{c\hbar}\right)^2$$

can be calculated using

$$m_e c^2 = 0.51 \times 10^6 \text{ eV}$$

$$e^2 = 23 \times 10^{-20} \text{ esu}^2 = 23 \times 10^{-20} \text{ dyn cm}^2,$$

$$\alpha = \frac{e^2}{c\hbar} = \frac{1}{137.036}.$$

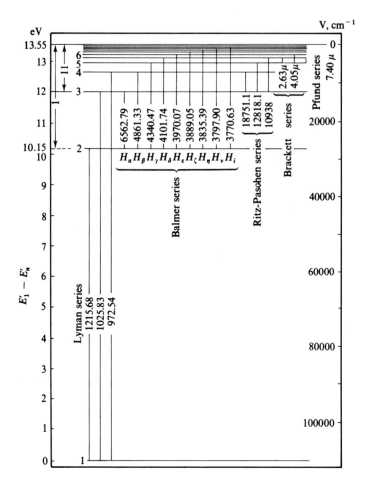

Figure 5.1   Diagram of the energy levels of hydrogen.

$R'' = m_e e^4/(2\hbar^2)$ is the Rydberg constant in energy units. Its numerical value is

$$R'' = 13.53 \text{ eV}. \tag{5.13}$$

The energy levels of the hydrogen atom are plotted in Figure 5.1, where details in the experimental energy spectrum have not been taken into account. The numbers on the left give $E_1' - E_n'$, which is measured in eV. As one sees, (5.12) is experimentally very well fulfilled. The energy levels are spaced closer and closer to one another as $n$ increases, and they finally converge to $E_\infty = 0$ for $n \to \infty$. $E > 0$ was excluded by our restriction to the subspace where $-2H$ is positive definite; this is the region in which the quantum-mechanical Kepler problem does not describe the hydrogen atom, but decribes an electron that is scattered by a proton (scattering states). Thus if we choose as our physical system the hydrogen atom only, $\mathscr{R}$ of (5.2) is the space of physical states. If we consider as our physical system the system that consists of the hydrogen atom, electron, and proton, then the space of physical states is larger than $\mathcal{R}$, and $\mathcal{R}$ is the subspace of discrete energy eigenstates.[5] $E = E_\infty = 0$ is the energy value at which this physical system ceases to be a hydrogen atom; beyond this point the electron energy can take any value and one can also prepare approximate momentum eigenstates of the electron. This means that if one starts with hydrogen atoms in the ground state, one can prepare the physical system in an approximate momentum eigenstate of the electron if one transfers to it an energy of at least $-E_1$, and that the system is no longer a hydrogen atom. The continuous energy states correspond to the hyperbolic orbits of the classical Kepler problem.

The numbers on the right-hand side of the diagram give the (negative) energy in units of cm$^{-1}$, i.e., the wave numbers $v_n$ that are connected with $E'_n$:

$$v_n = \frac{E'_n}{2\pi\hbar c}. \tag{5.14}$$

Inserting (5.12) gives for the wave numbers

$$v_n = -\frac{e^4 m_e}{\hbar^3 4\pi c} \frac{1}{n^2}. \tag{5.15}$$

The quantities

$$R' = \frac{1}{2\pi\hbar} R'' = \frac{e^4 m_e}{4\pi\hbar^3} = 3.29 \times 10^{15} \text{ sec}^{-1}, \tag{5.16}$$

$$R = \frac{1}{2\pi\hbar c} R'' = \frac{e^4 m_e}{4\pi\hbar^3 c} = 1.097 \times 10^5 \text{ cm}^{-1} \tag{5.16'}$$

are the Rydberg constant in frequency and wave-number units respectively. (Originally it was $R$ that was called the Rydberg constant.)

The lines connecting the different energy levels depict the transitions in the emission or absorption of electromagnetic interaction. According to (III.1.1), the frequency and wave number of the absorbed or emitted light quantum are

$$v'_{nm} = \left| R' \left( \frac{1}{n^2} - \frac{1}{m^2} \right) \right| \quad (\text{in sec}^{-1}), \tag{5.17}$$

$$v_{nm} = \left| R \left( \frac{1}{n^2} - \frac{1}{m^2} \right) \right| \quad (\text{in cm}^{-1}). \tag{5.17'}$$

---

[5] It should be remarked that we are discussing the electron states only, i.e., describing the electron under the external influence of the nucleus. If we wanted to describe the combined system of electron and nucleus (proton), we would have to take as the space of physical states the direct-product space of the electron space $\mathcal{R}$ and the proton space $\mathcal{H}_p: \mathcal{R} \otimes \mathcal{H}_p$ (Axiom $IV$), or $\mathcal{R} \otimes \mathcal{H}_{C.M.}$, where $\mathcal{H}_{C.M.}$ is the space for the center-of-mass motion. Neglecting the proton structure (e.g., proton spin), $\mathcal{H}_p$ is one-dimensional for protons at rest.

The number on each transition line in Figure 5.1 gives the wavelength of the light in angstroms:

$$\lambda = 1/v'_{nm} = c/v'_{nm},$$

with $1 \text{ Å} = 10^{-8} \text{ cm} = 10^{-4} \mu$.

All transitions terminating in the same lower state form a spectral series and each series of the hydrogen atom has a historical name. The series of lines corresponding to transitions to the level $n = 1$ is called the Lyman series. The frequencies of the Lyman series are

$$v'_{1m} = R'\left(\frac{1}{1} - \frac{1}{m^2}\right) \qquad (m = 2, 3, 4, \ldots). \tag{5.18}$$

The line with the longest wavelength is $c/v'_{12} = 1215.68 \text{ Å}$ This line lies in the ultraviolet region (cf. Figure III.1.2). The series with $n = 2$,

$$v'_{2m} = R'\left(\frac{1}{2^2} - \frac{1}{m^2}\right) \qquad (m = 3, 4, 5, \ldots) \tag{5.19}$$

is called the Balmer series. It lies in the visible region of the spectrum, and (5.19) was first obtained empirically by Balmer in 1885. It played an exceptionally important role in the conjecture of the quantum theory of atoms. The other series in Figure 5.1 with $n = 3$ (Paschen), $n = 4$ (Brackett), and $n = 5$ (Pfund) lie in the infrared region.

To each energy level with energy value $E_n$ corresponds the energy eigenspace $\mathcal{R}(n)$ of dimension $n^2$. The energy value $E$ is $n^2$-fold degenerate. Thus if the energy of the hydrogen atom is measured as $E_n$, then the state is in general a mixture (except for $n = 1$) and the (normalized) statistical operator of this state is

$$W = (\operatorname{Tr} \Lambda_n)^{-1}\Lambda_n, \tag{5.20}$$

where $\Lambda_n$ is the projection operator on the $n^2$-dimensional subspace $\mathcal{R}(n)$. Energy measurements on the hydrogen atom do not reveal the value of the angular momentum; they just give an upper limit to the possible angular momentum values $l = 0, 1, \ldots, n - 1$.

Then $n^2$-fold degeneracy of each energy level is a peculiarity of the hydrogen atom, or the $1/r$ potential. This is connected with the existence of the constant of motion $A_i$ in addition to the angular momenta $L_i$. In general one has only a $(2l + 1)$-fold degeneracy of the energy values as a consequence of $[L_i, H] = 0$.

However, the degeneracy for the hydrogen atom is only approximate, and Figure 5.1 is only a rough picture of the experimental situation. The above theoretical description, like every theoretical model, gives a description of the situation in nature only up to a certain accuracy. Finer details of the hydrogen spectrum are shown in Figure 5.2 for the low-lying energy values.

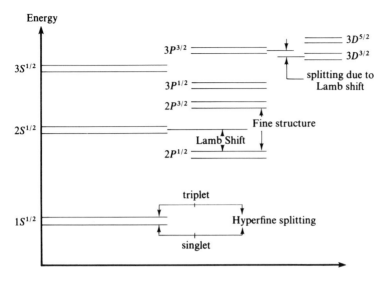

Figure 5.2   Low-lying energy levels of atomic hydrogen. (Not to scale.)

The first number by each energy level gives the value of $n$. The symbol $S, P, D$ stands for the angular-momentum value $l = 0, 1, 2$ respectively (e.g., $2P$ means $n = 2$, $l = 1$, and $m$ can be anything such that $-l \leq m \leq +l$). According to (5.12) all $2S$ and $2P$ states should have the same energy, as shown in Figure 5.1. This is only approximately the case. The $2P$ states split into two subsets $2P^{3/2}$ and $2P^{1/2}$ with

$$E(2P^{3/2}) - E(2P^{1/2}) = 0.453 \times 10^{-4} \, \text{eV} = 1095 \times 10^7 \, \text{sec}^{-1}. \quad (5.21)$$

Comparing this with the numbers of (5.12), (5.13), we see that our model indeed gives an excellent approximation to the hydrogen spectrum; the deviation (5.21) is only a fraction of one percent. The splitting (5.21) (fine structure) and the $3P^{3/2}$ to $3P^{1/2}$ and $3D^{5/2}$ to $3D^{3/2}$ splittings can be explained if a new observable, the electron spin, is introduced. The superscripts $^{3/2}, ^{1/2}$, etc. are connected with this new quantum number, the spin.

The splitting between the levels $2S^{1/2}$ and $2P^{1/2}$ cannot be explained even after the spin has been introduced. It is again an order of magnitude smaller than (5.21):

$$
\begin{aligned}
E(2S^{1/2}) - E(2P^{1/2}) &= 0.0353 \, \text{cm}^{-1} \\
&= 0.437 \times 10^{-5} \, \text{eV} \\
&= 1058 \times 10^6 \, \text{sec}^{-1}
\end{aligned}
$$

and is called the *Lamb shift*. The Lamb shift is a consequence of the interaction of the electron with the fluctuations of the radiation field, and its explanation requires a new theory called *quantum electrodynamics*.

The splitting of each term $1S^{1/2}$, $2S^{1/2}$, $2P^{1/2}$, etc. into two levels, the *hyperfine splitting*, is another order of magnitude smaller than the Lamb shift:

$$E(2S^{1/2} \text{ upper}) - E(2S^{1/2} \text{ lower}) = 0.0059 \text{ cm}^{-1},$$

$$E(2P^{1/2} \text{ upper}) - E(2P^{1/2} \text{ lower}) = 0.0020 \text{ cm}^{-1}.$$

In order to explain this splitting one needs to include observables that describe the internal structure of the nucleus (in particular, nucleon spin).

Summarizing, we have seen that our simple theoretical model describes the features of the hydrogen spectrum very well. However, to explain the fine details requires the introduction of new observables and even completely new theories (quantum electrodynamics and nuclear physics). Any given theory has only a limited domain of applicability, and this example of the hydrogen spectrum serves very well to illustrate that behind each new digit in an experimental number may be hidden a completely new domain of physics.

## Problem

1.  Fill in the missing steps in the treatment of the hydrogen atom: Let $L_i$ be the components (3.1) of the orbital-angular-momentum operator, and let $H$ [as given by (VI.3.5)] be the energy operator of the hydrogen atom. Let $\tilde{A}_i$ denote the Lenz vector given by (3.6).

(a)  Show that as a consequence of the canonical commutation relations the angular momentum and the Lenz vector are constants of the motion, i.e., satisfy (3.8):
$$[L_i, H] = 0 \quad \text{and} \quad [\tilde{A}_i, H] = 0.$$

(b)  Define $A_i$ by (3.9). Show that $A_i$ is Hermitian and a constant of the motion:
$$A_i^\dagger = A_i$$

and

$$[A_i, H] = 0. \tag{3.10}$$

(c)  Show that $L_i$ and $A_j$ fulfill the commutation relations of (3.11) and (3.12):
$$[A_i, L_j] = i\varepsilon_{ijk} A_k,$$
$$[A_i, A_j] = i\varepsilon_{ijk} L_k.$$

(d)  Define the Casimir operators $C_1$ and $C_2$ by Eqs. (3.13) and (3.14). Derive (3.17) and (3.18):
$$C_1 = a^2(-2H)^{-1} - I,$$
$$C_2 = 0.$$

The following fact is useful in carrying out the above derivations: If two operators $A$ and $B$ commute, $[A, B] = 0$, then $[f(A), B] = 0$ for any well-defined function of the operator $A$.

# CHAPTER VII

# Alkali Atoms and the Schrödinger Equation of One-Electron Atoms

In Section VII.1 the concept of perturbation theory is explained using the example of the alkali atoms. Section VII.2 represents an algebraic calculation of the matrix elements of $Q^{-\nu}$ ($\nu = 1, 3, 4, \ldots$); the results are used for the calculation of the energy values of the alkali atoms. Section VII.3 gives a brief description of the solution of the Schrödinger equation for the hydrogen atom, which is used for an alternative computation of the matrix elements of $Q^{-\nu}$ and an evaluation of the alkali energy values. It also lists some properties of the spherical harmonics which are used in the second part of this book.

## VII.1 The Alkali Hamiltonian and Perturbation Theory

The alkali spectra are very similar to the hydrogen spectrum. This is suggested by their classical model, according to which the alkali atom consists of one electron—the "valence electron"—that moves in the Coulomb field of the nucleus and in the average field of the other electrons that are in the orbits closer to the nucleus.

The classical potential $V_A(r)$ for this "outer" electron has, for large values of $r$, the form

$$V_A(r) \approx -e^2/r,$$

because the $Z - 1$ electrons in the inner orbits screen the charge of the nucleus $Ze$, so that the effective charge will be $e$. For very small $r$ (smaller than the radius of the inner orbits)

$$V_A(r) \approx -Ze^2/r.$$

Therefore $V_A(r)$ can be written

$$V_A(r) = -\frac{e^2}{r} + V(r), \tag{1.1}$$

where $V(r)$ is negative and different from zero only in the neighborhood of the nucleus. The classical Hamiltonian function of the outer electron is then

$$h = h_{\text{Hydr}} + V(r), \tag{1.2}$$

where $V(r)$ is a small perturbation of the Hamiltonian $h_{\text{Hydr}}$.

The exact expression for $V(r)$ is not known since it depends upon the unknown charge distribution of the core, which consists of the nucleus and the inner electrons. If $V_A(r)$ is spherically symmetric, then $V_A(r)$ may be expanded in the form

$$V_A(r) = -c_0 \frac{e^2}{r} - c_1 \frac{e^2}{r^2} - c_2 \frac{e^2}{r^3} - \cdots, \tag{1.2'}$$

where $c_0 e$ is the charge, $c_1 e = e\mathbf{d} \cdot \mathbf{x}/r$ is the dipole contribution, $c_2 e = eq_{ij}x_i x_j/r^2$ is the quadrupole contribution, etc., of the core. For the alkali atoms the extension of the core is small compared to the distance of the outer electron, so that $c_1$ is small and contributions by the quadrupole moment $q_{ij}$ and by higher multipole moments are negligible.

The Hamiltonian operator for the quantum-mechanical system is therefore, in correspondence to (1.2),

$$H = K + V, \tag{1.3}$$

where $K$ is the Hamiltonian operator of the hydrogen atom,

$$K = \frac{\mathbf{P}^2}{2} - \frac{a}{Q}, \tag{1.3'}$$

and

$$V = V(Q) = -\frac{c_1 a}{Q^2} - \frac{c_2 a}{Q^3} - \cdots \tag{1.3''}$$

is the "small perturbation" of $K$. In (1.3), (1.3'), and (1.3'') we have used the units adopted in Section VI.3, i.e., we have made the replacements $Q_i/\hbar \to Q_i$ and $m_e H \to H$.

The problem is now to find the spectrum of the Hamiltonian operator $H$ and its expectation values in the physical states. This can no longer be done exactly. Also, we do not know the exact form of $V(r)$ or $V(Q)$, though we know some of its properties. We can only hope, therefore, that the spectrum of $H$ will not differ significantly from the spectrum of $K$, i.e., that $V$ causes only a small perturbation of the spectrum of $K$.

The determination of the alkali spectrum is a problem of *perturbation theory*, which will be discussed in more general terms in the next chapter. One major factor in this kind of problem is the choice of an appropriate basis. One starts with eigenstates of $K$, i.e., chooses a c.s.c.o. that contains $K$. If the other members of the c.s.c.o. do not commute with $V$, then $V$ applied to this eigenvector will not only change the eigenvalue of $K$ but also the eigenvalues of these other observables. Thus $V$ will perturb not only the spectrum of $K$ but also the spectrum of the other members of the c.s.c.o. Therefore the c.s.c.o. should be chosen so that as many of their members as possible commute with $V$.

For the alkali atoms we have

$$[V, L_k] = 0, \tag{1.4}$$

because

$$[Q, L_k] = 0.$$

Consequently, we have the ideal situation where $H$ as well as $K$ can be diagonalized together with $\mathbf{L}^2$ and $L_3$. Thus the basis $|n\,l\,m\rangle$ of the hydrogen-atom problem is a very appropriate basis to start with.

We again define

$$A_i = (-2K)^{-1/2}(\tfrac{1}{2}\epsilon_{ikl}\{P_1, L_k\} + Q_i Q^{-1}), \tag{1.5}$$

and we again have

$$C_1|n\,l\,m\rangle = (\mathbf{A}^2 + \mathbf{L}^2)|n\,l\,m\rangle = (n^2 - 1)|n\,l\,m\rangle. \tag{1.6}$$

Unlike the case of the hydrogen atom, however, it will not be true that the full Hamiltonian operator $H$ commutes with $A_i$, $\mathbf{A}^2$, and $C_1$:

$$[H, A_i] = [V, A_i] \neq 0,$$
$$[H, C_1] = [V, C_1] \neq 0. \tag{1.7}$$

Thus $H$ and $C_1$ (and consequently $H$ and $K$) cannot be diagonalized together, i.e., there are no vectors in $\mathcal{R}$ that are eigenstates of $H$ and also eigenstates of $C_1$. Let us denote the eigenstates of $H$ by $|\lambda\,l\,m\rangle$ and the eigenvalue by $E_\lambda$:

$$H|\lambda\,l\,m\rangle = E_\lambda|\lambda\,l\,m\rangle. \tag{1.8}$$

$H$ can be diagonalized together with $\mathbf{L}^2$, $L_3$ because of (1.4).]

Now the question arises: Which are the physical states, i.e., which are the states that the alkali atoms can be prepared in? According to the basic assumption III, the alkali atoms should be in an energy eigenstate (or a mixture of energy eigenstates) if an energy measurement has been performed on the system, i.e., the statistical operator should be

$$W_H = P_\lambda, \quad W_H^{(l)} = P_{\lambda l} \quad \text{or} \quad W_H^{(lm)} = P_{\lambda lm} \tag{1.9}$$

if only the energy, or only the energy and angular momentum, or the energy, angular momentum, and $z$-component of the angular momentum have been measured, respectively. Here $P_\lambda$ is the projector on the space of eigenvectors of $H$ with eigenvalue $E_\lambda$, $P_{\lambda l}$ projects on the space spanned by $|\lambda\,l\,m\rangle$ $(m = -l, -l + 1, \ldots, +l)$, and $P_{\lambda lm}$ projects on the space spanned by $|\lambda\,l\,m\rangle$.

If the state of the alkali atom has not been prepared by an energy measurement but by a measurement of the observable $C_1$ or $K$, then, according to the basic assumption III, this physical system should be in an eigenstate of $C_1$, i.e., the statistical operator should be

$$W_K = P_n, \qquad W_K^{(l)} = P_{nl}, \qquad \text{or} \qquad W_K^{(lm)} = P_{nlm}, \qquad (1.10)$$

where $P_n$ is the projection operator on the space of eigenvectors of $C_1$ with eigenvalue $n^2 - 1$.

Although at this stage we cannot exclude the possibility that states of atoms are prepared by a $C_1$ measurement, this possibility appears very unlikely, as $C_1$ and $K$ have only an auxiliary meaning (except for the hydrogen atom) and the physical observable is $H$. (Also, in Chapter XII we shall see that states that do not change in time—and the states that correspond to the energy levels of the atoms are likely to be well described by this property—have to be eigenstates of the energy operator.) Thus, the occurence of physical states given by (1.10) would make the formulation of the theory very unsatisfactory and would cause doubts about the appropriateness of III. Therefore it will be interesting to see how the predictions of (1.9) and (1.10) differ. In this chapter we shall perform the calculation for the alkali atoms and see that an answer to the above question cannot be obtained from the alkali spectra. In the later chapter on two-electron atoms (helium atom) we shall, however, see that experimental data require (1.9), which reassures us of the correctness of our basic assumption III—and the appropriateness of the meaning of stationary states (cf. Chapter XII).

The values predicted for the measurement of the energy operator for the state $W_K^{(lm)}$ of (1.10) are, according to the basic assumption II,

$$
\begin{aligned}
\langle H \rangle_{W_K^{(lm)}} &= \mathrm{Tr}(W_K^{(lm)} H) = \mathrm{Tr}(P_{nlm} H) = \mathrm{Tr}(P_{nlm} K) + \mathrm{Tr}(P_{nlm} V(Q)) \\
&= \langle n\,l\,m | K | n\,l\,m \rangle + \langle n\,l\,m | V(Q) | n\,l\,m \rangle \\
&= -\frac{a^2}{2n^2} + \langle n\,l\,m | V(Q) | n\,l\,m \rangle \\
&= -\frac{a^2}{2n^2} + \epsilon(n, l) \equiv E(n, l).
\end{aligned}
\qquad (1.11)
$$

For the case where the state is $W_H^{(lm)}$ of (1.9), we have

$$\langle H \rangle_{W_H^{(lm)}} = \mathrm{Tr}(W_H^{(lm)} H) = \mathrm{Tr}(P_{\lambda lm} H) = \langle \lambda\,l\,m | H | \lambda\,l\,m \rangle = E_{\lambda l}. \qquad (1.12)$$

The term $\epsilon(n, l)$ in (1.11) is the matrix element of the small perturbing Hamiltonian $V$:

$$\epsilon(n, l) = \langle n\,l\,m | V(Q) | n\,l\,m \rangle. \qquad (1.13)$$

As $|n\,l\,m\rangle$ is known, this matrix element can be computed if $V(Q)$ is known. The value $\epsilon(n, l)$ does not depend upon $m$, because of (1.4). Therefore, for the expectation value in the state $P_{nl}$ one obtains

$$\langle H \rangle_{W_K^{(l)}} = \frac{\mathrm{Tr}(W_K^{(l)} H)}{\mathrm{Tr}\, W} = \frac{1}{2l+1} \sum_{m=-l}^{+l} \langle n\,l\,m | H | n\,l\,m \rangle = E(n, l). \qquad (1.14)$$

In contrast to the expectation value $E(n, l)$, $E_{\lambda l}$ is an element of the spectrum of $H$. The spectrum of $H$ and the states $|\lambda\, l\, m\rangle$ need to be determined. The spectrum of $K$ and the states $|n\, l\, m\rangle$ are known from the hydrogen-atom problem, and in order to calculate $E(n, l)$ it only remains to calculate the "small" matrix elements $\epsilon(n, l)$.

The eigenvalues of $H$ and the eigenvectors are determined by a perturbation-theoretical calculation. It will turn out that $E(n, l)$ of (1.11) will be the first-order approximation $E_{\lambda l}^{(1)}$ for the perturbation-theoretical calculations of the eigenvalue $E_{\lambda l}$. This is the reason we cannot distinguish between (1.9) and (1.10). As $E(n, l)$ cannot be calculated exactly [because $V(Q)$ is not known precisely], we can never tell from comparison with the experimental energy spectrum of the alkali atoms whether $E(n, l)$ is the exact energy value or only a first approximation.

## VII.2  Calculation of the Matrix Elements of the Operator $Q^{-\nu}$

To obtain the correction term $\epsilon(n, l) = \langle n\, l\, m|V(Q)|n\, l\, m\rangle$, we calculate the matrix elements of the operators $Q^{-\nu}$ for any integer $\nu$. These matrix elements will also be used for the evaluation of the fine structure splitting in Section IX.4 below.

We proceed in the following way. We first derive some relations between the operators $Q^{-\nu}$, $\mathbf{L}^2$, and

$$K = \frac{\mathbf{P}^2}{2} - \frac{a}{Q}. \tag{2.1}$$

The matrix element of one of these relations will give a recursion relation for the matrix elements $\langle n\, l\, m|Q^{-\nu}|n\, l\, m\rangle$.

We introduce the operator

$$P_r \equiv \frac{1}{2}\left\{\frac{Q_i}{Q}, P_i\right\} = \frac{Q_i}{Q}P_i - \frac{i}{Q}. \tag{2.2}$$

As a consequence of the Heisenberg commutation relation it follows that

$$[Q, P_r] = iI. \tag{2.3}$$

[Equation (2.3) follows from

$$\left[Q, \frac{1}{2}\left\{\frac{Q_i}{Q}, P_i\right\}\right] = \frac{1}{2}\left\{\frac{Q_i}{Q}, [Q, P_i]\right\} - \frac{1}{2}\left\{P_i, \left[\frac{Q_i}{Q}, Q\right]\right\}$$

by use of

$$[Q, P_i] = iQ_i/Q. \tag{2.4}$$

The proof of (2.4) proceeds as follows:

$$[Q^2, P_i] = [Q_j Q_j, P_i] = Q_j[Q_j, P_i] + [Q_j, P_i]Q_j = 2iQ_i.$$

On the other hand

$$[Q^2, P_i] = [QQ, P_i] = Q[Q, P_i] + [Q, P_i]Q = \{Q, [Q, P_i]\},$$

and consequently

$$\{Q, [Q, P_i]\} = 2iQ_i. \tag{2.5}$$

Furthermore,

$$\begin{aligned}
[Q^2, [Q, P_i]] &= (QQ[Q, P_i] + Q[Q, P_i]Q) \\
&\quad - (Q[Q, P_i]Q + [Q, P_i]Q^2) \\
&= [Q, Q[Q, P_i]] + [Q, [Q, P_i]Q] \\
&= [Q, \{Q, [Q, P_i]\}] = [Q, 2iQ_i] = 0,
\end{aligned}$$

where (2.5) has been used. However, if $Q^2$ commutes with $[Q, P_i]$, then $Q = (Q^2)^{1/2}$ commutes with $[Q, P_i]$ also (by the definition of the square root of an operator). Therefore it follows from (2.5) that

$$2Q[Q, P_i] = 2iQ_i,$$

which in turn gives (2.4).⟧

As a consequence of the definition of $\mathbf{L}$, $L_i = \epsilon_{ijk} Q_j P_k$, and the definition (2.2), it follows by a straightforward calculation that

$$\mathbf{P}^2 = P_r^2 + \frac{1}{Q^2} \mathbf{L}^2. \tag{2.6}$$

With this the energy operator $K$ can be written

$$K = \frac{P_r^2}{2} + \frac{\mathbf{L}^2}{2Q^2} - \frac{a}{Q}. \tag{2.7}$$

$P_r^2/2$ is the operator of the kinetic energy of radial motion, and $P_r$, the conjugate to the radius operator $Q = \sqrt{Q_i Q_i}$, is often called the *radial momentum operator*.

The radial momentum operator fulfills the commutation relation

$$[Q^{-\nu}, P_r] = -i\nu Q^{-(\nu+1)}. \tag{2.8}$$

⟦We prove (2.8) by induction on $\nu$. First note that

$$[I, P_r] = 0 = [QQ^{-1}, P_r] = Q[Q^{-1}, P_r] + [Q, P_r]Q^{-1}.$$

Then use of (2.3) gives $[Q^{-1}, P_r] = -iQ^{-2}$. Consequently (2.8) is true for $\nu = 0, \pm 1$. Assuming (2.8) is true for $\nu = n$, calculate that

$$\begin{aligned}
[Q^{-n-1}, P_r] &= Q^{-1}[Q^{-n}, P_r] + [Q^{-1}, P_r]Q^{-n} \\
&= Q^{-1}(-in)Q^{-(n+1)} - iQ^{-n-2} = -i(n+1)Q^{-(n+2)}.
\end{aligned}$$

Thus if (2.8) is true for $v = n$, it is true for $v = n + 1$. In the same way, one shows (2.8) is true for $v = -(n + 1)$ given that it is true for $v = -n$. Therefore (2,8) is generally true.]

As a consequence of (2.8) and (2.7) it follows by a straightforward calculation that

$$[K, Q^{-1(v-1)}] = \tfrac{1}{2}(v - 1)(2Q^{-v}iP_r - vQ^{-(v+1)}). \qquad (2.9)$$

A special case of (2.9) is

$$[K, Q] = -iP_r. \qquad (2.9a)$$

The relation

$$[K, iP_r] = \mathbf{L}^2 Q^{-3} - aQ^{-2} \qquad (2.10)$$

can be checked by inserting (2.7) into the right-hand side and using (2.8).
    With (2.9) and (2.10) one calculates

$$[K, [K, Q^{-(v-1)}]] = (v - 1)\Big(-v(v + 1)Q^{-(v+2)}iP_r - vQ^{-(v+1)}P_r^2$$
$$+ \tfrac{1}{4}v(v + 1)(v + 2)Q^{-(v+3)} + \mathbf{L}^2 Q^{-(v+3)} - aQ^{-(v+2)}\Big).$$
$$(2.11)$$

Inserting (2.9a) and (2.9) and taking the commutator of the result with $K$ gives

$$[Q^{-v}[Q, K], K] - \tfrac{1}{2}v[Q^{-(v+1)}, K] + v[Q^{-(v+1)}, K]$$

$$= \frac{1}{v - 1} [[K, Q^{-(v-1)}], K] + v[Q^{-(v+1)}, K], \qquad (2.12)$$

where one term has been added to both sides.
    Inserting (2.11), (2.7), and (2.9) into the right-hand side of (2.12) gives

$$[Q^{-v}[Q, K], K] + \tfrac{1}{2}v[Q^{-(v+1)}, K] = 2vQ^{-(v+1)}K - (v + 1)Q^{-(v+3)}\mathbf{L}^2$$
$$+ (2v + 1)aQ^{-(v+2)} + \tfrac{1}{4}v(v + 1)(v + 2)Q^{-(v+3)}. \qquad (2.13)$$

The recursion relation for the matrix elements of $Q^{-v}$ is obtained if one takes the matrix elements of (2.13) between the vectors $|n\,l\,m\rangle$, for which

$$K|n\,l\,m\rangle = -\frac{a^2}{2n^2}|n\,l\,m\rangle,$$

$$\mathbf{L}^2|n\,l\,m\rangle = l(l + 1)|n\,l\,m\rangle,$$

$$L_3|n\,l\,m\rangle = m|n\,l\,m\rangle.$$

The left-hand side of (2.13) between the vectors $|n\,l\,m\rangle$ gives zero, so that we get as a recursion relation

$$0 = -v\frac{a^2}{n^2}\langle n\,l\,m|Q^{-(v+1)}|n\,l\,m\rangle + (2v + 1)a\langle n\,l\,m|Q^{-(v+2)}|n\,l\,m\rangle$$

$$+ [-(v + 1)l(l + 1) + \tfrac{1}{4}v(v + 1)(v + 2)]\langle n\,l\,m|Q^{-(v+3)}|n\,l\,m\rangle \qquad (2.14)$$

If one takes the matrix elements of (2.13) between states $|n\,l\,m\rangle$ and $|n'\,l'\,m\rangle$, one obtains recursion relations for the off-diagonal matrix elements.[1]
    For $\nu = -1$ Equation (2.14) gives

$$\langle n\,l\,m|Q^{-1}|n\,l\,m\rangle = a/n^2. \tag{2.15}$$

The matrix elements of $Q^{-2}$ cannot be obtained from (2.14); we do not give its derivation here, but just the result:

$$\langle n\,l\,m|Q^{-2}|n\,l\,m\rangle = \frac{a^2}{n^3(l+\frac{1}{2})}. \tag{2.16}$$

For $\nu = -2$ Equation (2.14) gives

$$\langle n\,l\,m|Q|n\,l\,m\rangle = \frac{n^2}{2a}\left(3 - \frac{l(l+1)}{n^2}\right).$$

Of all the matrix elements of $Q^{-\nu}$ that can be obtained from (2.14), we list the lowest ones:

$$\langle n\,l\,m|Q^{-3}|n\,l\,m\rangle = \frac{a}{l(l+1)}\langle n\,l\,m|Q^{-2}|n\,l\,m\rangle$$

$$= \frac{a^3}{n^3 l(l+1)(l+\frac{1}{2})}, \tag{2.17}$$

$$\langle n\,l\,m|Q^{-4}|n\,l\,m\rangle = \frac{a}{n^2}\frac{3n^2 - l(l+1)}{2l(l+1) - \frac{3}{2}}\langle n\,l\,m|Q^{-3}|n\,l\,m\rangle. \tag{2.18}$$

The constant $a$ has, in our units, the dimension of an inverse length. According to (VI.3.5),

$$a = m_e c\alpha. \tag{2.19}$$

The above $Q$ is, because of our choice of units, not measured in cm but rather in $(eV/c)^{-1}$ [cf. the discussion preceding Eq. (VI.3.4)], i.e., the above $Q$ is in fact $Q/\hbar$. If $Q$ is measured in its usual units of cm, then the above $a$ is to be replaced by $a/\hbar$. Thus for $Q$ measured in cm the above equations hold with $a$ replaced by

$$a' = \frac{m_e c\alpha}{\hbar} = \frac{1}{r_B} = \frac{1}{0.529 \times 10^{-8}\ \text{cm}}. \tag{2.20}$$

$r_B$ is called the *Bohr radius*.
    We can now give the energy value

$$E(n, l) = -\frac{a^2}{2n^2} + \langle n\,l\,m|V(Q)|n\,l\,m\rangle \tag{2.21}$$

[1] L. C. Biedenharn, N. V. Swamy, *Journal Math. Phys.* **11**, 1165 (1970).

for any $V$ that is a sum of positive and negative powers of $Q$. Equation (2.21) is—as we mentioned above and as we shall see in the next chapter—also equal to the first approximation of the perturbation expansion. We shall use for $V(Q)$ only the dipole term

$$V(Q) = -c_1 a/Q^2, \tag{2.22}$$

which—as was mentioned above—should be a sufficiently good approximation.

From (2.16) and (2.22) one obtains for $E(n, l)$

$$E(n, l) = -\frac{a^2}{2n^2}\left(1 + \frac{2c_1 a}{n(l + \frac{1}{2})}\right). \tag{2.23}$$

Thus the representation spaces $\mathscr{R}(n)$ of $\mathscr{E}(SO(4))$ no longer have only one energy value, for $H$ does not commute with $A_i$ [In group-theoretical language this is expressed as "$V(Q)$ breaks the SO(4) symmetry."]

It is customary to write (2.23) in the form

$$E(n, l) = -\frac{2^2}{2n^{*2}} = -\frac{a^2}{2(n - \sigma)^2}, \tag{2.24}$$

where $n^*$ is called the *effective quantum number* and $\sigma = \sigma(n, l)$ is called the *quantum defect*. Comparing (2.24) with (2.23) and using the fact that $c_1$ is small compared to the Bohr radius $1/a$, one obtains that the quantum defect for the alkali atoms is given by

$$\sigma = \frac{c_1 a}{l + \frac{1}{2}}. \tag{2.25}$$

The energy levels for lithium, sodium, and cesium are given in the diagrams of Figures 2.1, 2.2, and 2.3, respectively. The energy levels are labeled in the standard notation. The first number stands for the principal quantum number $n$, and the letters $s$, $p$, $d$, $f$, ... denote the angular-momentum quantum-number values $l = 0, 1, 2, 3, \ldots$, respectively; e.g., $3p$ means $n = 3$, $l = 1$. The lines connecting different energy levels indicate dipole transitions. The numbers in the lines give the wavelengths in angstroms. These transitions fulfill the selection rule $\Delta l = \pm 1$ [cf. (V4.33)]. Thick lines indicate transitions of higher intensity.

Comparing these energy diagrams with the energy diagram of Figure VI.5.1 for hydrogen, we observe the $l$-dependence of the correction term in (2.23), which leads to a splitting of the levels with the same principal quantum number. Thus (2.23) gives a good quantitative approximation to the experimental energy spectrum. We observe further that some of the energy levels split in contradiction to (2.23); e.g., there are two energy levels $3p_1$ and $3p_2$ of sodium, whereas according to (2.23) there should be only one. According to (2.14) through (2.18), even if one takes for $V(Q)$ the general spherically symmetric expansion (1.3″), the energy values should depend upon $n$ and $l$ only. The splitting of these levels with the same value of the quantum

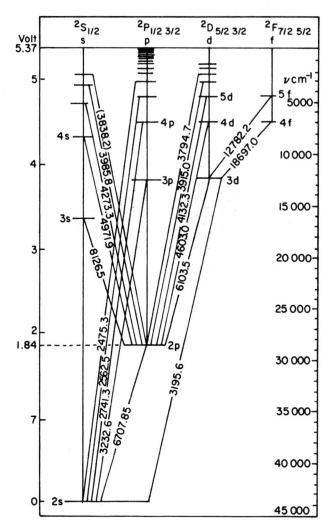

Figure 2.1   Energy levels of lithium atoms.

numbers $n$, $l$ will be explained by the introduction of the new observable *spin*, which will be discussed in Chapter IX.

We also observe in the experimental energy diagrams of Figures 2.1–2.3 that for lithium there are no energy levels with $n = 1$ or 2, that for potassium (which is not depicted here) there are no energy levels with $n = 1$, 2, or 3, and that for cesium there are no energy levels with $n = 1$, 2, 3, 4, or 5. This means that our physical system, which is the outer electron in the electrostatic potential $V_A(r) = -e^2/r + V(r)$, cannot be in a state with principal quantum number $n = 1$ (lithium), $n = 1$ or 2 (sodium), etc. In Chapter X we shall see that this can be explained as a consequence of a new basic assumption of quantum mechanics, the *Pauli exclusion principle*.

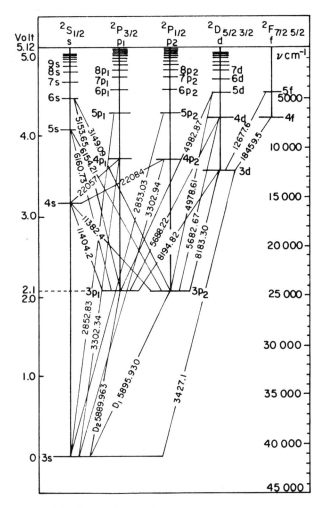

Figure 2.2    Energy levels of sodium atoms.

## VII.3  Wavefunctions and Schrödinger Equation of the Hydrogen Atom and the Alkali Atoms

The Solution of the Schrödinger equation for the hydrogen atom is so familiar that it need not be discussed in detail here. It does not lead to any new observable results, which have not already been discussed in the previous chapter. The new quantities that it provides are the wave functions $\psi_{nlm}(\mathbf{x}) = \langle x_1, x_2, x_3 | l\, n\, m \rangle = \langle \mathbf{x} | n\, l\, m \rangle$. The quantity $|\psi(\mathbf{x})|^2$ represents, according to the results in Chapters I and II, the probability density for the observation of the electron in the state $|n\, l\, m\rangle$ at the position $\mathbf{x}$. Such a position measurement, however, cannot be performed. Nevertheless, solving the Schrödinger equation is the most common method of calculating the energy values,

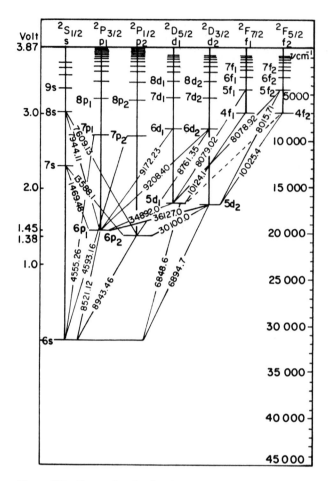

Figure 2.3   Energy levels of cesium atoms.

and it is often a very convenient method. Furthermore, the wave functions will be of physical importance for scattering problems. We give here only a brief description of this method, since detailed calculations are given in many good standard textbooks on quantum mechanics.

The Schrödinger equation for the one-electron atom is obtained by taking the "scalar product" of the energy eigenvalue equation

$$H|n\,l\,m\rangle = E|n\,l\,m\rangle$$

with the generalized eigenvector $|\mathbf{x}\rangle = |x_1 x_2 x_3\rangle$. Using (2.7) one obtains then:

$$\langle \mathbf{x}| \left( \frac{P_r^2}{2} + \frac{L^2}{2Q^2} + V_A(Q) \right) |n\,l\,m\rangle = E\langle \mathbf{x}|n\,l\,m\rangle. \qquad (3.1)$$

For the hydrogen atom $V_A(Q) = V_{\text{Hydr}}(Q) = -a/Q$. Equation (3.1) also applies to any other one-particle problem in three-dimensional space if

$V_A(Q)$ is replaced by the appropriate potential-energy operator, and the $|n\,l\,m\rangle$ by the appropriate basis vectors.

In the basis of generalized eigenvectors $|\mathbf{x}\rangle$ of the $Q_i$, the position operators $Q_i$ act like multiplication by $x_i$:

$$\langle\mathbf{x}|Q_i|\psi\rangle = x_i\langle\mathbf{x}|\psi\rangle. \tag{3.2a}$$

The momentum operators $P_i$, according to (II.8.59), act like differentiation with respect to $x_i$:[2]

$$\langle\mathbf{x}|P_i|\psi\rangle = \frac{1}{i}\frac{\partial}{\partial x_i}\langle\mathbf{x}|\psi\rangle. \tag{3.2b}$$

We label the generalized position eigenvectors $|\mathbf{x}\rangle$ by the polar coordinates $(r, \theta, \phi)$ instead of the Cartesian coordinates:

$$x_1 = r\sin\theta\cos\phi, \qquad x_2 = r\sin\theta\sin\phi, \qquad x_3 = r\cos\theta. \tag{3.3}$$

From (2.2) and (3.2) it then follows that

$$\langle\mathbf{x}|P_r|\psi\rangle = \frac{1}{i}\left(\frac{\partial}{\partial r} + \frac{1}{r}\right)\langle\mathbf{x}|\psi\rangle,$$

$$\langle\mathbf{x}|P_r^2|\psi\rangle = -\frac{1}{r^2}\frac{\partial}{\partial r}\left(r^2\frac{\partial}{\partial r}\langle\mathbf{x}|\psi\rangle\right). \tag{3.4}$$

The matrix element of the angular momentum operator $L_i = \epsilon_{ijk}Q_jP_k$ in the basis of generalized eigenvectors of the $Q_i$ is, according to (3.2), given by

$$\langle\mathbf{x}|L_i|\psi\rangle = \frac{1}{i}\epsilon_{ijk}x_j\frac{\partial}{\partial x_k}\langle\mathbf{x}|\psi\rangle. \tag{3.5}$$

If the generalized position eigenvectors are labeled by the polar coordinates (3.3), the differential operators on the right-hand side of (3.5) have to be

---

[2] Note that the derivation of (II.8.59) required, in addition to the Heisenberg commutation relation, the assumption that $P^2 + Q^2$ had at least one eigenvector (or equivalently $P^2 + Q^2$ was essentially self-adjoint). Thus (3.2b) is not a consequence of (VI.3.4) only, but requires an additional assumption. This assumption is natural for the case of the three-dimensional oscillator, because then $\mathbf{P}^2 + \mathbf{Q}^2$ represents the observable energy. For the hydrogen atom the energy is given by (2.1), and $\mathbf{P}^2 + \mathbf{Q}^2$ need not correspond to any observable for the hydrogen atom. Thus (3.2b) is an additional assumption, and the description of the hydrogen atom given in Chapter VI is not completely equivalent to the description by the Schrödinger equation. $\mathscr{R}$ given by (VI.5.2), considered as a Hilbertian direct sum, is a complete Hilbert space; it is a representation space of a unitary representation of SO(4) and of an irreducible unitary representation of the groups SO(4, 1) and SO(4, 2). In $\mathscr{R}$ the operator $C_1$ given by (VI.3.13)—and therewith, by (VI.3.17), the energy operator of the hydrogen atom—is self-adjoint. The space of square-integrable solutions of the Schrödinger equation (3.1) with discrete eigenvalues (the space of bound states) is not a complete Hilbert space. Remarkably, these distinct topological properties do not make any difference for the observable quantities.

transformed from Cartesian to spherical coordinates; the results of that somewhat lengthy transformation are

$$\langle \mathbf{x}|L_1|\psi\rangle = \frac{1}{i}\left(-\sin\phi\,\frac{\partial}{\partial\theta} - \cot\theta\cos\phi\,\frac{\partial}{\partial\phi}\right)\langle r\,\theta\,\phi|\psi\rangle, \qquad (3.6_1)$$

$$\langle \mathbf{x}|L_2|\psi\rangle = \frac{1}{i}\left(+\cos\phi\,\frac{\partial}{\partial\theta} - \cot\theta\sin\phi\,\frac{\partial}{\partial\phi}\right)\langle r\,\theta\,\phi|\psi\rangle, \qquad (3.6_2)$$

$$\langle \mathbf{x}|L_3|\psi\rangle = \frac{1}{i}\frac{\partial}{\partial\phi}\langle r\,\theta\,\phi|\psi\rangle. \qquad (3.6_3)$$

From this one calculates

$$\langle \mathbf{x}|\mathbf{L}^2|\psi\rangle = -\left(\frac{1}{\sin^2\theta}\frac{\partial^2}{\partial\phi^2} + \frac{1}{\sin\theta}\frac{\partial}{\partial\theta}\left(\sin\theta\,\frac{\partial}{\partial\theta}\right)\right)\langle r\,\theta\,\phi|\psi\rangle. \quad (3.6)$$

For $|\psi\rangle = |n\,l\,m\rangle$ the equations $(3.6_3)$ and $(3.6)$ become

$$\frac{1}{i}\frac{\partial}{\partial\phi}\langle r\,\theta\,\phi|n\,l\,m\rangle = m\langle r\,\theta\,\phi|n\,l\,m\rangle \qquad (3.6_3')$$

and

$$-\left(\frac{1}{\sin^2\theta}\frac{\partial^2}{\partial\phi^2} + \frac{1}{\sin\theta}\frac{\partial}{\partial\theta}\left(\sin\theta\,\frac{\partial}{\partial\theta}\right)\right)\langle r\,\theta\,\phi|n\,l\,m\rangle = l(l+1)\,\langle r\,\theta\,\phi|n\,l\,m\rangle.$$

$$(3.6')$$

Inserting (3.4) into (3.1) gives

$$-\frac{1}{2}\frac{1}{r^2}\frac{\partial}{\partial r}\left(r^2\frac{\partial}{\partial r}\langle r\,\theta\,\phi|n\,l\,m\rangle\right) + \frac{l(l+1)}{2r^2}\langle r\,\theta\,\phi|n\,l\,m\rangle + V(r)\langle r\,\theta\,\phi|n\,l\,m\rangle$$

$$= E\langle r\,\theta\,\phi|n\,l\,m\rangle. \quad (3.7)$$

Using the expansion

$$|n\,l\,m\rangle = \int d^3x\,|\mathbf{x}\rangle\langle\mathbf{x}|n\,l\,m\rangle$$

$$= \int r^2\sin\theta\,dr\,d\theta\,d\phi|r\,\theta\,\phi\rangle\langle r\,\theta\,\phi|n\,l\,m\rangle \qquad (3.8)$$

in the orthonormality condition

$$\langle n'\,l'\,m'|n\,l\,m\rangle = \delta_{nn'}\,\delta_{ll'}\,\delta_{mm'} \qquad (3.9)$$

gives the normalization condition for the transition coefficients:

$$\int r^2\sin\theta\,dr\,d\theta\,d\phi\,\langle n'\,l'\,m'|r\,\theta\,\phi\rangle\langle r\,\theta\,\phi|n\,l\,m\rangle = \delta_{nn'}\,\delta_{ll'}\,\delta_{mm'}. \quad (3.10)$$

Equation (3.7) for a given value of $l$ is a differential equation in the variable $r$ only, while (3.6$_3$) and (3.6′) are differential equations in the angular variables $\theta$, $\phi$ only. Therefore the transition coefficients are written

$$\langle r\,\theta\,\phi | n\,l\,m \rangle = R_{nl}(r)Y_{lm}(\theta, \phi). \tag{3.11}$$

Equations (3.7), (3.6′), and (3.6$_3$) then lead to

$$\left\{ -\frac{1}{2r^2}\frac{\partial}{\partial r}\left( r^2\frac{\partial}{\partial r} \right) + \frac{l(l+1)}{2r^2} + V_A(r) \right\} R_{nl}(r) = E(n, l)R_{nl}(r), \tag{3.12}$$

$$\left\{ -\frac{1}{\sin^2\theta}\frac{\partial^2}{\partial\phi^2} + \frac{1}{\sin\theta}\frac{\partial}{\partial\theta}\left( \sin\theta\frac{\partial}{\partial\theta} \right) \right\} Y_{lm}(\theta, \phi) = l(l+1)Y_{lm}(\theta, \phi), \tag{3.13}$$

and

$$\left\{ \frac{1}{i}\frac{\partial}{\partial\phi} \right\} Y_{lm}(\theta, \phi) = mY_{lm}(\theta, \phi). \tag{3.14}$$

For the hydrogen atom $V_A(r) = -a/r$ and $E(n, l) = -a^2/(2n^2)$. The normalization condition (3.10) gives the following normalization conditions for $R_{nl}$ and $Y_{lm}$:

$$\int r^2\,dr\,R_{n'l}(r)R_{nl}(r) = \delta_{nn'}, \tag{3.15}$$

and

$$\int \sin\theta\,d\theta\,d\phi\,\overline{Y}_{l'm'}(\theta, \phi)Y_{lm}(\theta, \phi) = \delta_{ll'}\,\delta_{mm'}. \tag{3.16}$$

The normalized solutions of (3.13) and (3.14) are the *spherical harmonics*

$$Y_{lm}(\theta, \phi) = \frac{(-1)^l}{2^l l!}\sqrt{\frac{2l+1}{4\pi}\frac{(l+m)!}{(l-m)!}}\,e^{im\phi}\sin^{-m}\theta\left( \frac{d}{d\cos\theta} \right)^{l-m}\sin^{2l}\theta. \tag{3.17}$$

It should be noted that different authors use different phase conventions in defining the $Y_{lm}(\theta, \phi)$. When $m = 0$, (3.17) becomes

$$Y_{l0}(\theta, \phi) = \sqrt{\frac{2l+1}{4\pi}}\,P_l(\cos\theta), \tag{3.18}$$

where $P_l$ is the *Legendre polynomial*.

It is frequently convenient to relate (3.17) to the *associated Legendre function*

$$P_l^{|m|}(\theta) = \frac{(-1)^{m+l}}{2^l l!}\frac{(l+|m|)!}{(l-|m|)!}\sin^{-|m|}\theta\left( \frac{d}{d\cos\theta} \right)^{l-|m|}\sin^{2l}\theta.$$

For $m \geq 0$ the desired relation is seen to be

$$Y_{lm}(\theta, \phi) = (-1)^m\frac{2l+1}{4\pi}\frac{(l-m)!}{(l+m)!}\,e^{im\phi}P_l^m(\theta); \tag{3.19}$$

for negative $m$ one uses

$$Y_{l,-m}(\theta, \phi) = (-1)^m \overline{Y}_{lm}(\theta, \phi). \tag{3.19'}$$

Occasionally we shall require the explicit form of the spherical harmonics. In tabulating the formulas it is convenient to write $Y_{lm}(\theta, \phi) = \Theta_{lm} e^{im\phi}/\sqrt{2\pi}$. Working from (3.17), one then finds

$$\Theta_{11}(\theta) = -\sqrt{\tfrac{3}{4}}\sin\theta, \qquad \Theta_{10}(\theta) = \sqrt{\tfrac{3}{2}}\cos\theta,$$

$$\Theta_{22}(\theta) = \sqrt{\tfrac{15}{16}}\sin^2\theta, \qquad \Theta_{21} = -\sqrt{\tfrac{15}{4}}\cos\theta\sin\theta, \tag{3.20}$$

$$\Theta_{20}(\theta) = \sqrt{\tfrac{5}{8}}(2\cos^2\theta - \sin^2\theta).$$

Finally we quote three important relations involving the spherical harmonics. The first of these states that $\{Y_{lm}\}$ is a complete set of single-valued functions on the unit sphere:

$$\sum_{l=0}^{\infty} \sum_{m=-l}^{l} Y_{lm}(\theta, \phi) \overline{Y}_{lm}(\theta', \phi') = \frac{\delta(\theta - \theta')\delta(\phi - \phi')}{\sin\theta}. \tag{3.21}$$

The other relationship is the *addition theorem*: Let $\mathbf{n}_1$ and $\mathbf{n}_2$ be unit vectors with orientations specified by $(\theta_1, \phi_1)$ and $(\theta_2, \phi_2)$, respectively; then

$$P_l(\mathbf{n}_1 \cdot \mathbf{n}_2) = \frac{4\pi}{2l+1} \sum_{m=-l}^{l} Y_{lm}(\theta_1, \phi_1)\overline{Y}_{lm}(\theta_2, \phi_2). \tag{3.22a}$$

The last relation is the *coupling rule*:[3]

$$Y_{l_1 m_1}(\theta, \phi) Y_{l_2 m_2}(\theta, \phi) = \sum_l \left[\frac{(2l_1+1)(2l_2+1)}{4\pi(2l+1)}\right]^{1/2}$$

$$\times \langle l_1 m_1 l_2 m_2 | lm \rangle \langle l_1 0 l_2 0 | l0 \rangle Y_{lm}(\theta, \phi). \tag{3.22b}$$

The normalized solutions of Equation (3.12) for $V_A(r) = -a/r$ and $E(n, l) = -a^2/(2n^2)$ are

$$R_{nl}(r) = (2na)^{3/2} \sqrt{\frac{n-l-1}{2n((n+l)!)^3}} (2nar)^l L_{n+l}^{2l+1}(2nar)e^{-nar}, \tag{3.23}$$

where $L_{n+l}^{2l+1}(\rho)$ are the *Laguerre polynomials*. $a = 1/r_{\text{Bohr}}$ is the inverse Bohr radius in units eV$/c$ [cf. (2.19)]; in the usual units in which $r$ and $r_{\text{Bohr}}$ have the dimension cm, $a$ is to be replaced by $a'$ of (2.20).

Figure 3.1b shows $R_{nl}(r)$ for some of the lower values of $n$ and $l$. Figure 3.1a gives the probability distribution for finding the electron at $r$.

---

[3] See, e.g., Rose (1957, Section 14).

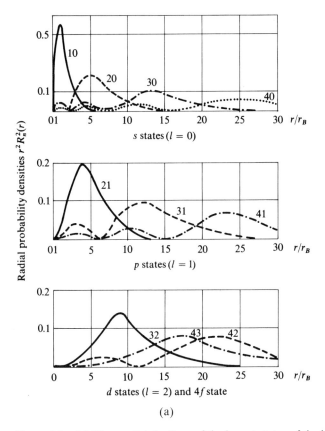

Figure 3.1   (a) Charge distributions of the lowest states of the hydrogen atom. The radial distance $r$ is in units of the Bohr radius. The curves are labeled by the numbers $nl$.

Knowledge of $\langle \mathbf{x} | n\,l\,m \rangle$ allows an alternative computation of the matrix element of $V(Q)$ to the one given in the previous section; $\epsilon(n, l) = \langle n\,l\,m | V(Q) | n\,l\,m \rangle$ can now be calculated by inserting the complete generalized basis $|\mathbf{x}\rangle$:

$$\epsilon(n, l) = \int d^3x \int d^3x' \, \langle n\,l\,m | \mathbf{x} \rangle \langle \mathbf{x} | V(Q) | \mathbf{x}' \rangle \langle \mathbf{x}' | n\,l\,m \rangle$$

$$= \int d^3 x \, |\langle n\,l\,m | \mathbf{x} \rangle|^2 V(r).$$

Multiplying (3.23) by $Y_{lm}(\theta, \phi)$, we obtain, according to (3.11), the transition coefficient $\langle r\,\theta\,\phi | n\,l\,m \rangle$. Substitution of $\langle r\,\theta\,\phi | n\,l\,m \rangle$ into the above equation and use of (3.16) to perform the $\theta, \phi$ integrations then yield

$$\epsilon(n, l) = (2na)^3 \frac{(n - l - 1)!}{(2n(n + l)!)^3} \int r^2 \, dr \, V(r)(2nar)^{2l} |L_{n+l}^{2l+1}(2nar)|^2 \, e^{-2anr}.$$

$$(3.24)$$

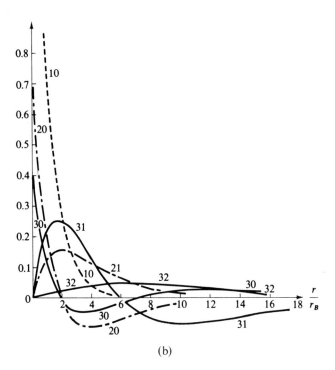

(b)

Figure 3.1   (b) The radial wave functions $R_{nl}(r)$ for hydrogenic atoms
for $n = 1, 2, 3$. Each curve is labeled by $nl$. Note the effect of the
centrifugal force in "pushing out" the wave function from the nucleus.
Note also that the functions have $n - l - 1$ nodes. [Part (b) from
Dicke & Wittke (1960), with permission.]

These integrals can be evaluated for the various functions $V(r) = -c/r^2$,
$V(r) = -c/r^3$, $V(r) = -c/r^4, \ldots$ and lead to the same result as given in
Equation (2.14)–(2.18). From Figure 3.1a we can make some qualitative
observations regarding the values $\epsilon(n, l)$. For a given value of $n$ the probability
distribution moves towards larger values of $r$ as $l$ increases. Therefore, for a
given value of $n$, $-\epsilon(n, l)$ is the largest, so that $E(n, l = 0)$ will be the lowest
energy value, and with increasing $l$ the $E(n, l)$ will approach the value
$-a^2/(2n^2)$. That this is indeed the case can be seen from the experimental
energy diagrams of the alkali atoms in the previous section.

## Problem

   1.   Calculate the matrix elements of the position operator $Q_\kappa$ ($\kappa = -1, 0, +1$)
between the SO(4) states $|n\,l\,m\rangle$ of the hydrogen atom, i.e., calculate

$$\langle n'\,l'\,m'|Q_\kappa|n\,l\,m\rangle.$$

It is sufficient to do this for the third component $Q_0$. (*Hint:* Use the matrix elements of
the modified Lenz vector $A_\kappa$ and algebraic relations between the $Q_\kappa$ and the $A_\kappa$.)

# CHAPTER VIII

# Perturbation Theory

In this chapter perturbation theory is developed in a general form that is easily adaptable to the discrete as well as the continuous spectrum. At the end of Section 1 the Wigner–Brillouin and the Rayleigh–Schrödinger perturbation expansions are obtained as special cases. Section 2 applies this general procedure to the continuous spectrum and results in the Lippman–Schwinger equation.

## VIII.1 Perturbation of the Discrete Spectrum

This section will develop the general procedure called *perturbation theory* for the evaluation of the eigenvalues $E_\lambda$ and eigenvectors $|\lambda\rangle$,

$$H|\lambda\rangle = E_\lambda|\lambda\rangle, \tag{1.1}$$

of an operator

$$H = K + V, \tag{1.2}$$

which differs by a small *perturbation* $V$ from an operator $K$ whose spectrum $\epsilon_a$ and eigenvectors $|a\rangle$,

$$K|a\rangle = \epsilon_a|a\rangle, \tag{1.3}$$

are known. To have a concrete case in mind, one may assume that $H$, $K$, and $V$ are the operators of Section VII.1, that

$$|\lambda\rangle = |\lambda\,l\,m\rangle, \qquad |a\rangle = |n\,l\,m\rangle$$

for any fixed values of $l, m$ (which is possible, since $[L_i, V] = 0$), and that[1]

$$\epsilon_{a=(nlm)} = -\frac{(m_e c \alpha)^2}{2 m_e n^2} = -\frac{m_e e^4}{2 \hbar^2} \frac{1}{n^2}.$$

In this kind of perturbation-theoretical treatment it is assumed that to every eigenvector $|a\rangle$ of $K$ there corresponds an eigenvector $|\lambda(a)\rangle$ of $H$ such that $E_{\lambda(a)}$ is given by $\epsilon_a$ plus a small correction term, i.e., such that if $V$ is zero, then $E_{\lambda(a)} = \epsilon_a$. In the case that $\epsilon_a$ is degenerate (i.e., that there are other quantum numbers $\eta$ besides $a$ or $\lambda$), $E_{\lambda(a)}$ will in general depend upon these additional quantum numbers even though $\epsilon_a$ does not.

Although at the moment we have in mind discrete values for $\epsilon_a$ and $a$, we do not want to exclude the possibility that $K$ has a continuous spectrum;[2] in that case the summation $\sum_a$, in the equations of this section, is to be replaced by an integration over the continuous spectrum (cf. Section VIII.2.)

We define the operator

$$G(E) \equiv (EI - H)^{-1} \qquad (I = \text{unit operator}), \qquad (1.4)$$

which depends upon the parameter $E$. $G(E)$ is called the "operator for the Green's function for the Hamiltonian" by the physicists, and is called the "resolvent of the operator $H$" by the mathematicians. It is (intuitively) clear that the task of finding the eigenvalues $E_\lambda$ of $H$ is equivalent to that of finding the "singularities" of $G(E)$.

We write $G$ as the product of two operators

$$G = Fg, \qquad (1.5)$$

where

$$g(E) \equiv \sum_a |a\rangle\langle a| G_a(E), \qquad (1.6)$$

$$G_a(E) \equiv \langle a| G(E) |a\rangle, \qquad (1.7)$$

and $F$ is defined by (1.5). If (1.5) is inserted into the identity[3]

$$(E - H)G(E) = I, \qquad (1.8)$$

one obtains

$$(E - K)Fg = I + VFg = I + Rg, \qquad (1.9)$$

where the operator $R(E)$ is defined by

$$R \equiv VF, \qquad (1.10)$$

and its diagonal matrix element is denoted by

$$R_a = \langle a| VF |a\rangle. \qquad (1.11)$$

---

[1] We revert to the usual cgs units in this chapter.

[2] The following treatment of perturbation theory is based on the Goldberger–Watson Collision Theory (1964, Section 8.1). It has been chosen because it displays the connection between the cases of the discrete and the continuous spectrum.

[3] $E$ stands for $E \cdot I$; the unit operator in a product with a number will be omitted.

$R$ is called the *level-shift operator* and $R_a$ the *level shift*, for the following reason: If one takes the diagonal matrix elements of (1.9) one obtains

$$(E - \epsilon_a)\langle a|F|a\rangle G_a = 1 + R_a G_a. \tag{1.12}$$

It is easily seen from the definitions (1.5) and (1.6) that

$$\langle a|F|a\rangle = 1. \tag{1.13}$$

Consequently,

$$G_a(E) = \frac{1}{E - \epsilon_a - R_a(E)}. \tag{1.14}$$

As the eigenvalues $E_\lambda$ are the singularities of $G(E)$ and consequently of its matrix element $G_a(E)$, one obtains from (1.14)

$$E_\lambda = \epsilon_a + R_a(E_\lambda). \tag{1.15}$$

Though (1.15) is valid for any $E_\lambda$ and $\epsilon_a$, it is used in practice to determine that value of $E_\lambda$ which is "closest" to $\epsilon_a$, $i^q E_{\lambda(a)}$. Thus $R_a(E_\lambda)$ represents the level shift from the "unperturbed" energy value $\epsilon_a$ to the corresponding $E_{\lambda(a)}$.

As already stated above, the assumption[4] that forms the basis of this perturbation theory is that to every $|a\,\eta\rangle$ there corresponds a $|\lambda(a)\,\eta\rangle$ (i.e., there are as many $|a\,\eta\rangle$ as there are $|\lambda(a)\,\eta\rangle$), but $E_{\lambda(a)}$ may be a function of $\eta$ because $R_a(E_\lambda)$ does in general depend upon $\eta$—e.g., in the example of Section VII.1

$$R_a(E_\lambda) = \langle a|R(E_\lambda)|a\rangle = \langle n\,l\,m|R(E_\lambda)|n\,l\,m\rangle \tag{1.16}$$

—and will in general depend upon $n, l, m$. In the particular case $[R, L_i] = 0$ it follows that $R_a(E_\lambda)$ does not depend upon $m$, but it will depend upon $l$ in

---

[4] This assumption, which underlies perturbation theory, can be shown to be fulfilled for certain $V$, as the following theorem states:
Let

$$H = H(\kappa) = K + \kappa K^{(1)} + \kappa^2 K^{(2)} + \cdots$$

with

$$\|K^{(k)}f\| \le \frac{M}{r^{k-1}}(\|f\| + \|Hf\|) \qquad (k = 1, 2, \ldots),$$

where the numbers $M$ and $r$ are positive constants. Then for $|\kappa| < r$ the eigenvalues $E(\kappa)$ of $H$ can be written

$$E(\kappa) = E^{(0)} + \kappa E^{(1)} + \kappa^2 E^{(2)} + \cdots,$$

and the eigenvectors $|E(\kappa)\rangle$ of $H$ can be written

$$|E(\kappa)\rangle = |E^{(0)}\rangle + \kappa|E^{(1)}\rangle + \kappa^2|E^{(2)}\rangle + \cdots.$$

For $\kappa \to 0$ it follows that

$$H = K,$$
$$E(\kappa = 0) = E^{(0)} = \epsilon_a,$$
$$|E(\kappa = 0)\rangle = |E^{(0)}\rangle = |a\rangle.$$

addition to its dependence upon $n$. Thus all the $|a\,\eta\rangle$ belong to one energy level $\epsilon_a$ of the unperturbed system, and under the influence of the perturbation this level splits into sublevels $E_{\lambda(a)\eta}$ and the state $|a\,\eta\rangle$ of the unperturbed system goes into the state $|\lambda(a)\,\eta\rangle$ of the perturbed system.

The level-shift operator $R$, and thus $R_a$ and $E_\lambda$, are evaluated by various sequences of approximations. These various sequences are obtained in the following way: (1.9) can be written

$$(E - K - O)Fg = I + (V - O)Fg, \tag{1.17}$$

where $O$ may be any operator. We restrict $O$ to be of the form

$$O(E) = \sum_a |a\rangle\langle a| O_a(E), \tag{1.18}$$

where $O_a(E)$ are numbers (i.e., $O$ is assumed to have the property $[O, K] = 0$). From (1.17) one obtains

$$F = \frac{1}{E - K - O}\frac{1}{g} + \frac{1}{E - K - O}(V - O)F. \tag{1.19}$$

This equation is not defined on every vector, not even on every vector $|a\rangle$, for any value $E$. That is, the first term is only defined on those $|a\rangle$ and for those $E$ for which

$$(E - \epsilon_a - O_a(E))G_a(E) \neq 0.$$

As discussed above, to every eigenvector $|a\rangle$ of $K$ there corresponds an eigenvector $|\lambda(a)\rangle$ of $H$. The operator $F(E)$ has been constructed such that $F(E_{\lambda(a)})$ is the operator that transforms $|a\rangle$ into the corresponding $|\lambda(a)\rangle$. To show this we calculate $F(E)|a\rangle$ at the value $E = E_{\lambda(a)}$. Applying (1.19) to $|a\rangle$ and making use of (1.6) and (1.14) gives

$$F(E)|a\rangle = \frac{1}{E - K - O(E)}(E - \epsilon_a - R_a(E))|a\rangle$$

$$+ \frac{1}{E - K - O(E)}(V - O(E))F(E)|a\rangle \tag{1.20}$$

$O_a(E)$ will be chosen such that this expression becomes the simplest possible relation for $F(E_{\lambda(a)})|a\rangle$; this can be done if at the value $E = E_{\lambda(a)}$, $F(E_{\lambda(a)})|a\rangle$ is written as $|a\rangle$ plus a correction term. Thus $O_a(E)$ is to be chosen so that the first term becomes $|a\rangle$.

Two choices of $O(E_{\lambda(a)})$ lead to particularly well-known perturbation expansions:

$$O(E_{\lambda(a)}) = R_a(E_{\lambda(a)})|a\rangle\langle a|, \tag{1.21}$$

or

$$O_{a'}(E_{\lambda(a)}) = R_{a'}(E_{\lambda(a)})\delta_{a'a}; \tag{1.21a}$$

and

$$O(E_{\lambda(a)}) = R_a(E_{\lambda(a)})|a\rangle\langle a| + (E_{\lambda(a)} - \epsilon_a)\sum_{a' \neq a}|a'\rangle\langle a'|, \tag{1.22}$$

or

$$O_a(E_{\lambda(a)}) = R_a(E_{\lambda(a)}) \qquad O_{a'}(E_{\lambda(a)}) = E_{\lambda(a)} - \epsilon_a \quad \text{for } a' \neq a. \quad (1.22a)$$

In the case (1.21),

$$(E - K - O(E))_{E=E_{\lambda(a)}}^{-1} = |a\rangle\langle a|(E - \epsilon_a - R_a(E))_{E=E_{\lambda(a)}}^{-1}$$
$$+ \sum_{a' \neq a} |a'\rangle\langle a'|(E - \epsilon_{a'})_{E=E_{\lambda(a)}}^{-1}; \quad (1.23)$$

and in the case (1.22),

$$(E - K - O(E))_{E=E_{\lambda(a)}}^{-1} = |a\rangle\langle a|(E - \epsilon_a - R_a(E))_{E=E_{\lambda(a)}}^{-1}$$
$$+ \sum_{a' \neq a} |a'\rangle\langle a'|(\epsilon_a - \epsilon_{a'})^{-1}. \quad (1.24)$$

Thus in both cases (1.21) and (1.22) the first term on the left-hand side of (1.20) becomes $|a\rangle$. Inserting (1.23) into (1.20) gives

$$F(E_{\lambda(a)})|a\rangle = |a\rangle + \frac{1}{E_{\lambda(a)} - \epsilon_a - R_a(E_{\lambda(a)})} |a\rangle\langle a|(V - O(E_{\lambda(a)}))F(E_{\lambda(a)})|a\rangle$$

$$+ \sum_{a' \neq a} \frac{1}{E_{\lambda(a)} - \epsilon_{a'}} |a'\rangle\langle a'|(V - O(E_{\lambda(a)}))F(E_{\lambda(a)})|a\rangle. \quad (1.25)$$

The second term on the right-hand side is zero because

$$\langle a|(V - O(E_{\lambda(a)}))F(E_{\lambda(a)})|a\rangle = 0, \quad (1.26)$$

which follows from (1.21), (1.10), and (1.13).
Inserting (1.24) into (1.20) gives

$$F(E_{\lambda(a)})|a\rangle = |a\rangle + \frac{1}{E_{\lambda(a)} - \epsilon_a - R_a(E_{\lambda(a)})} |a\rangle\langle a|(V - O(E_{\lambda(a)}))F(E_{\lambda(a)})|a\rangle$$

$$+ \sum_{a' \neq a} \frac{1}{\epsilon_a - \epsilon_{a'}} |a'\rangle\langle a'|(V - O(E_{\lambda(a)}))F(E_{\lambda(a)})|a\rangle. \quad (1.27)$$

The second term on the right-hand side is again zero because again (1.26) holds as a result of (1.22), (1.10), (1.13). Thus the cases (1.21) and (1.22) both lead to a very similar form for the equation of $F(E_{\lambda(a)})|a\rangle$. It is now easily established that

$$HF(E_{\lambda(a)})|a\rangle = E_{\lambda(a)}F(E_{\lambda(a)})|a\rangle. \quad (1.28)$$

Multiplying (1.25) by $(E_{\lambda(a)} - K - O(E_{\lambda(a)}))$ and using (1.21) gives

$$(E_{\lambda(a)} - K - O(E_{\lambda(a)}))F(E_{\lambda(a)})|a\rangle$$

$$= \sum_{a'} \frac{E_{\lambda(a)} - \epsilon_{a'}}{E_{\lambda(a)} - \epsilon_{a'}} |a'\rangle\langle a'|(V - O(E_{\lambda(a)}))F(E_{\lambda(a)})|a\rangle. \quad (1.29)$$

In the $\sum_{a'}$ it is irrelevant whether the term with $a' = a$ is included or excluded, because of (1.26). Consequently (1.29) reads

$$(E_{\lambda(a)} - K - O(E_{\lambda(a)}))F(E_{\lambda(a)})|a\rangle = (H - K - O(E_{\lambda(a)}))F(E_{\lambda(a)})|a\rangle, \quad (1.30)$$

which establishes (1.28). In a similar way (1.28) can be established for (1.27).

Equations (1.25) and (1.27) can be simplified. Inserting (1.21) into (1.25) gives

$$F(E_{\lambda(a)})|a\rangle = |a\rangle + \sum_{a' \neq a} \frac{1}{E_{\lambda(a)} - \epsilon_{a'}} |a'\rangle\langle a'|(VF(E_{\lambda(a)})|a\rangle. \quad (1.31)$$

This is an equation for $F(E_{\lambda(a)})|a\rangle$, and can be iterated by successive substitution of the left-hand side into the right-hand side:

$$F(E_{\lambda(a)})|a\rangle = |a\rangle + \sum_{a' \neq a} \frac{1}{E_{\lambda(a)} - \epsilon_{a'}} |a'\rangle\langle a'|V|a\rangle$$

$$+ \sum_{\substack{a' \neq a \\ a'' \neq a}} \frac{1}{E_{\lambda(a)} - \epsilon_{a'}} \frac{1}{E_{\lambda(a)} - \epsilon_{a''}} |a'\rangle\langle a'|V|a''\rangle\langle a''|V|a\rangle + \cdots.$$

$$(1.32)$$

The level shift (1.10) and therefore, by (1.15), the eigenvalue of $H$ is obtained from (1.32) as

$$E_{\lambda(a)} = \epsilon_a + \langle a|V|a\rangle + \sum_{a' \neq a} \frac{1}{E_{\lambda(a)} - \epsilon_{a'}} \langle a|V|a'\rangle\langle a'|V|a\rangle + \cdots. \quad (1.33)$$

This provides an equation for $E_{\lambda(a)}$ in terms of the unperturbed energy value $\epsilon_a$ and the matrix element of the perturbation $V$ between the unperturbed states $|a\rangle$. Equations (1.32) and (1.33), derived from the choice (1.21) for $O(E_{\lambda(a)})$, constitute the *Wigner–Brillouin* form of the perturbation expansion.

The choice (1.22) of $O(E_{\lambda(a)})$ leads to the *Rayleigh–Schrödinger* perturbation series. Inserting (1.22) into (1.27) leads to

$$F(E_{\lambda(a)})|a\rangle = |a\rangle + \sum_{a' \neq a} \frac{1}{\epsilon_a - \epsilon_{a'}} |a'\rangle\langle a'|(V - \delta E)F(E_{\lambda(a)})|a\rangle, \quad (1.34)$$

where $\delta E$ is defined to be the difference between the perturbed and the unperturbed energy:

$$\delta E \equiv E_{\lambda(a)} - \epsilon_a. \quad (1.35)$$

From (1.34) one obtains with (1.10) and (1.15) an equation for $\delta E$:

$$\delta E = R_a(E_{\lambda(a)}) = R_a(\delta E + \epsilon_a). \quad (1.36)$$

We iterate (1.34) and (1.36) to the first and second order, respectively. We denote by $\Delta_a^{(n)}$ the $n$th-order approximation of $\delta E$, and by $F^{(n)}|a\rangle$ the $n$th-order approximation of $F|a\rangle$. In zeroth order,

$$\Delta_a^{(0)} = E_a^{(0)} - \epsilon_a = 0 \qquad F^{(0)}|a\rangle = |a\rangle. \quad (1.37)$$

In first order,

$$\Delta_a^{(1)} = R_a(\Delta_a^{(0)} + \epsilon_a) = R_a(\epsilon_a) = \langle a | V F^{(0)} | a \rangle = \langle a | V | a \rangle, \quad (1.38)$$

$$F^{(1)} | a \rangle = | a \rangle + \sum_{a' \neq a} \frac{1}{\epsilon_a - \epsilon_{a'}} | a' \rangle \langle a' | (V - \Delta_{a'}^{(0)}) F^{(0)} | a \rangle$$

$$= | a \rangle + \sum_{a' \neq a} \frac{1}{\epsilon_a - \epsilon_{a'}} | a' \rangle \langle a' | V | a \rangle. \quad (1.39)$$

In second order,

$$\Delta_a^{(2)} = R_a(\Delta_a^{(1)} + \epsilon_a) = \langle a | V F^{(1)} | a \rangle$$

$$= \langle a | V | a \rangle + \sum_{a' \neq a} \frac{1}{\epsilon_a - \epsilon_{a'}} \langle a | V | a' \rangle \langle a' | V | a \rangle. \quad (1.40)$$

## VIII.2  Perturbation of the Continuous Spectrum—The Lippman–Schwinger Equation[5]

In the previous section it was nowhere stated that $\epsilon_a$ is a discrete eigenvalue and $| a \rangle$ a proper eigenvector of $K$, though all equations were written as if this were the case. The assumption of a discrete spectrum is in fact unnecessary; we shall now discuss in more detail the case of the continuous spectrum, which, with the tools of Section II.8, is easily obtained by replacing the proper vectors $| a \rangle$ with the generalized vectors, and the discrete sums by integrals.

For the following discussions, we have to make some assumptions regarding the properties of the operators $K$ and $H = K + V$. We shall assume that

$$\text{spectrum } K \subset \text{spectrum } H. \quad (2.1)$$

This is a general enough assumption to incorporate the situations that occur for physical problems. Usually for scattering and decay processes $K$ has only a continuous spectrum that is positive; the continuous spectrum of $H$ agrees with the continuous spectrum of $K$, and in addition $H$ has a set of negative discrete eigenvalues (and sometimes also positive discrete eigenvalues lying in the continuous spectrum) that are bounded from below.

In this case of a continuous spectrum the level shift $R_a$ is not an observable quantity. However (1.15) still holds, for (as remarked in the previous section) it is true for any value $E_\lambda$ in the spectrum of $H$ and any value $\epsilon_a$ in the spectrum of $K$.

---

[5] The results of this section will not be used until Chapter XIV.

We shall consider (1.20) at the value $E_{\lambda(a)} = \epsilon_a$, i.e., we shall calculate $F(E_{\lambda(a)})|a\rangle$ for that value $E_{\lambda(a)} \in$ spectrum $H$ which is identical to the eigenvalue $\epsilon_a \in$ spectrum $K$ corresponding to the generalized eigenvector $|a\rangle$. Such a value $E_{\lambda(a)}$ exists in the spectrum of $H$ because of the assumption (2.1). We call this value $E_{\lambda(a)} = E_a$. For $E_a$ (1.15) becomes

$$E_a - \epsilon_a = R_a(E_a) = 0. \tag{2.2}$$

If we again choose (1.21) for the operator $O(E_a)$ we will again arrive at (1.31), which in the continuous form is written

$$F(E_a)|a\rangle = |a\rangle + {}^{\text{P}}\!\!\int da' \, \frac{1}{E_a - E_{a'}} |a'\rangle\langle a'|VF(E_a)|a\rangle. \tag{2.3}$$

${}^{\text{P}}\!\!\int$ means the *principal-value integral*, i.e., the integral over all $a'$ with the omission of an "infinitesimally small interval" around $E_{a'} = E_a$. $\int da \ldots$ is a symbolic way of expressing summation over the discrete label and integration with respect to a suitable weight function over the continuous label. The label $a$ in $|a\rangle$ stands for the energy $E_a$ and some additional labels $\eta_a$, which may be discrete or continuous: $|a\rangle = |E_a\eta_a\rangle$. If the $|a\rangle$ are "normalized" so that

$$\langle \eta_a' E_a'|E_a\eta_a\rangle = \rho^{-1}(E_a)\delta(E_a - E_{a'})\delta_{\eta_a\eta_a'},$$

then

$$\int da = \sum_{\eta a} \int \rho(E_a) \, dE_a$$

[cf. (II.8.24) and Chapter XIV]. Again (1.28) is fulfilled, i.e., $F(E_a)|a\rangle$ is a generalized eigenvector of $H$ with eigenvalue $E_a$. This can also be seen directly from (2.3): Multiplying (2.3) by $(E_a - K)$ and using

$$K|a\rangle = E_a|a\rangle$$

and

$${}^{\text{P}}\!\!\int da' \, |a'\rangle\langle a'| = I - |a\rangle\langle a|,$$

one obtains

$$(I - |a\rangle\langle a|)(H - E_a)F(E_a)|a\rangle = 0. \tag{2.4}$$

In the case that we have (2.2) and (1.21), Equation (1.26) becomes

$$\langle a|(H - K)F(E_a)|a\rangle = 0. \tag{1.26'}$$

This together with (2.4) leads to

$$(H - E_a)F(E_a)|a\rangle = 0. \tag{1.28'}$$

We now make use of a well-known relation from the theory of generalized functions,[6]

$$\frac{1}{x \pm i0} = \lim_{\eta > 0+} \frac{1}{x \pm i\eta} = P\frac{1}{x} \mp i\pi\delta(x). \tag{2.5}$$

Then (2.3) can be written

$$F(E_a)|a\rangle = |a\rangle + \int da' \frac{1}{E_a - E_{a'} \pm i0} |a'\rangle\langle a'|VF(E_a)|a\rangle$$

$$\pm i\pi \int da' \, \delta(E_a - E_{a'})|a'\rangle\langle a'|VF(E_a)|a\rangle. \tag{2.6}$$

Integrating and using (2.2), we see that the last term in (2.6) vanishes:

$$\pm i\pi|a\rangle\langle a|VF(E_a)|a\rangle = \pm i\pi R_a(E_a)|a\rangle = 0.$$

Because $K|a'\rangle = E_{a'}|a'\rangle$, Equation (2.6) may be written in the more usual form

$$F(E_a)|a\rangle = |a\rangle + \frac{1}{E_a - K \pm i0} V(F(E_a)|a\rangle). \tag{2.7}$$

This is called the *Lippman–Schwinger equation*. It is an equation for the generalized eigenvectors

$$|\alpha\rangle = |\alpha(a)\rangle = F(E_a)|a\rangle \tag{2.8}$$

of the operator $H$ with eigenvalue $E_a$; $E_a$ is equal to the eigenvalue $\epsilon_a$ of $K$ belonging to the generalized eigenvector $|a\rangle$. Thus

$$H|\alpha\rangle = H|\alpha(a)\rangle = E_a|\alpha\rangle \qquad K|a\rangle = E_a|a\rangle. \tag{2.9}$$

The Lippman–Schwinger equation is often written

$$|\alpha^{\pm}\rangle = |\alpha^{\pm}(a)\rangle = |a\rangle + \frac{1}{E_a - K \pm i0} V|\alpha^{\pm}\rangle. \tag{2.7'}$$

Besides the solutions of the Lippman–Schwinger equation $|\alpha^{\pm}\rangle$, there are other basis vectors of interest. These are obtained from the form (2.3) of the

[6] The proof of (2.5) may be found in Gelfand et al. (1964, Vol. 1, Chapter I, Section 2.4). Further discussion of generalized functions of the type $(x \pm i0)^{\lambda}$ may be found in Chapter I, Sections 3 and 4 of the same volume. Relations between generalized functions (distributions) like (2.5) can be understood in the conventional sense when they are multiplied with a well-behaved function $\phi(x)$ and integrated. Thus (2.5) means

$$\int dx \, \frac{1}{x \pm i0} \phi(x) = \overset{P}{\int} dx \, \frac{1}{x} \phi(x) \mp i\pi \int dx \, \delta(x)\phi(x)$$

for all well-behaved functions $\phi(x)$. Analogously, relations between generalized eigenvectors like (2.6) can be understood in the conventional sense when the scalar product is taken with a well-behaved vector $\phi \in \Phi$ (cf. Section II.10).

integral equation for $F(a)|a\rangle$, which in operator form is written as

$$F(E_a)|a\rangle = |a\rangle + \frac{P}{E_a - K} V F(E_a)|a\rangle, \qquad (2.10)$$

where P indicates that the principle-value integral is to be taken. To distinguish the solution of this equation from the two solutions of the Lippman–Schwinger equation we denote it by $|\alpha^P\rangle$ and write (2.10):

$$|\alpha^P\rangle = |a\rangle + \frac{P}{E_a - K} V |\alpha^P\rangle. \qquad (2.10')$$

## Problems

1. Let the energy operator be given by

$$H = K + \alpha Q^3 + \beta Q^4,$$

where

$$K = \frac{1}{2m} P^2 + \frac{m\omega^2}{2} Q^2$$

(anharmonic oscillator).

   (a) Calculate the eigenvalues to first order in the perturbation expansion.
   (b) Calculate for $\beta = 0$ the eigenvalues of $H$ to the second order of the Rayleigh–Schrödinger perturbation series. Compare the result with that of (a).
   (c) Calculate for $\beta = 0$ the eigenvalues of $H$ to the second order of the Wigner–Brillouin perturbation series. Compare the result with those of (a) and (b).

2. According to Equation (VIII.1.20):

$$F(E)|a\rangle = \frac{1}{E - K - O(E)} (E - K - R_a(E))|a\rangle$$

$$+ \frac{1}{E - K - O(E)} (V - O(E))F(E)|a\rangle,$$

which for $O_a(E) = R_a(E)$ can be written

$$F(E)|a\rangle = |a\rangle + \frac{1}{E - K - O(E)} (H - K - O)F(E)|a\rangle.$$

Further according to Equation (VIII.1.28) is for $E = E_{\lambda(a)}$ the vector $F(E)|\rangle$ an eigenvector of $H$:

$$HF(E_{\lambda(a)})|a\rangle = E_{\lambda(a)}F(E_{\lambda(a)})|a\rangle.$$

Consequently the above equation becomes at $E = E_\lambda$

$$F(E_\lambda)|a\rangle = |a\rangle + \frac{1}{E_\lambda - K - O(E_\lambda)} (E_\lambda - K - O(E_\lambda))F(E_\lambda)|a\rangle$$

$$= |a\rangle + F(E_\lambda)|a\rangle,$$

from which it follows that $|a\rangle = 0$. Give your comment of the above arguments.

3.   Show that [Equation (1.28')

$$HF(E_a)|a\rangle = E_a F(E_a)|a\rangle,$$

where according to Equation (VIII.2.3)

$$F(E_a)|a\rangle = |a\rangle + \int^P da' \frac{1}{E_a - E_{a'}} |a'\rangle\langle a'| V F(E_a)|a\rangle$$

is the continuous-spectrum analogue to (VIII.1.31).

# CHAPTER IX

# Electron Spin

In Section IX.2 the doublet finestructure splitting of one-electron atoms is explained as a spin effect. To obtain quantitative results the interaction Hamiltonian for the spin interaction is determined in Section IX.3. In Section IX.3a the magnetic moment of the electron is determined by classical arguments; in Section IX.3b the magnetic field, which acts on the electron magnetic moment in an atom, is presented. In Section IX.4 these results are used to calculate the finestructure splitting. In Section IX.5 selection rules for dipole transitions are derived, and the chapter closes with some general remarks concerning the visualization of quantum systems.

## IX.1 Introduction

The existence of *electron spin* was suggested by the fine structure in atomic spectra. (See the description of the experimental situation at the end of Chapter VI.) The electron spin cannot be expressed in terms of the position and momentum operators of the electron. If the electron is considered as a physical object with translational and rotational degrees of freedom, then spin is an observable that corresponds to the rotational degrees of freedom in the same way that momentum corresponds to the translational degrees of freedom. Thus the electron is an elementary rotator (Section V.1) with a translational degree of freedom. In the preceding chapters we have ignored the rotational degrees of freedom of the electron, as their contributions to the energy of the electrons bound in atoms is small. We have shown in Chapter III

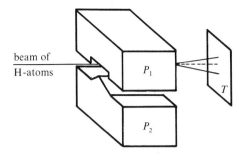

Figure 1.1 Splitting of a particle beam after traversing a Stern–Gerlach magnet.

that angular momenta of any integral and half-integral value may exist and that there is nothing peculiar about half-integral angular momenta. It turns out that the spin (*intrinsic angular momentum*) of the electron is $\frac{1}{2}$.

A direct verification of the electron-spin hypothesis came from the Stern–Gerlach experiment. The experimental arrangement is shown in Figure 1.1. A strongly inhomogeneous magnetic field was produced between pole pieces $P_1$ and $P_2$, of which $P_2$ has a sharp edge. A beam of hydrogen atoms[1] in the ground state was sent closely past the sharp edge of pole piece $P_2$, eventually hitting plate $T$. With no field the beam formed a narrow line on the plate (Figure 1.1 dashed line). When the magnetic field was turned on the line divided into two lines. (Figure 1.1 solid lines).

Without going into the details for the description of this experiment (see Chapter XIII), one can already see that the model of the hydrogen atom described in Chapter VI cannot explain such a separation. According to that model the ensemble of hydrogen atoms with lowest energy is in a pure state

$$W = \Lambda_{\mathscr{R}(n=1)}, \tag{1.1}$$

because $\mathscr{R}(n = 1)$ is a one-dimensional space. According to the model of Chapter VI, it would thus be impossible to separate the ensemble of the original beam into two subensembles as has been done by the magnetic field. The experimental separation into two subensembles shows that instead of $\mathscr{R}(n = 1)$ we must have (at least) a two-dimensional space. If we assume that the ensembles of the separated beams are in pure states, then we must have an exactly two-dimensional space. Within a certain degree of accuracy (including only the electron structure and disregarding possible structure of the nucleus) this assumption has so far been upheld.

As the separation of the beam was caused by a magnetic field, one conjectures that the new observable, whose existence has been demonstrated by the Stern–Gerlach experiment, must be connected with the magnetic moment. As the magnetic moment is connected with rotating charges, one would suspect that this new observable must be angular momentum.

---

[1] The original experiment used silver atoms. The experiment has since been repeated with other atoms, including hydrogen.

## IX.2 The Fine Structure—Qualitative Considerations

The angular momentum that causes the separation of the beam in the Stern–Gerlach experiment cannot be the orbital angular momentum $L_i = \epsilon_{ijk} Q_j P_k$, because this is zero for the ground state of the hydrogen atom. Furthermore, there are no two-dimensional spaces of states for the algebra of orbital angular momentum as $l = 0, 1, 2, \ldots$ (Problem III.1). The two-dimensional representation space of the algebra of spin angular momentum is $\mathscr{R}^{j=1/2}$ (Section III.3), containing the two one-dimensional spaces of states with opposite helicity. Denoting this two-dimensional space by $i^s$, we may write

$$i^s = \mathscr{R}^{j=1/2} \xrightarrow[\mathscr{E}(SO(2)_{j3})]{} \mathscr{R}^{j=1/2}_{j_3=-1/2} \oplus \mathscr{R}^{j=1/2}_{j_3=+1/2}. \tag{2.1}$$

We thus conjecture that the electron is an elementary rotator and the electron in the hydrogen atom is a combination of the physical systems of an orbiting electron and this elementary rotator (spinning electron), the space of states of which we call $i^s$. Then the space of physical states of lowest energy of the hydrogen atom is

$$\mathscr{H}(n = 1) = \mathscr{R}(n = 1) \otimes i^s, \tag{2.2}$$

and in general the $n$th space of the hydrogen atom is given by

$$\mathscr{H}(n) = \mathscr{R}(n) \otimes i^s. \tag{2.3}$$

The orbital observables (functions of $Q_i$, $P_i$) $A_i$ and $L_i$, given by (VI.3.1), (VI.3.3), and (VI.3.9), act in the space $\mathscr{R}(n)$, and the new angular momenta, which are called spin and denoted by $S_i$, act in the space $i^s$.

To check whether our conjecture is in agreement with experiment, we reduce the space $\mathscr{H}(n)$ with respect to angular-momentum spaces. According to (VI.4.4), each space $\mathscr{R}(n)$ is the direct sum $\mathscr{R}(n) = \sum_{l=0}^{n-1} \oplus \mathscr{R}^l$. Then $\mathscr{H}(n)$ given by Equation (2.3) is the direct sum

$$\mathscr{H}(n) = \left( \sum_{\substack{L_i \\ l=0}}^{n-1} \oplus \mathscr{R}^l \right) \otimes i^s = \sum_{l=0}^{n-1} \oplus (\mathscr{R}^l \otimes i^s) \tag{2.4}$$

$\mathscr{R}^l \otimes i^s$ is the space of physical states of the combination of two elementary rotators with angular momenta $l$ and $s = \frac{1}{2}$. The orbital angular momenta $L_i = \epsilon_{ijk} Q_j P_k$ act in $\mathscr{R}^l$, and the spin angular momenta $S_i$ act in $i^s$. The total angular momentum of the combined elementary rotators is then, according to Equation (V.2.3),

$$J_i = L_i \otimes I + I \otimes S_i. \tag{2.5}$$

Using (V.2.38) we can reduce $\mathscr{R}^l \otimes i^s$ to the sum of total-angular-momentum eigenspaces

$$\mathscr{R}^l \otimes i^s = \begin{cases} \mathscr{R}^{j=1/2} & \text{if } l = 0, \\ \mathscr{R}^{j=l+1/2} \oplus \mathscr{R}^{j=l-1/2} & \text{otherwise.} \end{cases} \tag{2.6}$$

Whether the $\mathscr{R}^j$ or the $\mathscr{R}^l \otimes \imath^s$ are the spaces of physical states depends, of course, upon the preparability of these states. If physically preparable states are eigenstates of the total angular momentum, then the $\mathscr{R}^j$ are the spaces of physical states; if the physically preparable states are eigenstates of the orbital and spin angular momentum, then the $\mathscr{R}^l \otimes \imath^s$ are the spaces of physical states. The experimental data show that the physical spaces are the $\mathscr{R}^j$, for each orbital-angular-momentum level (except the $l = 0$ level) splits into two sublevels as required by (2.4) and (2.6).

From (2.4) and (2.6) it follows that

$$\mathscr{H}(n) = \mathscr{R}^{1/2}_{(l=0)} \oplus (\mathscr{R}^{1/2}_{(l=1)} \oplus \mathscr{R}^{3/2}_{(l=1)}) \oplus (\mathscr{R}^{3/2}_{(l=2)} \oplus \mathscr{R}^{5/2}_{(l=2)})$$
$$\oplus \cdots \oplus (\mathscr{R}^{n-3/2}_{(l=n-1)} \oplus \mathscr{R}^{n-1/2}_{(l=n-1)}). \quad (2.7)$$

For $n = 1$,

$$\mathscr{H}(n = 1) = \mathscr{R}^{1/2}_{(l=0)}, \quad (2.8)$$

and there is one energy level corresponding to this two-dimensional space. The two-dimensionality of $\mathscr{H}(n = 1)$ allows us to explain the separation of the beam in the Stern–Gerlach experiment into two subensembles, one corresponding to the spaces $\mathscr{R}^{1/2}_{(l=0) \, j_3 = +1/2}$ and the other to $\mathscr{R}^{1/2}_{(l=0) \, j_3 = -1/2}$.

For $n = 2$,

$$\mathscr{H}(n = 2) = \mathscr{R}^{1/2}_{(l=0)} \oplus \mathscr{R}^{1/2}_{(l=1)} \oplus \mathscr{R}^{3/2}_{(l=1)}, \quad (2.9)$$

and there are three energy levels, each of which corresponds to one of the spaces in (2.9):

$$\mathscr{R}^{1/2}_{(l=0)} \leftrightarrow 2s^{1/2} \quad \mathscr{R}^{1/2}_{(l=1)} \leftrightarrow 2p^{1/2} \quad \text{and} \quad \mathscr{R}^{3/2}_{(l=1)} \leftrightarrow 2p^{3/2}. \quad (2.10)$$

For $n = 3$,

$$\mathscr{H}(n = 3) = \mathscr{R}^{1/2}_{(l=0)} \oplus \mathscr{R}^{1/2}_{(l=1)} \oplus \mathscr{R}^{3/2}_{(l=1)} \oplus \mathscr{R}^{3/2}_{(l=2)} \oplus \mathscr{R}^{5/2}_{(l=2)}. \quad (2.11)$$

which belong to the energy levels

$$3s^{1/2} \quad 3p^{1/2} \quad 3p^{3/2} \quad 3d^{3/2} \quad 3d^{5/2}$$

respectively. For $n = 4, 5, \ldots$ the same arguments hold. These results reflect precisely (if hyperfine splitting is neglected) the experimental data shown in Figure VI.5.2 for hydrogen and Figures 2.1, 2.2, 2.3 of Chapter VII for the alkali atoms. Thus to each energy level there corresponds an eigenspace of the total angular momentum; eigenspaces of $\mathbf{J}^2$ are also eigenspaces of $H$, and

$$[H, J_i] = 0. \quad (2.12)$$

This equation which we obtained here from the empirical facts, expresses the rotational invariance of $H$ and can theoretically be obtained from general symmetry arguments.

We shall denote by $\mathscr{H}^{\text{orb}}$ the space $\mathscr{R}$ given by (VI.5.2):

$$\mathscr{H}^{\text{orb}} = \sum_{n=1}^{\infty} \oplus \mathscr{R}(n). \quad (2.13)$$

The operator $H$ given by (VI.3.2) or (VI.3.5) is not the total-energy operator when spin is taken into account; let us rename this approximate energy operator $H_0$:

$$H_0 = \frac{\mathbf{P}^2}{2} - \frac{a}{Q} \qquad (2.14)$$

(or

$$H_0 = \frac{\mathbf{P}^2}{2} - \frac{a}{Q} + V(Q) \qquad (2.14')$$

for the alkali atom). The space of physical states of the hydrogen atom in a description that takes spin into account is then

$$\mathcal{H} = \mathcal{H}^{\text{orb}} \otimes \imath^s = \left( \sum_{n=1}^{\infty} \oplus \mathcal{R}(n) \right) \otimes \imath^s = \sum_{n=1}^{\infty} \mathcal{H}(n). \qquad (2.15)$$

This description is the combination of the spinless hydrogen atom and an elementary rotator. A basis system in this space is thus

$$|n\, l\, m\rangle \otimes |s = \tfrac{1}{2} s_3\rangle, \qquad (2.16)$$

where the

$$|n\, l\, m\rangle\ (n = 1, 2, 3, \ldots; l = 0, 1, 2, \ldots, n-1; m = -l, -l+1, \ldots, +l)$$

form a basis system for $\mathcal{H}^{\text{orb}}$, and the $|s = \tfrac{1}{2} s_3\rangle$ $(s_3 = -\tfrac{1}{2}, +\tfrac{1}{2})$ form a basis system for $\imath^s = \mathcal{R}^{1/2}$. Each observable $A$ in $\mathcal{H}$ is written, according to the basic assumption IV,

$$A = \sum_{ij} A^{\text{orb}}_{(i)} \otimes A^{\text{spin}}_{(j)}, \qquad (2.17)$$

where the $A^{\text{orb}}_{(i)}$ act only in $\mathcal{H}^{\text{orb}}$ and the $A^{\text{spin}}_{(j)}$ act only in $\imath^s$ [cf. (III.5.7)]. Orbital angular momentum and spin are special examples of this:

$$L_i = L_i^{\text{orb}} \otimes I^{\text{spin}} = (\epsilon_{ijk} Q_j^{\text{orb}} P_k^{\text{orb}}) \otimes I^{\text{spin}} = L_i \otimes I = (\epsilon_{ijk} Q_j P_k) \otimes I, \qquad (2.18)$$

$$S_i = I^{\text{orb}} \otimes S_i^{\text{spin}} = I \otimes S_i. \qquad (2.19)$$

(For typographical convenience we shall usually omit the superscripts "orb" and "spin" if it is clear from the context in which space an operator acts.) The total angular momentum of the combined system is the sum of these two angular momenta,

$$J_i = K_i + S_i = L_i \otimes I + I \otimes S_i. \qquad (2.5)$$

The energy operator $H$ can also be written in the form (2.17). Since $H_0 \otimes I$ already gives a very good approximation to the energy spectrum (cf. Figures VI.5.2, VII.2.1, VII.2.2, and VII.2.3), we shall write the energy operator as

$$H = H_0 + H_1, \qquad (H_0 = H_0 \otimes I), \qquad (2.20)$$

where $H_1$ acts on the entire space $\mathcal{H} = \mathcal{H}^{\text{orb}} \otimes \imath^s$ and is small in the sense that its contributions to the energy values are small compared with the values of $H_0$. $H_1$ cannot be of the form $H_1 \otimes I$. If it were, no fine-structure splitting between levels with the same $l$ could occur, for $H_1^{\text{orb}}$ could just be added to $H_0^{\text{orb}}$ and the resultant $H = (H_0 + H_1) \otimes I$ (for a given value of $l$) would only lead to a shift of the energy levels. Neither can $H_1$ be of the form $I \otimes H_1$. For $\imath^s$ is a very simple two-dimensional space: Every operator in $\imath^s$ can be written as a linear combination of the four operators $I, S_1, S_2$, and $S_3$; thus $H_1$ would have to be $H_1 = \alpha^0 I + \sum_{i=1}^{3} \alpha^i S_i \, (\alpha^i \in \mathbb{C})$. The first term would only lead to an overall shift of all energy values. The second term $(\sum_{i=1}^{3} \alpha^i S_i)$, and consequently $H_1$, does not commute with $J_k$, because

$$[J_i, S_j] = i\epsilon_{ijk} S_k \tag{2.21}$$

as a consequence of (2.5) and of

$$[S_i, S_j] = i\epsilon_{ijk} S_k. \tag{2.21a}$$

Thus it would follow that $H_1$ and consequently $H = H_0 + H_1$ is not rotationally invariant, in contradiction to (2.12). Hence $H_1$ must be of the form (2.17). Now the most general possible form of $H_1$ is

$$H_1 = A \otimes I + \sum_{i=1}^{3} B_i \otimes S_i, \tag{2.22}$$

where $A$ and $B_i$ are operators in $\mathcal{H}^{\text{orb}}$, i.e., functions of the operators $Q_i, P_i$. $A$ is a scalar operator, while the $B_i$ are the components of a vector operator with respect to the $L_i$. The former follows from the fact that every operator in $\imath^s = \mathcal{R}^{1/2}$ can be written as a linear combination of the operators $I, S_1, S_2$, and $S_3$. The latter follows from the requirement that

$$[H_1, J_i] = 0.$$

which in turn follows from (2.12) (2.20) and $[H_0, L_i] = 0$.

To see this we calculate

$$\sum_{k=1}^{3} [B_k \otimes S_k, J_i] = [B_k, L_i] \otimes S_k + i\epsilon_{kij} B_k \otimes S_j,$$

which is zero only if

$$[B_k, L_i] = i\epsilon_{kij} B_j, \tag{2.22b}$$

i.e., if $B_k$ is a vector operator with respect to the $L_i$.

As

$$[H_1, L_i] \neq 0, \tag{2.22a}$$

which follows immediately from (2.22b), the basis vectors given by (2.16) cannot be eigenvectors of $H$. Physically preparable states appear to be always energy eigenstates or mixtures of energy eigenstates (cf. Chapter XII). Therefore we should use a basis of eigenvectors of the energy operator

$H$. Because of (2.12), eigenvectors of the total-angular-momentum operators $\mathbf{J}^2$, $J_3$ can be eigenvectors of $H$. Therefore we use (V.2.12) to form the new basis vectors

$$|n \, l \, s = \tfrac{1}{2} j j_3\rangle = \sum_{l_3 s_3} |n \, l \, l_3\rangle \otimes |s = \tfrac{1}{2} s_3\rangle \langle l \, l_3 \, s = \tfrac{1}{2} s_3 | j j_3\rangle, \quad (2.23)$$

where $\langle l l_3 \tfrac{1}{2} s_3 | j j_3\rangle$ are the Clebsch–Gordan coefficients given in Table V.2.1. The basis vectors (2.23) are eigenvectors of the c.s.c.o.

$$H_0, \mathbf{L}^2, \mathbf{S}^2, \mathbf{J}^2, \quad J_3. \quad (2.24')$$

As $\mathbf{S}^2$ is $\tfrac{1}{2}(\tfrac{1}{2} + 1)$ on the whole space $\mathscr{H}$, we can ignore it and have in

$$H_0, \mathbf{L}^2, \mathbf{J}^2, \quad J_3 \quad (2.24)$$

a c.s.c.o., of which the basis vectors of (2.23) are eigenvectors. The basis (2.23) need not be a basis of eigenvectors of $H$. If the basis vectors (2.23) are not eigenvectors of $H$, then eigenvectors of $H$ can be obtained from (2.23) by perturbation-theoretical calculations that affect only the quantum number $n$ [cf. the remark on perturbation theoretical calculations preceeding Equation (VII.1.4)]. The reason for this is that the system

$$H, \mathbf{L}^2, \mathbf{J}, \quad J_3 \quad (2.25)$$

is also a c.s.c.o., the eigenvectors of which we call

$$|E \, l \, j \, j_3\rangle. \quad (2.26)$$

The difference between (2.23) and (2.26) is that

$$H_0 |n \, l \, j \, j_3\rangle = E_n^0 |n \, l \, j \, j_3\rangle \quad (2.27)$$

where $E_n^0 = -a/(2n^2)$, and

$$H |E \, l \, j \, j_3\rangle = E |E \, l \, j \, j_3\rangle \quad (2.28)$$

where $E$ is a yet to be determined value. Thus the basis systems (2.23) and (2.26) have all but one quantum number in common.

To prove that (2.25) is indeed a c.s.c.o. one needs only to show that

$$[H, \mathbf{L}^2] = 0. \quad (2.29)$$

This cannot be derived from the general form (2.22), but constitutes a condition on the $A$ and $B_i$. The justification of (2.29) comes from the considerations in Section V.4 on parity, also expressed by (VI.5.9), and a new assumption about $H$:

$$[H, U_P] = 0, \quad \text{or} \quad U_P H U_P^{-1} = H. \quad (2.30)$$

This assumption is called *parity invariance* (of $H$). $U_P$ in the spin space is given by

$$U_P |s = \tfrac{1}{2} s_3\rangle = \pi_S |s = \tfrac{1}{2} s_3\rangle, \quad (2.31)$$

where $|\pi_s| = 1$. Equation (2.31) is (V.4.20) for the special case that $j = s = \frac{1}{2}$ and follows from (V.4.3) for the $S_i$, i.e., from

$$U_P S_I U_P^{-1} = S_i. \tag{2.32}$$

Equation (V.4.20) and (V.4.21) for the orbital angular momentum ($j = l$) give

$$U_P |\xi \, l \, l_3\rangle = (-1)^l \eta_{\text{orb}} |\xi \, l \, l_3\rangle; \tag{2.33}$$

here $\xi$ may be any additional quantum number whose operator commutes with $U_P$, e.g., $\xi = n$ or $\xi = E$. From (2.31) and (2.32) it follows that

$$U_P |\xi \, l \, l_3\rangle \otimes |s = \tfrac{1}{2} s_3\rangle = (-1)^l \eta |\xi \, l \, l_3\rangle \otimes |s = \tfrac{1}{2} s_3\rangle. \tag{2.34}$$

($\eta = \eta_{\text{orb}} \pi_s$). It then follows from (2.23) that

$$U_P |\xi \, l \, j \, j_3\rangle = (-1)^l \eta |\xi \, l \, j \, j_3\rangle. \tag{2.35}$$

Consequently the eigenvalue of $\mathbf{L}^2$ is connected with the eigenvalue of $U_P$, and the quantum number $l$ in the basis (2.23) may be considered the parity quantum number. [The parity can have only the two values $+1$ and $-1$; thus for given $n, j, j_3$ there exist two states of opposite parity. This has already been expressed in (2.7) in the form that to every $j$ there belong two $l$'s, $l = j + \frac{1}{2}$ and $l = j - \frac{1}{2}$.] Thus if (2.30) holds, $\xi$ in (2.35) may be $E$, the eigenvalue of $H$; from (2.30) it follows that $E, l, j, j_3$ may label the states, and consequently that (2.25) is a c.s.c.o., one which has now been shown to be identical with the c.s.c.o.

$$H, U_P, \mathbf{J}^2, \quad J_3. \tag{2.36}$$

Now that (2.29) has been established, it can be used to obtain some new conditions on the operators $A$ and $B_i$ in (2.22). From (2.22) follows

$$[H_1, \mathbf{L}^2] = 0 \tag{2.37}$$

which does not give any new restrictions on $A$, but, along with (2.22b), does restrict the $B_i$:

$$0 = [B_k, \mathbf{L}^2] = \{L_i, [B_k, L_i]\} = i\epsilon_{kij}\{L_i, B_j\}.$$

From this follows the restriction

$$B_j = f(Q, P)L_j, \tag{2.38}$$

where $[f(Q, P), L_j] = 0$, i.e., $f(Q, P)$ is a scalar operator with respect to the $L_j$. We will later determine $B_j$ by correspondence from some physical considerations with the classical case, and will then see what functions of $Q$ and $P$ $f$ is.

Although we have denoted the basis vectors (2.23) and (2.26) differently, it may happen that they are identical (up to a phase). We have to distinguish the two possibilities

$$[H_1, H_0] = 0, \quad \text{which implies } [H, H_0] = 0, \tag{2.39}$$

and

$$[H_1, H_0] \neq 0, \quad \text{which implies} \quad [H, H_0] \neq 0. \tag{2.40}$$

If (2.39) holds, then the basis vectors (2.23) are also eigenvectors of $H$ and

$$H|n\,l\,j\,j_3\rangle = E_{nlj}|n\,l\,j\,j_3\rangle; \tag{2.41}$$

the eigenvalue of $H$ may in general depend upon $n$, $l$, and $j$, and is given by

$$E_{nlj} = E_n^0 + \epsilon_{nlj}, \qquad E_n^0 = -\frac{a}{2n^2}, \tag{2.42}$$

where $\epsilon_{nlj}$ is the eigenvalue of the operator $H_1$ in the state $|n\,l\,j\,j_3\rangle$:

$$H_1|n\,l\,j\,j_3\rangle = \epsilon_{nlj}|n\,l\,j\,j_3\rangle. \tag{2.43}$$

If (2.40) holds, then (2.23) and (2.26) are different vectors, and (2.26) can be determined from (2.23) by perturbation theory. The first-order perturbation expression $E_{nlj}^{(1)}$ for the eigenvalue of $H$ is then given, according to (VIII.1.33) or (VIII.1.38), by

$$E_{nlj}^{(1)} = E_n^0 + \epsilon_{nlj}^{(1)}, \tag{2.44}$$

where $\epsilon_{nlj}^{(1)}$ is the expectation value of $H_1$ in the states (2.23),

$$\epsilon_{nlj}^{(1)} = \langle n\,l\,j\,j_3|H_1|n\,l\,j\,j_3\rangle. \tag{2.45}$$

## IX.3  Fine-Structure Interaction

The calculation of quantitative results requires knowledge of $H_1$; $H_1$ can be conjectured by some plausible physical arguments from the corresponding classical situation. The fine-structure term in the energy operator consists of two contributions: (1) a contribution caused by the interaction between the internal magnetic moment of the electron and the magnetic field in the electron rest frame due to the motion of the proton charge as seen in this frame, and (2) a contribution caused by the variation of the mass of the electron with its velocity. Both terms arise from relativistic kinematics; although the last term is of the same order of magnitude as the magnetic-moment term, it does not contribute to the splitting of the levels with the same values of $n$ and $l$. Besides the fine-structure terms there are other contributions to the energy operator, which cause the Lamb shift and the hyperfine structure, but we shall not consider those here.

## IX.3a  The Magnetic Moment of a Spinning
##         Particle in Classical Physics[2]

The energy of a magnetic dipole of magnetic moment **m** in a magnetic field of strength **H** is

$$E_1^{(m)} = -\mathbf{m} \cdot \mathbf{H}. \tag{3.1}$$

[2] Corben (1968).

So we need to know the magnetic moment **m** of the spinning electron and the magnetic field **H** in the electron's rest frame. In this subsection we shall determine **m**, and in Section IX.3b we shall discuss **H**.

Before we consider the magnetic moment of a spinning particle, let us recall the connection between the orbital angular momentum and the magnetic moment caused by the revolution of a charged spinless particle around a center. According to classical electromagnetic theory, the magnetic dipole moment of a point charge at position **x** with mass $m_e$ and charge $(-e)$ moving with a velocity **v** is given by

$$\mathbf{m} = \frac{1}{2c}(-e)\mathbf{x} \times \mathbf{v} = -\frac{1}{2m_e c}\,e\mathbf{l}. \tag{3.2}$$

(Note that we consider negatively charged revolving particles like the electron, and that $e = +4.803 \times 10^{-10}$ esu.) The connection between the magnetic moment of a spinning charged particle and its internal angular momentum (spin) **s** differs from (3.2) by a factor of $g_s \approx 2$, the Landé factor, i.e., the gyromagnetic ratio is $e/(m_e c)$ and not $e/(2m_e c)$ as in (3.2). In order to derive this factor we consider the spinning point particle.

A classical spinning particle is a physical system with two different dynamical variables, the momentum **p** and the intrinsic angular momentum **s**. If **x** denotes the position, then the total angular momentum **j** is given by

$$\mathbf{j} = \mathbf{l} + \mathbf{s} = \mathbf{x} \times \mathbf{p} + \mathbf{s}. \tag{3.3}$$

The external force **F** acting on the particle is given as the time rate of change of the momentum of the particle,

$$\mathbf{F} = \frac{dp}{dt}, \tag{3.4}$$

and the moment **M** about the origin of the forces acting on the particle is defined as

$$\mathbf{M} = \mathbf{x} \times \mathbf{F} + \mathbf{T}, \tag{3.5}$$

the first term representing the orbital contribution to this moment from the resultant force **F**, and the second term describing an extra applied torque **T**. This torque may arise, for example, from the effect of an externally applied magnetic field upon any intrinsic magnetic moment the particle may possess. We now postulate that

$$\mathbf{M} = \frac{d\mathbf{j}}{dt}, \tag{3.6}$$

so that, from Equation (3.3), (3.4), (3.5), and the definition $\mathbf{v} = d\mathbf{x}/dt$ of the velocity,

$$\frac{d\mathbf{s}}{dt} + \mathbf{v} \times \mathbf{p} = \mathbf{T}. \tag{3.7}$$

If the momentum $\mathbf{p}$ and velocity $\mathbf{v}$ are parallel, or if either of them vanishes, then

$$\frac{d\mathbf{s}}{dt} = \mathbf{T} \tag{3.8a}$$

and

$$\frac{d\mathbf{l}}{dt} = \mathbf{x} \times \mathbf{F}. \tag{3.8b}$$

Spin and orbital motions are then separately and respectively determined by the torque $\mathbf{T}$ and the moment of the force $\mathbf{F}$, so that the spin and orbital motions become uncoupled and each can be discussed without reference to the other. Since in nonrelativistic mechanics $\mathbf{p} = m\mathbf{v}$, it is possible in this approximation to discuss the orbital motion in terms of (3.8b) and the spin motion in terms of (3.8a). For example, in the equation describing the precession in a magnetic field $\mathbf{H}$ of a particle that possesses a magnetic moment $\mathbf{m}$, the torque is

$$\mathbf{T} = \mathbf{m} \times \mathbf{H}; \tag{3.9}$$

consequently

$$\frac{d\mathbf{s}}{dt} = \mathbf{m} \times \mathbf{H}. \tag{3.10}$$

Let us now assume that we have a particle that has *no intrinsic magnetic moment*, i.e., $\mathbf{m} = 0$; consequently, by (3.9), $\mathbf{T} = 0$, where $\mathbf{T}$ is interpreted as the torque acting on any intrinsic moment the particle may possess. *But let us assume that* $\mathbf{v} \times \mathbf{p} \neq 0$, i.e., that we do not have the strictly nonrelativistic relation $\mathbf{p} = m\mathbf{v}$ or the relativistic relation for free spinless particles $\mathbf{p} = m\mathbf{v}/(1 - v^2/c^2)^{1/2}$.[3] Then (3.7) becomes

$$\frac{d\mathbf{s}}{dt} = -\mathbf{v} \times \mathbf{p}, \tag{3.11}$$

Suppose the particle is moving in a uniform constant magnetic field $\mathbf{H}$. The Lorentz force on the particle is then

$$\mathbf{F} = \frac{d\mathbf{p}}{dt} = \frac{-e}{c} \mathbf{v} \times \mathbf{H}, \tag{3.12}$$

so that

$$\frac{1}{2}\frac{d}{dt}(\mathbf{p}^2) = \mathbf{p} \cdot \frac{d\mathbf{p}}{dt} = -\frac{e}{c}\mathbf{p} \cdot \mathbf{v} \times \mathbf{H}. \tag{3.13}$$

But, using (3.11) together with the supposition that $\mathbf{H}$ is a constant field ($d\mathbf{H}/dt = 0$),

$$\frac{e}{c}\frac{d}{dt}(\mathbf{s} \cdot \mathbf{H}) = \frac{e}{c}\frac{d\mathbf{s}}{dt} \cdot \mathbf{H} = -\frac{e}{c}\mathbf{v} \times \mathbf{p} \cdot \mathbf{H}$$

$$= +\frac{e}{c}\mathbf{p} \times \mathbf{v} \cdot \mathbf{H} = \frac{e}{c}\mathbf{p} \cdot \mathbf{v} \times \mathbf{H}. \tag{3.14}$$

---

[3] $\mathbf{v} \times \mathbf{p} \neq 0$ will occur if the time components $s_{01}, s_{02}, s_{03}$, which together with $s_{ij} = \epsilon_{ijk}s_k$ make up the relativistic spin tensor $s_{\mu\nu}$, are not constant in time. This is always the case for an observer for whom the particle is moving with a velocity $\mathbf{v}$.

Adding (3.13) and (3.14), we conclude

$$\tfrac{1}{2}\mathbf{p}^2 + \frac{e}{c}\mathbf{s}\cdot\mathbf{H} = \text{const.} \tag{3.15}$$

Hence, supposing the particle possess a constant mass $m_e$, it follows that

$$\frac{\mathbf{p}^2}{2m_e} - \boldsymbol{\mu}\cdot\mathbf{H} = \text{const,} \tag{3.16}$$

where

$$\boldsymbol{\mu} = -\frac{e}{m_e c}\mathbf{s}. \tag{3.17}$$

Equation (3.13) tells us that the kinetic energy $\mathbf{p}^2/(2m_e)$ of the particle is not a constant of the motion unless two of the vectors $\mathbf{v}$, $\mathbf{p}$, and $\mathbf{H}$ are parallel. In general, the particle has an extra energy $-\boldsymbol{\mu}\cdot\mathbf{H}$ in the magnetic field $\mathbf{H}$ [cf. (3.1)], i.e., it behaves as if it had a magnetic moment $\boldsymbol{\mu}$ in the direction of its spin, with gyromagnetic ratio $e/(m_e c)$. Equation (3.17) is similar to (3.2), but with an additional factor of 2 (the value of the Landé factor $g_s$) on the right-hand side. We are thus led to the conclusion that all charged spinning particles automatically possess a magnetic moment, given by (3.17), which is of relativistic origin. [The sign in Equation (3.12)–(3.15) and (3.17) reverses for positively charged particles.]

If the electron is a quantum-mechanical particle with charge $-e$ and *without intrinsic magnetic moment*, then its magnetic-moment operator should be given by the quantum-mechanical analogue of (3.17), i.e., by

$$\mathbf{M}_s = -\frac{e}{m_e c}\mathbf{S} = -g_s\frac{e}{2m_e c}\mathbf{S} \quad (g_s = 2), \tag{3.18}$$

where $\mathbf{S}$ is the spin operator. It turns out that (3.18) indeed gives a very accurate description of the electron magnetic moment. The deviation from the $g_s = 2$ value for the electron comes from the radiative corrections of quantum electrodynamics and is of the same order as, and of analogous origin to, the Lamb shift. Including these corrections up to the second order gives

$$g_s = 2\left(1 + 0.5\frac{\alpha}{\pi} - 0.328\left(\frac{\alpha}{\pi}\right)^2\right),$$

where $\alpha = e^2/(\hbar c)$ is the fine-structure constant; this value for $g_s$ agrees with experimental values up to the eighth decimal place. Thus electrons—and also muons—are particles without an intrinsic magnetic structure.

The value $g_s = 2$ was first established as far back as 1915 by an experiment of Einstein and de Haas, and was incorporated in the spin hypothesis put forward around 1926. The existence of the radiative correction terms was first discovered experimentally by Rabi and collaborators in 1947, and was calculated within the framework of quantum electrodynamics by Schwinger in 1948. The value $g_s = 2$ is also obtainable from the assumption of "minimal coupling to the electromagnetic field," first used in the Dirac relativistic wave equation for the electron and considered one of the great achievements of the Dirac equation. The above argument shows that it is already a consequence of classical considerations.

There are elementary particles with intrinsic structure; e.g., the proton has a magnetic moment

$$\mathbf{M}_p = 5.59 \frac{e}{2m_p c} \mathbf{S} = \frac{e}{2m_p c}(2 + 3.59)\mathbf{S}, \qquad (3.18p)$$

i.e., a $g$-value of $g_p = 5.59$. The value of the magnetic moment in excess of the value $2(e/(2m_p c))\mathbf{S}$ given by (3.17) is called the anomalous magnetic moment; it is $1.79e/(m_p c)$ for the proton. The neutron has no charge, and therefore the magnetic moment given by (3.17) is also zero; however it has an anomalous magnetic moment

$$\mathbf{M}_n = \frac{e}{2m_n c}(0 - 1.92)\mathbf{S}. \qquad (3.18n)$$

## IX.3b  The Spin–Orbit Interaction Term

The determination of the magnitude of the magnetic field $\mathbf{H}$ must take into account relativistic kinematical effects. We go into a coordinate system that moves with the electron around the proton. In this system the electron is at rest, and the proton charge moves with a velocity $\mathbf{v}$ that is equal in magnitude but opposite in direction to the electron velocity. This movement constitutes a current. The magnetic field caused by a current of a single charge $+e$ moving with a velocity $\mathbf{v}$ is, according to the Biot–Savart law, given by

$$\mathbf{H}(x) = +e \frac{\mathbf{v} \times \mathbf{x}}{cr^3} \qquad (3.19)$$

($e$ is measured in esu, $H$ in gauss), where $\mathbf{x}$ is the vector of magnitude $r$ from the moving charge to the observation point. The angular momentum of the electron is

$$\mathbf{l} = \mathbf{x} \times (-m_e \mathbf{v}).$$

Consequently the magnetic field at the position of the electron caused by the rotating proton is

$$\mathbf{H} = + \frac{e}{m_e cr^3} \mathbf{l}. \qquad (3.20)$$

Both (3.19) and (3.20) neglect relativistic effects. From (3.1) we thus obtain

$$E_1^{(m)}\big|_{\text{r.f.}} = - \frac{e}{m_e c} \frac{1}{r^3} \mathbf{l} \cdot \mathbf{m}_s \qquad (3.21)$$

for the energy of the magnetic moment in this field. The subscript "r.f." denotes that we have used a rotating frame. If the frame rotates, there is an extra contribution to the energy that reduces (3.21) by a factor of $\frac{1}{2}$. [This factor is known as the "Thomas factor," and is caused by the "Thomas precession"; a detailed calculation may be found in Jackson (1975, p. 364). Thus the energy of the moving spin magnetic moment in the magnetic field of the proton is given by

$$E_1^{(m)} = -\frac{e}{2m_e c}\frac{1}{r^3}\mathbf{l}\cdot\mathbf{m}_s. \tag{3.22}$$

We obtain the quantum-mechanical expression corresponding to (3.22) by the usual procedure of replacing the classical quantities $\mathbf{l}$ and $\mathbf{m}_s$ by the quantum-mechanical observables $\mathbf{L}$ amd $\mathbf{M}_s$ given by (VI.3.1) and by (3.18):

$$H_1^{(m)} = +g_s\frac{e^2}{4m_e^2 c^2}\frac{1}{Q^3}\mathbf{L}\cdot\mathbf{S} = \frac{1}{2}\frac{e^2}{m_e^2 c^2}\frac{1}{Q^3}L_i S_i \quad \text{(usual cgs units)}$$

$$= g_s\mu_B^2\frac{1}{Q^3}L_i S_i \quad (\hbar = 1). \tag{3.23}$$

The quantity

$$\mu_B = \frac{eh}{2m_e c} = \frac{(4.8\times 10^{-10}\text{ esu})(6.6\times 10^{-16}\text{ eV sec})}{2(3.0\times 10^{10}\text{ cm/sec})(9.1\times 10^{-28}\text{ g})}$$

$$= 0.579\times 10^{-8}\frac{\text{esu eV sec}^2}{\text{g cm}} = 0.579\times 10^{-8}\text{ eV}\frac{\text{cm}^{1/2}\text{ sec}}{\text{g}^{1/2}}$$

$$= 0.579\times 10^{-8}\text{ eV/gauss} = 9.27\times 10^{-21}\text{ erg/gauss}$$

is called the Bohr magneton. The operator $\mathbf{L}\cdot\mathbf{S}$ of (3.23) is easily calculated from (2.5),

$$\mathbf{J}^2 = (\mathbf{L}+\mathbf{S})^2 = \mathbf{L}^2 + \mathbf{S}^2 + 2\mathbf{L}\cdot\mathbf{S},$$

and the fact that

$$\mathbf{S}^2|n\,l\,j\,j_3\rangle = s(s+1)|n\,l\,j\,j_3\rangle = \tfrac{3}{4}|n\,l\,j\,j_3\rangle.$$

We then obtain for (3.23) that (in the usual cgs units)

$$H_1^{(m)} = \frac{1}{2}\frac{e^2}{m_e^2 c^2}Q^{-3}\tfrac{1}{2}(-\mathbf{L}^2 - \tfrac{3}{4}I + \mathbf{J}^2). \tag{3.24}$$

## IX.3c  The Kinematical Correction Term

We obtain the contribution arising from the relativistic mass effect by expanding the relativistic expression for the kinetic energy

$$p_0 = \sqrt{(m_e c^2)^2 + c^2 p^2}$$

of the free electron in powers of $p/(m_e c)$ $[p = (\mathbf{p}^2)^{1/2}]$:

$$p_0 = m_e c^2 \sqrt{1 + \left(\frac{p}{m_e c}\right)^2} = m_e c^2 \left(1 + \frac{1}{2}\left(\frac{p}{m_e c}\right)^2 - \frac{1}{8}\left(\frac{p}{m_e c}\right)^4 + \cdots\right)$$

$$= m_e c^2 + \frac{p^2}{2m_e} - \frac{1}{2}\left(\frac{p^2}{2m_e}\right)^2 \frac{1}{m_e c^2} + \cdots.$$

The rest energy $m_e c^2$ of the electron is ignored because nonrelativistically the energy is determined only up to an additive constant (the rest energy would shift the energy levels by the same fixed amount). Thus classically the kinetic energy with the first-order relativistic correction is

$$E_{\text{kin}} = \frac{p^2}{2m_e} - \frac{1}{2}\left(\frac{p^2}{2m_e}\right)^2 \frac{1}{m_e c^2},$$

in contrast to the usual expression

$$E_{\text{kin}} = \frac{p^2}{2m_e}.$$

Going to quantum mechanics, we replace the numbers $p_i$ by the operators $P_i$ and obtain

$$H_1^{(k)} = -\frac{1}{2m_e c^2}\left(\frac{\mathbf{P}^2}{2m_e}\right)^2 = -\frac{1}{2m_e c^2}\left(H_0 + \frac{e^2}{Q}\right)^2 \qquad (3.25)$$

for the kinematical correction to

$$H_0 = \frac{\mathbf{P}^2}{2m_e} - \frac{e^2}{Q}. \qquad (2.14')$$

[Both this equation for $H_0$ and (3.25) are in the usual cgs units, as opposed to the units described above (VI.3.4).] The total-energy operator with the two corrections is then

$$H = H_0 + H_1^{(m)} + H_1^{(k)}, \qquad (3.26)$$

where $H_1^{(m)}$ and $H_1^{(k)}$ are given by (3.24) and (3.25), respectively.

## IX.4  Fine Structure of Atomic Spectra

As $H_1 = H_1^{(m)} + H_1^{(k)}$ contains the operators $Q^k$ $(k = -1, -2, -3)$, $H_1$ and therefore $H$ do not commute with $H_0$. It therefore appears that we have the situation described by (2.40), (2.44), (2.45), that $|n\,l\,j\,j_3\rangle$ are not the physical eigenvectors and that $E_{nlj}$ given by (2.44) and (2.45) is only a first approximation. Without further justification (which comes from a relativistic theory), we state that the hydrogen-atom energy operator that includes fine-structure effects has eigenvectors that are not eigenvectors of the operator $H_0$ but that are again labelled by $n$, $l$, $j$, and $j_3$, thus the eigenvectors (2.26) are also labeled by $n$. The values $E_{nlj}^{(1)}$ given by (2.44) and (2.45) are identical to the eigenvalues of this relativistic energy operator.

In order to calculate the matrix elements (2.45) we make use of (VII.2.15), (VII.2.16), (VII.2.17) with (VII.2.20). [Note that for $l = 0$ Equation (VII.2.17) is singular; for $l = 0$ the matrix element of $\mathbf{j} \cdot \mathbf{s}$ is also zero because

$$\langle n \, l \, j \, j_3 | \mathbf{L} \cdot \mathbf{S} | n \, l \, j \, j_3 \rangle \propto l/2$$

—cf. (4.2) below—so that the matrix element of $\mathbf{L} \cdot \mathbf{S}/Q^3$ is always well defined.] We calculate the matrix element of the spin-orbit term $H_1^{(m)}$ and find

$$\langle n \, l \, j \, j_3 | H_1^{(m)} | n \, l \, j \, j_3 \rangle = \frac{e^2}{2m_e^2 c^2} \langle n \, l \, j \, j_3 | Q^{-3} | n \, l \, j \, j_3 \rangle$$

$$\times \begin{cases} l/2 & \text{for } j = l + \tfrac{1}{2}, \\ -(l+1)/2 & \text{for } j = l - \tfrac{1}{2}, \end{cases} \tag{4.1}$$

because

$$\tfrac{1}{2}(\mathbf{J}^2 - \mathbf{L}^2 - \tfrac{3}{4}I)|n \, l \, j \, j_3\rangle = \tfrac{1}{2}(j(j+1) - l(l+1) - \tfrac{3}{4})|n \, l \, j \, j_3\rangle$$

$$= \tfrac{1}{2}|n \, l \, j \, j_3\rangle \begin{cases} l & \text{for } j = l + \tfrac{1}{2}, \\ -(l+1) & \text{for } j = l - \tfrac{1}{2}. \end{cases} \tag{4.2}$$

According to (2.23),

$$\langle n \, l \, j \, j_3 | Q^{-3} | n \, l \, j \, j_3 \rangle = \sum_{l_3' s_3'} \sum_{l_3 s_3} \langle j \, j_3 | l \, l_3' \, s = \tfrac{1}{2} \, s_3' \rangle \langle l \, l_3' \, s = \tfrac{1}{2} s_3' | j \, j_3 \rangle$$

$$\times \langle s_3 | s_3' \rangle \langle n \, l \, l_3 | Q^{-3} | n \, l \, l_3' \rangle.$$

Using the orthogonality relation (V.2.15) of the Clebsch–Gordan coefficients, and the fact that the last matrix element is zero unless $l_3' = l_3$, which we write as

$$\langle n \, l \, l_3' | Q^{-3} | n \, l \, l_3 \rangle = \delta_{l_3 l_3'} \langle n \, l | Q^{-3} | n \, l \rangle,$$

we obtain

$$\langle n \, l \, j \, j_3 | Q^{-3} | n \, l \, j \, j_3 \rangle = \langle n \, l | Q^{-3} | n \, l \rangle.$$

Thus from (4.1) and (VII.2.17) together with (VII.2.20) it follows that

$$\langle n \, l \, j \, j_3 | H_1^{(m)} | n \, l \, j \, j_3 \rangle = \frac{1}{2} \left( \frac{1}{m_e c} \right)^2 e^2 \frac{m_e^3 e^6}{1} \frac{1}{n^3 (l+1)(l + \tfrac{1}{2})l}$$

$$\times \frac{1}{2} \begin{cases} l & \text{for } j = l + \tfrac{1}{2}, \\ -(l+1) & \text{for } j = l - \tfrac{1}{2}. \end{cases} \tag{4.3}$$

Recalling that

$$E_n^0 = -\frac{m_e e^4}{1}\frac{1}{2n^2} = -\left(\frac{e^2}{c}\right)^2 m_e c^2 \frac{1}{2n^2}, \tag{4.4}$$

(4.3) can be written as

$$\langle n\, l j j_3 | H_1^{(m)} | n\, l j j_3 \rangle = -E_n^0 \left(\frac{e^2}{c}\right)^2 \frac{1}{n(2l+1)}\begin{cases} 1/(l+1) & \text{for } j = l + \tfrac{1}{2}, \\ -1/l & \text{for } j = l - \tfrac{1}{2}. \end{cases}$$
$$\tag{4.5}$$

The matrix element of the kinetic-energy term $H_1^{(k)}$ of (3.25) is

$$\langle n\, l j j_3 | H_1^{(k)} | n\, l j j_3 \rangle$$

$$= -\frac{1}{2m_e c^2}\left\{E_n^{02} + 2E_n^0 e^2 \langle n\, l | Q^{-1} | n\, l \rangle + e^4 \langle n\, l | Q^{-2} | n\, l \rangle \right\}$$

$$= -E_n^0 \left\{ -\left(\frac{e^2}{c}\right)^2 \frac{1}{4n^2} + \frac{2e^2}{2m_e c^2}\frac{m_e e^2}{1}\frac{1}{n^2} + \frac{e^4 m_e^2 e^4}{2m_e c^2 E_n^0}\frac{1}{(l+\tfrac{1}{2})n^3}\right\}$$

$$= -E_n^0 \left(\frac{e^2}{c}\right)^2 \left\{ -\frac{1}{4n^2} + \frac{1}{n^2} - \frac{1}{n(l+\tfrac{1}{2})}\right\}$$

$$= -E_n^0 \left(\frac{e^2}{c}\right)^2 \frac{1}{n^2}\left(\frac{3}{4} - \frac{n}{l+\tfrac{1}{2}}\right). \tag{4.6}$$

In this calculation (VII.2.15) and (VII.2.16) together with (VII.2.20) have been used. Adding (4.5) and (4.6), we obtain for the spin-orbit interaction term

$$\langle n\, l j j_3 | H_1 | n\, l j j_3 \rangle = -E_n^0 \left(\frac{e^2}{c}\right)^2 \frac{1}{n^2}\left(\frac{3}{4} - \frac{n}{j+\tfrac{1}{2}}\right). \tag{4.7}$$

Therefore the matrix element of $H = H_0 + H_1$ is (with $\hbar$ restored)

$$E_{nlj} = \langle n\, l j j_3 | H | n\, l j j_3 \rangle = E_n^0\left(1 + \left(\frac{e^2}{\hbar c}\right)^2 \frac{1}{n^2}\left(\frac{n}{j+\tfrac{1}{2}} - \frac{3}{4}\right)\right). \tag{4.8}$$

The term in (4.8) that gives rise to the fine-structure splitting is the one proportional to $1/(j+\tfrac{1}{2})$. Recalling that $\alpha^2 = (e^2/(\hbar c))^2 \cong (1/137)^2$, we see that this fine-structure term is four orders of magnitude smaller than $E_n^0$ and gives the experimentally correct splitting between the energy levels of different $j$ (cf. Figure VI.5.2). Equation (4.8) is independent of $l$ and therefore does not describe the splitting between $2S^{1/2}$, $2P^{1/2}$, $3P^{3/2}$, $3D^{3/2}$, etc., i.e., between states with the same value of $j$ but different values of $l$ or different values of parity. As mentioned above, (4.8) is also the eigenvalue of the relativistic energy operator, which does not include corrections for the Lamb shift and nucleon structure.

## IX.5 Selection Rules

Selection rules for dipole transitions have been discussed previously, in particular in Section V.4 for general angular momentum states and also briefly in Section VI.5 for the hydrogen atom without spin. The selection rules, i.e., the rules that tell when the matrix elements $\langle n\, l\, j\, j_3 | Q_i | n\, l\, j\, j_3 \rangle$ are zero, follow from the property that $Q_i$ is a proper vector operator, i.e., that it fulfills the relations

$$[L_i, Q_j] = i\epsilon_{ijk} Q_k, \tag{5.1a}$$

$$[J_i, Q_j] = i\epsilon_{ijk} Q_k, \tag{5.1b}$$

and

$$U_P Q_i U_P = -Q_i. \tag{5.2}$$

Equation (5.1b) expresses the fact that $Q_j$ is not only a vector operator with respect to the orbital angular momentum (an **L**-vector operator) but also a vector operator with respect to total angular momentum (a **J**-vector operator). Equation (5.1b) follows from $J_i = L_i + S_i$ together with (5.1a).

As a consequence of (5.1b), the selection rules for dipole transitions between the physical states $| n\, l\, j\, j_3 \rangle$ for the hydrogen atom (and all other one-electron atoms) are:

$$\langle n'\, l'\, j'\, j_3' | Q_i | n\, l\, j\, j_3 \rangle = 0 \tag{5.3}$$

unless $j' = j + 1, j$, or $j - 1$. From (5.2) it follows that

$$\langle n'\, l'\, j'\, j_3' | Q_i | n\, l\, j\, j_3 \rangle = 0 \tag{5.4}$$

unless $\pi(l)\pi(l') = -1$ [i.e., unless $(-1)^{l'+l} = -1$, as $\pi(l) = (-1)^l$]. From (5.3) and the fact that $j' = l' \pm \frac{1}{2}$ and $j = l \pm \frac{1}{2}$, it follows that the matrix element is zero unless $l' = l + 2, l + 1, l, l - 1, l - 2$. Thus (5.4) can be written

$$\langle n'\, l'\, j'\, j_3' | Q_i | n\, l\, j\, j_3 \rangle = 0 \tag{5.5}$$

unless $l' = l \pm 1$. Equations (5.3), (5.4) or (5.5) give the selection rules for dipole transitions in the hydrogen atom and all other one-electron atoms.

## IX.6 Remarks on the State of an Electron in Atoms

We should close this part on the hydrogen atom with a remark. To conjecture the algebraic structure that is the mathematical image of the hydrogen atom, we made use of the classical particle picture of the Kepler problem, i.e., we had as the point of departure for our conjecture the picture of an electron as a particle that moves in closed orbits around a center. And when we discussed spin, we even implied some analogy of the spin with the rotation of this particle around its own axis [in the comparision of (3.17) with (3.18), for instance]. This was the classical picture, which is described by the mathematical relations between the observables when they are not represented by operators but by numbers. In quantum mechanics these mathematical relations are operator relations, and are not the mathematical image of a point particle revolving around a center and rotating around its own axis. The electron in the hydrogen atom is not a particle, and the spin is not a rotation around the particle's axis. A quantum-mechanical object was called a particle when it was approximately a generalized eigenstate of the position operator, because localization is the characteristic property of a particle. The electron in the hydrogen atom is not in a generalized position eigenstate. Equally wrong is the view (which originates from the misinterpretation of the solution of the Schrödinger wave equation) that the electron in the hydrogen atom is a standing plane wave, i.e., the classical picture complementary to the particle picture. A quantum-mechanical object was called a wave when it was a generalized eigenstate of the momentum operator (Section II.11), because this state has the characteristic property of a wave (wave motion over all space). The electron in the hydrogen atom is not in a generalized momentum eigenstate. Thus neither of the complementary classical pictures of the electron is applicable to the electron in the hydrogen atom. In the hydrogen atom the electron does not appear as a wave or as a particle, but in a form different from both, namely, as an angular-momentum and energy eigenstate, in which it has neither a definite position (particle) nor a definite momentum (wave), but does have a definite angular momentum (rotator). The classical picture that comes closest to this is that of a standing spherical wave.

## Problems

1. (a) Show that every operator on $\boldsymbol{\ell}^s$ is a linear combination of the operators $I, S_1, S_2$, and $S_3$; as stated in the paragraph preceding equation (2.21).
   (b) Check to see whether there are other solutions of

$$i\epsilon_{ijk}\{L_k, B_j\} = 0$$

[the equation immediately preceeding (2.38)] besides (2.38),

$$B_j = f(Q, P)L_j.$$

2. The proton possess a magnetic moment

$$M_p = \frac{e}{2m_p c}(2 + 3.59)\,\mathbf{S}.$$

   (a) Determine the Hamiltonian that describes the interaction of the electron's spin with the proton's spin in the hydrogen atom (hyperfine interaction).
   (b) Treating the spin-spin interaction Hamiltonian in part (a) as a further perturbation term on the hydrogen-atom Hamiltonian, determine the matrix elements of this spin-spin term between states $|n\,l\,j\,j_3\rangle$ of $l = 0$ and $l = 1$. See how this correction compares with the fine-structure correction.

# CHAPTER X

# Indistinguishable Particles

In this chapter the basic assumption about the combination of identical physical systems is conjectured from the indistinguishability of these systems.

## X.1 Introduction

The quantum-mechanical systems that we have considered so far have consisted of only one constituent of the same kind. The hydrogen atom was considered as one electron in an electric field. The vibrating diatomic molecule was described as the problem of one oscillator. The rotating diatomic molecule was reduced to the problem of one system rotating around a center. And when the vibrating and rotating diatomic molecule was considered, it was described as the combination of one rotator and one oscillator, i.e., as one rotating oscillator. These systems are called *one-particle systems*. Thus one-particle systems are systems that consist of only one constituent of the "same kind."

We now want to consider *many-particle systems*, i.e., systems that are the combination of many ($N = 2, 3, \ldots$) one-particle systems of the same kind. Let $\mathscr{H}_1, \mathscr{H}_2, \ldots, \mathscr{H}_N$ be the space of physical states of the first, second, $\ldots$, $N$th one-particle system. (All of the spaces $\mathscr{H}_i$ are identical; the subscript serves only to identify a space with a particular particle.) We would expect from the basic assumption IVa that the $N$-particle system, which is the combination of these $N$ one-particle systems, will have as its space of states the direct product space

$$\mathfrak{H} = \mathscr{H}_1 \otimes \mathscr{H}_2 \otimes \cdots \otimes \mathscr{H}_N. \tag{1.1}$$

The algebra of the $N$-particle systems would be the direct product of the algebras of observables of the one-particle systems, i.e., the algebra is the set of all operators

$$A = \sum A_1 \otimes A_2 \otimes \cdots \otimes A_N, \tag{1.2}$$

where $A_i$ is an element of the algebra of observables in $\mathcal{H}_i$. (In particular, if these $N$ one-particle systems do not interact with each other, then all elements of the algebra of observables in $\mathfrak{H}$ are of the kind $A = A_1 \otimes I \otimes I \otimes \cdots \otimes I + I \otimes A_2 \otimes I \otimes \cdots \otimes I + \cdots + I \otimes I \otimes \cdots \otimes I \otimes A_N$.)

That the $N$-particle system would be described by (1.1) we would expect from classical considerations. Classical particles can, in principle, be numbered; i.e., we can imagine that each particle is given at some instant a label, and we can then follow the subsequent motion of each particle in its path and identify a given particle at any subsequent instant. This is not possible, however, for quantum-mechanical systems. Following the motion of each particle of a quantum-mechanical system means performing a series of position measurements. Each measurement—according to the axiom III—changes the state of the quantum-mechanical system in an uncontrollable fashion: If we have localized the particle in the neighborhood of a certain point, then we do not known what its momentum is and we do not known where it will go. The concept of path does not exist for a quantum-mechanical system (uncertainty principle). Thus in quantum mechanics it is, even in principle, impossible to follow each of a number of *identical particles* and thereby distinguish them. (By identical particles we mean particles that have the same observable values.) Thus: Identical quantum-mechanical particles are indistinguishable.

We shall now give a precise, mathematical formulation of indistinguishability and then formulate the consequences as another basic assumption of quantum mechanics. Let $|\xi_i\rangle_i$ denote a basis in the space $\mathcal{H}_i$, i.e., the one symbol $\xi_i$ stands for the full set of quantum numbers (eigenvalues of a c.s.c.o.) necessary to label the basis system of $\mathcal{H}_i$. [For the sake of definiteness, one may assume that the $N$ particles are electrons in a Coulomb field; then each $\mathcal{H}_i$ is the space $\mathcal{H}$ of (IX.2.15), and $|\xi_i\rangle_i = |n_i l_i j_i j_{i3}\rangle$.] The basis in $\mathfrak{H}$ of (1.1) is then given by

$$|\xi_1 \xi_2 \cdots \xi_N\rangle = |\xi_1\rangle_1 \otimes |\xi_2\rangle_2 \otimes \cdots \otimes |\xi_N\rangle_N. \tag{1.3}$$

Suppose we are given $N$ objects (elements) in a certain order,

$$(\xi_1, \xi_2, \ldots, \xi_N);$$

such an arrangement is called a *permutation*. These $N$ elements can be written in a different order, $(\eta_1, \eta_2, \ldots, \eta_N)$; this is called a permutation of the $N$ objects. There are $N!$ different permutations of $N$ objects. One particular permutation can be considered to be the "original" or "natural" or "standard" one. All other permutations can be obtained from this original one by changing the order in which the objects appear. It is clear that the *operation* of changing the order is specified by the resulting permutation;

therefore this operation is also called a permutation. Thus we may consider the permutation operation (or just "permutation") $P$ that changes $(\xi_1, \xi_2, \ldots, \xi_N)$ into $(\eta_1, \eta_2, \ldots, \eta_N)$. For example, we can consider the permutation $P_{12}$ that changes $(\xi_1, \xi_2, \ldots, \xi_N)$ into $(\eta_1 = \xi_2, \eta_2 = \xi_1, \eta_3 = \xi_3, \ldots, \eta_N = \xi_N)$. Or we can consider the permutation $P_{ij}$ that changes $(\xi_1, \ldots, \xi_i, \ldots, \xi_j, \ldots, \xi_N)$ into $(\xi_1, \ldots, \xi_j, \ldots, \xi_i, \ldots, \xi_N)$ by interchanging the $i$th and $j$th elements; such permutations, which consist of the exchange of the position of two elements, are called *transpositions*. Every permutation can be obtained by a finite number of transpositions; e.g., the permutation $(\xi_2, \xi_3, \xi_1, \xi_4, \ldots, \xi_N)$ may be obtained by the transposition $P_{13}$ followed by the transposition $P_{12}$: $(\xi_1, \xi_2, \xi_3, \xi_4, \ldots) \rightarrow (\xi_3, \xi_2, \xi_1, \xi_4, \ldots) \rightarrow (\xi_2, \xi_3, \xi_1, \xi_4, \ldots)$. While the decomposition of a permutation into successive transpositions is not unique, the number of transpositions will always be even or odd, depending upon the particular permutation. A permutation is *odd* (with respect to the original permutation) if it is obtained from the original permutation by an odd number of transpositions; it is *even* if it is obtained by an even number of transpositions.

Now let the $N$ objects be the $N$ sets of quantum numbers $(\xi_1, \xi_2, \ldots) = (n_1 l_1 j_1 j_{13}, n_2 l_2 j_2 j_{23}, \ldots)$. Each permutation $(\eta_1, \eta_2, \ldots, \eta_N)$, or each permutation operation $P : (\xi_1, \xi_2, \ldots, \xi_N) \mapsto (\eta_1, \eta_2, \ldots, \eta_N)$ can then be represented in the space $\mathfrak{H}$ by a linear operator $\mathbb{P}$ defined by

$$\mathbb{P}|\xi_1 \xi_2 \cdots \xi_N\rangle = |\eta_1 \eta_2 \cdots \eta_N\rangle. \tag{1.4}$$

For example, the transposition $P_{12}$ is represented by the operator $\mathbb{P}_{12}$ with

$$\mathbb{P}_{12}|\xi_1 \xi_2 \cdots \xi_N\rangle = |\xi_2 \xi_1 \cdots \xi_N\rangle. \tag{1.5}$$

The operators $\mathbb{P}$ will be chosen to be unitary.[1]

⟦The set $\{P\}$ of all permutations of $N$ objects forms a group known as the *symmetric* (or *permutation*) *group*, while the set $\{\mathbb{P}\}$ of all representing operators forms a *representation* of the permutation group. If all operators $\mathbb{P}$ are unitary, then the representation is called a *unitary representation* of the permutation group. Because the permutation group has a finite number of elements (is finite and consequently compact), every representation of the permutation group, according to a theorem, may be considered unitary. We shall make use of only one property of the permutation group, to be stated below, and shall not require any group theory.⟧

Let $\wedge_{|\xi\rangle}$ denote the projection operator onto the one-dimensional subspace spanned by $|\xi\rangle = |\xi_1 \xi_2 \cdots \xi_N\rangle$, and let $\wedge_{|\eta\rangle}$ denote the projector onto the one-dimensional subspace spanned by $|\eta\rangle = |\eta_1 \eta_2 \cdots \eta_N\rangle$. Because of (1.4), we have the connection

$$\wedge_{|\eta\rangle} = |\eta\rangle\langle\eta| = \mathbb{P}|\xi\rangle\langle\xi|\mathbb{P}^\dagger = \mathbb{P}\wedge_{|\xi\rangle}\mathbb{P}^\dagger. \tag{1.6}$$

---

[1] The reason for the requirement than $\mathbb{P}$ be unitary follows from the fact that $\mathbb{P}$ is a symmetry transformation. A brief justification of this will be given in Appendix to Section XIX.2.

Now, due to *indistinguishability*, $\wedge_{|\xi\rangle}$ and $\wedge_{|\eta\rangle}$ can represent a (pure) physical state only if

$$\wedge_{|\eta\rangle} = \wedge_{|\xi\rangle}, \tag{1.7}$$

or, using (1.6),

$$\wedge_{|\xi\rangle} = \mathbb{P}\wedge_{|\xi\rangle}\mathbb{P}^{\dagger}. \tag{1.8}$$

Thus of the one-dimensional spaces spanned by $|\xi_1\rangle \otimes |\xi_2\rangle \otimes \cdots \otimes |\xi_N\rangle$, only those for which (1.8) is fulfilled can represent physical states. From this we conclude that, because of indistinguishability, it is not the whole direct-product space $\mathfrak{H}$ of (1.1) but only a subspace of it that is the space of physical states. Also, it is not the whole direct-product algebra of operators given by (1.2) but only a subalgebra of it that is the algebra of observables. We now want to determine the physical subspace of (1.1). It is clear that, in general, the direct-product basis (1.3) is not a suitable basis [(1.8) means $\mathbb{P}|\eta\rangle \propto |\eta\rangle$, which cannot be fulfilled if all the quantum numbers are different]; and that, in general, the physical states are represented by linear combinations of (1.3).

Let $\psi$ be a vector of the physical subspace [in general, a linear combination of (1.3)], and let $\mathbb{P}$ be a permutation operator. Then, if the particles are indistinguishable, $\psi$ and $\chi = \mathbb{P}\psi$ [or $\wedge_{\psi}$ and $\wedge_{\chi} = \wedge_{P\psi}$] represent the same physical state. The expectation value of every observable $A$ must therefore be the same for $\psi$ and for $\chi$, i.e.,

$$\langle\psi|A|\psi\rangle = \langle\chi|A|\chi\rangle = \langle\psi|\mathbb{P}^{\dagger}A\mathbb{P}|\psi\rangle \tag{1.9}$$

for every observable $A$. As $\psi$ is an arbitrary vector of the physical subspace, we conclude that

$$A = \mathbb{P}^{\dagger}A\mathbb{P} \tag{1.10}$$

for every observable $A$. Hence for any $A \in \mathscr{A}$ and for any permutation operator $\mathbb{P}$,

$$[\mathbb{P}, A] = 0. \tag{1.11}$$

[Equation (1.11) follows immediately from (1.10) for $\mathbb{P}$ unitary, which, as mentioned above, can be assumed for the permutation group. For $\mathbb{P}$ nonunitary, one must take linear combinations $\psi_1 + i\alpha\psi_2$, $\psi_1 + \alpha\psi_2$ in order to deduce (1.11).]

Equation (1.11) *is the* mathematical formulation of the *statement that identical particles are indistinguishable*. From this mathematical formulation of indistinguishability, one can deduce that the vectors of the physical subspace of $\mathfrak{H}$ must fulfill either

$$\mathbb{P}\psi = +\psi \quad \text{for all } \mathbb{P} \tag{1.12}$$

or

$$\mathbb{P}\psi = (-1)^{p}\psi \quad \text{where } \begin{cases} p \text{ is even if } \mathbb{P} \text{ is even,} \\ p \text{ is odd if } \mathbb{P} \text{ is odd.} \end{cases} \tag{1.13}$$

The $\psi$ fulfilling (1.12) are called *symmetric*, while those fulfilling (1.13) are called *antisymmetric*.

To deduce this, we need the mathematical formulation of a physically obvious property of *the algebra $\mathscr{A} = \{A\}$ of observables $A$*. This can be formulated in the following way:[2] *$\mathscr{A} = \{A\}$ contains a complete set of commuting operators.* To justify this condition physically, we recall that we called a state "pure" (up to a certain accuracy) if there is no observable whose measurement allowed the separation of an ensemble in this state into two or more subensembles. If such a separation is possible (as in the case of the hydrogen atoms in the ground state), it leads to the introduction of a new quantum number and therewith to the introduction of a new observable (in the case of the hydrogen atom, spin). Consequently one must enlarge the algebra of observables to accomodate this new observable. What is a pure state with respect to a certain accuracy need not be a pure state with respect to a higher accuracy. But up to every desired (and observable) accuracy, every pure state is completely specified by a set of quantum numbers that are connected with observables. (In fact, the algebra of observables is conjectured from these quantum numbers and observables.) Each label of a vector is connected with an observable, and the observables whose eigenvalues label the vectors form a complete set. [As we have seen, the statement that a certain set of commuting operators is a complete system is a *physical* statement, conjectured from the physical properties of the system, and not a mathematical statement (cf. Chapter IV)]. Therefore the apparent occurrence of pure states not fulfilling (1.12) or (1.13) is always an indication that the available set of quantum numbers is not complete and that there exist other quantum numbers completing this set that have not yet been uncovered.

Let us denote by $A_1, A_2, \ldots, A_n$ a complete system of commuting operators of the algebra of observables, by $|a\rangle = |a_1 a_2 \cdots a_n\rangle$ the corresponding eigenvectors, and by $\wedge_{|a\rangle}$ the projectors onto the subspaces spanned by $|a\rangle$. $\wedge_{|a\rangle}$ is the observable whose expectation value gives the probability of obtaining $a_1, a_2, \ldots, a_n$ in a measurement of $A_1, A_2, \ldots, A_n$. As observables, the $\wedge_{|a\rangle}$ must commute with all permutations $\mathbb{P}$ in accord with (1.11):

$$[\mathbb{P}, \wedge_{|a\rangle}] = 0 \quad \text{for all } \mathbb{P}. \tag{1.14}$$

From (1.14) it follows that $|a\rangle$ must be an eigenvector of all $\mathbb{P}$.

*Proof*: Equation (1.14) applied to a vector $|\psi\rangle$ gives $\mathbb{P} \wedge_{|a\rangle} |\psi\rangle = \wedge_{|a\rangle} \mathbb{P} |\psi\rangle$. Using $\wedge_{|a\rangle} = |a\rangle\langle a|$, we see that $\mathbb{P}|a\rangle\langle a|\psi\rangle = |a\rangle\langle a|\mathbb{P}|\psi\rangle$, and hence that

$$\mathbb{P}|a\rangle = |a\rangle \frac{\langle a|\mathbb{P}|\psi\rangle}{\langle a|\psi\rangle}. \tag{1.15}$$

---

[2] It is this condition that is relaxed when parastatistics are allowed.

The possible eigenvalues, $\langle a|\mathbb{P}|\psi\rangle/\langle a|\psi\rangle$, follow from the following property (which we shall not derive here) of the permutation group: The permutation group has two one-dimensional representations: (1) the symmetrical representation in which all permutations are represented by the unit operator $I$, i.e.,

$$\mathbb{P}|\psi\rangle = +|\psi\rangle \quad \text{for all permutations } \mathbb{P}; \tag{1.16}$$

and (2) the antisymmetrical representation in which all even permutations are represented by the operator $I$ and all odd permutations are represented by the operator $-I$, i.e.,

$$\mathbb{P}|\psi\rangle = (-1)^p|\psi\rangle \quad \text{where} \begin{cases} p \text{ is even if } \mathbb{P} \text{ is even,} \\ p \text{ is odd if } \mathbb{P} \text{ is odd.} \end{cases} \tag{1.17}$$

We recall that an irreducible-representation space (ladder representation) of the algebra of $\{\mathbb{P}\}$ is a representation space that is obtained by applying all $\mathbb{P}$ to one element of the space. The fact that $|a\rangle$ is an eigenvector of all $\mathbb{P}$ means that $|a\rangle$ spans a one-dimensional irreducible-representation space. Therefore, according to the above property of the permutation group, we must have either

$$\mathbb{P}|a\rangle = +|a\rangle \quad \text{for all } \mathbb{P} \tag{1.18}$$

or

$$\mathbb{P}|a\rangle = (-1)^p|a\rangle \quad \text{for all } \mathbb{P}. \tag{1.19}$$

Since this is true for every basis vector $|a\rangle$, and as every vector $|\psi\rangle$ of the representation space of the algebra of observables (i.e., of the physical subspace of $\mathfrak{H}$) can be written as a linear combination of the basis vectors, we conclude that either (1.12) or (1.13) must be fulfilled.

Let us denote by $\mathscr{H}_+^N$ that subspace of $\mathfrak{H}$ which consists of the symmetric vectors,

$$\mathscr{H}_+^N = \{|\psi\rangle \in \mathfrak{H} : \mathbb{P}|\psi\rangle = |\psi\rangle\}, \tag{1.20}$$

and let us denote by $\mathscr{H}_-^N$ that subspace of $\mathfrak{H}$ which consists of the antisymmetric vectors,

$$\mathscr{H}_-^N = \{|\psi\rangle \in \mathfrak{H} : \mathbb{P}|\psi\rangle = (-1)^p|\psi\rangle\}. \tag{1.21}$$

We can then formulate the statement, which we have derived from indistinguishability (1.11), in the following way: The space of physical states of $N$ identical quantum-mechanical systems is either the antisymmetric space $\mathscr{H}_-^N$ or the symmetric space $\mathscr{H}_+^N$.

A basis of nonnormalized vectors in $\mathscr{H}_+^N$ is given by

$$|\xi\rangle_+ = |\xi_1\xi_2\cdots\xi_N\rangle_+ = \sum_{\mathbb{P}} \mathbb{P}|\xi_1\xi_2\cdots\xi_N\rangle, \tag{1.22}$$

where $|\xi_1 \xi_2 \cdots \xi_N\rangle$ is given by (1.3) and $\sum_\mathbb{P}$ is the sum over all permutations $\mathbb{P}$ of the $N$ objects $\xi_1, \xi_2, \ldots, \xi_N$. It is clear that the order of the quantum numbers $\xi_1, \xi_2, \ldots, \xi_N$ in $|\xi\rangle_+$ is irrelevant. To prove that the $|\xi\rangle_+$ are symmetric, we calculate for an arbitrary permutation $\mathbb{P}_1$ that

$$\mathbb{P}_1 |\xi\rangle_+ = \sum_\mathbb{P} \mathbb{P}_1 \mathbb{P} |\xi_1 \xi_2 \cdots \xi_N\rangle = \sum_{\mathbb{P}'} \mathbb{P}' |\xi_1 \xi_2 \cdots \xi_N\rangle = |\xi\rangle_+ \, ,$$

where we have set $\mathbb{P}' = \mathbb{P}_1 \mathbb{P}$ and where the sum over all $\mathbb{P}$ becomes a sum over all $\mathbb{P}'$. The latter follows from the fact that if $\mathbb{P}_1$ is fixed and $\mathbb{P}$ runs over all permutations, then $\mathbb{P}'$ runs over all permutations. Thus for any permutation $\mathbb{P}_1$,

$$\mathbb{P}_1 |\xi\rangle_+ = |\xi\rangle_+ \, . \tag{1.23}$$

A basis of nonnormalized vectors in $\mathcal{H}^N_-$ is given by

$$|\xi\rangle_- = |\xi_1 \xi_2 \cdots \xi_N\rangle_- = \sum_\mathbb{P} (-1)^p \mathbb{P} |\xi_1 \xi_2 \cdots \xi_N\rangle. \tag{1.24}$$

To prove that they are antisymmetric, we first note that $(-1)^{p'} = (-1)^{p+p_1} = (-1)^p (-1)^{p_1}$ and hence that $(-1)^p = (-1)^{p_1} (-1)^{p'}$. Then

$$\mathbb{P}_1 |\xi\rangle_- = \sum_\mathbb{P} (-1)^p \mathbb{P}_1 \mathbb{P} |\xi_1 \xi_2 \cdots \xi_N\rangle$$

$$= \sum_{\mathbb{P}'} (-1)^{p_1} (-1)^{p'} \mathbb{P}' |\xi_1 \xi_2 \cdots \xi_N\rangle = (-1)^{p_1} |\xi\rangle_- . \tag{1.25}$$

for any permutation $\mathbb{P}_1$. If the $|\xi_i\rangle_i$ are normalized vectors in $\mathcal{H}_i$, then the $|\xi_1 \xi_2 \cdots \xi_N\rangle$ are normalized in $\mathfrak{H}$, i.e., $\langle \xi_1 \xi_2 \cdots \xi_N | \xi_1 \xi_2 \cdots \xi_N \rangle = 1$. Consequently the $|\xi\rangle_+$ and the $|\xi\rangle_-$ are not normalized; the normalizing factors are calculated in Problem XI.1.

We state the result of the preceeding considerations as a new basic assumption (axiom) of quantum mechanics:

IVb    The space of physical states of $N$ identical quantum-mechanical systems (particles) is $\mathcal{H}^N_+$ if their angular momentum (spin) has an integral value, and is $\mathcal{H}^N_-$ if their angular momentum has a half-integral value.

Particles whose space of physical states is $\mathcal{H}^N_+$ and $\mathcal{H}^N_-$ are called *bosons* and *fermions*, respectively. Bosons are said to obey *Bose statistics*, while fermions are said to obey *Fermi statistics*. The above axiom IVb then states that half-integral-spin particles are fermions, that integral-spin particles are bosons, and that there are no other particles obeying some other "parastatistics" (which would have to belong to higher-dimensional representations of the permutation group). This axiom has been confirmed in all cases where it has been investigated (electrons, protons, neutrons are fermions; pions, photons, phonons, alpha particles are bosons). We remark that the axiom IVb has been derived, to a large extent, from indistinguishability (1.11), which in turn was deduced from previously formulated basic assumptions of quantum mechanics. The part that has not been deduced is the connection between spin and statistics.

The Pauli principle in its original form follows immediately from IVb. The Pauli exclusion principle states: The quantum numbers of two or more electrons can never entirely agree.

# CHAPTER XI

# Two-Electron Systems— The Helium Atom

The system with two electrons is studied in this chapter. Section XI.1 shows that the space of physical states of the helium atom is the sum of the parahelium and the orthohelium spaces. In Section XI.2 the ionization thresholds (i.e., the energy values $E_{n\infty}$ at which one electron is in the $n$th level and the other is just dissociating from the atom) are determined, and the energy levels below the first ionization threshold are discussed. Section XI.3 discusses the energy levels above the first ionization threshold without considering the interaction between these levels and the energy continuum of the $(\text{He}^+, e)$ system.

## XI.1 The Two Antisymmetric Subspaces of the Helium Atom

We shall illustrate the consequences of the basic assumption IVb with the example of two electrons in a Coulomb field. This is the simplest nontrivial case; although it does not demonstrate the full extent of IVb, it is mathematically simple and does not require the introduction of further properties of the representations of the permutation group.

The energy of two classical spinless particles of mass $m_e$ with charge $-e$ that move in the field of a central charge $Ze$ is given by

$$E = \frac{1}{2m_e}(\mathbf{p}_1^2 + \mathbf{p}_2^2) - \frac{Ze^2}{r_1} - \frac{Ze^2}{r_2} + \frac{e^2}{r_{12}}, \tag{1.1}$$

where $r_1$ and $r_2$ are the distances of the first and second charges $-e$ from the charge $Ze$, and where $r_{12}$ is the distance between these two charges.

245

This system is the classical analogue of the helium atom (or of any two-electron ion if $Z \neq 2$). We obtain the energy operator of the helium atom by the usual procedure of replacing the numbers $p_{\alpha i}$, $x_{\alpha i}$, $r_{\alpha} = (x_{\alpha}^2)^{1/2}$, and $r_{12} = ((\mathbf{x}_1 - \mathbf{x}_2)^2)^{1/2}$ by the operators $P_{\alpha i}$, $Q_{\alpha i}$, $Q_{\alpha} = (Q_{\alpha}^2)^{1/2}$, and $Q_{12} = ((\mathbf{Q}_1 - \mathbf{Q}_2)^2)^{1/2}$, respectively. In addition we have to add a term $H_1$ that describes the influence of the electron spin. Thus we have

$$H = \frac{1}{2m_e}(\mathbf{P}_1^2 + \mathbf{P}_2^2) - \frac{Ze^2}{Q_1} - \frac{Ze^2}{Q_2} + \frac{e^2}{Q_{12}} + H_1 \qquad (1.2)$$

for the energy operator of the helium atom ($Z = 2$), which we write in the form

$$H = H_0 + H_1. \qquad (1.3a)$$

$H_0$ is the Hamiltonian operator corresponding to the classical Hamiltonian given by (1.1):

$$H_0 = \frac{1}{m_e}(h_1 \otimes I + I \otimes h_2) + W = \frac{1}{m_e} H_{00} + W, \qquad (1.3b)$$

where

$$H_{00} = h_1 \otimes I + I \otimes h_2, \qquad (1.3c)$$

$$h_\alpha = \frac{\mathbf{P}_\alpha^2}{2} - \frac{a_Z}{Q_\alpha} \qquad (a_Z \equiv m_e e^2 Z/\hbar, \quad \hbar = 1), \qquad (1.3d)$$

and

$$W = \frac{e^2}{Q_{12}} = e^2 \left( \sum_{i=1}^{3} (Q_{1i} \otimes I - I \otimes Q_{2i})^2 \right)^{-1/2}. \qquad (1.3e)$$

The operator $H$ and all the other operators act in the space

$$\mathfrak{H} = \mathscr{H}_1 \otimes \mathscr{H}_2, \qquad (1.4)$$

where $\mathscr{H}_\alpha$ is the space of the system that consists of one electron in the Coulomb field of the charge $Ze$.

For two objects $\xi_1$ and $\xi_2$ there are only $2! = 2$ permutations, $(\xi_1, \xi_2)$ and $(\xi_2, \xi_1)$; therefore for a fixed set of quantum numbers $\xi_1$ and $\xi_2$ with $\xi_1 \neq \xi_2$ there are only two basis vectors in $\mathfrak{H}$: $|\xi_1 \xi_2\rangle$ and $|\xi_2 \xi_1\rangle$. The normalized symmetric and antisymmetric vectors for this fixed set of quantum numbers are, according to (X.1.22) and (X.1.24),

$$|\xi_1 \xi_2\rangle_+ = \frac{1}{\sqrt{2}}(|\xi_1 \xi_2\rangle + |\xi_2 \xi_1\rangle) \qquad (1.5)$$

and

$$|\xi_1 \xi_2\rangle_- = \frac{1}{\sqrt{2}}(|\xi_1 \xi_2\rangle - |\xi_2 \xi_1\rangle). \qquad (1.6)$$

Thus for a fixed set of quantum numbers $\xi_1$ and $\xi_2$ with $\xi_1 \neq \xi_2$ we have a two-dimensional space spanned alternatively by $|\xi_1 \xi_2\rangle, |\xi_2 \xi_1\rangle$ or by

$|\xi_1\xi_2\rangle_+, |\xi_1\xi_2\rangle_-$. If $\xi_1$ and $\xi_2$ are fixed with $\xi_1 = \xi_2$, we have a one-dimensional space spanned by $|\xi_1\xi_2\rangle = |\xi_1\xi_2\rangle_+$. The space $\mathfrak{H}$ is spanned by $|\xi_1\xi_2\rangle$, where $\xi_1$ and $\xi_2$ independently can take any of the possible sets of values $(nljj_3)$ $(n = 1, 2, \ldots, j = 0, 1, \ldots, n-1;\ j_3 = -j, -j+1, \ldots, j; l = j \pm \frac{1}{2})$.[1] The space $\mathcal{H}^2_-$ is spanned by all the vectors $|\xi_1\xi_2\rangle_-$, and the space $\mathcal{H}^2_+$ is spanned by all the vectors $|\xi_1\xi_2\rangle_+$. Consequently

$$\mathfrak{H} = \mathcal{H}^2_- \oplus \mathcal{H}^2_+, \tag{1.7}$$

i.e., the product space is the direct sum of the symmetric and antisymmetric subspaces.

⟦Equation (1.7) is a particular feature of the case $N = 2$; for $N > 2$, $\mathfrak{H}$ of (X.1.1) is not the direct sum of the symmetric and antisymmetric subspaces (X.1.20) and (X.1.21). Rather

$$\mathfrak{H} = \mathcal{H}^N_+ \oplus \mathcal{H}^N_- \oplus \mathcal{H}^N_{\sigma_1} \oplus \mathcal{H}^N_{\sigma_2} \oplus \cdots,$$

where there are a finite number of terms, as many as the ways in which one can write $N$ as a sum $\sum_i N_i$ of positive integers $N_i$. For $N = 2$ there is only one way possible $(2 = 1 + 1)$, and consequently there are only two different terms, as given by (1.7).⟧

According to the axiom IVb, or the Pauli principle, only the subspace $\mathcal{H}^2_-$ of (1.7) is the space of physical states for the two-electron system. (Were we considering a two-boson system, $\mathcal{H}^2_+$ would be the space of physical states.) To construct $\mathcal{H}^2_-$ and to find the properties of the algebra of observables in $\mathcal{H}^2_-$, we proceed in the following way: Each $\mathcal{H}_\alpha$ $(\alpha = 1, 2)$ is, according to (IX.2.15), written as

$$\mathcal{H}_\alpha = \mathcal{H}^{\mathrm{orb}}_\alpha \otimes \imath^s_\alpha, \tag{1.8}$$

where $\mathcal{H}^{\mathrm{orb}}_\alpha$ is the space in which the orbital observables (i.e., the observables that are obtained as functions of the $P_{\alpha i}$ and $Q_{\alpha i}$) of the $\alpha$th electron act, and where $\imath^s_\alpha$ is the space in which the spin observables of the $\alpha$th electron act. We now combine the orbital and spin spaces of the two electrons separately, i.e., we form

$$\mathcal{H}^{\mathrm{orb}2} = \mathcal{H}^{\mathrm{orb}}_1 \otimes \mathcal{H}^{\mathrm{orb}}_2 \tag{1.9}$$

and

$$\imath^{s^2} = \imath^s_1 \otimes \imath^s_2. \tag{1.10}$$

We then find the symmetric and antisymmetric subspaces of $\mathcal{H}^{\mathrm{orb}2}$ and $\imath^{s^2}$ separately by the same procedure as described above; the $\xi_\alpha$ in (1.5) and (1.6) stand for $\xi^{\mathrm{orb}}_\alpha = (n_\alpha l_\alpha l_{\alpha 3})$ when considering $\mathcal{H}^{\mathrm{orb}2}$ and stand for $\xi^s_\alpha = s_{\alpha 3}$ when considering $\imath^{s^2}$. In this way we arrive at

$$\mathcal{H}^{\mathrm{orb}2} = \mathcal{H}^{\mathrm{orb}2}_+ \oplus \mathcal{H}^{\mathrm{orb}2}_- \tag{1.11}$$

---

[1] $\xi_1 \neq \xi_2$ then means not all $(n_1 l_1 j_1 j_{13})$ agree with $(n_2 l_2 j_2 j_{23})$.

and

$$\imath^{s^2} = \imath^{s^2}_+ \oplus \imath^{s^2}_-. \tag{1.12}$$

The total space $\mathfrak{H}$ is then given by[2]

$$\mathfrak{H} = \mathscr{H}^{\mathrm{orb}^2} \otimes \imath^{s^2}$$

$$= (\mathscr{H}^{\mathrm{orb}^2}_+ \otimes \imath^{s^2}_+) \oplus (\mathscr{H}^{\mathrm{orb}^2}_+ \otimes \imath^{s^2}_-) \oplus (\mathscr{H}^{\mathrm{orb}^2}_- \otimes \imath^{s^2}_+) \oplus (\mathscr{H}^{\mathrm{orb}^2}_- \otimes \imath^{s^2}_-). \tag{1.13}$$

In Problem XI.2 it is shown that the symmetric subspace is

$$\mathscr{H}^2_+ = (\mathscr{H}^{\mathrm{orb}^2}_+ \otimes \imath^{s^2}_+) \oplus (\mathscr{H}^{\mathrm{orb}^2}_- \otimes \imath^{s^2}_-) \tag{1.14}$$

while the antisymmetric subspace is shown to be

$$\mathscr{H}^2_- = (\mathscr{H}^{\mathrm{orb}^2}_+ \otimes \imath^{s^2}_-) \oplus (\mathscr{H}^{\mathrm{orb}^2}_- \otimes \imath^{s^2}_+). \tag{1.15}$$

Thus *the space of physical states $\mathscr{H}^2$ is the direct sum of two spaces: one is the space of symmetric orbital states and antisymmetric spin states, and the other is the space of antisymmetric orbital states and symmetric spin states.*

As has already been discussed (Section IX.2) in the case of the hydrogen atom, the basis vectors that are eigenvectors of the (total) spin and of the (total) orbital angular momentum are not a physical basis, because it is the (total) angular momentum that is the physical observable, and not the spin or orbital angular momentum. Thus to obtain the physical states one has to form those linear combinations of the direct product states $|\zeta^{\mathrm{orb}}_1 \zeta^{\mathrm{orb}}_2\rangle_+ \otimes |s_{13}s_{23}\rangle_-$ (in $\mathscr{H}^{\mathrm{orb}^2}_+ \otimes \imath^{s^2}_-$) and $|\zeta^{\mathrm{orb}}_1 \zeta^{\mathrm{orb}}_2\rangle_- \otimes |s_{13}s_{23}\rangle_+$ (in $\mathscr{H}^{\mathrm{orb}^2}_- \otimes \imath^{s^2}_+$) that are eigenstates of the total angular momentum. Furthermore, it will turn out that the spaces $\mathscr{H}^{\mathrm{orb}^2}_+ \otimes \imath^{s^2}_-$ and $\mathscr{H}^{\mathrm{orb}^2}_- \otimes \imath^{s^2}_+$ are not eigenspaces of the energy operator $H$ of (1.3a); the reason is that $H_1$ does not commute with the operator of total spin $S_i = S_{1i} + S_{2i}$, but $\imath^{s^2}_+$ and $\imath^{s^2}_-$ are eigenspaces of $\mathbf{S}^2$. If the physical states are eigenstates of $H$ (as all experimental data confirm) then the physical state vectors are elements of neither $\mathscr{H}^{\mathrm{orb}^2}_+ \otimes \imath^{s^2}_-$ nor $\mathscr{H}^{\mathrm{orb}^2}_- \otimes \imath^{s^2}_+$, but are linear combinations with a small component in one of the spaces and a large component in the other. Thus the reduction of $\mathscr{H}^2_-$ into the direct sum given by (1.15) is only approximately physical; the subspaces $\mathscr{H}^{\mathrm{orb}^2}_+ \otimes \imath^{s^2}_-$ and $\mathscr{H}^{\mathrm{orb}^2}_- \otimes \imath^{s^2}_+$ are spaces of physical states only to the extent that the contribution of $H_1$ (the spin-orbit interaction) to $H$ can be neglected. As in the case of the hydrogen atom, this will turn out to be a very good approximation.

We shall now neglect $H_1$ and undertake a detailed construction of all four spaces on the right-hand side of (1.15). We start with the spin spaces $\imath^{s^2}_+$ and $\imath^{s^2}_-$, as these are much simpler than the orbital spaces. The space $\imath^{s^2} = \imath^s_1 \otimes \imath^s_2$ is the direct product of 2 two-dimensional spaces and is thus

---

[2] We wish to stress again that the appearance of only symmetric or antisymmetric subspaces is a particular feature of $N = 2$; for $N > 2$, higher-dimensional representations $\sigma_i$ of the permutation group in $\mathscr{H}^{\mathrm{orb}^N}$ and in $\imath^{s^N}$ also have to be considered. The antisymmetric space $\mathscr{H}^N_-$ then contains not only the spaces $\mathscr{H}^{\mathrm{orb}^N}_{\mp} \otimes \imath^{s^N}_{\pm}$ as in (1.15), but also contains all direct-product spaces of the form $\mathscr{H}^{\mathrm{orb}^N}_\sigma \otimes \imath^{s^N}_{\sigma'}$, where $\sigma'$ is the irreducible representation of the permutation group that is "associated" with the irreducible representation $\sigma$ in such a way that $\mathscr{H}^{\mathrm{or}^N}_\sigma \otimes \imath^{s^N}_{\sigma'}$ is an antisymmetric subspace of $\mathfrak{H}$.

four-dimensional. Its direct-product basis is given by the four vectors

$$|s_{13}\rangle_1 \otimes |s_{23}\rangle_2 \qquad (s_{13} = \pm\tfrac{1}{2}, \quad s_{23} = \pm\tfrac{1}{2}). \qquad (1.16)$$

One easily finds the antisymmetric and symmetric combinations of these four vectors. The symmetric ones are

$$|\tfrac{1}{2}\rangle_1 \otimes |\tfrac{1}{2}\rangle_2, \qquad |-\tfrac{1}{2}\rangle_1 \otimes |-\tfrac{1}{2}\rangle_2,$$

$$\text{and} \quad \frac{1}{\sqrt{2}}(|\tfrac{1}{2}\rangle_1 \otimes |-\tfrac{1}{2}\rangle_2 + |-\tfrac{1}{2}\rangle_1 \otimes |\tfrac{1}{2}\rangle_2), \qquad (1.17)$$

while the single antisymmetric vector is

$$\frac{1}{\sqrt{2}}(|\tfrac{1}{2}\rangle_1 \otimes |-\tfrac{1}{2}\rangle_2 - |-\tfrac{1}{2}\rangle_1 \otimes |\tfrac{1}{2}\rangle_2). \qquad (1.18)$$

The four vectors of (1.17) and (1.18) are orthonormal and therefore constitute a basis in $\iota^{s^2}$; consequently the three symmetric vectors span the symmetric space $\iota^{s^2}_+$, which is therefore three-dimensional, and the vector (1.18) spans the antisymmetric space $\iota^{s^2}_-$, which is therefore one-dimensional. $\iota^{s^2} = \mathscr{R}^{s_1 = 1/2} \otimes \mathscr{R}^{s_2 = 1/2}$ is the space of the combination of two elementary rotators. We can thus apply the results of Section V.2 and define the operator of total spin,

$$S_i = S_{1i} \otimes I + I \otimes S_{2i} \qquad (i = 1, 2, 3). \qquad (1.19)$$

Equation (V.2.33) then tells us that

$$\iota^{s^2} = \mathscr{R}^{s = 1} \oplus \mathscr{R}^{s = 0}, \qquad (1.20)$$

i.e., the total spin is $s = 1$ or $s = 0$, and we can introduce in $\iota^{s^2}$ the basis $|s\, s_3\rangle$ with

$$\left.\begin{array}{l} \mathbf{S}^2|s\, s_3\rangle = s(s + 1)|s\, s_3\rangle, \\ S_3|s\, s_3\rangle = s_3|s\, s_3\rangle \end{array}\right\} \qquad \left\{\begin{array}{l} s = 0, \quad s_3 = 0; \\ s = 1, \quad s_3 = -1, 0, 1. \end{array}\right. \qquad (1.21)$$

The observables $S_i$ commute with the permutation operators (in this case the transposition operator $\mathbb{P}_{12}$), since

$$\mathbb{P}_{12}S_i = \mathbb{P}_{12}(S_{1i} \otimes I + I \otimes S_{2i}) = (I \otimes S_{2i} + S_{1i} \otimes I)\mathbb{P}_{12} = S_i\mathbb{P}_{12}. \qquad (1.22)$$

The two subspaces $\iota^{s^2}_+$ and $\iota^{s^2}_-$ are eigenspaces of $\mathbb{P}_{12}$ corresponding to the eigenvalues $+1$ and $-1$, respectively. Because of (1.22), $S_i$ cannot transform out of either $\iota^{s^2}_+$ or $\iota^{s^2}_-$, i.e., $S_i$ leaves $\iota^{s^2}_+$ and $\iota^{s^2}_-$ invariant. Thus $S_i$ leaves invariant $\mathscr{R}^{s = 1}$ and $\mathscr{R}^{s = 0}$ on the one hand, and leaves invariant $\iota^{s^2}_+$ and $\iota^{s^2}_-$ on the other hand. Also, $\mathscr{R}^{s = 1}$ is three-dimensional, as is $\iota^{s^2}_+$, and $\mathscr{R}^{s = 0}$ is one-dimensional, as is $\iota^{s^2}_-$; consequently $\mathscr{R}^{s = 1} = \iota^{s^2}_+$ and $\mathscr{R}^{s = 0} = \iota^{s^2}_-$. We can therefore write (1.15) as

$$\mathscr{H}^2_- = (\mathscr{H}^{\text{orb}^2}_+ \otimes \mathscr{R}^{s = 0}) \oplus (\mathscr{H}^{\text{orb}^2}_- \otimes \mathscr{R}^{s = 1}). \qquad (1.23)$$

The space of physical states (neglecting the spin-orbit interaction) is thus the direct sum of a space in which the total spin is zero (space of singlet states) and a space in which the total spin is one (space of triplet states). This, as we shall discuss below, is the explanation for para- and orthohelium, first given by Heisenberg in 1926.

## XI.2  Discrete Energy Levels of Helium

We will now investigate the structure of the orbital spaces in (1.11). In each $\mathscr{H}_\alpha^{\text{orb}}$ ($\alpha = 1, 2$) we have a reducible representation of the algebra $\mathscr{E}(SO(4))$ of the orbital angular momentum

$$L_{\alpha i} = \varepsilon_{ijk} Q_{\alpha j} P_{\alpha k} \tag{2.1}$$

and the Lenz vector

$$A_{\alpha i} = (-2h_\alpha)^{-1/2} \left( \tfrac{1}{2}\varepsilon_{ikl} \{ P_{\alpha l}, L_{\alpha k} \} + \frac{a_Z Q_{\alpha i}}{Q_\alpha} \right), \tag{2.2}$$

which, according to (IX.2.13), is given by

$$\mathscr{H}_\alpha^{\text{orb}} = \sum_{n=1}^{\infty} \oplus \mathscr{R}_\alpha(n). \tag{2.3}$$

In the direct-product space

$$\mathscr{H}^{\text{orb}2} = \mathscr{H}_1^{\text{orb}} \otimes \mathscr{H}_2^{\text{orb}} \tag{2.4}$$

we have a representation of the algebra of orbital angular momentum and the Lenz vector given by

$$\begin{aligned} L_i &= L_{1i} \otimes I + I \otimes L_{2i}, \\ A_i &= A_{1i} \otimes I + I \otimes A_{2i}. \end{aligned} \tag{2.5}$$

The operators of (2.5) are defined in analogy to the definition (V.2.3) of the total angular momentum of the combined system of two elementary rotators. It is easy to see that $L_i$ and $A_i$ obey the same commutation relations as $L_{\alpha i}$ and $A_{\alpha i}$, i.e., the commutation relations of $\mathscr{E}(SO(4))$:

$$[L_i, L_j] = i\varepsilon_{ijk} L_k, \qquad [L_i, A_j] = i\varepsilon_{ijk} A_k, \qquad [A_i, A_j] = i\varepsilon_{ijk} L_k. \tag{2.6}$$

We can define the operator

$$C_1 \equiv C_{11} \otimes I + I \otimes C_{21}, \tag{2.7}$$

where, similarly to (VI.3.13) and (VI.3.17),

$$C_{\alpha 1} = \mathbf{A}_\alpha^2 + \mathbf{L}_\alpha^2 = a_Z^2(-2h_\alpha)^{-1} - I, \, a_Z = \frac{Zm_e e^2}{\hbar} \tag{2.8}$$

We can further define the operator

$$C_2 \equiv A_i L_i \tag{2.9}$$

in analogy to (VI.3.14) for the hydrogen atom. The operators $C_1$ and $C_2$ commute with $A_i$ and $L_i$. However, these operators do not fulfill the same relations as the operators $C_{\alpha 1}$ and $C_{\alpha 2}$. In particular, $C_1$ is not related to the energy operator $H_0$ by as simple a relation as (VI.3.17), and $C_2$ is no longer zero. To see the latter, insert (2.5) into (2.9); one then calculates that

$$C_2 = L_{1i} A_{1i} \otimes I + I \otimes L_{2i} A_{2i} + A_{1i} \otimes L_{2i} + L_{1i} \otimes A_{2i},$$

which gives [making use of (VI.3.18)]

$$C_2 = A_{1i} \otimes L_{2i} + L_{1i} \otimes A_{2i}. \tag{2.10}$$

This is not, in general, identically zero.

The energy operator of the helium atom in the approximation in which the influence of the spin is neglected is given by (1.3b):

$$H_0 = \frac{1}{m_e} H_{00} + W. \tag{2.11}$$

It is customary to consider first the term (1.3c):

$$H_{00} = h_1 \otimes I + I \otimes h_2. \tag{2.12}$$

$H_{00}$ is the energy operator for a system of two noninteracting electrons in the (nuclear) Coulomb field. As it is very unrealistic to neglect the Coulomb interaction between the two electrons (which is of the same "strength" as the Coulomb interaction between each electron and the nucleus), $H_{00}$ is a very poor approximation to the energy operator of the helium atom. Thus we cannot expect that the spectrum of $H_{00}$ will give a good approximation of the energy spectrum of the helium atom. As we will see later, it happens that the qualitative features of $H_{00}$ agree with those of the energy operator $H_0 = (1/m_e)H_{00} + W$; this justifies the usual treatment of first considering $H_{00}$ separately.

Using (2.8) and (2.12), $H_{00}$ can be written

$$H_{00} = -\frac{a_Z^2}{2} \left( \frac{1}{C_{11} + I} \otimes I + I \otimes \frac{1}{C_{21} + I} \right). \tag{2.13}$$

The spectrum of $H_{00}$ is easily found. We introduce into $\mathcal{H}^{\mathrm{orb}2} = \mathcal{H}_1^{\mathrm{orb}} \otimes \mathcal{H}_2^{\mathrm{orb}}$ the direct product basis

$$|n\,l\,l_3\rangle_1 \otimes |n'\,l'\,l_3'\rangle_2. \tag{2.14}$$

As, according to (1.23), we want to known the spectrum of $H_{00}$ in the symmetric subspace $\mathcal{H}_+^{\mathrm{orb}2}$ and in the antisymmetric subspace $\mathcal{H}_-^{\mathrm{orb}2}$, we introduce in $\mathcal{H}^{\mathrm{orb}2} = \mathcal{H}_+^{\mathrm{orb}2} \oplus \mathcal{H}_-^{\mathrm{orb}2}$ the basis system of symmetric and antisymmetric vectors

$$\frac{1}{\sqrt{2}} (|n\,l\,l_3\rangle_1 \otimes |n'\,l'\,l_3'\rangle_2 \pm |n'\,l'\,l_3'\rangle_1 \otimes |n\,l\,l_3\rangle_2). \tag{2.15$\pm$}$$

Equation (2.15$_+$) gives the basis system in $\mathcal{H}_+^{\mathrm{orb}2}$, while (2.15$_-$) gives the basis system in $\mathcal{H}_-^{\mathrm{orb}2}$. The vectors (2.15$_\pm$) are eigenvectors of $H_{00}$ and together form a complete basis system in $\mathcal{H}^{\text{orb}^2}$. The spectrum of $H_{00}$ is therefore obtained by applying (2.13) to (2.15±); the result is

$$\text{spectrum}\left(\frac{H_{00}}{m_e}\right) = E_{nn}^{00'} = -R_{He}''\left(\frac{1}{n^2} + \frac{1}{n'^2}\right), \qquad (2.16)$$

where

$$R_{He}'' = \frac{a_{Z=2}^2}{2m_e} = 4\left(\frac{m_e e^4}{2\hbar^2}\right) = 4R'' = 54.4 \text{ eV}. \qquad (2.17)$$

$R_{He}''$ differs from the Rydberg constant for the hydrogen atom by a factor of 4. It is the Rydberg constant for a one-electron system in the Coulomb field of a charge $Ze = 2e$, i.e., for the $He^+$ ion. Its value in $cm^{-1}$ (wave-number units or inverse wavelength units) is

$$R_{He} = \frac{1}{2\pi\hbar c}R_{He}'' = \frac{54.4 \text{ eV}}{12.40 \times 10^{-5} \text{ eV cm}} = 4.39 \times 10^5 \text{ cm}.$$

The basis vectors (2.15±) are not eigenvectors of the total orbital angular momentum $\mathbf{L}^2$ and $L_3$, and are not eigenvectors of the energy operator $H_0$ (recall that $H_0$ is the energy operator if the contribution of the spin $H_1$ is neglected). If the physical states are eigenstates of the energy operator $H$, they are very closely eigenstates of $H_0$. Eigenstates of $H_0$ can be eigenstates of the total orbital angular momentum $\mathbf{L}^2$ and $L_3$, but can be neither the direct-product states (2.14) of angular momentum $l$ and $l'$ nor the particular linear combinations (2.15±) of those direct product states. [The states (2.15±) are eigenstates of $L_1^2 + L_2^2$. But

$$[L_1^2 + L_2^2, Q_{12}^2] = -2[L_{1k}L_{2k}, (Q_{1i} - Q_{2i})(Q_{1i} - Q_{2i})]$$
$$= -4i\epsilon_{kij}(L_{1k}Q_{2j}Q_{1i} - Q_{1i}Q_{2j}L_{2k}) \neq 0.$$

Consequently the vectors (2.15±) are eigenvectors of an operator that does not commute with $H_0$.] One therefore has to couple the angular momenta $l$ and $l'$ in (2.15±), according to the rules of Section V.2, to obtain eigenvectors of the total orbital angular momentum. These eigenvectors, which are formed as linear combinations of (2.15±), are not yet $H_0$ eigenstates but still $H_{00}$ eigenstates; $H_0$ eigenstates can then be formed as linear combinations with the same value of total orbital angular momentum.

Instead of approximating the helium atom by a model system that consists of two noninteracting electrons that move in the Coulomb field of a doubly charged nucleus, one can try to approximate it by a model system that consists of one electron moving in the electric field that is formed by the doubly charged nucleus and the other electron. This model is certainly much more realistic if the one electron is—in the classical picture—far away from the nucleus and from the other electron, which are close together. To obtain this approximation we write

$$H_0 = \frac{1}{m_e}h_1 + H_2^{\text{el}} = \frac{1}{m_e}h_2 + H_1^{\text{el}}, \qquad (2.18)$$

where

$$H_\alpha^{el} = \frac{1}{m_e} h_\alpha + \frac{e^2}{Q_{12}}. \tag{2.19}$$

More precisely,

$$H_1^{el} = \frac{1}{m_e} h_1 \otimes I + \frac{e^2}{Q_{12}}$$

and

$$H_2^{el} = \frac{1}{m_e} I \otimes h_2 + \frac{e^2}{Q_{12}}.$$

Equation (2.18) is exact, but $H_\alpha^{el}$ is not an operator in $\mathscr{H}_\alpha^{orb}$; it would be an operator in $\mathscr{H}_\alpha^{orb}$ if we made the replacement

$$Q_{12} \to Q_\alpha. \tag{2.20}$$

In our classical picture this would mean that the electron that is close to the nucleus is really at the position of the nucleus. With the asymptotic replacement (2.20), the energy operator for the electron far away from the nucleus has the asymptotic form

$$H_\alpha^{el} = \frac{\mathbf{P}_\alpha^2}{2m_e} + \left( -\frac{2e^2}{Q_\alpha} + \frac{e^2}{Q_{12}} \right) \sim \frac{\mathbf{P}_\alpha^2}{2m_e} + \frac{e^2}{Q_\alpha} = \frac{1}{m_e} H_{\alpha\,\mathrm{Hydr}}, \tag{2.21}$$

i.e., the energy operator for the distant electron is the same as the energy operator of the electron in the hydrogen atom. In the approximation (2.21) the energy operator $H_0$ is approximated by

$$H_0 \to \tilde{H}_0 \equiv \frac{1}{m_e} h_1 + \frac{1}{m_e} H_{2\,\mathrm{Hydr}} = \frac{1}{m_e} h_2 + \frac{1}{m_e} H_{1\,\mathrm{Hydr}}, \tag{2.22}$$

where $H_{\alpha\,\mathrm{Hydr}}$ is the hydrogen-atom Hamiltonian corresponding to (VI.3.5). Letting $n' \geq n$, the spectrum of $\tilde{H}_0$ is then

$$\text{spectrum } \tilde{H}_0 = \tilde{E}_{nn'} = -R''_{\mathrm{He}} \frac{1}{n^2} - R'' \frac{1}{n'^2}$$

$$= -R'' \left( \frac{4}{n^2} + \frac{1}{n'^2} \right) \geq E_{nn'}^{00} = -R'' \left( \frac{4}{n^2} + \frac{4}{n'^2} \right). \tag{2.23}$$

A more realistic approximation for $H_0$ of the form (2.18) is to make the replacement

$$H_\alpha^{el} \to H_{\alpha(Z_{\mathrm{eff}})}^{el} = \frac{\mathbf{P}_\alpha^2}{2m_e} - \frac{Z_{\mathrm{eff}}^{(\alpha)} e^2}{Q_\alpha}, \tag{2.24}$$

where $Z_{\mathrm{eff}}^{(\alpha)}$ is a number between one and two that expresses the screening of the nuclear Coulomb field by the nearby electron. A possible approximation for $H_0$ is thus

$$H_0 \rightarrow H_{0(Z_{eff})} = H^{el}_{1(Z_{eff})} + H^{el}_{2(Z_{eff})} \tag{2.25}$$

The value of $Z^{(\alpha)}_{eff}$ should be different for different states of the helium atom. For the ground state one would expect $Z^{(1)}_{eff} = Z^{(2)}_{eff}$; for the state with one electron in the ground state and the other in a very high state one would expect, according to (2.22), $Z^{(1)}_{eff} \approx 2$ and $Z^{(2)}_{eff} \approx 1$. Thus for different subspaces we have different operators $H_{0(Z_{eff})}$.

Let us now consider the subspace of states of the helium atom that can be characterized—in the classical picture—in the following way: One of the electrons has just dissociated itself from the helium atom, i.e., the system consists of a $He^+$ ion together with an electron of zero relative energy. Let us call the subspace of these states $\mathscr{H}_\infty$; on this subspace $H_0$ can be approximated very well by

$$H_0 = \frac{1}{m_e} h_1 + H^{el}_2 = \frac{1}{m_e} h_2 + H^{el}_1, \tag{2.26}$$

where $H^{el}_\alpha$ is approximated by (2.21). The subspace $\mathscr{H}_\infty$ is then defined as that subspace of $\mathscr{H}^{orb^2}$ on which

$$H^{el}_\alpha = 0. \tag{2.27}$$

Let us denote the eigenvectors of $H^{el}_\alpha$ in $\mathscr{H}^{orb}_\alpha$ by

$$|n \, l \, l_3\}_\alpha, \tag{2.28}$$

in distinction to the eigenvectors $|n \, l \, l_3\rangle_\alpha$ of $h_\alpha$. These eigenvectors have the property [under the approximation (2.21)]

$$H^{el}_\alpha |n \, l \, l_3\}_\alpha = -R'' \frac{1}{n^2} |n \, l \, l_3\}_\alpha \tag{2.29}$$

and are eigenstates of $\mathbf{L}^2_\alpha$ and $L_{\alpha 3}$. The eigenvectors on which (2.27) is fulfilled are denoted by

$$|\infty \, l \, l_3\}_\alpha. \tag{2.30}$$

The direct-product basis in $\mathscr{H}_\infty$ is therefore given by

$$|\infty \, l' \, l'_3\}_1 \otimes |n \, l \, l_3\rangle_{2'} \qquad |n \, l \, l_3\rangle_1 \otimes |\infty \, l' \, l'_3\}_2. \tag{2.31}$$

We introduce in $\mathscr{H}_\infty$ the basis of symmetric and antisymmetric vectors

$$\frac{1}{\sqrt{2}} (|\infty \, l' \, l'_3\}_1 \otimes |n \, l \, l_3\rangle_2 \pm |n \, l \, l_3\rangle_1 \otimes |\infty \, l' \, l'_3\}_2). \tag{2.32±}$$

On $\mathscr{H}_\infty$ the energy operator (2.26) has, according to (2.27), the following spectrum:

$$E_{n\infty} = \text{spectrum } H_0|_{\mathscr{H}_\infty} = -\frac{4m_e e^4}{2\hbar^2} \frac{1}{n^2} = -R''_{He} \frac{1}{n^2}. \tag{2.33}$$

For $H_0$ of (2.26) applied to (2.32$\pm$) gives

$$\frac{1}{\sqrt{2}}\left(|\infty\, l'\, l_3'\}_1 \otimes \frac{h_2}{m_e} |n\, l\, l_3\rangle_2 + \underbrace{H_1^{\text{el}}|\infty\, l'\, l_3'\}_1 \otimes |n\, l\, l_3\rangle_2}_{=0}\right.$$

$$\left. \pm \frac{h_1}{m_e} |n\, l\, l_3\rangle_1 \otimes |\infty\, l'\, l_3'\}_2 + |n\, l\, l_3\rangle_1 \otimes \underbrace{H_2^{\text{el}}|\infty\, l'\, l_3'\}_2}_{=0}\right)$$

$$= -\frac{a_Z}{2m_e n^2}\frac{1}{\sqrt{2}}(|\infty\, l'\, l_3'\}_1 \otimes |n\, l\, l_3\rangle_2 \pm |n\, l\, l_3\rangle_1 \otimes |\infty\, l'\, l_3'\}_2).$$

The $E_{n\infty}$ are the energy values of the helium atom when one of the electrons has zero energy and are also the energy values at which one of the electrons dissociates from the atom (ionization thresholds). We expect that these are the highest energy values for a given quantum number $n$, i.e., for states in which one of the electrons has the principal quantum number $n$, because if the other electron has the principle quantum number $n'$, its relative energy with the He$^+$ ion will be negative. $E_{n\infty}$ is called the $n$th ionization threshold; it is the value at which one of the electrons is just dissociating itself from the atom while the other is in the $n$th energy level.

Comparing (2.33) with the spectrum of $H_{00}$ given by (2.16), we observe that

$$E_{nn'}^{00} \to E_{n\infty} \quad \text{from below} \quad \text{as } n' \to \infty. \tag{2.34}$$

Comparing it with (2.23), we observe

$$\tilde{E}_{nn'} \to E_{n\infty} \quad \text{from below} \quad \text{as } n' \to \infty. \tag{2.35}$$

From our above considerations we expect the eigenvalue $E_{nn'}$ of $H_0$ to lie between the eigenvalues of $H_{00}$ and $\tilde{H}_0$:

$$E_{nn'}^{00} \leq E_{nn'} \leq \tilde{E}_{nn'} < E_{n\infty}. \tag{2.36}$$

In the energy diagram of helium we first draw the energy levels $E_{n\infty}$ ($n = 1, 2, 3, \ldots$). The energy levels $E_{nn'}$ lie, according to (2.34), (2.35), and (2.36), below $E_{n\infty}$, and lie closer and closer together as $n'$ becomes higher.

The larger the values of $n'$, the closer are the eigenvalues of $H_{00}$ to the energy values of the helium atom. Consequently for large $n'$ and a given $n$, the space spanned by the vectors (2.15$+$) and (2.15$-$) represents to a good approximation the space of physical states. In the classical picture of the helium atom, increasing values of $n'$ mean that the "second" electron is further away from the nucleus and from the "first" electron, so that the influence of the interaction term $e^2/r_{12}$ is small; thus the classical picture supports the result described above. For decreasing values of $n'$ the interaction term $e^2/r_{12}$ becomes more important, and we expect larger deviations of the energy value of the helium atom from the value $E_{nn'}^{00}$. We expect the largest deviation for the states in which both electrons have the principal quantum number

$n = n' = 1$. In this case the interaction term is expected to be largest (both electrons are closest to the nucleus and to each other), and $E_{11}^{00}$ is a very poor approximation to the eigenvalue of $H_0$. In this case, therefore, the space spanned by (2.15+) with $n = n' = 1$ is far from being the space of (orbital) physical states. Thus for low values of $n$ and $n'$ the helium atom cannot be considered as consisting of two independent electrons (independent-particle approximation) in a Coulomb field. The physical states are not states in which one electron is in one particular state and the other electron is in another particular state; rather it is only the helium atom considered as a whole that is in the physical states.

We now consider the space of states (which we call $\hbar_1$) of the helium atom that have energy values below the value $E_{1\infty}$. The space of eigenstates of $H_0$ with eigenvalue $E_{1\infty}$ describes the state of the helium atom in which one electron is in the lowest state ($n' = 1$, $l' = 0$, $l'_3 = 0$) and the other electron is just dissociating from the helium atom. The eigenvectors of $H_{00}$ in this subspace $\hbar_1$ can be obtained from the independent-particle states (2.15±):

$$\psi_{nll_3}^0 = \frac{1}{\sqrt{2}} (|n' = 1\, l' = 0\, l'_3 = 0\rangle_1 \otimes |n\, l\, l_3\rangle_2$$

$$+ |n\, l\, l_3\rangle_1 \otimes |n' = 1\, l' = 0\, l'_3 = 0\rangle_2), \qquad (2.37)$$

$$\gamma_{nll_3}^0 = \frac{1}{\sqrt{2}} (|n' = 1\, l' = 0\, l'_3 = 0\rangle_1 \otimes |n\, l\, l_3\rangle_2$$

$$- |n\, l\, l_3\rangle_1 \otimes |n' = 1\, l' = 0\, l'_3 = 0\rangle_2). \qquad (2.38)$$

The $\psi_{nll_3}^0$ span the space of symmetric states of $\hbar_1$, which we call $\hbar_{1+}$; and the $\gamma_{nll_3}^0$ span the space of antisymmetric states of $\hbar_1$, which we call $\hbar_{1-}$. That is, $\psi_{nll_3}^0$ and $\gamma_{nll_3}^0$ are eigenstates of $H_{00}$ with eigenvalue $E_{nn'=1}^{00}$:

$$H_{00}\psi_{nll_3}^0 = -R_{\text{He}}'' \left(1 + \frac{1}{n^2}\right) \psi_{nll_3}^0, \qquad (2.39)$$

$$H_{00}\gamma_{nll_3}^0 = -R_{\text{He}}'' \left(1 + \frac{1}{n^2}\right) \gamma_{nll_3}^0. \qquad (2.40)$$

As $l' = 0$ and $l'_3 = 0$, $\psi_{nll_3}^0$ and $\gamma_{nll_3}^0$ are also eigenvectors of $\mathbf{L}^2$ and $L_3$ with eigenvalues $l(l + 1)$ and $l_3$, respectively. Let us denote by $\psi_{nll_3}$ and $\gamma_{nll_3}$ the eigenvectors of $H_0$, $\mathbf{L}^2$, and $L_3$ in $\hbar_{1+}$ and in $\hbar_{1-}$, respectively. The eigenvectors $\psi_{nll_3}$ and $\gamma_{nll_3}$ of $H_0$ can be obtained from the eigenvectors $\psi_{nll_3}^0$ and $\gamma_{nll_3}^0$ of $H_{00}$ by acting on them with a (unitary) operator that does not change the eigenvalues of $\mathbf{L}^2$, $L_3$, and $\mathbb{P}_{12}$:

$$\psi_{nll_3} = U^+(nl)\psi_{nll_3}^0, \qquad (2.41)$$

$$\gamma_{nll_3} = U^-(nl)\gamma_{nll_3}^0. \qquad (2.42)$$

This transformation $U^\pm(nl)$ may be different for different values of $n$ and $l$; $U^\pm(nl)$ does not depend upon $l_3$, as $[H_0, L_i] = 0$. Equations (2.41) and

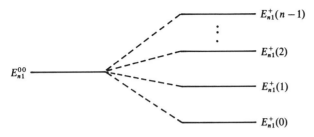

$E^{00}_{n1}$

$E^{+}_{n1}(n-1)$

$\vdots$

$E^{+}_{n1}(2)$

$E^{+}_{n1}(1)$

$E^{+}_{n1}(0)$

Figure 2.1   Splitting of the eigenvalues of $H_{00}$ under the influence of $W$.

(2.42) show that the eigenvectors of $H_0$ can still be characterized by the principal quantum number $n$, though they might be quite different from the eigenvectors of $H_{00}$ with eigenvalues $E^{00}_{nn'=1} = -R''_{\mathrm{He}}(1 + 1/n^2)$. For large values of $n$, $U^{\pm}(nl)$ is very close to the unit operator. The eigenvalues of $H_0$ in $\hbar_1$ we denote by $E^{+}_{n1}(l)$ and $E^{-}_{n1}(l)$; in general they will depend upon $l$:

$$H_0\psi_{nll_3} = E^{+}_{n1}(l)\psi_{nll_3}, \qquad (2.43)$$

$$H_0\gamma_{nll_3} = E^{-}_{n1}(l)\gamma_{nll_3}. \qquad (2.44)$$

We are not that much interested in the calculation of the exact values of $E^{\pm}_{n1}(l)$; as we want only to obtain a qualitative understanding of the energy spectrum of the helium atom. Therefore it is not necessary to know the operators $U^{\pm}(nl)$. (They are connected with the interaction term $W$ and can be calculated by approximation methods.)

The important conclusion for the qualitative understanding of the energy spectrum that we draw from (2.41) and (2.42) is that to each $n^2$-fold degenerate eigenvalue $E^{00}_{n1}$ of $H_{00}$ on $\hbar_{1+}$ there correspond the $n$ energy values $E^{+}_{n1}(l)$, the interaction term $W$ splits $E^{00}_{n1}$ into $n$ sublevels as shown in Figure 2.1. And to each $n^2$-fold degenerate eigenvalue $E^{00}_{n1}$ of $H_{00}$ on $\hbar_{1-}$ there correspond the $n$ energy values $E^{-}_{n1}(l)$. As the result of our above consideration we obtain the following energy spectrum of the helium atom below $E_{1\infty}$ (Figure 2.2 and 2.3): The lowest energy value is $E^{+}_{11}(l = 0)$ in the symmetric subspace $\mathscr{H}^{\mathrm{orb}^2}_{+}$, which is connected with $E^{00}_{11} = -R''_{\mathrm{He}}(1 + 1)$. There is no corresponding energy value in the antisymmetric subspace $\mathscr{H}^{\mathrm{orb}^2}_{-}$, because $\gamma_{1ll_3} = 0$. The values of $\tilde{E}_{11}$ and $E^{00}_{11}$ as calculated from (2.23) and (2.16) are

$$\tilde{E}_{11} = -68.05 \text{ eV} \quad \text{and} \quad E^{00}_{11} = -108.8 \text{ eV}.$$

The experimentally measured value of $E^{+}_{11}(l = 0)$, the double ionization energy of helium (minimum energy required to free both electrons), is:

$$E^{+}_{11}(l = 0) = -79.0 \text{ eV},$$

which shows that $H_{00}$ is really a very poor approximation to the energy operator $H_0$. The energy difference

$$E_{1\infty} - E^{+}_{11}(0) = -54.4 \text{ eV} + 79.0 \text{ eV} = 24.6 \text{ eV}$$

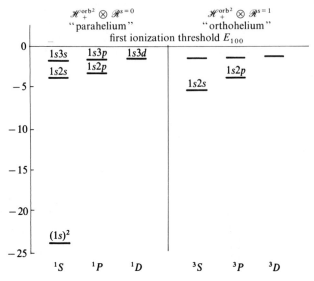

Figure 2.2   Energy levels of the Helium atom below the lowest ionization threshold.

is the energy that is necessary to dissociate one electron from the helium atom when it is in the ground state; it is called the "ionization energy" or "ionization potential." For $n = 2$ there are energy values $E_{12}^{+}(l)$ in $\mathcal{H}_{+}^{\mathrm{orb}^{2}}$ and $E_{12}^{-}(l)$ in $\mathcal{H}_{-}^{\mathrm{orb}^{2}}$. They correspond to the eigenvalues

$$E_{12}^{00} = -R_{\mathrm{He}}''(1 + \tfrac{1}{4}) = -68.0 \text{ eV}$$

of $H_{00}$ and

$$\tilde{E}_{12} = -R_{\mathrm{He}}'' \cdot 1 - R'' \cdot \tfrac{1}{4} = -57.8 \text{ eV}$$

of $\tilde{H}_0$. The experimentally measured values are

$$E_{12}^{-}(0) - E_{11}^{+}(0) = 19.8 \text{ eV} \quad \text{or} \quad E_{12}^{-}(0) = -59.2 \text{ eV},$$
$$E_{12}^{+}(0) - E_{11}^{+}(0) = 20.6 \text{ eV} \quad \text{or} \quad E_{12}^{+}(0) = -58.4 \text{ eV}$$

for $l = 0$, and

$$E_{12}^{-}(1) - E_{11}^{+}(0) = 20.9 \text{ eV} \quad \text{or} \quad E_{12}^{-}(1) = -58.1 \text{ eV},$$
$$E_{12}^{+}(1) - E_{11}^{+}(0) = 21.2 \text{ eV} \quad \text{or} \quad E_{12}^{+}(1) = -57.8 \text{ eV}$$

for $l = 1$. Thus for $n = 2$ the energy values are already closer to the eigenvalues of $H_{00}$, as we expect from our general discussion above. For $n = 3, 4, 5, \ldots$ the agreement improves further.

Figure 2.3 gives all known discrete energy levels of the helium atom. This energy spectrum is in agreement with our discussion above. To summarize our findings for the energy spectrum below $E_{1\infty}$, there is one distorted, shifted,

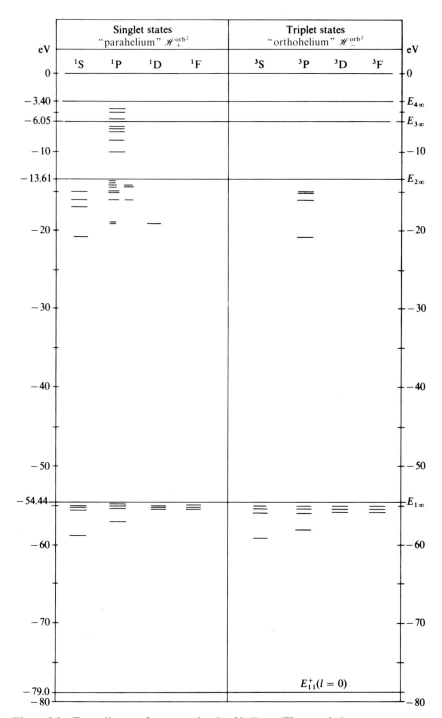

Figure 2.3  Term diagram for energy levels of helium. (The part below the first ionization threshold is again given in Fig. 2.2 which is the term diagram of the Helium that one usually finds.) The energy levels above the first ionization threshold $E_{100}$ lie in the continuous spectrum.

and split hydrogenlike spectrum of states in the space $\mathcal{H}_+^{\mathrm{orb}^2}$, and there is one distorted, shifted, and split hydrogenlike spectrum—without the lowest level—of states in the space $\mathcal{H}_-^{\mathrm{orb}^2}$. According to (1.23) the states in $\mathcal{H}_+^{\mathrm{orb}^2}$ are states with total spin zero, called singlet states; the states in $\mathcal{H}_-^{\mathrm{orb}^2}$ are states with total spin one, called triplet states. As we shall explain below, there are practically no transitions between the singlet and triplet states of the helium atom. (The $2^3S$ state is a quasistable state with a very long lifetime). Therefore the splitting of the energy diagram into two parts had in earlier days lead to the hypothesis that helium was a mixture of two elements, "parahelium," described by $\mathcal{H}_+^{\mathrm{orb}^2} \otimes \mathcal{R}^{s=0}$, and "orthohelium," described by $\mathcal{H}_-^{\mathrm{orb}^2} \otimes \mathcal{R}^{s=1}$.

For parahelium the total angular momentum is equal to the total orbital angular momentum $L = l$ because the total spin $S = 0$. Therefore the interaction caused by the spin $H_1$ cannot split the energy level $E_{1n}^+(l)$, but can only cause a small shift of the same order of magnitude as the fine structure in the hydrogen atom, which has been calculated by approximation methods (singlet).

For orthohelium the total spin $S = 1$. Therefore the direct-product vectors

$$\gamma_{nll_3} \otimes |s = 1\ s_3\rangle \tag{2.45}$$

in $h_{1-} \otimes \mathcal{R}^{s=1}$ are not eigenvectors of the total-angular-momentum operators $\mathbf{J}^2$ and $J_3$, where

$$J_i = L_i \otimes I + I \otimes S_i. \tag{2.46}$$

To obtain eigenstates of the total angular momentum we have to couple the orbital angular momentum and the spin in $h_{1-} \otimes \mathcal{R}^{s=1}$ according to the rules of Section V.2. Let $\mathcal{M}^l$ denote the subspace of $h_{1-} \subset \mathcal{H}_-^{\mathrm{orb}^2}$ with orbital angular momentum $l$; it is the space spanned by $\gamma_{nll_3}$ with a fixed value of $l$. The eigenspaces $\mathcal{M}^j$ of total angular momentum $j$ are then given by

$$\mathcal{M}^l \otimes \mathcal{R}^{s=1} = \begin{cases} \mathcal{M}^{k=l+1} \oplus \mathcal{M}^{j=l} \oplus \mathcal{M}^{j=l-1} & \text{if } l \neq 0, \\ \mathcal{M}^{j=1} & \text{if } l = 0. \end{cases} \tag{2.47}$$

So we see that for orthohelium, except for the $S$ states, all orbital-angular-momentum spaces are actually a triplet of total-angular-momentum spaces. Therefore the spin perturbation $H_1$ of the Hamiltonian $H = H_0 + H_1$ splits the $P\ (l = 1)$, $D\ (l = 2)$, and $F\ (l = 3)$ energy levels into triplets of fine-structure levels, each belonging to an eigenspace of total angular momentum $j = l + 1, l, l - 1$.

These results agree with the experimental energy spectrum: parahelium consists only of singlet terms, whereas the orthohelium consists (except for $l = 0$) only of triplet terms (these are so close to each other that they are shown by one line in Figures 2.2 and 2.3). To avoid labeling the states by their symmetry properties under permutation of the orbital part (ortho- and para-), one also calls the $S$ term of the orthohelium a "triplet" terms, $^3S_1$, though it consists in fact of only a singlet.

## XI.3 Selection Rules and Singlet–Triplet Mixing for the Helium Atom

The operator of the dipole moment for the helium atom is

$$\mathbf{d} = e(\mathbf{Q}_1 \otimes I + I \otimes \mathbf{Q}_2), \tag{3.1}$$

where $\mathbf{d}$ is a vector operator with respect to the total angular momentum operator

$$J_i = J_{1i} \otimes I + I \otimes J_{2i} = L_i \otimes I + I \otimes S_i \tag{3.2}$$

and also with respect to the total orbital angular momentum

$$L_i = L_{1i} \otimes I + I \otimes L_{2i}. \tag{2.5'}$$

Thus it obeys the commutation relations

$$[J_i, d_j] = i\epsilon_{ijk} d_k \tag{3.3}$$

$$[L_i, d_j] = i\epsilon_{ijk} d_k \tag{3.3'}$$

As a consequence, one obtains by (V.3.7) for dipole radiation the selection rules

$$J \to J - 1, \quad J, \quad J + 1 \tag{3.4}$$

$$L \to L - 1, \quad L, \quad L + 1 \tag{3.5'}$$

The physical states are total angular momentum eigenstates, therefore the selection rules (3.4) hold strictly for physical states. The selection rule (3.5') holds to a very good approximation, precisely to the extent to which $H_1$ commutes with $\mathbf{L}^2$.

The parity operator for the two-electron system is the direct product of the parity operators in the one-particle subspaces:

$$U_p = U_{p_1} \otimes U_{p_2}. \tag{3.6}$$

Therefore, for the vectors (2.41) and (2.42), one obtains

$$U_p \psi_{nll_3} = (-1)^L \eta_0 \psi_{nll} \qquad U_p \gamma_{nll_3} = (-1)^L \eta_1 \gamma_{nll_3} \tag{3.7}$$

because $L = l$ below the first ionization threshold. As (3.1) is a proper vector

$$U_p \mathbf{d} U_p^{-1} = -\mathbf{d},$$

it follows by the arguments of Section V.4 that the transition $L \to L$ is forbidden, so that with (3.5') one has the selection rule

$$L \to L + 1, \quad L - 1. \tag{3.5}$$

A further selection rule follows from the fact that

$$[\mathbf{d}, \mathbb{P}_{12}] = 0 \tag{3.8}$$

For every vector

$$\gamma \in \mathcal{H}_-^{\text{orb}^2} \otimes i_+^{s^2} \tag{3.9}$$

and

$$\psi \in \mathscr{H}_+^{\mathrm{orb}^2} \otimes \imath_-^{s^2} \tag{3.10}$$

it follows as a consequence of (3.8) that

$$\mathbf{d}\gamma = \gamma' \in \mathscr{H}_-^{\mathrm{orb}^2} \otimes \imath_+^{s^2} \tag{3.11}$$

$$\mathbf{d}\psi = \psi' \in \mathscr{H}_-^{\mathrm{orb}^2} \otimes \imath_-^{s^2} \tag{3.12}$$

Therefore,

$$(\psi, \mathbf{d}\gamma) = 0 \tag{3.13}$$

Consequently, there are no transitions between singlet and triplet states (the same arguments as above hold for the quadrupole and higher multipole transitions, too). As already mentioned in Section 2, this had led to the hypothesis that helium is a mixture of two elements—parahelium, possessing only singlet states with an energy diagram given by the left part of Figure 2.2, and orthohelium, possessing only triplet states with an energy diagram given by the right part of Figure 2.2. However, one then observed transitions of very weak intensity between energy levels of para- and orthohelium. These inter-multiplet transitions are due to "singlet–triplet" mixing"[3] caused by $H_1$. We shall describe the eigenstates of total angular momentum in the spaces $\mathscr{H}_+^{\mathrm{orb}^2} \otimes \mathscr{R}^{s=0}$, $\mathscr{H}_-^{\mathrm{orb}^2} \otimes \mathscr{R}^{s=1}$, which are obtained by coupling the spin and orbital angular momentum, according to the rules of Chapter V. For $\mathscr{H}_+^{\mathrm{orb}^2} \otimes \mathscr{R}^{s=0}$, we obtain:

$$\psi_{nl=J}^{JJ_3} = \psi_{nll_3 = J_3} \otimes \sum_{s_3's_3''} |s_3'\rangle_1 |s_3''\rangle_2 \langle \tfrac{1}{2}s_3', \tfrac{1}{2}s_3''|00\rangle \tag{3.14}$$

where $\psi_{nll_3}$ is given by (2.41) with (2.37) and the second factor on the right-hand side of (3.14) is the same vector as the one given by (1.18).

For $\mathscr{H}_-^{\mathrm{orb}^2} \otimes \mathscr{R}^{s=1}$ the total angular momentum eigenstates are given by

$$\gamma_{nl}^{JJ_3} = \sum_{l_3s_3} \gamma_{nll_3} \otimes \left( \sum_{s_3's_3''} |\tfrac{1}{2}s_3'\rangle_1 |\tfrac{1}{2}s_3''\rangle_2 \langle \tfrac{1}{2}s_3', \tfrac{1}{2}s_3''|1s_3\rangle \right) \langle ll_3 \, 1s_3|JJ_3\rangle, \tag{3.15}$$

where $\gamma_{nll_3}$ is given by (2.42) with (2.38). For a fixed $n$ and $l$, $\gamma_{nl}^{JJ_3}$ is a basis in the spaces $\mathscr{M}^J$ on the right-hand side of (2.47).

According to the considerations in Chapter VIII on perturbation theory, the eigenstates of $H = H_0 + H_1$ are in first order given by

$$\tilde{\psi}_{nl}^{JJ_3} = \psi_{nl}^{JJ_3} + \sum_{\substack{n'l' \\ n' \neq n \\ l' \neq l}} \left( \frac{1}{E_{nl}^+ - E_{n'l'}^+} \psi_{n'l'}^{JJ_3} \langle \psi_{n'l'}^{JJ_3}|H_1|\psi_{nl}^{JJ_3}\rangle \right.$$

$$\left. + \frac{1}{E_{nl}^- - E_{nl}^-} \gamma_{n'l'}^{JJ_3} \langle \gamma_{n'l'}^{JJ_3}|H_1|\psi_{nl}^{JJ_3}\rangle \right) \tag{3.16}$$

---

[3] Mixing is the common term used for this kind of phenomenon, which is a somewhat misleading nomenclature as compared with the meaning of the word mixture of states in quantum mechanics. The singlet–triplet mixed states are not a mixture of states but a linear combination of states.

and by

$$\tilde{\gamma}_{nl}^{JJ_3} = \gamma_{nl}^{JJ_3} + \sum_{\substack{n'l' \\ n' \neq n \\ l' \neq l}} \left( \frac{1}{E_{nl}^- - E_{n'l'}^-} \gamma_{n'l'}^{JJ_3} \langle \gamma_{n'l'}^{JJ_3} | H_1 | \gamma_{nl}^{JJ_3} \rangle \right.$$

$$\left. + \frac{1}{E_{nl}^+ - E_{n'l'}^+} \psi_{n'l'}^{JJ_3} \langle \psi_{n'l'}^{JJ_3} | H_1 | \gamma_{nl}^{JJ_3} \rangle \right), \tag{3.17}$$

where $E_{nl}^+ = E_{n1}^+(l)$ and $E_{nl}^- = E_{n1}^-(l)$ are the eigenvalues (2.43), (2.44) of $H_0$ in the singlet and triplet states, respectively.

In (3.16) and (3.17) we have already taken into account that $[H, J_i] = [H_1, J_i] = 0$. From parity conservation,

$$[H, U_p] = 0 \tag{3.18}$$

and, consequently, $[H_1, U_p] = 0$ then follows according to (3.7):

$$\langle \psi_{n'l'}^{JJ_3} | H_1 | \psi_{nl}^{JJ_3} \rangle = 0 \quad \text{unless } (-1)^{l'} = (-1)^l,$$
$$\langle \gamma_{n'l'}^{JJ_3} | H_1 | \gamma_{nl}^{JJ_3} \rangle = 0 \quad \text{unless } (-1)^{l'} = (-1)^l, \tag{3.19}$$

and

$$\langle \psi_{n'l'}^{JJ_3} | H_1 | \gamma_{nl}^{JJ_3} \rangle = 0 \quad \text{unless } (-1)^{l'} \eta_0 = (-1)^l \eta_1. \tag{3.20}$$

In the perturbation series of (3.16), (3.17), only states of $\mathscr{h}_{1+}$ or $\mathscr{h}_{1-}$ with energy below the first ionization threshold have been included. This should suffice for the qualitative considerations that we wish to present here.[4]

The matrix elements $\langle \gamma_{n'l'}^{JJ_3} | H_1 | \gamma_{nl}^{JJ_3} \rangle$ and $\langle \psi_{n'l'}^{JJ_3} | H_1 | \psi_{nl}^{JJ_3} \rangle$ give the fine structure contributions, i.e., the splitting of the terms $^3(L)_J$ into three energy levels and a small shift of the energy levels of $^1(L)_J$. Calculations show (Bethe and Salpeter, 1957, section 40) that the $J = 1$ and $J = 2$ levels happen to have almost the same energy.

The matrix elements $\langle \psi_{n'l'}^{JJ_3} | H_1 | \gamma_{nl}^{JJ_3} \rangle$ give the singlet–triplet mixing, which we shall discuss now.

$H_1$, like every observable commutes with the permutation operator $\mathbb{P}_{12}$, which represents the transposition of the two electrons. However, if we write $\mathbb{P}_{12}$ as a direct product of an orbital part $\mathbb{P}_{12}^{\text{orb}}$ and a spin part $\mathbb{P}_{12}^{\text{s}}$ according to the representation (1.23)

$$\mathbb{P}_{12} = \mathbb{P}_{12}^{\text{orb}} \otimes \mathbb{P}_{12}^{\text{s}}, \tag{3.21}$$

then

$$[H_1, \mathbb{P}_{12}^{\text{orb}}] \neq 0, \qquad [H_1, \mathbb{P}_{12}^{\text{s}}] \neq 0, \tag{3.22}$$

i.e., $H_1$ contains contributions that are not symmetrical in the orbital operators and spin operators separately. An example of such a contribution is the

---

[4] For more details, see Bethe and Salpeter (1957), in particular Chapter IIb.

interaction between the spin magnetic moment and orbital magnetic moment of each electron which would, according to (XI.3.23), be given by

$$H^1 = C(Q_{(1)})\mathbf{S}_{(1)} \cdot \mathbf{L}_{(1)} + C(Q_{(2)})\mathbf{S}_{(2)} \cdot \mathbf{L}_{(2)} \tag{3.23}$$

with

$$C_{(i)} = C(Q_{(i)}) = \frac{1}{2}\left(\frac{e^2}{m_e^2 c^2}\right)\frac{1}{Q_{(i)}^3}. \tag{3.24}$$

$H^1$ is only one term in $H_1$: other terms like the relativistic mass correction term and the interaction between the spin magnetic moment of one electron and the orbital magnetic moment of the other are of the same order of magnitude. As we want to discuss only the principle of singlet–triplet mixing, we will restrict our discussions to (3.23).

In order to calculate the matrix element $(\gamma_{n'L}^{JJ_3}|H^1|\psi_{nJ}^{JJ_3})$, we use (3.14) and (3.15) and obtain

$$
\begin{aligned}
(\gamma_{n'l'}^{JJ_3}|H^1|\psi_{nJ}^{JJ_3}) = &\sum_{l'_3 s_3} \sum_{\substack{s'_3 s''_3 \\ \tilde{s}'_3 \tilde{s}''_3}} ((\gamma_{n'l'l'_3}|C_{(1)}\mathbf{L}_{(1)}|\psi_{nJJ_3})\langle JJ_3|l'l'_3\,1s_3\rangle\langle 1s_3|s'_3 s''_3\rangle \\
&\times \langle s'_3|\mathbf{S}_{(1)}|\tilde{s}'_3\rangle\langle s''_3|\tilde{s}''_3\rangle\langle \tilde{s}\tilde{s}''_3|00\rangle \\
&+ (\gamma_{n'l'l'_3}|C_{(2)}\mathbf{L}_{(2)}|\psi_{nJJ_3})\langle JJ_3|l'l'_3\,1s_3\rangle\langle 1s_3|s'_3 s''_3\rangle\langle s'_3|\tilde{s}'_3\rangle \\
&\times \langle s''_3|\mathbf{S}_{(2)}|\tilde{s}''_3\rangle\langle \tilde{s}''_3\rangle\langle \tilde{s}'_3 \tilde{s}''_3|00\rangle)
\end{aligned} \tag{3.25}
$$

Changing in the second term the summation indices $s''_3 \to s'_3$, $\tilde{s}''_3 \to \tilde{s}'_3$ and making use of the property of the Clebsch–Gordan coefficients $\langle \frac{1}{2}s''_3 \frac{1}{2}s'_3|1s_3\rangle = \langle \frac{1}{2}s'_3 \frac{1}{2}s''_3|1s_3\rangle$; $\langle \frac{1}{2}s''_3 \frac{1}{2}s'_3\|00\rangle = (-1)\langle \frac{1}{2}s'_3 \frac{1}{2}s''_3|00\rangle$ one obtains

$$
\begin{aligned}
= &\sum_{l'_3 s_3} \sum_{\substack{s'_3 s''_3 \\ \tilde{s}'_3}} \langle JJ_3|l'l'_3\,1s_3\rangle\langle 1s_3|s'_3 s''_3\rangle\langle s'_3|\mathbf{S}|\tilde{s}'_3\rangle\langle \tilde{s}'_3 s''_3|00\rangle \\
&\times \{(\gamma_{n'l'l'_3}|C_{(1)}\mathbf{L}_{(1)}|\psi_{nJJ_3}) - (\gamma_{n'l'l'_3}|C_{(2)}\mathbf{L}_{(2)}|\psi_{nJJ_3})\}.
\end{aligned} \tag{3.26}
$$

Since $\psi_{nJJ_3}$ is symmetric and $\gamma_{nl'l'_3}$ is antisymmetric under $\mathbb{P}_{12}^{\text{orb}}$ one has

$$(\gamma_{n'l'l'_3}|C_{(2)}\mathbf{L}_{(2)}|\psi_{nJJ_3}) = -(\gamma_{nl'l'_3}|C_{(1)}\mathbf{L}_{(1)}|\psi_{nJJ_3}). \tag{3.27}$$

To calculate these matrix elements, one uses (2.41) and (2.42) with (2.37) and (2.38) and obtains four terms. Two of these terms are zero unless $l' = 0, J = 0$, and one of them is zero unless $l' = J$. In order to reduce the number of terms, we shall restrict ourselves to the case $l' \neq J = l$ and obtain for this case

$$
\begin{aligned}
(\gamma_{n'l'l'_3}|C_{(1)}\mathbf{L}_{(1)}|\psi_{nJJ_3}) &= -{}_i\langle n'l'l'_3|U^{-\dagger}(n'l')C_{(i)}\mathbf{L}_{(i)}U^+(nJ)|nJJ_3\rangle_i \\
&= -{}_i\langle n'-l'l'_3|C_{(i)}\mathbf{L}_{(i)}|n^+JJ_3\rangle_i
\end{aligned} \tag{3.28}
$$

where $i$ can be either 1 or 2 and $U^-(n'l')$ and $U^+(nJ)$ in here are the restrictions of the unitary operators of (2.42) and (2.41) to $\mathcal{H}_i^{\text{orb}}$. All we need to know of them here is that they are scalar operators (because $W$ is a scalar operator) and do not change the value of $l'l'_3$ and $JJ_3$. Thus the new vectors

$$|n^{\pm}ll_3\rangle_i = U^{\pm}(nl)|nll_3\rangle_i \qquad (3.29)$$

are again eigenvectors of $\mathbf{L}^2$ and $L_3$ with the same eigenvalue. Inserting (3.28) and (3.27) into (3.26), one obtains

$$(\gamma_{n'L}^{JJ_3}|H^1|\psi_{nJ}^{JJ_3})$$

$$= \sum_{l_3s_3} \sum_{\substack{s_3's_3'' \\ \tilde{s}_3}} \langle JJ_3|Ll_3\,1s_3\rangle\langle 1s_3|\tfrac{1}{2}s_3', \tfrac{1}{2}s_3''\rangle\langle s_3'|S_{(i)}|\tilde{s}_3\rangle\langle \tfrac{1}{2}\tilde{s}_3, \tfrac{1}{2}s_3''|00\rangle$$

$$\times (-1)_i\langle n'^-Ll_3|C_{(i)}\mathbf{L}_{(i)}|n^+JJ_3\rangle_i, \qquad (3.30)$$

where $C_{(i)}$ is an $\mathbf{L}_{(i)}$-scalar operator. $L_{(i)\kappa}$ $\kappa = 0 \pm 1$, are $\mathbf{L}_{(i)}$ vector operators. Thus

$$_i\langle n'^-Ll_3|C_{(i)}L_{(i)\kappa}|n^+JJ_3\rangle_i = \langle JJ_3\,1\kappa|Ll_3\rangle\langle n'^-L\|C_{(i)}L_{(i)}\|n^+J\rangle. \qquad (3.31)$$

The reduced matrix element can also be further calculated using (V.3.10):

$$\langle n'^-L\|C_{(i)}L_{(i)}\|n^+J\rangle = \delta_{LJ}\sqrt{J(J+1)}\langle n'^-\|C_{(i)}\|n^+\rangle_J. \qquad (3.32)$$

Further evaluation of the quantity $\langle n'^-\|C_{(i)}\|n^+\rangle_J$ requires actual calculation using perturbation theory with $W$ as the interaction Hamiltonian or other approximation methods for the determination of $|n^{\pm}ll_3\rangle$ and then calculation of the matrix element $\langle n'^-\|C_{(i)}\|n^+\rangle_J$ using (3.24). We shall not do this calculation here but just remark that these quantities are of the same order of magnitude as the fine-structure splitting terms which are given by $\langle n'^-\|C_{(i)}\|n^-\rangle$.

Inserting (3.31) and

$$_i\langle s_3'|S_{(i)\kappa}|\tilde{s}_3\rangle_i = \langle \tfrac{1}{2}\tilde{s}_3\,1\kappa|\tfrac{1}{2}s_3'\rangle\sqrt{\tfrac{1}{2}\tfrac{3}{2}} \qquad (3.33)$$

into (3.30), one obtains

$$\langle \gamma_{n'L}^{JJ_3}|H^1|\psi_{nl}^{JJ_3}\rangle = \sum_{\kappa} \sum_{l_3s_3} \sum_{s_3's_3''} \sum_{\tilde{s}_3} \langle JJ_3\,1\kappa|Ll_3\rangle\langle \tfrac{1}{2}s_3'', \tfrac{1}{2}\tilde{s}_3|00\rangle$$

$$\times \langle \tfrac{1}{2}s_3'', \tfrac{1}{2}s_3'|1s_3\rangle\langle \tfrac{1}{2}\tilde{s}_3\,1\kappa|\tfrac{1}{2}s_3'\rangle\langle Ll_3\,1s_3|JJ_3\rangle$$

$$\times \sqrt{\tfrac{3}{4}}\langle n'^-L\|C_{(i)}L_{(i)}\|n^+J\rangle. \qquad (3.34)$$

It is now a straightforward though slightly tedious calculation using the symmetry properties (V.2.48)–(V.2.53) and orthogonality relations (V.2.14), (V.2.15) of the Clebsch–Gordan coefficients to calculate (cf. Problem 4) that

$$\langle \gamma_{n'L}^{JJ_3}|H^1|\psi_{nl}^{JJ_3}\rangle = (-1)^{J-L+1}\sqrt{2}\delta_{LJ}\sqrt{J(J+1)}\langle n'^-\|C_{(i)}\|n^+\rangle_J. \qquad (3.35)$$

Therewith we have expressed the matrix element of the spin orbit interaction between singlet and triplet states in terms of a quantity which (can be only calculated numerically) is different from zero and of the same order as the fine structure splitting term.

Inserting this quantity into (3.16) and (3.17) we see that the eigenvectors of the operator $H = H_0 + H_1, \tilde{\gamma}_{nl}^{JJ_3}$ and $\tilde{\psi}_{nl}^{JJ_3}$, are not pure triplet and singlet states, but have a small component in the singlet and triplet space, respectively. Thus, an orthohelium state $\tilde{\gamma}$ is mainly triplet $\gamma$ with a small singlet component

$$\tilde{\gamma} = \gamma + \epsilon\psi, \qquad |\epsilon| \ll 1, \tag{3.17'}$$

and the parahelium state $\tilde{\psi}$ is mainly singlet $\psi$ with a small triplet component

$$\tilde{\psi}' = \psi' + \epsilon\gamma', \qquad |\epsilon| \ll 1 \tag{3.16'}$$

The dipole matrix element between para- and orthohelium is therefore

$$(\tilde{\psi}', \mathbf{d}\tilde{\gamma}) = (\psi'\mathbf{d}\gamma) + \epsilon(\psi', \mathbf{d}\psi) + \epsilon(\gamma'\mathbf{d}\gamma) \tag{3.36}$$

and even though the first term is zero according to (3.13), this matrix element is different from zero due to the "singlet–triplet mixing."

We had mentioned above in Section VII.1 that it was impossible to verify in the spectra of one-electron atoms whether the energy values were the expectation values of the energy operator in a state which is not an energy eigenstate or the eigenvalues of the energy operator. From the helium spectrum it follows that the latter is the case; the physical para- and orthohelium states are eigenstates of $H = H_0 + H_1$ and *not* pure singlet and triplet states with the energy values being the expectation values of $H$ between these pure singlet and triplet states. The existence of the transitions between para- and orthohelium states is an experimental verification that the physical states are eigenstates of the energy operator. In Chapter XII we will see that for theoretical reasons all stationary states have to be eigenstates of the energy operator.

## XI.4  Doubly Excited States of Helium

The energy levels $E_{1n'}(l)$ of the helium atom are experimentally very well known. They have been measured by light spectroscopy, i.e., by measuring the wave number (or frequency) of the emitted or absorbed light in a transition between two energy levels according to

$$\nu_{n'l,\tilde{n}\tilde{l}} = \frac{1}{2\pi\hbar c} (E_{1n'}(l) - E_{1\tilde{n}}(\tilde{l})).$$

The energy levels have also been measured in energy-loss experiments as described in Section II.4, and as depicted schematically in Figure II.4.1. The collision chamber was filled with helium gas, and the intensity of the electron current was measured as a function of the energy lost in collisions with the helium atoms. The results of this experiment are shown in Figure 4.1, where the intensity $I$ has been plotted as a function of the energy loss $E$. According to the interpretation of this experiment in Section II.4, the energy lost by the electrons is used to excite the atom and cause the transition from the ground state $1\,^1S$ to an excited state; therefore the intensity shows maxima at values of energy loss that agree with the energy difference between the ground state and one of the states excited by this experiment. The excitation

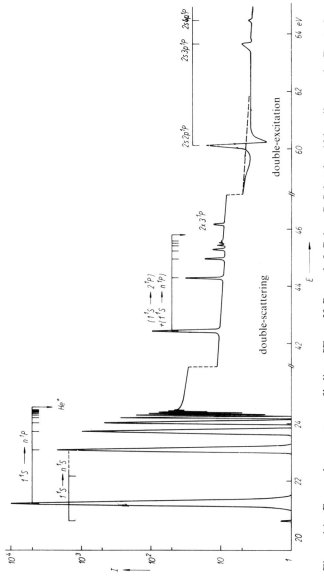

Figure 4.1  Energy-loss spectrum of helium. [From H. Boersch, J. Geiger, B. Schröder, Abhandlungen der Deutschen Akademie der Wissenschaften, Berlin (1967), No. 1, p. 15, with permission.]

**Table 4.1** Excitation energies of helium

|  | Excitation energy (eV) measured in | |
|---|---|---|
| Transition | light spectros-copy experiment | energy-loss experiment |
| $1\,^1S-2\,^1P$ | 21.212 | 21.213 |
| $3\,^1P$ | 23.081 | 23.084 |
| $4\,^1P$ | 23.736 | 23.739 |
| $5\,^1P$ | 24.039 | 24.034 |
| $6\,^1P$ | 24.205 | 24.209 |
| $7\,^1P$ | 24.304 | 24.308 |
| $8\,^1P$ | 24.369 | 24.374 |
| $9\,^1P$ | 24.413 | 24.419 |
| $10\,^1P$ | 24.445 | 24.451 |
| $11\,^1P$ | 24.469 | 24.474 |
| $12\,^1P$ | 24.487 | (24.493) |
| $2\,^1S$ | 20.610 | 20.612 |
| $2s2p\,^1P$ | 60.123 | 60.120 |
| $2s3p\,^1P$ | 63.651 | 63.651 |
| $2s4p\,^1P$ | 64.462 | (64.450) |

energies that are observed in this experiment are tabulated in Table 4.1, column 3, and are compared with the energy of the $n^1\,P$ levels measured by light spectroscopy (column 2); the first column gives the energy levels that participate in the transition. Not given in the table are the bumps between 42-eV and 46-eV energy loss. These bumps are the result of two successive excitations (double scattering); the electrons in the collision chamber scatter inelastically twice, losing energy to the transition $1\,^1S \rightarrow 2\,^1P$ and to the transition $1\,^1S \rightarrow n\,^1P$, or two times to the same transition $1\,^1S \rightarrow 3\,^1P$. From Figure 4.1 we see that there is a very weak transition $1\,^1S \rightarrow 2\,^1S$, and that the transitions of high intensity are the $^1S \rightarrow\,^1P$ transitions (dipole transitions). No transition between the $1\,^1S$ ground state and the $^1D$, $^1F$, ... states or between the $1\,^1S$ state and the triplet states are observed, which agrees with what one would expect for dipole excitations.

Figure 4.1 shows that there are further bumps above 60 eV; these bumps look different from the lower-lying bumps and have a typical asymmetric profile. They correspond to energy levels that are well above the ionization energy for one electron. Energy levels with such a high energy above the ground state level have also been observed as spectral lines in the optical (ultraviolet) absorption spectra. Figure 4.2 shows a photograph of the absorption spectrum obtained for helium in the region 160–215 Å. The observed discrete spectral lines are superimposed upon a continuous absorption background. The lowest-lying discrete spectral line in this region is at 206.21 Å, corresponding to an energy of 60.1 eV; this agrees well with the position of the first bump of this kind in the energy-loss spectrum of Figure 4.1 (cf. also Table 3.1). It is the lowest-lying line of a single long series

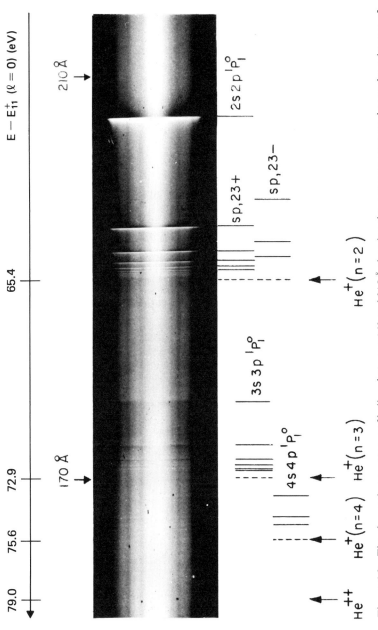

Figure 4.2  The absorption spectrum of helium between 160 and 215 Å showing the many resonances due to the existence of two-electron excitation states in neutral helium. The photograph is a positive print (black is absorption). The white portion of each resonance indicates a reduced-adsorption zone or "window" in the continuous photoionization background absorption. [From R. P. Madden, K. Codling: *Astrophysical Journal* **141**, 364 (1965), University of Chicago Press, with permission.]

of discrete spectral lines of outstanding intensity that converge to the energy value 65.4 eV. (Some of these lines have corresponding bumps in the energy-loss spectrum of Figure 4.1.) Three fainter lines indicate the existence of a second series that also converges towards this energy value. A number of fainter lines can be seen in Figure 4.2 lying to shorter wavelengths from the most prominent series. These lines are grouped into series having as limit points the energy values 72.9 eV and 75.6 eV. Thus we have a series of series of spectral lines. As these lines are observed in the absorption of electromagnetic radiation by helium atoms in the ground state, we can safely assume that (at least for the high-intensity lines and probably for the fainter lines too) they correspond to dipole transitions and that (according to the selection rules for dipole transitions) the corresponding excited energy levels belong to singlet states with orbital angular momentum 1, i.e., are $^1P$ states.

The threshold for double ionization of the helium atom—i.e., the energy value $E_{\infty\infty}$ at which both electrons have just dissociated from the helium nucleus and the physical system is in a state consisting of the nucleus and two electrons, both with relative energy 0—lies 79.0 eV above the ground state. We stated this experimental fact before by saying that the ground-state energy $E_{11}^+ = -79.0\,\text{eV}$. [The energy scale in Figure 3.2 gives $E(\text{level}) - E_{11}^+$.]

The observations in Figure 4.2 indicate that one has a series of series of $^1P$ energy levels, and it appears that the limit point of this series of series lies 79.0 eV above the ground-state energy, i.e., at the threshold for double ionization. The limit points for each single series are then seen to agree with the energy values $E_{n\infty}$ as calculated from (2.33):

$$E_{n\infty} - E_{11}^+ = -R''_{\text{He}}\frac{1}{n^2} + 79.0\text{ eV} = (-54.4\text{ eV})\frac{1}{n^2} + 79.0\text{ eV}; \quad (4.1)$$

hence

$$\begin{aligned}
E_{1\infty} - E_{11}^+ &= 24.6\text{ eV,} \\
E_{2\infty} - E_{11}^+ &= 65.4\text{ eV,} \\
E_{3\infty} - E_{11}^+ &= 72.9\text{ eV,} \\
E_{4\infty} - E_{11}^+ &= 75.6\text{ eV.}
\end{aligned} \quad (4.2)$$

Thus the observations in Figure 4.2 shows that the discrete $^1P$ energy levels converge to the $n$th ionization threshold of the helium atom, i.e., to the energy value at which one electron has the principal quantum number $n$ and the other has just dissociated itself from the helium atom. These experimentally measured $^1P$ levels above the first ionization level are depicted in Figure 2.3. Also depicted there are all the other known (1971) discrete energy levels of the helium atom. We see that the energy-level diagram below the second, third, etc. thresholds is qualitatively very similar to the energy diagram below the first threshold.

A rough explanation of the discrete energy levels above the first ionization threshold is now very simple:[5] Whereas the energy levels below $E_{1\infty}$ correspond to states in which one of the electrons is always in the ground state ($n = 1$), the energy levels below $E_{n\infty}$, $n > 1$, correspond to states in which one of the electrons is in the state with principal quantum number $n$ and the other is in a state with $n' \geq n$. Both electrons are excited, the atom is doubly excited. In the approximation where the interaction $W$ between the two electrons is neglected, these are the states that correspond to the vectors (2.15±) with $n' \geq n > 1$. For the case $n = 2$, (2.15±) reads

$$\psi^0_{2ll_3, n'l'l_3'} = \frac{1}{\sqrt{2}} (|2\, l\, l_3\rangle_1 \otimes |n'\, l'\, l_3'\rangle_2 + |n'\, l'\, l_3'\rangle_1 \otimes |2\, l\, l_3\rangle_2, \quad (4.3+)$$

$$\gamma^0_{2ll_3, n'l'l_3'} = \frac{1}{\sqrt{2}} (|2\, l\, l_3\rangle_1 \otimes |n'\, l'\, l_3'\rangle_2 - |n'\, l'\, l_3'\rangle_1 \otimes |2\, l\, l_3\rangle_2, \quad (4.3-)$$

where $l$ can take the two values $l = 0$ and $l = 1$; i.e., we have the singlet states $\psi^0_{211_3, n'l'l_3'}$ and $\psi^0_{200, n'l'l_3'}$ and the triplet states $\gamma^0_{211_3, n'l'l_3'}$ and $\gamma^0_{200, n'l'l_3'}$. If we want to explain the $^1P$ levels, we have to couple the angular momenta $l$ and $l'$ to obtain eigenvectors of total orbital angular momentum $\mathbf{L}^2 = (\mathbf{L}_1 + \mathbf{L}_2)^2$ with eigenvalue $L(L+1) = 1(1+1)$. As $L$ can be any of the values $l + l', l + l' - 1, \ldots, |l - l'|$, we can obtain $L = 1$ in the following cases:

$$
\begin{array}{llll}
2snp & l = 0 \quad l' = 1 & L = l + l' & \pi = -1 \\
2pns & l = 1 \quad l' = 0 & L = l + l' & \pi = -1 \\
2pnp & l = 1 \quad l' = 1 & L = l + l' - 1 & \pi = +1 \\
2pnd & l = 1 \quad l' = 2 & L = l + l' - 2 & \pi = -1
\end{array}
\quad (4.4)
$$

where $\pi$ is the parity, $\pi = (-1)^l (-1)^{l'}$. Therefore one has three possibilities to obtain $^1P$ states with parity $\pi = -1$ as linear combinations of the states (4.3±):

$$(n = 2, l = 0; n', l' = 1), \qquad (4.5p)$$

$$(n = 2, l = 1; n', l' = 0), \qquad (4.5s)$$

$$(n = 2, l = 1; n', l' = 2). \qquad (4.5d)$$

All of these linear combinations are eigenvectors of $H_{00}$ with the same eigenvalue $E^{00}_{nn'}$. The energy shift due to the interaction of the two electrons depends not only on $n'$ and $L$ but also on $l$ and $l'$; one would therefore expect three different energy values for every value of $n'$ (except for $n' = 2$). Thus one would expect there to be three series of energy levels converging to the $E_{2\infty}$ limit. Figure 4.2 shows only two series, one with a very high transition

[5] Extensive theoretical calculations of these energy levels by various approximation methods have been performed and agree with the observed values; see J. Macek, *J. Phys. B* **1**, 831 (1968) and references therein. See also U. Fano: Doubly Excited States of Atoms in Atomic Physics edited by V. W. Hughes et al., Plenum, New York (1969)

probability from the $1\,^1S$ ground state and the other with a very low one. This indicates the existence of an approximate selection rule that cannot be explained on the basis of the assumption that the spaces spanned by (4.5p), (4.5s), and by (4.5d) are good approximations to spaces of physical states. It is believed that linear combinations of (4.5p) and (4.5s) span, to a good approximation, the spaces of energy eigenstates that belong to the two observed series,[6] and that (4.5d) corresponds to a still fainter and therefore yet unobserved series.

The situation becomes even more complicated for $n = 3, 4, \ldots$. With similar arguments the $^3P$ series can be discussed.

We conclude the discussion of the double excited states with a remark concerning the broad profile of the spectral lines corresponding to these states, which is shown in Figure 4.1 and is also visible in the optical spectra. These energy levels lie above the first ionization threshold. Therefore, in addition to the doubly excited states with $n = 2$, $n' \geq 2$ there are states with the same energy as the energy of the doubly excited states which belong to the system that consist of the $He^+$ ion in the ground state $n = 1$ and one electron ($He^+ + e$). The system $He^+ + e$ can take any possible energy value $E \geq E_{1\infty}$, because the relative energy between $He^+$ and $e$ can take any value, i.e., the energy spectrum of $He^+(n = 1) + e$ is continuous above $E_{1\infty}$. Therefore each energy value of the doubly excited states coincides with one of the energy values in the continuous spectrum of the states of the system $He^+(n = 1) + e$. Thus there are two different states with the same energy: the discrete state $n = 2$, $n' \geq 2$, and the state of the continuum with $n = 1$ and a relative energy between $He^+(n = 1)$ and $e$. These two states mix and interact with each other, as a result of which the doubly excited state cannot persist and easily decays into the state of a $He^+(n = 1)$ and an electron. The lifetime of such doubly excited states is therefore much smaller, and the spectral line has the typical broad profile. Such states are often called resonances.

For the doubly excited states with $n = 3$, $n' \geq 3$, there exist two continuum states with the same energy, namely $He^+(n = 1) + e$ and $He^+(n = 2) + e$. For $n = 4$, $n' \geq 4$ there are three, and so forth.

## Problems

1.  Calculate the length of the vectors

$$|\xi_1 \xi_2 \cdots \xi_N\rangle_+ = \sum_{\mathbb{P}} \mathbb{P} |\xi_1\rangle \otimes |\xi_2\rangle \otimes \cdots \otimes |\xi_N\rangle$$

and

$$|\xi_1 \xi_2 \cdots \xi_N\rangle_- = \sum_{\mathbb{P}} (-1)^{\mathbb{P}} \mathbb{P} |\xi_1\rangle \otimes |\xi_2\rangle \otimes \cdots \otimes |\xi_N\rangle,$$

[6] J. W. Cooper, U. Fano, and F. Prats, *Phys. Rev. Lett.* **10**, 518 (1965).

where the $|\xi_i\rangle$ are normalized basis vectors in the space $\mathcal{H}_i$ and the sum runs over all permutations $\mathbb{P}$. The factor $(-1)^P$ is $+1$ or $-1$ according as $\mathbb{P}$ is an even or odd permutation.

2.   Show that the symmetric and antisymmetric parts of $\mathcal{H}^2$ are given by

$$\mathcal{H}^2_+ = (\mathcal{H}^{\text{orb}^2}_+ \otimes \imath^{s^2}_+) \otimes (\mathcal{H}^{\text{orb}^2}_- \otimes \imath^{s^2}_-),$$

$$\mathcal{H}^2_- = (\mathcal{H}^{\text{orb}^2}_+ \otimes \imath^{s^2}_-) \otimes (\mathcal{H}^{\text{orb}^2}_- \otimes \imath^{s^2}_+),$$

respectively.

3.   Check whether $H^1$ given by (3.23) commutes with the operator of total orbital angular momentum square $\mathbf{L}^2$.

4.   Using the properties of the Clebsch–Gordan coefficients given in Section V.2, perform the calculations that lead from (3.34) to (3.35).

# Time Evolution

In this chapter the basic assumption on time evolution is introduced. The Schrödinger picture, the Heisenberg picture, the Dirac picture, and their relations are discussed. Time-dependent external forces are only briefly mentioned at the end of the chapter.

## XII.1 Time Evolution

Until now we have not mentioned time and have completely ignored the time evolution of the state of a system and the time development of an observable. We have restricted ourselves to properties of systems measured at one instant. In general, one should specify not only the physical values obtained in a measurement but also the time at which the observation is made. In this chapter we consider the time evolution—the dynamics—of physical systems. Although the notion of one instant of time makes as little sense physically as the notion of a definite position (measurements always require a finite amount of time in the same way that they always require a finite amount of space), we make the idealizing assumption that time is a continuous real parameter $t$ that labels a sequence of states. (Note that we have called time a "parameter" as opposed to an "observable.") That is, we make the idealizing assumption that measurements can be made at a succession of arbitrarily small time intervals, and that at any instant of time the specification of a complete set of observables is possible.

274

Let us denote by $W(t_0)$ the state of a system that is prepared at the time $t = t_0$. We then expect that $W(t_0)$ will uniquely determine another state $W(t_1)$ at any later time $t = t_1 > t_0$. During the evolution the system may be subject to external influences, so that the manner of evolution will depend upon the interval from $t_0$ to $t_1$—its starting time $t_0$ as well as its length $t_1 - t_0$. We first consider systems that evolve undisturbed by external influences (isolated systems). For such systems the evolution

$$W(t_0) \rightarrow W(t_1) \tag{1.1}$$

does not depend upon $t_0$ but only upon the system itself, the initial state $W(t_0)$, and the length $\tau_1 = t_1 - t_0$ of the time interval from $t_0$ to $t_1$. Our intuitive understanding of the time evolution suggests that we impose the following requirements on the map (1.1):

1.  Expression (1.1) is a linear transformation of the set of states, continuous in a certain sense: If $W_1(t_0) \rightarrow W_1(t_1)$ and if $W_2(t_0) \rightarrow W_2(t_1)$, then for $a, b \in \mathbb{C}$,
$$aW_1(t_0) + bW_2(t_0) \rightarrow aW_1(t_1) + bW_2(t_1).$$

2.  Time evolution forms an additive semigroup, i.e., if $W(t_0) \rightarrow W(t_0 + \tau_1)$ for $\tau_1 = t_1 - t_0$ and if $W(t_1) \rightarrow W(t_1 + \tau_2)$ for $\tau_2 = t_2 - t_1$, then
$$W(t_0) \rightarrow W(t_0 + \tau_1 + \tau_2).$$

Requirement 1 expresses the physical statement that first mixing the states at time $t_0$ and then letting the system evolve in time leads to the same results as first letting the states evolve separately in time and mixing them afterwards at time $t_1$; the continuity of the transformation expresses the intuitive statement that time changes continuously. Requirement 2 is intuitively obvious; it is a statement of the fact that letting a physical state evolve first for a period of length $\tau_1$ and then letting the resultant state evolve for a period of length $\tau_2$ leads to the same result as letting the original state evolve for a period of length $\tau_1 + \tau_2$. Requirements 1 and 2 are very general and also include irreversible processes.

We are here concerned with the quantum mechanics of reversible processes, and for such processes we make the additional requirement

3.  The time evolution (1.1) is given by a unitary operator $U(\tau)$ satisfying

$$W(t_0) \rightarrow W(t_1) = U^\dagger(\tau_1)W(t_0)U(\tau_1), \tag{1.2}$$

$$U^\dagger(\tau) = U^{-1}(\tau), \tag{1.3}$$

$$U(0) = I, \tag{1.4}$$

$$U^{-1}(\tau) = U(-\tau), \tag{1.5}$$

$$U(\tau_1 + \tau_2) = U(\tau_1)U(\tau_2) \qquad (-\infty < \tau_1, \tau_2 < +\infty), \tag{1.6}$$

$U(\tau)$ is a continuous operator function[1] of the parameter $\tau$. $\tag{1.7}$

---

[1] See appendix to this chapter.

Equation (1.5) follows from (1.2) and (1.3)—specifically from

$$U(\tau_1)W(t_1)U^{-1}(\tau_1) = W(t_0) \tag{1.2'}$$

—and from the assumption that (1.2) is also valid for evolution backwards in time, i.e.,

$$W(t_0) = U^{-1}(-\tau_1)W(t_1)U(-\tau_1). \tag{1.8}$$

Equation (1.6) follows from requirement 2, which, when the time evolution is described by a unitary operator as in (1.2), is expressed as

$$U(t_2 - t_0) = U(t_1 - t_0)U(t_2 - t_1). \tag{1.6'}$$

Equation (1.6) follows for $\tau_1 = t_1 - t_0$, $\tau_2 = t_2 - t_1$. The linearity of the transformation (1.2) is obvious. The continuity of the transformation (1.1) is expressed by (1.7). We shall not give here the mathematical definition of a continuous operator function, but mention only that the formal operations, employed below, of differentiation, Taylor series expansion, etc.[2] can be rigorously defined. A continuous operator function satisfying (1.6) is called a *one-parameter group of operators*.

We differentiate $U(\tau)$ with respect to $\tau$ and define

$$A \equiv \left. \frac{dU(\tau)}{d\tau} \right|_{\tau=0}. \tag{1.9}$$

$A$ is called the *infinitesimal operator* or *generator* of the operator $U(\tau)$ of time evolution. $U(\tau)$ is not only "differentiable," but is differentiable an arbitrary number of times. We can calculate all "derivatives" using (1.6). Differentiating (1.6) with respect to $\tau_1$,

$$\frac{dU(\tau + \tau_1)}{d(\tau + \tau_1)} \frac{d(\tau + \tau_1)}{d\tau_1} = \frac{dU(\tau_1)}{d\tau_1} U(\tau),$$

and evaluating the resultant expression at $\tau_1 = 0$, we obtain

$$\frac{dU(\tau)}{d\tau} = AU(\tau).$$

We differentiate this again and obtain

$$\frac{d^2 U(\tau)}{d\tau^2} = A\frac{dU(\tau)}{d\tau} = A \cdot AU(\tau) = A^2 U(\tau).$$

We may continue to differentiate, obtaining for the $p$th derivative

$$\frac{d^p U(\tau)}{d\tau^p} = A^p U(\tau);$$

use of (1.4) then gives

$$\left. \frac{d^p U(\tau)}{d\tau^p} \right|_{\tau=0} = A^p.$$

[2] See the appendix.

The "Taylor series expansion" of $U(\tau)$ is therefore given by

$$U(\tau) = I + \tau A + \frac{\tau^2}{2!} A^2 + \cdots + \frac{\tau^n}{n!} A^n + \cdots; \qquad (1.10)$$

or, if we compare (1.10) with the series for $e^{\tau A}$, we have

$$U(\tau) = e^{\tau A}. \qquad (1.10')$$

In particular this means [according to our rule that $f(A)|a\rangle = f(a)|a\rangle$ for functions $f(A)$ of $A$] that

$$U(\tau)|a\rangle = e^{\tau a}|a\rangle$$

for (generalized) eigenvectors $|a\rangle$ of $A$ corresponding to the eigenvalue $a$. Taking the adjoint of (1.10) gives

$$U^\dagger(\tau) = I + \tau A^\dagger + \frac{\tau^2}{2!} A^{\dagger 2} + \cdots = e^{\tau A^\dagger}. \qquad (1.11)$$

If we use (1.3) and differentiate $U^\dagger(\tau)U(\tau) = I$, we obtain

$$\frac{dU^\dagger(\tau)}{d\tau} U(\tau) + U^\dagger(\tau) \frac{dU(\tau)}{d\tau} = 0;$$

setting $\tau = 0$, we obtain

$$A^\dagger = -A. \qquad (1.12)$$

An operator that fulfills (1.12) is called *skew-Hermitian*. As $\tau$ is the time, $A$ has the dimensions of frequency, i.e., of energy/$\hbar$. The operator

$$H = -i\hbar A = -i\hbar \left.\frac{dU(\tau)}{d\tau}\right|_{\tau=0}, \qquad (1.13)$$

often also called the generator of time evolution, is therefore a Hermitian operator with the dimension of energy. $H$ is the Hamiltonian operator or energy operator of the system, and, from (1.10'),

$$U(\tau) = e^{iH\tau/\hbar}. \qquad (1.14)$$

   What $H$ is depends, of course, upon the particular physical system under consideration, and how it is found in particular cases has been discussed previously. $H$ is an element of the algebra of observables of the particular system, and its properties are determined by the defining algebraic relations of the particular physical system. We remark that (if one uses a precise mathematical formulation—something that has not been done here) one can show that on the space of physical states there is a one-to-one correspondence between $U(\tau)$ and $H$; one can therefore consider $H$ as a quantity derived [using (1.13)] from the more fundamental time-evolution operator $U(\tau)$, or one can consider $H$ as the more fundamental quantity and the time-evolution operator $U(\tau)$ as a quantity derived [using (1.14)] from $H$. It is in the spirit of our preceeding formulation to consider $H$, an element of the algebra of observables, as the fundamental quantity, and to consider $U(\tau)$ as the derived quantity.

We formulate the result of our preceeding plausibility argument as the basic assumption of quantum mechanics for time evolution—the basic dynamical law for quantum-mechanical systems:

**Va** (The Schrödinger Picture). A (conservative) physical system has a generator $H$ of time translation, which is a Hermitian element of the algebra of observables and which is characteristic of the physical system. The time evolution of the state of the physical system is given by

$$W(t) = U^\dagger(t)W_0 U(t), \tag{1.15}$$

where $W_0$ is the state of the system at some initial time $t_0 = 0$ and where $U(t)$ is given by

$$U(t) = e^{itH/\hbar}. \tag{1.16}$$

According to the axiom II, the expectation value of an observable $A$ is given by

$$\langle A \rangle = \text{Tr}(AW). \tag{1.17}$$

The observable that is represented by the operator $A$ is defined by the prescription for its measurement, i.e., how to set up a measurement apparatus and how to take the measurement. This prescription is given for all time and does not change with time. The experimentally measured value changes in time; if we make a measurement at a time $t_0$, the expectation value $\langle A \rangle_{t=t_0}$ will, in general, be different from the expectation value $\langle A \rangle_{t=t_1}$ at a different time $t_1$. This experimental fact is represented by a description of the change of the state $W(t)$ according to (1.15). Inserting (1.15) into (1.17), we obtain

$$\langle A \rangle_t = \text{Tr}(AW(t)) = \text{Tr}(AU^\dagger(t)W_0U(t)). \tag{1.18}$$

This can be written

$$\langle A \rangle_t = \text{Tr}(U(t)AU^\dagger(t)W_0), \tag{1.19}$$

since the trace of a product of operators is invariant under a cyclic permutation of the order of the factor operators. We can now define a new operator $A(t)$ by

$$A(t) = U(t)AU^\dagger(t) \tag{1.20}$$

and write (1.19) as

$$\langle A \rangle_t = \langle A(t) \rangle = \text{Tr}(A(t)W_0). \tag{1.21}$$

In (1.21) the change of the experimentally measured value $\langle A \rangle_t$ with time is described in a form that can be interpreted in the following way: The state of the system is described by an operator $W_0$; this statistical operator $W_0$ does not change with time, i.e., the whole history of the state of the physical system is described by the one operator $W_0$. The observable develops with time according to (1.20), i.e., the prescription for the measurement must include instructions as to when to take the measurement, because the observable—the experimental setup—is different for different times. Thus instead of the basic assumption Va we can formulate the time evolution by the following basic assumption:

**Vb** (The Heisenberg Picture). A (conservative) physical system has a generator $H$ of time translation, which is a Hermitian element of the algebra of observables and which is characteristic of the physical system. The time development of every observable $A(t)$ of the physical system is given by

$$A(t) = U(t)AU^\dagger(t), \tag{1.22}$$

where $A$ describes the observable at some initial time $t_0 = 0$ and where $U(t)$ is given by

$$U(t) = e^{itH/\hbar}. \tag{1.23}$$

These two descriptions of the change in time of the physically measured value $\langle A \rangle_t$ are, of course, completely equivalent. In the form Va the change in time is described as a change of the state of the system. The description of the change in time in this form is called the "Schrödinger picture." In the form Vb the state is held fixed and the change in time is described by the change of the observable. The description in this form is called the "Heisenberg picture."

The time development of an observable as given by (1.22) is often written in a different form, which is obtained from (1.22) by differentiating:

$$\frac{dA(t)}{dt} = \frac{i}{\hbar} H e^{itH/\hbar} A e^{-itH/\hbar} + e^{itH/\hbar} A \left( -\frac{i}{\hbar} H \right) e^{-itH/\hbar}$$

$$= \frac{i}{\hbar}(HA(t) - A(t)H),$$

or

$$i\hbar \frac{dA(t)}{dt} = [A(t), H]. \tag{1.24}$$

This is the "Heisenberg equation of motion." An observable is a *constant of the motion* if its does not depend upon $t$. From (1.24) it follows that all constants of the motion commute with $H$. In particular, $H$ itself is a constant of the motion. Furthermore, every member of any c.s.c.o. that contains $H$ is a constant of the motion.

Consider the particular observables momentum $P_i$ and position $Q_i$ for a physical system in which neither of them is a constant of the motion. It is easily seen that $[P_i(t), Q_j(t)]$ has its usual form for all times $t > 0$ given that $[P_i, Q_n] = (\hbar/i)\delta_{ij}I$ at $t = 0$:

$$P_i(t)Q_j(t) - Q_j(t)P_i(t) = U(t)P_iU^\dagger(t)U(t)Q_iU^\dagger(t) - U(t)Q_iU^\dagger(t)U(t)P_iU^\dagger(t)$$

$$= U(t)(P_iQ_j - Q_jP_i)U^\dagger(t)$$

$$= U(t)U^\dagger(t)\frac{\hbar}{i}\delta_{ij}I = \frac{\hbar}{i}\delta_{ij}I,$$

i.e.,

$$[P_i(t), Q_j(t)] = \frac{\hbar}{i}\delta_{ij}I. \tag{1.25}$$

We are now in a position to calculate the equations of motion for the momentum and for the position and to show that these equations have the same form as those for the classical quantities. For a system whose Hamiltonian is given by

$$H = \frac{1}{2m} P_i P_i + V(\mathbf{Q}),$$

where

$$Q_i = Q_i(t) \quad \text{and} \quad P_i = P_i(t),$$

we calculate that

$$[Q_k, H] = \frac{1}{2m} [Q_k, P_i P_i] = \frac{1}{2m} \{[Q_k, P_i], P_i\}$$

$$= \frac{i\hbar}{2m} \delta_{ki} \{I, P_i\} = \frac{i\hbar}{m} P_k.$$

Inserting this into (1.24), we obtain

$$\frac{dQ_k(t)}{dt} = \frac{P_k(t)}{m}. \tag{1.26}$$

This is the quantum-mechanical analogue of the classical relation between momentum and the time derivative of position.

In classical physics the force is obtained from the potential function $V(\mathbf{x})$ by

$$f_k = -\frac{\partial V(\mathbf{x})}{\partial x_k}.$$

In analogy to this we define the "force operator" by

$$F_k \equiv -\frac{i}{\hbar} [P_k, V(\mathbf{Q})] = -\frac{"\partial V(\mathbf{Q})"}{\partial Q_k} \tag{1.27}$$

where the last symbol is defined below. Inserting this into (1.24) *for* $A(t) = P_k(t)$, we obtain

$$\frac{dP_k(t)}{dt} = \frac{1}{i\hbar} [P_k, H] = -\frac{i}{\hbar} [P_k, V(\mathbf{Q})] = F_k(\mathbf{Q}(t)). \tag{1.28}$$

This is the quantum-mechanical analogue of the classical relation between the time derivative of the momentum and the force. Combining (1.26) and (1.28) gives the quantum-mechanical Newton's equation,

$$m \frac{d^2 Q_k(t)}{dt^2} = F_k(\mathbf{Q}(t)). \tag{1.29}$$

["$\partial V(\mathbf{Q})/\partial Q_k$" is a symbolic way of indicating the following procedure: Replace $Q_i$ by the numbers $x_i$ in $V(\mathbf{Q})$ [recall that this is just the inverse to the way $V(\mathbf{Q})$ was obtained from the corresponding classical expression], differentiate $V(\mathbf{x})$ with respect to $x_k$, and replace $x_i$ by $Q_i$ in $\partial V(\mathbf{x})/\partial x_k$. That "$\partial V(\mathbf{Q})/\partial Q_k$" is in fact $(i/\hbar)$ $[P_k, V(\mathbf{Q})]$ can be seen in the following ways:

1. Write

$$V(\mathbf{Q}) = C + C_i Q_i + C_{ij} Q_i Q_j + \cdots$$
$$+ C_{i_1 \cdots i_n} Q_{i_1} \cdots Q_{i_n} + \cdots, \qquad (1.30)$$

where the $C_{i_1 \cdots i_n}$ are symmetric, i.e.,

$$C_{i_1 \cdots i_p \cdots i_q \cdots i_n} = C_{i_1 \cdots i_q \cdots i_p \cdots i_n}, \qquad (1.31)$$

because the $Q_i$ commute with each other. Using the canonical commutation relations (1.23), calculate that

$$\frac{i}{\hbar} [P_k, V(\mathbf{Q})] = C_k + 2 C_{ki} Q_i + \cdots$$
$$+ n C_{k i_2 i_3 \cdots i_n} Q_{i_2} Q_{i_3} \cdots Q_{i_n} + \cdots, \qquad (1.32)$$

because

$$C_{i_1 \cdots i_n} [P_k, Q_{i_1} \cdots Q_{i_n}]$$
$$= C_{i_1 \cdots i_n} \sum_{p=1}^{n} [P_k, Q_{i_p}] Q_{i_1} \cdots Q_{i_{p-1}} Q_{i_{p+1}} \cdots Q_{i_n}$$
$$= C_{i_1 \cdots i_n} \sum_{p=1}^{n} \frac{\hbar}{i} \delta_{k i_p} Q_{i_1} \cdots Q_{i_{p-1}} Q_{i_{p+1}} \cdots Q_{i_n}$$
$$= n \frac{\hbar}{i} C_{k i_2 i_3 \cdots i_n} Q_{i_2} Q_{i_3} \cdots Q_{i_n},$$

where (1.31) has been used. The right-hand side of (1.32) is identical to "$\partial V(\mathbf{Q})/\partial Q_k$."

2. Calculate $[P_k, V(\mathbf{Q})]$ in the position representation:

$$\langle \mathbf{x} | [P_k, V(\mathbf{Q})] | \psi \rangle = \langle \mathbf{x} | P_k (V(\mathbf{Q}) | \psi \rangle) - \langle \mathbf{x} | V(\mathbf{Q}) P_k | \psi \rangle$$
$$= \frac{\hbar}{i} \frac{\partial}{\partial x_k} \langle \mathbf{x} | V(\mathbf{Q}) | \psi \rangle - V(\mathbf{x}) \frac{\hbar}{i} \frac{\partial}{\partial x_k} \langle \mathbf{x} | \psi \rangle$$
$$= \frac{\hbar}{i} \frac{\partial}{\partial x_k} (V(\mathbf{x}) \langle \mathbf{x} | \psi \rangle) - V(\mathbf{x}) \frac{\hbar}{i} \frac{\partial}{\partial x_k} \langle \mathbf{x} | \psi \rangle$$
$$= \frac{\hbar}{i} \frac{\partial V(\mathbf{x})}{\partial x_k} \langle \mathbf{x} | \psi \rangle. \qquad ]$$

As the Heisenberg commutation relations do not change with time, all algebraic relations that are derived from them are also valid at any time, even if the operators in these algebraic relations do not commute with $H$. That is, any algebra of observables whose elements are functions of $P_i$, $Q_j$ does not change in time, even though some of the elements may not commute with $H$.

A *symmetry transformation* of a set of elements is a transformation of the elements that leaves the relations among them unchanged. Thus, the time evolution of a physical system whose algebra of observables is derived from the Heisenberg commutation relations is a symmetry transformation of the algebra of observables. If time evolution is a symmetry transformation, then the mathematical structure (in particular the algebraic relations) of the algebra of observables does not change in time; this means that the physical structure is indistinguishable at two different points in time. Our experience shows that there are physical systems that have this property, and in fact it is this property that defines the isolated physical systems. Thus isolated physical systems do not age, an absolute value of time has no meaning for these systems, and only time differences are accessible to measurement. Irreversible processes do not take place in isolated physical systems defined as above. These considerations lead us to another formulation of the basic assumption V:

**Vc.** The time evolution of a conservative physical system is given by a continuous symmetry transformation of its algebra of observables.

Let us return to the Schrödinger picture. The time evolution (1.15) can also be expressed in differential form. Differentiating (1.15), one obtains

$$\frac{dW(t)}{dt} = \frac{i}{\hbar}(W(t)H - HW(t)) = \frac{i}{\hbar}[W(t), H]. \tag{1.33}$$

We consider now the special case where the state $W_0$ is a pure state, i.e.,

$$W_0 = \Lambda_{|\psi_0\rangle} = |\psi_0\rangle\langle\psi_0|, \tag{1.34}$$

where $\Lambda_{|\psi_0\rangle}$ is the projection operator onto the one-dimensional space spanned by the vector $|\psi_0\rangle$. For this case (1.15) reads

$$W(t) = U^\dagger(t)\Lambda_{|\psi_0\rangle} U(t) = U^\dagger(t)|\psi_0\rangle\langle\psi_0|U(t).$$

We define the vector $|\psi(t)\rangle$ by

$$|\psi(t)\rangle \equiv U^\dagger(t)|\psi_0\rangle \tag{1.35}$$

and have

$$W(t) = U^\dagger(t)\Lambda_{|\psi_0\rangle} U(t) = |\psi(t)\rangle\langle(t)| = \Lambda_{|\psi(t)\rangle}. \tag{1.36}$$

We obtain the differential form of (1.36) by differentiating (1.35):

$$\frac{d}{dt}|\psi(t)\rangle = \frac{dU^\dagger(t)}{dt}|\psi_0\rangle = -\frac{i}{\hbar}HU^\dagger(t)|\psi_0\rangle = -\frac{i}{\hbar}H|\psi(t)\rangle.$$

This is the *Schrödinger equation* for the state vector $|\psi(t)\rangle$:

$$i\hbar \frac{d}{dt} |\psi(t)\rangle = H |\psi(t)\rangle. \tag{1.37}$$

For the Hamiltonian given by

$$H = \frac{1}{2m} P_i P_i + V(\mathbf{Q}),$$

the Schrödinger equation in the position representation reads

$$i\hbar \frac{d}{dt} \langle \mathbf{x} | \psi(t)\rangle = \frac{1}{2m} \langle \mathbf{x} | P_i P_i | \psi(t)\rangle + V(\mathbf{x})\langle \mathbf{x} | \psi(t)\rangle.$$

Using

$$\langle \mathbf{x} | P_i P_i | \psi(t)\rangle = \frac{\hbar}{i} \frac{\hbar}{i} \frac{\partial}{\partial x_i} \frac{\partial}{\partial x_i} \langle \mathbf{x} | \psi(t)\rangle$$

(see Equation (II.8.64)), the Schrödinger equation may be written in the conventional form

$$i\hbar \frac{d}{dt} \psi(\mathbf{x}, t) = -\frac{\hbar^2}{2m} \nabla^2 \psi(\mathbf{x}, t) + V(\mathbf{x})\psi(\mathbf{x}, t), \tag{1.38}$$

where we have defined the time-dependent wave function

$$\psi(\mathbf{x}, t) \equiv \langle \mathbf{x} | \psi(t)\rangle. \tag{1.39}$$

Summarizing, the distinction between the Heisenberg picture and the Schrödinger picture is the following: In the Heisenberg picture the observables are transformed by $U(t)$ and the states are kept fixed, while in the Schrödinger picture the observables are kept fixed and states are transformed in the "opposite direction," i.e., by $U^\dagger(t) = U^{-1}(t)$. The time dependence of $\text{Tr}(AW)$ is the same in both cases. Expressed in terms of vectors: In the Heisenberg picture the eigenvectors $|a\,t\rangle$ of the observables $A(t)$,

$$A(t)|a\,t\rangle = a|a\,t\rangle, \tag{1.40}$$

are "rotated" by $U(t)$:

$$U(t)|a\,0\rangle = |a\,t\rangle,$$

whereas the state vectors $|\psi_0\rangle$ are kept constant. [Note that in (1.40) no time dependence occurs on the part of the eigenvalue $a$. This is easily shown: $A(t)|a\,t\rangle = A(t)U(t)|a\,0\rangle = U(t)A(0)|a\,0\rangle = U(t)a(0)|a\,0\rangle = a(0)|a\,t\rangle$.]In the Schrödinger picture the eigenvectors $|a\,0\rangle$ are kept constant and the state vectors are "rotated" in the opposite direction:

$$U^\dagger(t)|\psi_0\rangle = |\psi(t)\rangle.$$

The time dependence of the physically measurable quantities representing the probabilities is in both cases the same:

$$|\langle a\, t | \psi_0 \rangle|^2 = |(U(t)|a\,0\rangle, |\psi_0\rangle)|^2$$
$$= |\langle a\,0 | U^\dagger(t) | \psi_0 \rangle|^2 = |\langle a\,0 | \psi(t) \rangle|^2.$$

A state is called a *stationary state* if it does not change with time:

$$W(t) = U^\dagger(t) W_0 U(t) = W_0 = W. \tag{1.41}$$

This terminology is justified by the fact that every observable measured in this state gives a value independent of the time of the measurement; for from (1.41) it follows that the expectation value of any observable $A$ is

$$\langle A \rangle_t = \mathrm{Tr}(A W(t)) = \mathrm{Tr}(A U^\dagger(t) W U(t)) = \mathrm{Tr}(A W) = \langle A \rangle_{t=0}.$$

Stationary states are always mixtures of eigenstates of the energy operator $H$. This follows from (1.41) in the form

$$W e^{itH/\hbar} = e^{itH/\hbar} W,$$

or equivalently,

$$WH - HW = 0. \tag{1.42}$$

In particular, for a pure stationary state,

$$W = \Lambda_{|\psi_0\rangle} = |\psi_0\rangle\langle\psi_0|,$$

(1.42) reads

$$|\psi_0\rangle\langle\psi_0| H = H |\psi_0\rangle\langle\psi_0|,$$

or

$$H|\psi_0\rangle = |\psi_0\rangle \frac{\langle\psi_0|H|\psi_0\rangle}{\langle\psi_0|\psi_0\rangle}. \tag{1.43}$$

Thus pure stationary states are represented by eigenvectors of the energy operator. The time dependence of a stationary state vector is given by

$$|\psi(t)\rangle = U^\dagger(t)|\psi_0\rangle = e^{-itE/\hbar}|\psi_0\rangle. \tag{1.44}$$

*Stationary systems* are systems that have only stationary physical states. Stationary systems can therefore only be prepared in states that are mixtures of eigenstates of the energy operator, i.e., in states that fulfill (1.42). And pure physical states of stationary systems must always be represented by energy eigenvectors. In the preceeding chapters all physical systems considered were stationary physical systems, and we have now found an "explanation" for the fact that these systems always appeared in eigenstates of the energy operator.

To what extent a given physical system may be considered stationary depends upon the desired accuracy of the description. If one ignores details, atoms and molecules are stationary physical systems. If a measurement process has prepared them to be in a particular state, which has to be in an energy eigenstate, then they will for "all time" remain in that state. "All time" means fairly long on atomic scales: $10^{-2}$ sec (for infrared transitions) to $10^{-9}$ sec (for ultraviolet transitions) for the excited states of atoms and molecules. If one wants to describe not only energy levels but also details of the decay process of excited states, then one cannot consider the atom as a stationary system. This will be discussed in later chapters.

The Heisenberg picture and the Schrödinger picture are the limiting cases of a more general picture, which is called the "Dirac picture" or the "interaction picture." One writes the Hamiltonian operator $H$ as the sum of two parts,

$$H = H_0 + H_1. \tag{1.45}$$

How this splitting is performed, i.e., how one makes the choice as to which part of $H$ is to be $H_0$ and which part is to be $H_1$, depends upon the particular situation. In principle any part of $H$ can be called $H_0$ and the rest then called $H_1$. In practice one chooses for $H_0$ an operator whose eigenvalues are known or can easily be obtained. Often $H_0$ will represent the energy of two or more combined systems when the interaction between the systems is ignored; $H_1$ then represents the interaction energy.

Assuming that $H$ is not explicitly time-dependent (a conservative system), we define, in accordance with (1.16),

$$U(t) = e^{itH/\hbar}. \tag{1.46}$$

The expectation value of an observable $A$ at time $t$ is then given by

$$\langle A \rangle_t = \mathrm{Tr}(U(t)AU^\dagger(t)W_0), \tag{1.47}$$

where $A$ represents the observable at some initial time $t_0 = 0$, and where $W_0$ is the state of the system at that initial time. If we define operators

$$U_0(t) \equiv e^{itH_0/\hbar} \tag{1.48}$$

and

$$U_1(t) \equiv U(t)U_0^\dagger(t) = U(t)e^{-itH_0\hbar}, \tag{1.49}$$

then (1.46) may be rewritten

$$\begin{aligned}
\langle A \rangle_t &= \mathrm{Tr}(U_1(t)U_0(t)AU_0^\dagger(t)U_1^\dagger(t)W_0) \\
&= \mathrm{Tr}(U_0(t)AU_0^\dagger(t)U_1^\dagger(t)W_0 U_1(t)) \\
&= \mathrm{Tr}(A^D(t)W^D(t)), 
\end{aligned} \tag{1.50}$$

where

$$A^D(t) \equiv U_0(t)AU_0^\dagger(t), \tag{1.51}$$

and where

$$W^D(t) \equiv U_1^\dagger(t)W_0 U_1(t). \tag{1.52}$$

Note the similarity between the time development (1.22) of $A$ in the Heisenberg picture and the time development (1.51) in the Dirac picture. In the former the evolution operator $U(t) = e^{itH/\hbar}$ is generated by the full Hamiltonian $H$, while in the latter $U_0(t) = e^{itH_0/\hbar}$ is generated by only part of the Hamiltonian, namely $H_0$. In the extreme case that $H_0 = H$ and $H_1 = 0$ we have $U_0(t) = U(t)$ and $U_1(t) = I$, so that (1.50), (1.51), and (1.52) reduce to the Heisenberg picture. Alternatively, there is a similarity between the time development (1.15) of the statistical operator $W$ in the Schrödinger picture and the time development (1.52) in the Dirac picture. This similarity is even more pronounced if

$$[H_0, H_1] = 0, \qquad (1.53)$$

for then (1.49) becomes

$$U_1(t) = e^{itH/\hbar}e^{-itH_0/\hbar} = e^{itH_1/\hbar}. \qquad (1.54)$$

In the case (1.15), then, $U(t) = e^{itH/\hbar}$ is generated by the full Hamiltonian $H$, while $U_1(t) = e^{itH_1/\hbar}$ is generated by the interaction part of the Hamiltonian, namely $H_1$. In the extreme case $H_0 = 0$ and $H_1 = H$, we have $U_0(t) = I$ and $U_1(t) = U(t)$, and we thereby regain the Schrödinger picture.

Suppose $W_0$ describes a pure state,

$$W_0 = |\psi_0\rangle\langle\psi_0|; \qquad (1.55)$$

then we can define a vector

$$|\psi(t)\rangle_D \equiv U_1^\dagger(t)|\psi_0\rangle \qquad (1.56)$$

such that, according to (1.51), $W^D(t)$ is expressed as

$$W^D(t) = |\psi(t)\rangle_{DD}\langle\psi(t)|. \qquad (1.57)$$

Differentiating (1.56), one obtains for the most common case that (1.53) holds:

$$ih\frac{d}{dt}|\psi(t)\rangle_D = H_1|\psi(t)\rangle_D. \qquad (1.58)$$

Thus the vectors representing the state in the Dirac picture obey the "Schrödinger" equation with the interaction Hamiltonian $H_1$.

In the interaction picture the time dependence of the physically measurable quantity $\langle A\rangle_t$ is described by letting both observable and state vary with time. The observables are rotated in one direction with a transformation generated by part of the Hamiltonian, and the states are rotated in the opposite direction by a transformation generated by the other part (the interaction part) of the Hamiltonian. Expressed in terms of vectors: The eigenvectors $|a\,t\rangle_D$ of the observable $A$,

$$A^D(t)|a\,t\rangle_D = a(t)|a\,t\rangle_D,$$

are given by a rotation by $e^{itH_0/\hbar}$ of $|a\,0\rangle_D$,

$$|a\,t\rangle_D = e^{itH_0/\hbar}|a\,0\rangle_D;$$

one obtains the state vectors $|\psi(t)\rangle_D$ by rotation in the opposite direction by $U_1^\dagger(t)$ $(= e^{-itH_1/\hbar}$ if $[H_0, H_1] = 0)$ of $|\psi_0\rangle$:

$$|\psi(t)\rangle_D = U_1^\dagger(t)|\psi_0\rangle.$$

The time dependence of the probabilities is the same as in the case of the Heisenberg and Schrödinger pictures:

$$\begin{aligned} |\langle a\,t|\psi_0\rangle|^2 &= |\langle a\,0|U^\dagger(t)|\psi_0\rangle|^2 \\ &= |\langle a\,0|U_0^\dagger(t)U_1^\dagger(t)|\psi_0\rangle|^2 \\ &= |_D\langle a\,t|\psi(t)\rangle_D|^2. \end{aligned}$$

Isolated systems are an idealization, because a physical system cannot be completely isolated from the rest of the world. However, in many cases these idealizations give a sufficiently good approximation to the physical situation. If the isolation between a physical system and some part of the rest of the world is not negligible, one can enlarge the physical system by this part of the rest of the world and then consider the combined system. If that is not sufficiently isolated, one can continue to enlarge the physical system and will ultimately end up with an isolated system to which the previous considerations about time development apply. In many situations, however, this procedure is not practical. The enlarged system might be too complicated to be described by a sufficiently simple mathematical structure (if such a structure exists at all). In such a case it is useful to consider nonconservative systems, in particular systems with time-dependent external forces.

For such systems the physical structure changes with time, and so does the mathematical structure that describes it. The observables, in particular the Hamiltonian, will depend explicitly upon time; and thus the dynamical laws will change with time. For such systems it is not possible to give the time evolution by a unitary operator, as in (1.15) and (1.16) or as in (1.22) and (1.23). But one can still try to express it in a differential form in analogy to (1.24) or in analogy to (1.33). The Hamiltonian operator, which is no longer the energy operator of the system, depends explicitly upon time. In this case the time dependence may be formulated in the following form, which is suggested from correspondence to classical mechanics:

$$\frac{dA(t)}{dt} = \frac{1}{i\hbar}[A(t), H(t)] + \frac{\partial A(t)}{\partial t}, \tag{1.59}$$

where $\partial A(t)/\partial t$ is the derivative of $A(t)$ with respect to its explicit time dependence. If one assumes that every observable $A$ is a function $A = A(Q_i(t), P_j(t), t)$ of $Q_i(t)$ and $P_j(t)$ with an additional explicit dependence upon $t$, then (1.59) is obtained from the corresponding equation in non-relativistic classical mechanics by replacing the Poisson bracket by $1/i\hbar$ times the commutator. Equation (1.59) is a generalization of (1.24) and, in the case of an isolated physical system, goes over to (1.24), which corresponds to the classical expression for conservative systems. For systems that have quantum-mechanical observables that cannot be expressed as functions of

$Q_i(t)$ and $P_j(t)$, it is not clear whether a Hamiltonian operator can be found such that the time development can be expressed by (1.59). Equation (1.59) can no longer be integrated into a form analogous to (1.22); even for the case that $A$ does not depend explicitly upon time $[\partial A(t)/\partial t = 0]$, there does not exist a general integration theory for the solution of

$$\frac{dA(t)}{dt} = \frac{1}{ih} [A(t), H(t)]. \tag{1.60}$$

## XII.A Mathematical Appendix: Definitions and Properties of Operators That Depend upon a Parameter

Let $\mathfrak{S}$ be a linear space in which a limiting process is defined; such a space is called a *linear topological space*. For example, in the *Hilbert space* $\mathfrak{S} = \mathcal{H}$ a limiting process is defined in the following way: A sequence $\phi_1, \phi_2, \phi_3, \ldots$ is said to converge to $\phi \in \mathcal{H}$, denoted

$$\phi_n \overset{\mathcal{H}}{\to} \phi,$$

iff

$$\|\phi_n - \phi\| \to 0 \quad \text{as } n \to \infty. \tag{A.1}$$

The Hilbert space $\mathcal{H}$ is not the only linear topological space; if one has a linear space, one can equip it with all kinds of topologies, i.e., one can give all kinds of meanings to the convergence of a sequence. For example, instead of one norm—as in the Hilbert space—one can define in a linear space $\mathfrak{S}$ a countable sequence of norms $\| \ \|_p$,

$$\|\phi\|_0 \leq \|\phi\|_1 \leq \|\phi\|_2 \leq \cdots \quad \text{for all } \phi \in \mathfrak{S},$$

and define convergence in the following way: A sequence of elements $\{\phi_n\}$ converges to $\phi$ with respect to the topology given by the countable sequence of norms, denoted

$$\phi_n \overset{\mathfrak{S}}{\to} \phi,$$

iff for every $p = 0, 1, 2, \ldots$

$$\|\phi_n - \phi\|_p \to 0 \quad \text{as } n \to \infty. \tag{A.2}$$

Such a linear topological space $\mathfrak{S}$ is called a *countably normed space*. In quantum mechanics one usually uses the definition of convergence (A.1), i.e., one uses the Hilbert space rather than any other linear topological space. No physical justification for this choice (made by J. von Neumann) exists.

The notions of "continuity," "differentiability," etc. are defined with respect to a certain topology or meaning of convergence. Hence continuity in a Hilbert space has a meaning different from continuity in a countably normed space or continuity in any space with another kind of topology. For a given linear topological space $\mathfrak{S}$, then, we say that an operator function

$A(t)$ (i.e., an operator which depends upon a parameter $t$) is *continuous at* $t = t_0$ iff for any $\phi \in \mathfrak{S}$

$$A(t)\phi \overset{\mathfrak{S}}{\to} A(t_0)\phi \quad \text{as } t \to t_0. \tag{A.3}$$

$A(t)$ is said to be *continuous* iff (A.3) holds for all values of $t_0$. If one assumes that in quantum mechanics the space of physical states is the Hilbert space, then the vague statement made about the continuity of the time-evolution operator $U(t)$ has the precise meaning given by (A.3), where

$$\text{``} \overset{\mathfrak{S}}{\to} \text{''} \quad \text{means} \quad \text{``} \overset{\mathscr{H}}{\to} \text{''},$$

i.e., $U(t)$ has the property

$$\|U(t)\phi - U(t_0)\phi\| \to 0 \quad \text{as } t \to t_0. \tag{A.4}$$

As mentioned in several places in particular in section II.10 there is no physical reason for which the Hilbert-space convergence is preferred over any other definition of convergence, and therefore there is also no reason to give the vague physical meaning of continuity the precise mathematical definition (A.4). One could as well have chosen any of the other topologies, i.e., used the definition (A.3) with any other mathematical definition of convergence in $\mathfrak{S}$. In the numerous derivations of the Hilbert space for quantum mechanics, it is usually at this point (i.e., the choice of (A.4) for the vague physical meaning of continuity) that the Hilbert space is sneaked in.

An operator function $A(t)$ is called *differentiable at* $t = t_0$ (with respect to the topology on $\mathfrak{S}$) *in a subspace* $\mathscr{D} \subseteq \mathfrak{S}$ if for any $\phi \in \mathscr{D}$

$$\frac{A(t_0 + \Delta t) - A(t_0)}{\Delta t}\phi \overset{\mathfrak{S}}{\to} \psi \in \mathfrak{S} \quad \text{as } \Delta t \to 0, \tag{A.5}$$

i.e., if the limit of $((A(t_0 + \Delta t) - A(t_0))/\Delta t)\phi$ exists in $\mathfrak{S}$. The vector $\psi$ defines a linear operator which we shall denote by $dA(t)/dt|_{t=t}$,

$$\psi = \frac{dA(t)}{dt}\bigg|_{t=t_0} \phi,$$

and which is called the *derivative* (with respect to the topology on $\mathfrak{S}$) *of* $A(t)$ at $t = t_0$. Thus the derivative $dA(t)/dt|_{t=t_0}$ is defined by

$$\frac{dA(t)}{dt}\bigg|_{t=t_0} \phi = \lim_{\Delta t \to 0}^{(\mathfrak{S})} \frac{A(t_0 + \Delta t) - A(t_0)}{\Delta t}\phi. \tag{A.6}$$

if one chooses the Hilbert-space convergence, then one writes

$$\frac{dA(t)}{dt}\bigg|_{t=t_0} \phi = \lim_{\Delta t \to 0}^{(\mathscr{H})} \frac{A(t_0 + \Delta t) - A(t_0)}{\Delta t}\phi. \tag{A.7}$$

The action of the derivative $dA(t)/dt|_{t=t_0}$ is defined on all vectors $\phi \in \mathscr{D} \subseteq \mathfrak{S}$, i.e., on all vectors for which the limit (A.5) exists.

An operator function $A(t)$ that satisfies

$$A(t_1 + t_2) = A(t_1)A(t_2) \tag{A.8}$$

and is continuous with respect to the Hilbert-space topology is called a *one-parameter group of operators* with respect to the Hilbert-space topology. For such a one-parameter group of operators the following properties can be proved:

1. $A(t)$ is differentiable on a "dense" subspace $\mathscr{D} \subseteq \mathscr{H}$. This justifies (1.9), not for all vectors, but for all vectors in $\mathscr{D}$.
2. On $\mathscr{D}$, $A(t)$ can be differentiated an arbitrary number of times.
3. The sequence $\{\phi_n\}$, where

$$\phi_n = \sum_{k=0}^{n} \frac{t^k}{k!} \frac{d^k A(t)}{dt^k}\bigg|_0 \phi,$$

converges (with respect to the Hilbert-space topology) for all elements $\phi$ of a dense subspace $\tilde{\mathscr{D}} \subseteq \mathscr{D} \subseteq \mathscr{H}$. This justifies (1.10).

## Problems

1.   $H$, $a$, and $a^\dagger$ are the operators as defined by (II.3.1) and (II.2.1b). Using the axiom of time development ($V$), calculate $da/dt$ and $da^\dagger/dt$.

2.   The Hamiltonian $H$ for a harmonic oscillator subject to an external force $K(t)$ is given by

$$H = \frac{p^2}{2m} + \frac{m\omega^2}{2} Q^2 - QK(t) = H_0 - QK(t).$$

Here $H_0$, not the Hamiltonian $H$, is the total energy.
   (a)   Express $H$ in terms of the operators $a$, $a^\dagger$ and the identity operator.
   (b)   Using the axiom on time development, calculate $da/dt$.
   (c)   Solve the differential equation obtained in part (b) and find the explicit expression for $a(t)$.

3.   For the harmonic oscillator under the external force $K(t)$ in Problem 2, the orthonormal eigenvectors $\psi_n$ of $N = a^\dagger a$ can be constructed in a manner analogous to that used in the force-free case. One finds

$$\psi_n(t) = \frac{1}{\sqrt{n!}} a^\dagger(t)\psi_0(t),$$

where $\psi_0(t)$ represents the ground state. At the time $t = 0$, the energy of the harmonic oscillator is measured to be $\hbar\omega/2$. Denote the ground state at $t = 0$ by $\phi_0 = \psi_0(t = 0)$.
   (a)   What is the statistical operator for the harmonic oscillator after the energy measurement?
   (b)   Show that $w_n = |(\psi_n, \phi_0)|^2$ is the probability for measuring a value of the energy $E = \hbar\omega(n + \frac{1}{2})$ at some later time $t$.
   (c)   Show

$$w_n = \frac{|F(t)|^{2n}}{n!} e^{-|F(t)|^2},$$

where

$$F(t) = \frac{i}{\sqrt{2m\hbar\omega}} \int_0^t K(\tau)e^{i\omega\tau}\, d\tau.$$

(d)   Show that the expectation value for the position operator immediately after the energy measurement at $t = 0$ is

$$\langle Q \rangle = \frac{1}{m\omega} \int_0^t K(\tau) \sin(t - \tau) \, d\tau.$$

4.   The harmonic oscillator under the influence of an external force

$$K(t) = \begin{cases} K(t) & \text{if } t > 0, \\ 0 & \text{if } t < 0 \end{cases}$$

is at the time $t = 0$ in a state in which the energy is $\hbar\omega/2$. The property that the energy at a time $t$ is equal or smaller than $\hbar\omega(m + \frac{1}{2})$ is characterized by the operator $\Lambda = \sum_{n=0}^{m} \Lambda_{\psi_n(t)}$, where $\Lambda_{\psi_n(t)}$ are the projection operators on the one-dimensional subspace spanned by $\psi_n(t)$.

(a)   What is the probability for the measurement of $\Lambda$.
(b)   What is the statistical operator $W_2$ after the measurement of $\Lambda$ with a positive result?
(c)   Show that

$$W_2 \psi_{n'} = \begin{cases} \sum_{n=0}^{m} \psi_n e^{i(n'-n)\omega t} \dfrac{1}{\sqrt{n!n'!}} (F)^n |F^*|^{n'} e^{-|F|^2} & \text{if } n' \le m, \text{ if } n' > m. \end{cases}$$

# Change of the State by the Dynamical Law and by the Measuring Process— The Stern–Gerlach Experiment

As an illustration of the characteristic features by which quantum mechanics differs from the classical theories, this chapter discusses the change of the state by a measurement. In a *Gedanken* experiment with the Stern–Gerlach apparatus it is shown that a polarized beam (a pure state) cannot be split by the magnetic field and that the splitting of such a beam is a consequence of the measurement.

## XIII.1 The Stern–Gerlach Experiment

In the preceding chapter we have formulated how the state of a system changes in time as a consequence of the dynamical law. In Section II.5 we have described another kind of change in time, the change of the state by a measurement. These are two completely different processes of time change. In this chapter we shall illustrate these processes by considering in detail a Gedanken experiment with the Stern–Gerlach apparatus, which was briefly described in Section IX.1.

In the Stern–Gerlach experiment (Figure 1.1), a beam of hydrogen atoms in the ground state passes through a strongly inhomogeneous magnetic field **H**. Under the conditions of this experiment the hydrogen atoms may be considered to be the combination of two elementary physical systems (cf. the basic assumption IVa of Section III.5): Physical system I is the elementary rotator with angular momentum $\frac{1}{2}$, and describes the spinning electron; physical system II is the elementary particle that describes the motion of the structureless hydrogen atom in the experimental setup.

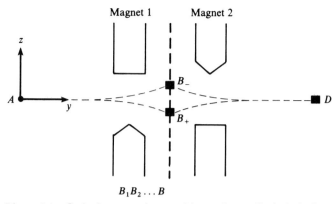

Figure 1.1    Gedanken experiment with two Stern–Gerlach devices.

The space of physical states of system I is the two-dimensional spin space $\mathscr{H}_1 = i^s = \mathscr{R}^{1/2}$ (cf. Section III.3). As a basis for $\mathscr{H}_1$ we choose the eigenvectors $|+\rangle$ and $|-\rangle$ of the spin component $S_3$ in the direction of the magnetic field $\mathbf{H}$, with[1]

$$S_3|+\rangle = +\tfrac{1}{2}|+\rangle \quad \text{and} \quad S_3|-\rangle = -\tfrac{1}{2}|-\rangle. \tag{1.1}$$

The magnetic moment of this system is [Equation (IX.3.18)]

$$\mathbf{M} = -2\frac{e}{2m_e c}\mathbf{S} = -2\mu_B\mathbf{S}. \tag{1.2}$$

The energy operator is a constant; because of our freedom of choice in defining the zero of energy, we may set this constant equal to zero: $H_1 = 0$. The space of physical states of system II is the space $\mathscr{H}_{II}$ spanned by the generalized eigenvectors $|\mathbf{p}\rangle$ of the momentum operator $\mathbf{P}$ or by the generalized eigenvectors $|\mathbf{x}\rangle$ of the position operator $\mathbf{Q}$; $\mathbf{Q}$ and $\mathbf{P}$ are the position and momentum of the hydrogen atom considered as a whole. The energy operator for system II is just the kinetic-energy operator

$$H_{II} = \mathbf{P}^2/2M, \tag{1.3}$$

where $M$ is the mass of the hydrogen atom.

System I and system II are coupled by the external magnetic field $\mathbf{H}$;[1a] according to Equation (IX.3.1) and (IX.3.18) the interaction energy operator that represents this coupling is

$$H_{\text{int}} = -\mathbf{M}\cdot\mathbf{H} = -(\mathbf{M}\otimes I)\cdot(I\otimes\mathbf{H}) = 2\mu_B S_3 H_3(\mathbf{Q}) = 2\mu_B S_3 H_3(Q_3). \tag{1.4}$$

---

[1] We are again using units where $\hbar \equiv 1$.

[1a] In the following, we use the same letter $H$ for the components $H_i$ of the magnetic field as for the Hamiltonians $H, H_1, H_{II}$, etc.

In (1.4) we have approximated the magnetic field **H** as having only a $z$-component and as being a function of $z = x_3$ alone.[2] Consequently the energy operator of the physical combination of these two systems, i.e., of the hydrogen atom in the ground state with electron spin in the magnetic field, is

$$H = H_1 + H_{II} + H_{int} = \frac{1}{2M} \mathbf{P}^2 + 2\mu_B \mathbf{S} \cdot \mathbf{H}(\mathbf{Q})$$

$$= \frac{1}{2M} \mathbf{P}^2 + 2\mu_B S_3 H_3(Q_3). \tag{1.5}$$

In choosing this combination of I and II for the description of the hydrogen beam in the magnetic field we have neglected (1) the magnetic moment of the proton, (2) the influence of the magnetic field **H** upon the electron-proton interaction in the hydrogen atom, and (3) the difference between the electron position and the position of the hydrogen atom as a whole. The justification of (1) and (2) may be seen by an examination of the orders of magnitude involved: (1) The magnetic moment of the proton is [cf. (IX.3.18$_p$)] $\mathbf{M}_p = (2 + 3.59)(e/2m_p c)\mathbf{S} = 5.59(m_e/m_p)\mu_B \mathbf{S}$. Since the mass of the proton is much greater than that of the electron ($m_p = 1836 m_e$), it follows that the magnetic moment of the proton is almost three orders of magnitude smaller than the magnetic moment of the electron; the interaction energy of the proton magnetic moment in the magnetic field is therefore negligible when compared to (1.4). (2) The Bohr magnetron $\mu_B = 5.795 \times 10^{-9}$ eV/gauss is sufficiently small that even in a fairly strong field of $10^3$ gauss the magnetic interaction energy $\mu_B H$ is six orders of magnitude smaller than the difference between the lower energy levels of the hydrogen atom. Thus the internal states of the hydrogen atoms do not change significantly; and the operator for the internal energy of the hydrogen atom [essentially the energy operator (VI.3.2) of the Kepler problem] may be considered a constant for the present problem, which we set equal to zero. Instead of the above described space $\mathscr{H}_{II}$, the space of states of system II should be $\mathscr{H}'_{II} = \mathscr{H}_{II} \otimes \mathscr{R}$, where $\mathscr{R}$ is the space of states for the Kepler problem [cf. (VI.5.2)]. But as we are keeping the hydrogen atom in the ground state [i.e., in the one-dimensional subspace $\mathscr{R}(n = 1) \subset \mathscr{R}$], the space of states for system II is in fact $\mathscr{H}_{II} \otimes \mathscr{R} (n = 1)$, i.e., is $\mathscr{H}_{II}$.

The experimental arrangement is such that at $t = 0$ a pulse of hydrogen atoms enters the region of the magnetic field at the point $A$ (cf. Figure 1.1) and moves with an average momentum $\tilde{p}$ in the $y$ direction. Idealizing, we assume that in this experimental arrangement a pure state of system II, has been prepared. This state is described by the state vector (statistical operator)

$$|\psi_0\rangle \qquad (W_{II} = |\psi_0\rangle\langle\psi_0|) \tag{1.6}$$

in the Heisenberg picture for all times $t \geq 0$, or by

$$|\psi_0\rangle \qquad (W_{II}(0) = |\psi_0\rangle\langle\psi_0|) \tag{1.7}$$

---

[2] The Maxwell equation $\mathbf{V} \cdot \mathbf{H} = \partial H_i/\partial x_i = 0$ tells us that the variation of $H_1$ with $x$ is just as great as the variation of $H_3$ with $z$. We shall return to the subject of this approximation in the appendix to this chapter.

in the Schrödinger picture at time $t = 0$. System I may be in a pure state with statistical operator

$$W_1 = |\phi\rangle\langle\phi| \quad \text{where} \quad |\phi\rangle = \alpha|+\rangle + \beta|-\rangle \qquad (\alpha, \beta \in \mathbb{C}) \qquad (1.8)$$

if the beam has been polarized before it reaches point $A$. (The state $|\phi\rangle$ will be an eigenstate of $\hat{\mathbf{n}} \cdot \mathbf{S}$ for some direction $\hat{\mathbf{n}}$ determined by the choice of $\alpha$ and $\beta$.) Alternately the system I state may be a mixture, e.g.,

$$W_1 = \Lambda^{1/2} = \tfrac{1}{2}(|+\rangle\langle+| + |-\rangle\langle-|) = \tfrac{1}{2}I, \qquad (1.9)$$

where $\Lambda^{1/2}$ is one-half the identity operator in the spin space $\mathscr{H}_1 = \imath^s$. The state of the combined physical system is described by

$$W = W_1 \otimes W_{II}, \quad \text{or} \quad |\chi\rangle = |\phi\rangle \otimes |\psi_0\rangle, \qquad (1.10)$$

the second description in terms of the state vector $|\chi\rangle$ being an alternative in the case of a pure system I state to the description in terms of the statistical operator $W$.[2a]

In considering the time development of the pulse of hydrogen atoms we shall use both the Heisenberg and the Schrödinger picture. By use of (XII.1.59), (1.5), and the canonical commutation relations, we see that the Heisenberg equations of motion for the position and momentum operators are

$$\frac{d\mathbf{Q}(t)}{dt} = \frac{1}{i}[\mathbf{Q}(t), H(t)] = \frac{1}{M}\mathbf{P}(t), \qquad (1.11)$$

$$\frac{d\mathbf{P}(t)}{dt} = \frac{1}{i}[\mathbf{P}(t), H(t)] = \frac{1}{i}2\mu_B S_3[\mathbf{P}(t), H_3(Q_3(t))] = \left(0, 0, -2\mu_B S_3 \frac{\partial H_3}{\partial Q_3(t)}\right). \qquad (1.12)$$

[The meaning of $\partial/\partial Q_i$ was given in the paragraph following Equation (XII.1.29).] Let us first investigate the case that system I is in the spin state $|+\rangle$ or $|-\rangle$, so that the state of the combined system is

$$|\chi\pm\rangle \equiv |\pm\rangle \otimes |\psi_0\rangle. \qquad (1.13)$$

The motion of the expectation value of the momentum operator in these states is then given by

$$\frac{d}{dt}\langle\chi\pm|\mathbf{P}(t)|\chi\pm\rangle = \left(0, 0, -2\mu_B\langle\pm|S_3|\pm\rangle\langle\psi_0|\frac{\partial H_3}{\partial Q_3(t)}|\psi_0\rangle\right)$$

$$= \left(0, 0, \mp\mu_B\langle\psi_0|\frac{\partial H_3}{\partial Q_3(t)}|\psi_0\rangle\right). \qquad (1.14)$$

[2a] Remark on the state of a combination of physical systems: Recall that the operators in the direct-product space are given by Equation (III.5.7). A general statistical operator in $\mathscr{H}_1 \otimes \mathscr{H}_{II}$ is therefore not given simply by the direct product of statistical operators in $\mathscr{H}_1$ and in $\mathscr{H}_{II}$. However, if either of the states $W_1$ or $W_{II}$ is a pure state, then $W_1$ and $W_{II}$ uniquely determine the state $W$ of the combined system as the product $W = W_1 \otimes W_{II}$. This is the situation in the problem under investigation. $W$ is also given by $W = W_1 \otimes W_{II}$ if the state of the combined system is determined by measurements that have been performed upon systems I and II separately. If, however, measurements are performed that measure correlated properties of systems I and II, then in general $W$ is not factorable into the form $W_1 \otimes W_{II}$.

Thus if the state has spin up, $s_3 = +\frac{1}{2}$ (spin down, $s_3 = -\frac{1}{2}$), i.e., if the state is $|+\rangle$ ($|-\rangle$), then the time development of the expectation value of the momentum's $z$-component fulfills

$$\frac{d}{dt}\langle \chi \pm |P_3(t)| \chi \pm \rangle = \mp \mu_B \left\langle \psi_0 \left| \frac{\partial H_3}{\partial Q_3(t)} \right| \psi_0 \right\rangle$$

$$= \mp \mu_B \int d^3x \frac{\partial H_3}{\partial x_3} \langle \psi_0 | \mathbf{x}\, t\rangle \langle \mathbf{x}\, t | \psi_0 \rangle, \quad (1.15\pm)$$

where the $|\mathbf{x}\, t\rangle$ are the time-dependent eigenvectors of $\mathbf{Q}$ in the Heisenberg picture. [The upper (lower) sign in $(1.15\pm)$ refers to the spin-up (spin-down) case.] The right-hand side of $(1.15\pm)$ shows that the change of the momentum depends upon the particular properties of the magnetic field; let us assume for simplicity that

$$\frac{\partial H_3}{\partial x_3} = \text{const} > 0. \tag{1.16}$$

Then

$$\frac{d}{dt}\langle \chi \pm |P_3(t)| \chi \pm \rangle = \mp \kappa \langle \psi_0 | \psi_0 \rangle = \mp \kappa, \quad (1.17\pm)$$

where

$$\kappa = \mu_B \frac{\partial H_3}{\partial x_3}. \tag{1.18}$$

Using the initial condition that the particle has momentum

$$\langle \chi \pm |\mathbf{P}(t = 0)| \chi \pm \rangle = (0, \tilde{p}, 0) \quad \text{at } t = 0,$$

we may integrate $(1.17\pm)$; the expectation value of the momentum operator at time $t$ is then

$$\langle \chi \pm |\mathbf{P}(t)| \chi \pm \rangle = (0, \tilde{p}, \mp \kappa t). \quad (1.19\pm)$$

If we take the expectation value between $|\chi \pm\rangle$ of $(1.11)$ and use the initial condition $\langle \chi \pm |\mathbf{Q}(t = 0)| \chi \pm \rangle = 0$ together with $(1.19\pm)$, then the expectation value of the position operator is

$$\langle \chi \pm |\mathbf{Q}(t)| \chi \pm \rangle = \frac{1}{M}(0, \tilde{p}t, \mp \kappa t^2/2). \quad (1.20\pm)$$

Thus the expectation value of the position operator, i.e., the position of the pulse of hydrogen atoms, moves on a parabolic orbit that bends downward (upward) in the case of a pure spin-up (spin-down) initial state, as illustrated in Figure 1.1. [This presumes that $\kappa > 0$, i.e., presumes $(1.16)$.]

Let us now describe the same situation in the Schrödinger picture, where the state changes in time. At the time $t = 0$ the state is given by

$$|\chi \pm (t = 0)\rangle = |\chi \pm\rangle = |\pm\rangle \otimes |\psi_0\rangle, \quad (1.21\pm)$$

which describes the pulse of hydrogen atoms with spin up (spin down) at the position $A$. According to (XII.1.33), the time development of this state is

$$|\chi \pm (t)\rangle = U^\dagger(t)|\pm\rangle \otimes |\psi_0\rangle = e^{-it(H_{II} + 2\mu_B S_3 H_3)}|\pm\rangle \otimes |\psi_0\rangle$$
$$= e^{-it(H_{II} \pm \mu_B H_3)}|\pm\rangle \otimes |\psi_0\rangle = |\pm\rangle \otimes e^{-it(H_{II} \pm \mu_B H_3)}|\psi_0\rangle.$$

$$(1.22\pm)$$

Thus the initial pure state with spin up will move through the magnetic field according to

$$|\chi + (t)\rangle = |+\rangle \otimes e^{-it(H_{II} + \mu_B H_3)}|\psi_0\rangle = |+\rangle \otimes |\psi + (t)\rangle, \quad (1.23+)$$

in contrast to the different motion

$$|\chi - (t)\rangle = |-\rangle \otimes e^{-it(H_{II} - \mu_B H_3)}|\psi_0\rangle = |-\rangle \otimes |\psi - (t)\rangle \quad (1.23-)$$

of the pure state with spin down. In (1.23$\pm$) we have defined

$$|\psi \pm (t)\rangle \equiv e^{-it(H_{II} \pm \mu_B H_3)}|\psi_0\rangle. \qquad (1.24\pm)$$

Now the relations connecting the momentum operator $\mathbf{P}$, the position operator $\mathbf{Q}$, and the state vectors $|\chi \pm (t)\rangle$ of the Schrödinger picture with $\mathbf{P}(t)$, $\mathbf{Q}(t)$, and $|\chi \pm\rangle$ of the Heisenberg picture are

$$\mathbf{P} = \mathbf{P}(t = 0) = U^\dagger(t)\mathbf{P}(t)U(t), \qquad (1.25a)$$

$$\mathbf{Q} = \mathbf{Q}(t = 0) = U^\dagger(t)\mathbf{Q}(t)U(t), \qquad (1.25b)$$

$$|\chi \pm (t)\rangle = U^\dagger(t)|\chi \pm (t = 0)\rangle = U^\dagger(t)|\chi \pm\rangle, \qquad (1.25c)$$

where $U(t) = e^{itH} = \exp(it(H_{II} + 2\mu_B S_3 H_3(Q_3)))$. We may write the Schrödinger-picture analogue to (1.14) by taking the expectation value of (1.12) between the state vectors $|\chi \pm\rangle$ and shifting the time dependence from the observables to the state vectors:

$$\frac{d}{dt}\langle \chi \pm (t)|P_3|\chi \pm (t)\rangle = \frac{d}{dt}\langle \chi \pm |P_3(t)|\chi \pm\rangle$$

$$= \langle \chi \pm |\frac{dP_3(t)}{dt}|\chi \pm\rangle = \langle \chi \pm |-2\mu_B S_3 \frac{\partial H_3}{\partial Q_3(t)}|\chi \pm\rangle$$

$$= -2\mu_B \langle \chi \pm |e^{it(H_{II} + 2\mu_B S_3 H_3)}S_3 \frac{\partial H_3}{\partial Q_3}e^{-it(H_{II} + 2\mu_B S_3 H_3)}|\chi \pm\rangle$$

$$= -2\mu_B \langle \chi \pm |e^{it(H_{II} \pm \mu_B H_3)}S_3 \frac{\partial H_3}{\partial Q_3}e^{-it(H_{II} \pm \mu_B H_3)}|\chi \pm\rangle$$

$$= -2\mu_B \langle \pm |S_3|\pm\rangle\langle \psi_0|e^{it(H_{II} \pm \mu_B H_3)}\frac{\partial H_3}{\partial Q_3}e^{-it(H_{II} \pm \mu_B H_3)}|\psi_0\rangle$$

$$= \mp\mu_B \langle \psi \pm (t)|\frac{\partial H_3}{\partial Q_3}|\psi \pm (t)\rangle.$$

(Note that $[S_3, H] = 0$ for the Hamiltonian $H$ of (1.5); the matrix element $\langle \pm | S_3 | \pm \rangle$ thus has no time dependence.) If we again assume (1.16), then

$$\langle \psi \pm (t) | \frac{\partial H_3}{\partial Q_3} | \psi \pm (t) \rangle = \int d^3x \frac{\partial H_3}{\partial x_3} \langle \psi \pm (t) | \mathbf{x} \rangle \langle \mathbf{x} | \psi \pm (t) \rangle$$

$$= \frac{\partial H_3}{\partial x_3} \langle \psi \pm (t) | \psi + (t) \rangle = \frac{\partial H_3}{\partial x_3}.$$

Consequently

$$\frac{d}{dt} \langle \chi \pm (t) | \mathbf{P} | \chi \pm (t) \rangle = \left(0, 0, \mp \mu_B \frac{\partial H_3}{\partial x_3}\right) = (0, 0, \mp \kappa). \quad (1.26)$$

The expectation value

$$\frac{d}{dt} \langle \chi \pm | \mathbf{Q}(t) | \chi \pm \rangle = \frac{1}{M} \langle \chi \pm | \mathbf{P}(t) | \chi \pm \rangle$$

of (1.11) easily transforms to the Schrödinger-picture analogue

$$\frac{d}{dt} \langle \chi \pm (t) | \mathbf{Q} | \chi \pm (t) \rangle = \frac{1}{M} \langle \chi \pm (t) | \mathbf{P} | \chi \pm (t) \rangle. \quad (1.27)$$

Equations (1.26) and (1.27) together with the initial conditions

$$\langle \chi \pm (t = 0) | \mathbf{P} | \chi \pm (t = 0) \rangle = (0, \tilde{p}, 0)$$

and

$$\langle \chi \pm (t = 0) | \mathbf{Q} | \chi \pm (t = 0) \rangle = (0, 0, 0)$$

lead to the Schrödinger-picture analogue of (1.20$\pm$):

$$\mathbf{x}_\pm(t) \equiv \langle \chi \pm (t) | \mathbf{Q} | \chi \pm (t) \rangle = \frac{1}{M}\left(0, \tilde{p}t, \mp \frac{\kappa t^2}{2}\right). \quad (1.28\pm)$$

As we would expect, the result is the same as in the Heisenberg picture.

The probability distribution for the position is the "expectation value" of the projection operator $I \otimes |\mathbf{x}\rangle\langle\mathbf{x}|$. More precisely, the probability of obtaining a value in the volume $(\mathbf{x} - \boldsymbol{\epsilon}, \mathbf{x} + \boldsymbol{\epsilon})$ in a position measurement is the expectation value of the operator

$$I \otimes \Lambda(\mathbf{x}, \boldsymbol{\epsilon}) = I \otimes \int_{(\mathbf{x} - \boldsymbol{\epsilon})}^{\mathbf{x} + \boldsymbol{\epsilon}} d^3x' \, |\mathbf{x}'\rangle\langle\mathbf{x}'| \quad (1.29)$$

[cf. Equation (II.10.8)]. Thus the position probability distribution in the state $|\chi + (t)\rangle = | + \rangle \otimes |\psi + (t)\rangle$ is given by

$$w_+(\mathbf{x}, t) = \langle + | + \rangle\langle\psi + (t)|\mathbf{x}\rangle\langle\mathbf{x}|\psi + (t)\rangle = \langle\psi + (t)|\mathbf{x}\rangle\langle\mathbf{x}|\psi + (t)\rangle.$$
$$(1.30+)$$

In particular, the probability of obtaining a value in the interval $(\mathbf{x} - \boldsymbol{\epsilon}, \mathbf{x} + \boldsymbol{\epsilon})$ at time $t$ is

$$w_+(\mathbf{x} - \boldsymbol{\epsilon}, \mathbf{x} + \boldsymbol{\epsilon}, t) = \int_{\mathbf{x} - \boldsymbol{\epsilon}}^{\mathbf{x} + \boldsymbol{\epsilon}} d^3x' \, \langle \psi + (t)|\mathbf{x}'\rangle\langle\mathbf{x}'|\psi + (t)\rangle. \quad (1.31+)$$

The probability distribution moves with time. The exact motion of $w_+(\mathbf{x}, t)$ may be obtained by solving the Schrödinger equation for $\langle\mathbf{x}|\psi + (t)\rangle$. From our considerations concerning the expectation value $(1.28_+)$, however, we have a rough idea of how $w_+(\mathbf{x}, t)$ moves: At $t = 0$, $w_+(\mathbf{x} - \boldsymbol{\epsilon}, \mathbf{x} + \boldsymbol{\epsilon}, t)$ and $w_+(\mathbf{x}, t)$ will be essentially zero everywhere except in a small volume around the point $A = (0, 0, 0)$. From $A$ the pulse moves along the lower parabola, given by $(1.28+)$; at the time $t_B = M y_B / \tilde{p}$ (i.e., at the time $t_B$ at which the pulse has the $y$-coordinate $y_B$), $w_+(\mathbf{x}, t)$ will be essentially zero everywhere except in a small volume around the point $B_+ = (0, \tilde{p} t_B / M, -\kappa t_B^2/2M)$.

Similarly, the probability distribution

$$w_-(\mathbf{x}, t) = \langle \psi - (t)|\mathbf{x}\rangle\langle\mathbf{x}|\psi - (t)\rangle \quad (1.30-)$$

and the probability

$$w_-(\mathbf{x} - \boldsymbol{\epsilon}, \mathbf{x} + \boldsymbol{\epsilon}, t) = \int_{\mathbf{x} - \boldsymbol{\epsilon}}^{\mathbf{x} + \boldsymbol{\epsilon}} d^3x' \langle \psi - (t)|\mathbf{x}'\rangle\langle\mathbf{x}'|\psi - (t)\rangle \quad (1.31-)$$

of finding the system in the volume $(\mathbf{x} - \boldsymbol{\epsilon}, \mathbf{x} + \boldsymbol{\epsilon})$ are the ones for the state $|\chi - (t)\rangle = |-\rangle \otimes |\psi - (t)\rangle$, which was a pure state with spin down at $A$. Then $w_-(\mathbf{x}, t)$ and $w_-(\mathbf{x} - \boldsymbol{\epsilon}, \mathbf{x} + \boldsymbol{\epsilon}, t)$ move along the upper parabola $(1.28-)$, and will be significantly nonzero at time $t_B$ only around the point $B_- = (0, \tilde{p} t_B / M, +\kappa t_B^2/2M)$.

These facts enable us to perform a measurement of the polarization—a measurement of the system-I observable $S_3$—by measuring the system-II observable $\mathbf{Q}$. If we put a detector plate at $B$ (at a distance $y_B = \tilde{p} t_B / M$ from $A$), then it will record the presence of hydrogen atoms at $B_+ = (0, \tilde{p} t_B / M, -\kappa t_B^2/2M)$ or at $B_- = (0, \tilde{p} t_B / M, +\kappa t_B^2 2M)$ only if one started at $A = (0, 0, 0)$ with the pure spin-up state $|\chi + (t = 0)\rangle = |+\rangle \otimes |\psi_0\rangle$ or with the pure spin-down state $|\chi - (t = 0)\rangle = |-\rangle \otimes |\psi_0\rangle$, respectively. Thus a measurement of $Q_3$ on system II with result $z = z_+(t_B) = -\kappa t_B^2/2$ constitutes a measurement of $S_3$ on system I with result $s_3 = +\frac{1}{2}$, and a measurement of $Q_3$ on system II with result $z = z_-(t_B) = +\kappa t_B^2/2$ constitutes a measurement of $S_3$ on system I with result $s_3 = -\frac{1}{2}$.

We now investigate the case where the system-I state is the mixture (1.9), where the coefficients $\frac{1}{2}$ before each of the projectors $|+\rangle\langle+|$ and $|-\rangle\langle-|$ represent the equal a priori probabilities of the system being in the states $|+\rangle$ or $|-\rangle$. At $t = 0$ the beam is described by the statistical operator (1.10) with $W_I(t = 0)$ given by (1.9) and $W_{II}(t = 0)$ given by (1.7):

$$\begin{aligned} W(0) = W_I(0) \otimes W_{II}(0) &= (\tfrac{1}{2}|+\rangle\langle+| + \tfrac{1}{2}|-\rangle\langle-|) \otimes (|\psi_0\rangle\langle\psi_0|) \\ &= \tfrac{1}{2}|+\rangle \otimes |\psi_0\rangle\langle+| \otimes \langle\psi_0| + \tfrac{1}{2}|-\rangle \otimes |\psi_0\rangle\langle-| \otimes \langle\psi_0| \\ &= \tfrac{1}{2}|\chi + (0)\rangle\langle\chi + (0)| + \tfrac{1}{2}|\chi - (0)\rangle\langle\chi - (0)|. \quad (1.32) \end{aligned}$$

The time development of $W$ is

$$W(t) = \tfrac{1}{2}|+\rangle \otimes |\psi + (t)\rangle \langle + | \otimes \langle \psi + (t)|$$
$$+ \tfrac{1}{2}|-\rangle \otimes |\psi - (t)\rangle \langle - | \otimes \langle \psi - (t)|$$
$$= \tfrac{1}{2}|\chi + (t)\rangle \langle \chi + (t)| + \tfrac{1}{2}|\chi - (t)\rangle \langle \chi - (t)|. \qquad (1.33)$$

The probability distribution $w(\mathbf{x}, t) = \mathrm{Tr}(I \otimes |\mathbf{x}\rangle \langle \mathbf{x}| W(t))$ for the position is then

$$w(\mathbf{x}, t) = \tfrac{1}{2}\langle \mathbf{x}|\psi + (t)\rangle \langle \psi + (t)|\mathbf{x}\rangle + \tfrac{1}{2}\langle \mathbf{x}|\psi - (t)\rangle \langle \psi - (t)|\mathbf{x}\rangle \quad (1.34)$$

$$= \tfrac{1}{2}w_+(\mathbf{x}, t) + \tfrac{1}{2}w_-(\mathbf{x}, t), \qquad (1.35)$$

where $w_+(\mathbf{x}, t)$ and $w_-(\mathbf{x}, t)$ are the probability distributions of $(1.30_\pm)$, and their (significantly) nonzero parts move downward and upward along the parabolic trajectories $(1.28+)$ and $(1.28-)$, respectively. Thus *the unpolarized beam is split into two beams*.

This was in fact what Stern and Gerlach observed, in contradiction to the predictions of classical physics.[3] Classically an unpolarized beam contains atoms with magnetic moments equally distributed in all directions. The deviation would be proportional to the magnitude of the $z$-component of the magnetic moment, and hence one would expect the narrow beam to just smear out in the $z$-direction.

We can now add (at least in a *Gedanken* experiment) a second magnet that reverses the splitting effect of the first magnet. The two beams are then brought together again; and at $D$ at time $t_D = My_D/\bar{p}$ we again have the same state as at $A$ at time $t = 0$, i.e., a mixture described by (1.32). If this pulse is now passed through another Stern–Gerlach apparatus with the magnetic field in the $x$-direction, it will split into two beams. This is easily seen by representing $W_I$ as

$$W_I = \Lambda^{1/2} = \tfrac{1}{2}I = \tfrac{1}{2}|s_1 = +\tfrac{1}{2}\rangle\langle s_1 = +\tfrac{1}{2}| \quad + \quad \tfrac{1}{2}|s_1 = -\tfrac{1}{2}\rangle\langle s_1 = -\tfrac{1}{2}|,$$
$$(1.36)$$

where $|s_1 = +\tfrac{1}{2}\rangle$ and $|s_1 = -\tfrac{1}{2}\rangle$ are the basis vectors of $\mathscr{H}_1$ satisfying

$$S_1|s_1 = \pm\tfrac{1}{2}\rangle = \pm\tfrac{1}{2}|s_1 = \pm\tfrac{1}{2}\rangle; \qquad (1.37)$$

the above arguments are then repeated with the $z$-component exchanged with the $x$-component.

We will now analyze the Stern–Gerlach experiment for the case that the state at $A$ is a pure state, polarized in any direction $\mathbf{n}$ other than the $z$-direction (this case we have already treated), e.g., the $x$-direction.[4] At time $t = 0$ the state is then described by the statistical operator

$$W(0) = |\chi(0)\rangle\langle\chi(0)| = |\phi\rangle\langle\phi| \otimes |\psi_0\rangle\langle\psi_0| \qquad (1.38)$$

---

[3] The original Stern–Gerlach experiment (1922) used silver atoms rather than hydrogen atoms.

[4] Use of the Pauli spin matrices easily gives $|s_1 = \pm\tfrac{1}{2}\rangle = (1/\sqrt{2})(|+\rangle \pm |-\rangle)$ (up to an arbitrary phase factor). For the Pauli spin Matrices see problem 7 of Chapter III.

or by the vector

$$|\chi(0)\rangle = |\phi\rangle \otimes |\psi_0\rangle = \alpha|+\rangle \otimes |\psi_0\rangle \;+\; \beta|-\rangle \otimes |\psi_0\rangle, \quad (1.38')$$

where $|\phi\rangle$ is given by (1.8). This state develops in time according to (XII.1.35); using the above results $(1.22_+)$, $1.23_+)$, and $(1.24_+)$ one obtains

$$|\chi(t)\rangle = \alpha|+\rangle \otimes |\psi + (t)\rangle + \beta|-\rangle \otimes |\psi - (t)\rangle. \quad (1.39)$$

Now in a unitary time development a pure state goes into a pure state. The system will therefore remain in a pure state while it passes through the magnetic field; *there is no splitting into two subensembles*. If the beam reaches $D$ after having passed through both magnets, the pulse is again in the pure state with the same polarization that it had at $A$. For example, if we start at $A$ with a beam polarized in the $x$-direction, it will again be polarized in the $x$-direction at $D$ at the time $t_D = My_D/\tilde{p}$; if it is then sent through the second Stern–Gerlach apparatus with the field in the $x$-direction, *it will not split*, but will just bend.

Suppose now that we perform a position measurement on this pure state at $B$, i.e., we place at $B$ a measuring device (screen) that lets the beam pass through but that registers the presence of hydrogen atoms. Recall that the probability of finding the hydrogen atom in the volume $(\mathbf{x} - \boldsymbol{\epsilon}, \mathbf{x} + \boldsymbol{\epsilon})$ around $\mathbf{x}$ is given by the expectation value of the operator $I \otimes \Lambda(\mathbf{x}, \boldsymbol{\epsilon})$ of (1.29) with the statistical operator

$$
\begin{aligned}
W(t) = |\chi(t)\rangle\langle\chi(t)| = {}& |\alpha|^2| +\rangle \otimes |\psi + (t)\rangle\langle + | \otimes \langle \psi + (t)| \\
& + |\beta|^2| -\rangle \otimes |\psi - (t)\rangle\langle - | \otimes \langle \psi - (t)| \\
& + \alpha\bar{\beta}| +\rangle \otimes |\psi + (t)\rangle\langle - | \otimes \langle \psi - (t)| \\
& + \bar{\alpha}\beta| -\rangle \otimes |\psi - (t)\rangle\langle + | \otimes \langle \psi + (t)|
\end{aligned}
$$
$$(1.40)$$

corresponding to the pure state (1.39); thus

$$
\begin{aligned}
w_{|\chi(t)\rangle}(\mathbf{x} - \boldsymbol{\epsilon}, \mathbf{x} + \boldsymbol{\epsilon}) = {}& \mathrm{Tr}(W(t)I \otimes \Lambda(\mathbf{x}, \boldsymbol{\epsilon})) \\
= {}& |\alpha|^2\,\mathrm{Tr}_\mathrm{I}(| +\rangle\langle + |)\,\mathrm{Tr}_\mathrm{II}(|\psi + (t)\rangle\langle\psi + (t)|\Lambda(\mathbf{x}, \boldsymbol{\epsilon})) \\
& + |\beta|^2\,\mathrm{Tr}_\mathrm{I}(| -\rangle\langle - |)\,\mathrm{Tr}_\mathrm{II}(|\psi - (t)\rangle\langle\psi - (t)|\Lambda(\mathbf{x}, \boldsymbol{\epsilon})) \\
= {}& |\alpha|^2 w_+(\mathbf{x} - \boldsymbol{\epsilon}, \mathbf{x} + \boldsymbol{\epsilon}, t) + |\beta|^2 w_-(\mathbf{x} - \boldsymbol{\epsilon}, \mathbf{x} + \boldsymbol{\epsilon}, t),
\end{aligned}
$$
$$(1.41)$$

where $w_\pm(\mathbf{x} - \boldsymbol{\epsilon}, \mathbf{x} + \boldsymbol{\epsilon}, t)$ is given by $(1.31\pm)$. The probability distribution for the position, i.e., the "expectation value" of the "operator" $I \otimes |\mathbf{x}\rangle\langle\mathbf{x}|$, is obtained in the same manner, and is given by

$$w_{|\chi(t)\rangle}(\mathbf{x}) = |\alpha|^2 w_+(\mathbf{x}, t) + |\beta|^2 w_-(\mathbf{x}, t). \quad (1.42)$$

For the case that the hydrogen atoms were originally polarized in the $x$-direction (i.e., $|\alpha|^2 = \frac{1}{2}$ and $|\beta|^2 = \frac{1}{2}$),

$$w_{|\chi(t)\rangle}(\mathbf{x}) = \tfrac{1}{2}w_+(\mathbf{x}, t) + \tfrac{1}{2}w_-(\mathbf{x}, t).$$

Thus the position probability distribution for the pure state (1.8) is the same as the probability for a position measurement in the mixture with the system-I statistical operator

$$W_I = |\alpha|^2 |+\rangle\langle+| \quad + \quad |\beta|^2 |-\rangle\langle-|. \tag{1.43}$$

From our discussion of the motions of $w_+(\mathbf{x}, t)$ and $w_-(\mathbf{x}, t)$ we conclude that the hydrogen atoms pass through the screen at $B_+$ and at $B_-$ with relative probabilities $|\alpha|^2$ and $|\beta|^2$, respectively. Now Equations (1.40), (1.41), and (1.42) depend upon the assumption that the system is in the pure state $|\chi(t)\rangle$ given by (1.39). This assumption is valid only for $t < t_B$, for after the position has been measured at $B$ the state is no longer given by $|\chi(t)\rangle$; it is changed to a new state by the measurement process. If the screen has openings only around $B_+$ and $B_-$, then, according to the basic assumption IIIb. [Equation (II.5.1)], the state is given at time $t = t_B$ by

$$W^B(t_B) = \Lambda(\mathbf{x}_{B+}, \epsilon)|\chi(t_B)\rangle\langle\chi(t_B)|\Lambda(\mathbf{x}_{B+}, \epsilon)$$
$$+ \Lambda(\mathbf{x}_{B-}, \epsilon)|\chi(t_B)\rangle\langle\chi(t_B)|\Lambda(\mathbf{x}_{B-}, \epsilon). \tag{1.44}$$

Now

$$\Lambda(\mathbf{x}_{B\pm}, \epsilon)|\psi \pm (t_B)\rangle\langle\psi \pm (t_B)|\Lambda(\mathbf{x}_{B\pm}, \epsilon)$$

$$= \int_{\mathbf{x}_{B\pm} + \epsilon}^{(\mathbf{x}_{B\pm} + \epsilon)} d^3x' \, d^3x'' \, |\mathbf{x}'\rangle\langle\mathbf{x}'|\psi \pm (t_B)\rangle\langle\psi \pm (t_B)|\mathbf{x}''\rangle\langle\mathbf{x}''|$$

Since $\langle\mathbf{x}'|\psi - (t_B)\rangle = 0$ for $\mathbf{x}'$ near $\mathbf{x}_{B+}$ and since $\langle\psi - (t_B)|\mathbf{x}''\rangle = 0$ for $\mathbf{x}''$ near $\mathbf{x}_{B+}$, it follows that

$$\Lambda(\mathbf{x}_{B+}, \epsilon)|\psi - (t_B)\rangle\langle\psi - t_B|\Lambda(\mathbf{x}_{B+}, \epsilon) = 0.$$

Similarly

$$\Lambda(\mathbf{x}_{B-}, \epsilon)|\psi + (t_B)\rangle\langle\psi + (t_B)|\Lambda(\mathbf{x}_{B-}, \epsilon) = 0.$$

Expansion of (1.44) by means of (1.39) and use of the last three equations allows us to rewrite (1.44) as

$$W^B(t_B) = |\alpha|^2 |+\rangle\langle+| \otimes \Lambda(\mathbf{x}_{B+}, \epsilon)|\psi + (t_B)\rangle\langle\psi + (t_B)|\Lambda(\mathbf{x}_{B+}, \epsilon)$$
$$+ |\beta|^2 |-\rangle\langle-| \otimes \Lambda(\mathbf{x}_{B-}, \epsilon)|\psi - (t_B)\rangle\langle\psi - (t_B)|\Lambda(\mathbf{x}_{B-}, \epsilon). \tag{1.45}$$

We thus see that the performance of a measurement changes the situation drastically. If the beam in the pure state (1.39) passes $B$ without a measurement being performed, then the system remains in this pure state $|\chi(t)\rangle$. If *a measurement is performed* as the beam passes $B$, i.e., if it is registered that the hydrogen atom has passed through the opening at $B_+$ or at $B_-$, *then the system becomes a mixture* of the electron spin-up and spin-down states. For example, if the openings in the screen are symmetrical with respect to the $xy$ plane,

$$w_+(\mathbf{x}_{B+} - \epsilon, \mathbf{x}_{B+} + \epsilon, t_B) = w_-(\mathbf{x}_{B-} - \epsilon, \mathbf{x}_{B-} + \epsilon, t_B), \tag{1.46}$$

and are big enough to let all of the particles in the beam pass through, then

$$\Lambda(\mathbf{x}_{B\pm}, \epsilon)|\psi \pm (t_B)\rangle\langle\psi \pm (t_B)|\Lambda(\mathbf{x}_{B\pm}, \epsilon) = |\psi \pm (t_B)\rangle\langle\psi \pm (t_B)|. \tag{1.47}$$

The state after the measurement then becomes

$$
\begin{aligned}
W^B(t) &= |\alpha|^2 \, |+\rangle\langle+| \otimes |\psi+(t)\rangle\langle\psi+(t)| \\
&\quad + |\beta|^2 \, |-\rangle\langle-| \otimes |\psi-(t)\rangle\langle\psi-(t)| \\
&= |\alpha|^2 \, |+\rangle \otimes |\psi+(t)\rangle\langle+| \otimes \langle\psi+(t)| \\
&\quad + |\beta|^2 \, |-\rangle \otimes |\psi-(t)\rangle\langle-| \otimes \langle\psi-(t)| \\
&= |\alpha|^2 \, |\chi+(t)\rangle\langle\chi+(t)| + |\beta|^2 \, |\chi-(t)\rangle\langle\chi-(t)| = W_M(t) \\
&\quad \text{for} \quad t > t_B.
\end{aligned}
\tag{1.48}
$$

If the state (1.48) now passes through the second magnet, it will arrive at $D$ in a mixture of spin-up with spin-down states. A subsequent Stern–Gerlach apparatus with magnetic field in the $x$-direction will then split the beam.

In summary, we have found that the pure state (1.38) *is not split by the inhomogeneous magnetic fields* of the first and second magnets unless a position measurement is made at the plane $B$. *If* such *a measurement is made,* however, *the pure state splits at $B$ into two subensembles of spin up and spin down,* and consequently into two subensembles of spin right ($s_1 = +\frac{1}{2}$) and spin left ($s_1 = -\frac{1}{2}$), by the second Stern–Gerlach apparatus, whose field is in the $x$-direction. This concrete example illustrates the content of the basic assumption IIIa.

## XIII.A  Appendix

In (1.4) and (1.12) we have neglected the force upon the particle except for the force in the $z$-direction. This requires a justification, because due to the equations of electrodynamics

$$
\mathbf{V} \cdot \mathbf{H} = 0 \qquad \mathbf{V} \times \mathbf{H} = 0.
\tag{A.1}
$$

Therefore a large $\partial \mathscr{H}_3 / \partial x_3$ will at least require also a large

$$
\frac{\partial H_1}{\partial x_1} = - \frac{\partial H_3}{\partial x_3}
\tag{A.2}
$$

and therewith a force in the $x$-direction. We shall describe in the following the simplest theoretical arrangement,[5] in which the momentum acquired by the particle in the $x$-direction never becomes large. This can be achieved if one chooses a large magnetic field in the $z$-direction. This field causes a Larmor precession of the magnetic moment, as a consequence of which the time average of the force in the $x$-direction, $\bar{F}_x$ is small—not the force $F_x$ itself.

Choosing (A.2), we neglect any variation of the magnetic field with $x_2$ (edge effects). From (A.2) it then follows that

$$
\begin{aligned}
H_3 &= \hbar x_3 + \phi(x_1), \\
H_1 &= \hbar x_1 + \psi(x_3),
\end{aligned}
\tag{A.3}
$$

[5] Based on a homework problem by J. W. Alred.

where $\phi(x_1)$ and $\psi(x_3)$ are arbitrary functions except for the requirement that **H** fulfill (A.1). The second equation of (A.1) requires

$$\frac{\partial H_2}{\partial x_3} = \frac{\partial H_3}{\partial x_2} = 0, \qquad \frac{\partial H_2}{\partial x_1} = \frac{\partial H_1}{\partial x_2} = 0, \qquad \text{(A.4)}$$

and

$$\frac{\partial H_3}{\partial x_1} = \frac{\partial H_1}{\partial x_3} \qquad \text{(A.5)}$$

Equation (A.4) follows from the neglect of any variation with $x_2$. Equations (A.5) and (A.3) then lead to the condition

$$\frac{\partial \phi(x_1)}{\partial x_1} = \frac{\partial \psi(x_3)}{\partial x_3}, \qquad \text{(A.6)}$$

which is fulfilled if we choose

$$\phi(x_1) = H_0, \qquad \psi(x_3) = 0. \qquad \text{(A.7)}$$

Therewith we have a magnetic field which can have a strong component in the $z$-direction and is inhomogeneous in the $x$- and the $z$-direction:

$$\mathbf{H} = \hbar x_1 \mathbf{e}_1 + (H_0 - \hbar x_3)\mathbf{e}_3. \qquad \text{(A.8)}$$

This is the simplest possibility compatible with the Maxwell equations; in an actual experimental setup the field will, of course, be more complicated.

The torque on a magnetic moment **M** in a magnetic field **H** is $\mathbf{M} \times \mathbf{H}$, which leads to a change in angular momentum (IX.3.10):

$$\frac{d\mathbf{S}}{dt} = \mathbf{M} \times \mathbf{H}. \qquad \text{(A.9)}$$

And as for the electron, spin and magnetic moment are connected by (1.2):

$$\frac{d\mathbf{M}}{dt} = -2\mu_B \mathbf{M} \times \mathbf{H}. \qquad \text{(A.10)}$$

If we choose for the magnetic field (A.8) with

$$H_0 \gg |\hbar x_1| \approx |\hbar x_3|, \qquad \text{(A.11)}$$

then

$$\frac{d\mathbf{M}}{dt} \approx -2\mu_B \mathbf{M} \times H_0 \mathbf{e}_3. \qquad \text{(A.12)}$$

This means that the magnetic moment operator **M** performs a precession around the $z$-axis $\mathbf{e}_3$ with frequency

$$\omega = 2\mu_B \mathscr{H}_0. \qquad \text{(A.13)}$$

This can be seen by differentiating (A.12) again and writing it in component form:

$$\frac{d^2 M_1}{dt^2} = -\omega^2 M_1, \qquad \frac{d^2 M_2}{dt^2} = -\omega^2 M_2, \qquad \frac{d^2 M_3}{dt^2} = 0. \quad \text{(A.14)}$$

The solution of (A.14) is

$$M_{1,2}(t) = M_{1,2}(0)e^{-i\omega t}, \quad \text{(A.15)}$$

whereas the component $M_3$ does not change in time.

Instead of (1.4) one has now

$$H_{\text{int}} = -M_1(t)\hbar Q_1 - M_3 H_0 - M_3(\hbar Q_3), \quad \text{(A.16)}$$

and instead of (1.12) one has for the force operator

$$\frac{d\mathbf{P}(t)}{dt} = -\frac{1}{i}[\mathbf{P}(t), Q_1]M_1(t)\hbar - \frac{1}{i}[\mathbf{P}(t), Q_3]M_3 \hbar. \quad \text{(A.17)}$$

The second term on the right-hand side is the same as the term in (1.12), except that we have here the special case

$$\frac{dH_3}{dQ_3(t)} = \hbar. \quad \text{(A.8')}$$

As a consequence the expectation value of $P_3(t)$ is given by (1.19).

The first term in (A.17) results in a component of the force operator in the $x$-direction:

$$\frac{dP_1(t)}{dt} = M_1(t)\hbar = M_1(0)\hbar e^{-i\omega t},$$

which oscillates with the frequency (A.13). Though its expectation value may be of the same order as (1.17), it oscillates rapidly under the condition (A.11). And so does the expectation value of $P_1(t)$. The time average of the force as well as the momentum in the $x$-direction is zero. In an experiment, which always measures the average value of the expectation value over a finite time interval, a deviation from zero in the $x$-direction will not be observed.

In the conventional Stern–Gerlach experiment[6] the magnetic field is not given by the idealization (A.8) but is usually of the form

$$\mathbf{H} = H_1(x_1, x_3)\mathbf{e}_1 + (H_0 + H_3(x_1, x_3))\mathbf{e}_3$$

with $H_1(x_1, x_3)$ and $H_3(x_1, x_3)$ fulfilling (A.5) and (A.2) and $|H_0| \gg |H_1| \approx |H_2|$. Clearly, the same arguments as above apply, and (1.4) suffices to describe the observable effects.

---

[6] For other arrangements of a Stern–Gerlach experiment see Myer Bloom and Karl Erdman, *Can. J. Phys.* **40**, 179 (1962).

# CHAPTER XIV

# Transitions in Quantum Physical Systems—Cross Section

This is the fundamental chapter on scattering theory. Here the concept of transition probabilities is introduced and general formulas for the cross section are derived from the basic assumptions of quantum mechanics. In Section XIV.2, a general formula for the transition rate is obtained, which will be used in Section XIV.5 for the derivation of the cross-section formula and in Chapter XXI for the derivation of the decay rate. Section XIV.3 introduces the concept of cross section, and Section XIV.4 gives a brief description of the different ways to relate the cross section to fundamental physical observables. The main section of this chapter is Section XIV.5, where the cross-section formulas for very general physical situations are derived. At the end of section XIV.5 the physical state of the scattering system is further specified to obtain the well-known expressions for the cross section. For a superficial understanding of the material of this chapter one may omit parts of Section XIV.2 and Section XIV.5. For this purpose we have discussed special results of Section XIV.5 at the end of Section XIV.4.

## XIV.1 Introduction

In the present chapter we collect some material that will be used in subsequent chapters to describe collision and decay processes. We derive the transition probability, introduce the notion of transition rate, and then define the cross section and derive the formula that relates it to the transition probability.

The general situation that we shall describe, and of which collision and decay processes are just two particular cases, is the change in time of a nonstationary state $W(t)$. We shall use the basic assumptions of quantum mechanics, in particular the axiom V. We shall make the further simplifying assumption that the generator $H$ of time translation can be split into two parts:

$$H = K + V, \tag{1.1}$$

where $K$ is the energy operator of the isolated physical system with stationary states and is assumed to be well defined. This means that it is assumed to make sense to consider an approximate description of the physical system by a stationary physical system. There are many examples for which this is possible. Thus $K$ may be the energy operator of an atom or of a molecule. In the approximate description the states belonging to an energy level are stationary, but in reality they are only quasistationary and decay into the ground state. This transition is then caused by $V = H - K$. Another example is the combination of two physical systems that can be spatially separated far apart from each other, a situation which one usually encounters in collision processes. If they are far apart, the energy operator of the combined system is $K$; and $V = H - K$, which describes the interaction between the two subsystems, has a finite range.[1]

The observables that are measured in these processes are assumed to commute with the operator $K$. For example, a detector may register all the ground states of atoms that were initially in excited states (by registering, for example, all emitted light of a particular frequency). The observable measured is thus the projection operator on that subspace of the space of physical states which contains the ground states of the atoms, i.e., the observable measured is the property of being in the atomic ground state (cf. Section II.4). In the other example of collision experiments the detector is placed far away from the target and therefore detects eigenstates or mixtures of eigenstates of the operator $K$. The property $B$ measured by the detector is described by a projection operator $\Lambda_B$, which projects onto an (in general continuous) direct sum of energy eigenspaces of $K$. Usually $B$ is a more specific property. For example, the detectors may not be placed all around the target, but only at a particular angle. $\Lambda_B$ is then the projection operator onto that particular subspace of the above direct sum of energy subspaces that contains states whose momentum vectors are directed into the particular solid angle.

Thus the problem that we shall discuss (in the Schrödinger picture) is the following: The state $W(t)$ changes in time according to

$$W(t) = e^{-iHt/h}We^{+iHt/h} \qquad (W \equiv W(0)). \tag{1.2}$$

---

[1] It should be remarked that the assumption (1.1) is not really necessary for the description of collision processes; it suffices to assume only that asymptotic direct-product states do exist.

What is the expectation value of an observable $\Lambda$ (i.e., what is the probability of measuring the property $\Lambda$) for which

$$[\Lambda, K] = 0? \qquad (1.3)$$

This problem is often interpreted as the transition between different stationary states of a physical system caused by an interaction $V$. This is based on the assumed existence of a $K$ eigenstate $W^{\text{in}}(t = -\infty)$ in the remote past and of a $K$ eigenstate $W^{\text{out}}(t = +\infty)$ in the far future that equals or contains $\Lambda$.

## XIV.2  Transition Probabilities and Transition Rates

Denote the state of the physical system by $W$, and the observable to be measured by $\Lambda$. Then, according to the basic assumption II, the expectation value $\langle \Lambda \rangle$ of $\Lambda$ is given by

$$\langle \Lambda \rangle = \text{Tr}(\Lambda W). \qquad (2.1)$$

Since $\Lambda$ is a projector, $\langle \Lambda \rangle$ is the probability that the system is in the subspace $\Lambda \mathscr{H}$. $\langle \Lambda \rangle$ is called the *transition probability* into the state $\Lambda$.

The transition probability is a function of time. At the beginning of the process $\langle \Lambda \rangle$ will in general be zero, and will then increase with time. According to the results of Chapter XII, this time dependence can be described in various pictures. We shall use the Schrödinger picture first, because it appeals best to our intuitive understanding of a scattering experiment with a pulsed beam or of a decay process. In the Schrödinger picture the ensemble of physical systems is described by the time-dependent statistical operator $W(t)$, which describes the beam moving towards the detector, or the excited state of an atom decaying into the ground state.

The observable $\Lambda$, which represents the apparatus, does not change in time. [We assume that $\Lambda$ does not depend explicitly upon time ($\partial \Lambda / \partial t = 0$) i.e., that the apparatus is not influenced from the outside after the scattering experiment has begun.] Thus the transition probability is given by

$$\langle \Lambda \rangle_t = \text{Tr}(\Lambda W(t)), \qquad (2.2)$$

where

$$W(t) = U^\dagger(t) W U(t) \qquad (W \equiv W(0));$$

$W$ is the state of the system at some conveniently chosen time, which we call $t = 0$, and $U(t) = e^{iHt/\hbar}$ is the time-evolution operator of Chapter XII.

The *transition rate* is defined to be the time rate of change of the transition probability: $d\langle \Lambda \rangle_t / dt$. Differentiation of (2.2) and use of (XII.1.33) give

$$\frac{d}{dt} \langle \Lambda \rangle_t = \text{Tr}\left(\Lambda \frac{dW(t)}{dt}\right) = -i \, \text{Tr}(\Lambda[H, W(t)]). \qquad (2.3)$$

Substitution of $H = K + V$ into (2.3) yields

$$\frac{d}{dt} \langle \Lambda \rangle_t = -i \, \text{Tr}(\Lambda[V, W(t)]) - i \, \text{Tr}(\Lambda[K, W(t)]). \qquad (2.4)$$

With the use of (1.3) the second term in (2.4) is easily shown to vanish:

$$
\begin{aligned}
\mathrm{Tr}(\Lambda[K, W(t)]) &= \mathrm{Tr}(\Lambda[K, W(t)] + [K, \Lambda]W(t)) \\
&= \mathrm{Tr}(\Lambda K W(t) - \Lambda W(t)K + K\Lambda W(t) - \Lambda K W(t)) \\
&= -\mathrm{Tr}(\Lambda W(t)K) + \mathrm{Tr}(K\Lambda W(t)) = 0.
\end{aligned}
$$

Let $\{|b\rangle\}$ be a basis for the subspace $\Lambda \mathscr{H}$ onto which $\Lambda$ projects. Then

$$
\mathrm{Tr}(\Lambda W(t)V) = \sum_b \langle b| W(t)V |b\rangle = \sum_b \langle b| V W(t)|b\rangle^*
$$

$$
= (\mathrm{Tr}(\Lambda V W(t)))^*.
$$

The right-hand side of (2.4) may then be written

$$
\frac{d}{dt}\langle \Lambda \rangle_t = -i\,\mathrm{Tr}(\Lambda V W(t)) + i\,\mathrm{Tr}(\Lambda V W(t)))^*. \tag{2.5}
$$

But $i(\alpha^* - \alpha) = 2\,\mathrm{Im}\,\alpha$; consequently[1a]

$$
\hbar\frac{d}{dt}\langle \Lambda \rangle_t = 2\,\mathrm{Im}\,\mathrm{Tr}(\Lambda V W(t)). \tag{2.6}
$$

In order to express the transition rate in a form more easily used, we use a basis of generalized eigenvectors of $K$ and of $H$:

$$
K|a\rangle = E_a|a\rangle, \tag{2.7a}
$$

and

$$
H|a^\pm\rangle = E_a|a^\pm\rangle. \tag{2.7b}
$$

Recall that the generalized eigenvectors $|a^\pm\rangle$ of $H$ are related to the generalized eigenvectors $|a\rangle$ of $K$ by means of the Lippman–Schwinger equation [Equation (VIII.2.7′)]:

$$
|a^\pm\rangle = |a\rangle + \frac{1}{E_a - K \pm i0} V|a^\pm\rangle. \tag{2.8}
$$

On occasion we shall make the labeling on $|a\rangle$ and $|a^\pm\rangle$ more explicit by writing

(a) $|a\rangle = |E_a \hat{a}\rangle$   and   (b) $|a^\pm\rangle = |E_a \hat{a}^\pm\rangle$,

where $\hat{a} = (a_1, a_2, \ldots, a_k)$ are labels needed in addition to the energy to specify the generalized vectors. These labels (quantum numbers) are eigenvalues of additional operators $A_1, A_2, \ldots, A_k$, which together with $K$ form a c.s.c.o. For example, $\hat{a}$ may consist of the angular-momentum quantum numbers $l$ and $l_3$ together with other, internal quantum numbers $\eta$; or $\hat{a}$ may consist of the direction $\mathbf{p}/|\mathbf{p}|$ of the momentum $\mathbf{p}$ together with other quantum numbers $\eta$. If the operators $A_i$ ($i = 1, 2, \ldots, k$) all commute with $H$, then the $|E_a \hat{a}^\pm\rangle$ are generalized eigenvectors of the c.s.c.o. $\{H, A_1, \ldots, A_k\}$; otherwise the additional labels $\hat{a}$ just serve to indicate *that* $|E_a \hat{a}^\pm\rangle$ *is obtained*

---

[1a] We again use the units with $\hbar = 1$ and introduce only in a few important formulas the correct factors of $\hbar$.

*from* $|E_a\hat{a}\rangle$ *by way of the Lippman–Schwinger equation*, and do not indicate that $|E_a\hat{a}^{\pm}\rangle$ is an eigenvector of all the $A_i$'s (for more details cf. Section XV.1).

We shall assume that the $K$ eigenvectors $|a\rangle$ are normalized according to

$$\langle a|a'\rangle = \langle E_a\hat{a}|E_{a'}\hat{a}'\rangle = \rho(E_a)^{-1}\delta_{\hat{a}\hat{a}'}\,\delta(E_a - E_{a'}). \tag{2.9a}$$

$[\rho(E_a)$ is a weight function that is arbitrary but fixed; a convenient choice for $\rho(E_a)$ will be made later.] With such a normalization the summation $\sum_a$ is really an abbreviation for the more explicit

$$\sum_a = \int\rho(E_a)\,dE_a\sum_{\hat{a}}. \tag{2.10}$$

The generalized eigenvectors $|a^{\pm}\rangle$ of $H$ have the same normalization

$$\langle a^{\pm}|a'^{\pm}\rangle = \langle E_a\hat{a}^{\pm}|E_{a'}\hat{a}'^{\pm}\rangle = \rho(E_a)^{-1}\delta_{\hat{a}\hat{a}'}\,\delta(E_a - E_{a'}) \tag{2.9b}$$

as the corresponding generalized eigenvectors $|a\rangle$ of $K$. The derivation of (2.9b) from (2.9a) is given in Appendix A of Chapter XV.

In general the spectrum of $K$ and the spectrum of $H$ are not the same. Usually $K$ has only a continuous spectrum $\{E_a\}$ starting at a particular value, say $E = 0$, and going to infinity. But the spectrum $\{E_\alpha\}$ of $H$ is the combination of a continuous spectrum $\{E_a\}$, which is usually the same as the continuous spectrum of $K$, and a discrete spectrum $\{E_n\}$, which may be negative but is bounded from below. Physically the continuous spectrum corresponds to the scattering states and the discrete spectrum to the bound states of the projectile and target. There may be discrete eigenvalues in the continuous spectrum, but for physical reasons the energy eigenvalues cannot be arbitrary large—in particular, not arbitrarily large and negative. We may thus choose the generalized eigenvectors of $K$:

$$\{|a\rangle = |E_a\hat{a}\rangle\}$$

as a basis for the space of physical states, or we may choose the set

$$\{|\alpha\rangle\} = \{|a^{\pm}\rangle = |E_a\hat{a}^{\pm}\rangle\} \cup \{|\alpha_n\rangle\}$$

of discrete eigenvectors $|\alpha_n\rangle$ and generalized eigenvectors $|a^{\pm}\rangle$ of $H$ as a basis. The completeness property of these bases may be expressed as

$$I = \sum_a|a\rangle\langle a| = \int\rho(E_a)\,dE_a\sum_{\hat{a}}|E_a\hat{a}\rangle\langle E_a\hat{a}| \tag{2.11a}$$

and as

$$I = \sum_\alpha|\alpha\rangle\langle\alpha| = \int\rho(E_a)\,dE_a\sum_{\hat{a}}|E_a\hat{a}^{\pm}\rangle\langle E_a\hat{a}^{\pm}| + \sum_n|\alpha_n\rangle\langle\alpha_n|. \tag{2.11b}$$

The trace that appears on the right-hand side of (2.6) may now be written

$$\mathrm{Tr}\,(\Lambda VW(t)) = \sum_b\langle b|VW(t)|b\rangle$$

$$= \sum_b\sum_\alpha\langle b|V|\alpha\rangle\langle\alpha|W(t)|b\rangle.$$

Use of the Schrödinger-picture time development (1.2) of $W(t)$ and insertion of another complete set of $H$ eigenvectors $|\alpha'\rangle$ then gives

$$\mathrm{Tr}(\Lambda V W(t)) = \sum_b \sum_{\alpha \alpha'} \langle b|V|\alpha\rangle\langle\alpha|e^{-iHt}We^{+iHt}|\alpha'\rangle\langle\alpha'|b\rangle$$

$$= \sum_b \sum_{\alpha \alpha'} \langle b|V|\alpha\rangle\langle\alpha|W|\alpha'\rangle\langle\alpha'|b\rangle e^{-i(E_\alpha - E_{\alpha'})t}. \quad (2.12)$$

Suppose $H$ does have discrete eigenvalues, corresponding to bound states $|\alpha_n\rangle$. Since bound states cannot evolve from free states, we assume that the state $W$ contains no contributions from bound states; mathematically this assumption is expressed by

$$\langle\alpha_n|W|\alpha_{n'}\rangle = 0. \quad (2.13)$$

We may therefore restrict the summations over $\alpha$ and $\alpha'$ to the continuous spectrum of $H$.

Since $[\Lambda, K] = 0$, the basis $\{|b\rangle\}$ of the subspace $\Lambda\mathscr{H}$ may be chosen to consist of generalized eigenvectors of $K$:

$$K|b\rangle = E_b|b\rangle \quad (|b\rangle = |E_b, \hat{b}\rangle). \quad (2.14)$$

If we take the inner product of the Lippman–Schwinger equation (2.8) with $|b\rangle$, we obtain with (2.14)

$$\langle a'^\pm|b\rangle = \langle a'|b\rangle + \lim_{\epsilon\to 0\pm} \langle a'^\pm|V \frac{1}{E_{a'} - K \mp i\epsilon}|b\rangle$$

$$= \langle a'|b\rangle + \lim_{\epsilon\to 0\pm} \langle a'^\pm|V|b\rangle \frac{1}{E_{a'} - E_b \mp i\epsilon}. \quad (2.15_\pm)$$

Use of $(2.15_+)^2$ in (2.12) yields

$$\mathrm{Tr}(\Lambda V W(t)) = \sum_b \sum_{aa'} e^{-i(E_a - E_{a'})t} \frac{\langle b|V|a^+\rangle\langle a'^+|V|b\rangle}{E_{a'} - E_b - i0}\langle a^+|W|a'^+\rangle + \mathrm{II},$$

$$(2.16)$$

where

$$\mathrm{II} = \sum_b \sum_{aa'} e^{-i(E_a - E_{a'})t}\langle b|V|a^+\rangle\langle a'|b\rangle\langle a^+|W|a'^+\rangle. \quad (2.17)$$

In Appendix XV.A we shall show that the second term II vanishes; we content ourselves here with an intuitive argument: The factor $\langle a'|b\rangle$ in the second term does not describe a transition, but describes the probability of observing the configuration of the initial state. But in a scattering experiment (except for absorption measurements, which do not measure $\langle\Lambda\rangle$ directly), the detectors are placed at an angle to the direction of the incident beam so

---

[2] That it is more natural to choose $\varepsilon \to 0+$ rather than $\varepsilon \to 0-$ can be seen from Section XV.3, where we discuss in detail the significance of the choice $\varepsilon \to 0+$. However the derivation could have as well been continued with $(2.15_-)$ and would lead to the same results.

that they are not flooded by the incident beam. Thus $|E_A \hat{A}\rangle \notin \Lambda \mathcal{H}$ for all eigenvectors $|E_A \hat{A}\rangle$ that give a nonzero contribution to the initial state, and the term II does not contribute to the probability of observing $\Lambda$.[3]

The transition rate is then obtained by inserting the first term of (2.16) into (2.5). By using the Hermiticity properties

$$\langle a^+ | W | a'^+ \rangle^* = \langle a'^+ | W | a^+ \rangle \quad \text{and} \quad \langle b | V | a^+ \rangle^* = \langle a^+ | V | b \rangle$$

of $W$ and $V$, and by exchanging the summation indices $a$ and $a'$ in the second term arising from (2.5), one obtains

$$\frac{d}{dt} \langle \Lambda \rangle_t = -i \sum_b \sum_{aa'} e^{-i(E_a - E_{a'})t} \langle b | V | a^+ \rangle \langle a'^+ | V | b \rangle \langle a^+ | W | a'^+ \rangle$$

$$\times \left( \frac{1}{E_{a'} - E_b - i0} - \frac{1}{E_a - E_b + i0} \right). \tag{2.18}$$

We will make use of this last result both in Section XIV.5 when we calculate cross section and in Section XX when we calculate the decay rate.

## XIV.3  Cross Sections

In a scattering experiment a beam is directed towards a small target. A detector is located at a large distance from the target and at a (nonzero) angle $\Omega = (\theta, \phi)$ with respect to the incident beam.

In classical physics this beam can be a beam of particles or a beam of radiation. For the case of particle scattering, for example, a beam of $N_B$ particles of type $B$ with mass $m_B$ and velocity $\mathbf{v}_0$ is, during an interval of time $\Delta t$, sent towards a target consisting of $N_T$ particles of mass $m_T$. The number of particles per second that are scattered and reach the detector, $N/\Delta t$—the counting rate—is proportional to both the number $N_T$ of particles in the target and to the incoming flux (i.e., to the number of particles per second per unit area perpendicular to $\mathbf{v}_0$). The incoming flux is given by

$$\text{incident flux} = \rho_B \mathbf{v}_0, \tag{3.1}$$

where $\rho_B$ is the density of particles in the beam. Thus the counting rate $N/\Delta t$ is

$$N/\Delta t = \sigma N_T \rho_B \mathbf{v}_0, \tag{3.2}$$

where the proportionality constant $\sigma$ is called the *cross section*. The cross section for the scattering of particles by particles is thus defined by

$$\sigma = \frac{N/N_T}{\rho_B \mathbf{v}_0 \, \Delta t}. \tag{3.3}$$

---

[3] Note that II is the term that comes from the first term in the Lippman–Schwinger Equation.

Figure 3.1   Schematic diagram of a scattering experiment.

It has the dimensions of an area. [If the target is not at rest, then the relative speed $v = |\mathbf{v}_0 - \mathbf{v}_T|$ should be used in place of $v_0$ in (3.2).]

Different kinds of cross section are given different names. If the detector detects all particles of a third kind that leave the target area in all directions, then $\sigma$ is called the *production cross section*. If the detector detects particles of type B, the incident type, after they leave the target area, then $\sigma$ is called the *scattering cross section*. In particular $\sigma$ is called the *total scattering cross section* if the detector detects all particles B leaving the target area without regard to the direction in which the particles leave. The *differential cross section* $d\sigma$ is obtained if only those $dN$ particles which are directed into a specific cone of (infinitesimal) solid angle $d\Omega = \sin\theta \, d\theta \, d\phi$ are detected cf. Figure 3.1):

$$d\sigma = \frac{dN/\Delta t}{\rho_B N_T v_0} \quad \text{or} \quad \frac{d\sigma}{d\Omega} = \frac{dN}{d\Omega} \frac{1/\Delta t}{\rho_B N_T v_0}. \tag{3.4}$$

The other example of a scattering experiment in classical physics is the scattering of a beam of radiation incident on an obstacle (target). The detector, which is located outside the incident beam at a distance large compared with the wavelength of the radiation and with the size of the obstacle, measures the flux scattered in a given direction. The ratio of the scattered power $dI_{\text{scat}}$ to the incident power per unit area (flux) $I_{\text{incid}}$ is the differential cross section

$$d\sigma = \frac{dI_{\text{scat}}}{I_{\text{incid}}}. \tag{3.5}$$

In quantum physics the beam is a beam of quantum physical systems, and the measured quantities are probabilities. The detector measures the probability of a transition from the incident state. This transition probability is proportional to the probability that a particle before scattering is passing through a surface of unit area perpendicular to the incident velocity. The proportionality constant is the quantum-mechanical cross section:

$$\sigma = \frac{\text{transition probability}}{\text{incident probability per unit area}}. \tag{3.6}$$

As in the classical case, there are different kinds of cross section. If the detector registers all states that emerge from the target, then $\sigma$ is the total cross section. If the detector registers only a subset of states (e.g., only those states which are in a state $B$), then $\sigma = \sigma_{BA}$ is called the partial cross section. Thus

$$\sigma_{BA} = \frac{\text{transition probability from a state } A \text{ into a state } B}{\text{incident probability per unit area}}. \qquad (3.7)$$

Suppose $B$ is the property that the system's momentum vector is directed into a specific cone of solid angle $d\Omega$ around the direction $\Omega = (\theta, \phi)$; then

$$d\sigma_{\Omega A} = \frac{\begin{array}{c}(\text{transition probability from a state } A \text{ into} \\ \text{all states with momentum vector directed into } d\Omega)\end{array}}{\text{incident probability per unit area}}, \qquad (3.8)$$

or more often $d\sigma_{\Omega A}/d\Omega$ is called the differential cross section.

In typical scattering experiments the projectile is chosen to be as structureless as possible (e.g., electrons), whereas the target consists of more complicated objects (e.g., atoms). The experimental cross sections then reveal information about the structure of the target particles. The collision of the projectile with the target can be elastic, as is the case when the energy of the projectile is smaller than the difference between the internal energy levels of the target; or the collision can be inelastic, as is the case when the internal energy level of the target is changed by the collision. To fix the nomenclature we shall use "total collision cross section" for the cross section $\sigma$ of a collision that involves any elastic or inelastic process. The cross section $\sigma_{\text{elas}}$ for all elastic collisions is called the *total elastic cross section* while the cross section $\sigma_\eta$ for inelastic collisions into the $\eta$th internal energy level will be called the *total inelastic partial cross section for the $\eta$th level*. The *total inelastic cross* section is $\sum_\eta \sigma_\eta$. The total cross section $\sigma$ is then the sum of the total elastic cross section with the total inelastic cross sections:

$$\sigma = \sigma_{\text{elas}} + \sum_\eta \sigma_\eta. \qquad (3.9)$$

Corresponding notation is used for the differential cross section.

## XIV.4 The Relation of Cross Sections to the Fundamental Physical Observables

Cross sections are the observable quantities. Their measurement follows in principle from their definition, given in the previous section, in terms of counting rates and flux.[4] The cross sections are related to the fundamental observables which describe the structure of the physical system consisting of the combination of projectile and target. We shall now establish the connection of the cross sections to these quantum-mechanical observables.

If the theory of the physical system (i.e., the mathematical structure that describes the system) is known, then one can calculate the cross sections in

---

[4] The actual measurement of cross sections is described in books on experimental physics. See, for example, Massey et al. (1969, Vol. 1).

terms of the known fundamental observables and predict the outcome of a collision experiment. More often a physicist meets the reverse situation: The cross sections have been measured experimentally; from this information conjectures are made as to what the fundamental observables are, how they behave, and what mathematical structure best describes such behavior. This task is least complicated if one of the colliding subsystems (called the projectile) has a known structure that is as simple as possible. Such a projectile serves to test the structure of the target system and to provide information for a theoretical model of the target system. Some examples are electrons as projectiles with atoms or molecules as targets, protons as projectiles with nucleons as targets, and electrons or other leptons as projectiles with hadrons as targets.

To relate the cross sections to fundamental observables one may proceed in two different ways. The first and conventional way is based upon the time-development axiom V of Chapter XII. It is assumed that there exists an energy operator $H$ in the system's algebra of operators that develops the system in a continuous fashion. It is usually further assumed that the energy operator $H$ can be split into parts,

$$H = K + V, \qquad K = K_B \otimes I + I \otimes K_T, \qquad (4.1)$$

where $K$ is the energy operator of the combination of the physical systems in the absence of the interaction $V$ between them, $K_B$ is the energy operator for the beam system, and $K_T$ is the energy operator for the target system. $K$ is thus the energy operator when the projectile and target are far apart.[5]

The existence of a continuous unitary time development (1.2), generated by an operator $H$ according to the basic assumption V, is a rather questionable assumption in particle physics, where a fundamental length is believed to exist. Therefore a second way was suggested by Heisenberg (1943). It is based on the realization that $W(t)$, which describes the systems while they are interacting, is really not a measurable quantity. Measurable quantities are the initial state $W^{in}(t \to -\infty)$ that describes the systems before the interaction when they are prepared for the collision process, and the final state $W^{out}(t \to +\infty)$ that describes the systems after the interaction when they are detected. Each initial state is transformed into a final state by the interaction. The operator that describes this transformation

$$W^{in}(t) \to W^{out}(t) = S \, W^{in}(t) \, S^{-1},$$

or, for pure states,

$$\Psi^{in}(t) \to \Psi^{out}(t) = S \, \Psi^{in}(t),$$

is called the $S$-operator ("$S$-matrix").

---

[5] The assumption (4.1) that $H$ can be split is not really essential. One can derive expressions for the cross sections from the assumption of the existence of a time-development generator $H$ alone. See, for example, Goldberger and Watson (1964, Chapter 5).

$S$ is to have the following properties: (1) It transforms superpositions $a\Phi^{in} + b\Psi^{in}$ into superpositions $a\Phi^{out} + b\Psi^{out}$; therefore it must be a linear operator. (2) It transforms every normalized initial state uniquely into a normalized final state ("conservation of probability"); this, together with the assumption that the set of initial states as well as the set of final states spans the space of physical states, requires that $S$ be unitary: $SS^{\dagger} = S^{\dagger}S = 1$.

The cross section can then be related to the matrix elements of the operator $S$ ($S$-matrix). Thus in this second approach the existence of an unitary $S$-operator is taken as the fundamental postulate in place of the basic assumption V of a continuous unitary time development.

If the Hamiltonian $H$ exists and is the generator of a continuous time development, then the $S$-matrix may be expressed in terms of $H$.[6] The $S$-operator is still a meaningful concept, however, even if this is not the case. Heisenberg's assumption that it is the $S$-operator (and not the Hamiltonian) that is the fundamental physical quantity has become the basis for an attempt at a new formulation of elementary-particle theory known as *S-matrix theory*.

The first way thus assumes more and allows one to relate the observable quantities, such as cross sections, to fundamental observables that reveal more of the physical system's structure. The second way denies the possibility of this deeper insight into the structure of the physical system. For atomic scattering experiments in nonrelativistic quantum mechanics, the existence of an elementary length of less than $10^{-13}$ cm ($=10^{-5}$ Å) can be ignored. We can therefore connect the cross section to our basic assumptions for quantum mechanics (in particular to the axiom V) by proceeding in the first way, along which we shall meet the $S$ operator as a derived quantity.

In the next chapter we shall examine in more detail the concept of incoming and outgoing states and their relationship to the $S$-matrix.

In Section XIV.5 we shall give a derivation of the cross section from the basic assumptions of quantum mechanics under very general conditions. Here we list a few special results for the benefit of the reader who does not want to go through the tedious derivations of the following section.

The cross section is expressed in terms of the matrix elements of the interaction Hamiltonian $V$ or the $T$-matrix. The matrix elements of $V$ and $T$ are related by

$$\langle \eta_b \Omega E | T | E\Omega' \eta_A \rangle = \langle \eta_b \Omega E | V | E\Omega' \eta_A^+ \rangle \tag{5.41'}$$

Here $E = p^2/2m + E_\eta$ denotes the energy of the target-projectile system, $E_\eta$ denotes the internal energy (which may be a function of the internal quantum numbers $\eta$), and $\Omega = (\theta\phi)$ denotes the direction of the projectile momentum, $\mathbf{p} = p\Omega$. $|E\Omega\eta^+\rangle$ is the eigenvector of the exact operator $H$, and $|E\Omega\eta\rangle$ is the eigenvector of the free-energy operator $K = H - V$. If a Hamiltonian time development does not exist, then the $T$-matrix is con-

---

[6] In this context the $S$-matrix was first introduced by J. A. Wheeler in 1937.

nected with the $S$-matrix by

$$\langle \eta \Omega E | S | E' \Omega' \eta' \rangle = \langle \eta \Omega E | E' \Omega' \eta' \rangle - 2\pi i \delta(E - E') \langle \eta \Omega E | T | E \Omega' \eta' \rangle. \quad (XV.3.36')$$

If the generalized eigenvectors are normalized according to

$$\langle E \Omega \eta | E' \Omega' \eta' \rangle = \langle \mathbf{p} \eta | \mathbf{p}' \eta' \rangle = \delta_{\eta \eta'} \delta^3(\mathbf{p} - \mathbf{p}'),$$

then the differential cross section for the scattering of a projectile with mass $m_A$ and momentum $\mathbf{p}_A$ on a target in a state $\eta_A$ into a particle with mass $m_b$ going into the direction $\Omega_b = \mathbf{p}_b/p_b$ and leaving the target in the new state $\eta_b$, is given by

$$\frac{d\sigma}{d\Omega}(\Omega_b E_A \eta_b \leftarrow p_A \eta_A) = \frac{(2\pi)^4 \hbar^2 m_A m_b p_b(E_A)}{p_A} |\langle E_A \Omega_b \eta_b | T | \mathbf{p}_A \eta_A \rangle|^2, \quad (5.61')$$

where

$$p_b(E_A) = \sqrt{2m_b(E_A - E_{\eta_b})}, \qquad E_A = \frac{p_A^2}{2m_A} + E_{\eta_A}.$$

For elastic scattering, $m_A = m_b$, $\eta_A = \eta_b$, this goes into (5.62) at the end of Section XIV.5, which can be expressed in terms of the elastic-scattering amplitude (5.63) given by (5.64).

## XIV.5[6a] Derivation of Cross-Section Formulas for the Scattering of a Beam off a Fixed Target

We shall now give the precise expression for the quantum-mechanical definition of the cross section in terms of the state and the transition probability. We shall then proceed to derive various cross-section formulas.

The transition probability per unit time, corresponding to $d(N/N_T N_B)/dt$ for classical particle scattering, is in quantum mechanics given by $d\langle \Lambda \rangle_t/dt$. The probability that the system will make a transition during the time period $\Delta t$ taken by the experiment is $(d\langle \Lambda \rangle_t/dt) \Delta t$.

The probability density for the beam position, corresponding to $\rho_B/N_B$ for classical particle scattering, is given in quantum mechanics by $\langle \mathbf{x} | W_B^{in}(t) | \mathbf{x} \rangle$, where $W_B^{in}(t)$ is the statistical operator describing the state of the incident beam system. If we assume a small spread in the velocities of the beam's constituent particles, the incident probability per unit area per unit time for a beam that is incident in the $z$-direction with an average velocity $v_0$ is $\langle \mathbf{x} | W_B^{in}(t) | \mathbf{x} \rangle v_0$.

According to its definition (3.7) and corresponding to its classical expression (3.3), the cross section for an experiment that runs over the period

---

[6a] To follow these derivations may require special effort. We have therefore presented a few special results needed for the following chapters at the end of section XIV.4 and this section may be omitted in first reading.

$\Delta t$ is then given as the proportionality factor $\sigma$ in the equation

$$\frac{d\langle\Lambda\rangle_t}{dt}\,\Delta t = \sigma\langle\mathbf{x}|\,W_B^{\text{in}}(t)|\mathbf{x}\rangle v_0\,\Delta t. \qquad (5.1)$$

If the experiment extends over a long time period, the defining equation (5.1) goes over into the equation

$$\int_{-\infty}^{+\infty} dt\,\frac{d\langle\Lambda\rangle_t}{dt} = \int_{-\infty}^{+\infty} dt\,\sigma\langle\mathbf{x}|\,W_B^{\text{in}}(t)|\mathbf{x}\rangle v_0. \qquad (5.2)$$

If the beam does not consist of structureless quantum physical systems but has internal degrees of freedom described by a set of quantum numbers $\lambda$, then the beam's position probability density is $\sum_\lambda \langle\mathbf{x}\lambda|\,W_B^{\text{in}}(t)|\mathbf{x}\lambda\rangle$, where $|\mathbf{x}\lambda\rangle$ is a basis of generalized position eigenvectors for the space of physical states of the beam system. The quantum numbers $\lambda$ usually describe the polarization (spin state), and we shall therefore generally refer to $\lambda$ as the polarization, even though in the case of more complicated projectiles (e.g., atoms, molecules, or ions), $\lambda$ could stand for a whole set of quantum numbers upon which the energy eigenvalues of the beam system may depend.

If there are polarizations, then one has

$$\int_{-\infty}^{+\infty} dt\,\frac{d\langle\Lambda\rangle_t}{dt} = \int_{-\infty}^{+\infty} dt\,\sigma v_0 \sum_\lambda \langle\mathbf{x}\lambda|\,W_B^{\text{in}}(t)|\mathbf{x}\lambda\rangle \qquad (5.3)$$

instead of (5.2).

We assume in this section that the target is fixed in position; $v_0$ is thus the average relative speed between projectile (beam) and target. We further assume that in the absence of the interaction $V$ between the component systems, the statistical operator $W^{\text{in}}(t)$ for the combined projectile-target system would be factorable into the product

$$W^{\text{in}}(t) = W_B^{\text{in}}(t) \otimes W_T^{\text{in}}(t) \qquad (5.4)$$

of a statistical operator $W_B^{\text{in}}(t)$ describing the state of the projectile with a statistical operator $W_T^{\text{in}}(t)$ describing the state of the target. $W_B^{\text{in}}(t)$ acts in a space of states $\mathcal{H}_B$, $W_T^{\text{in}}(t)$ acts in a space $\mathcal{H}_T$, and $W^{\text{in}}(t)$ acts in the total space $\mathcal{H} = \mathcal{H}_B \otimes \mathcal{H}_T$. The justification for these assumptions lies in the initial conditions.

$W^{\text{in}}(t)$ can be thought of as the operator which describes the system before the interaction becomes effective (i.e., for $t \to -\infty$), for it is then that the beam and target system are prepared. Developing in time according to

$$W^{\text{in}}(t) = e^{-iKt}W^{\text{in}}e^{iKt}$$

where $K$ is the interaction free-energy operator, it describes a fictitious state in which the beam passes through the target without interaction. The preparation of the component systems is by way of measurements of properties which are uncorrelated at the time of preparation, with the consequence that at this time $W^{\text{in}}(t)$ is factorable into the form (5.4).[7] But then it remains so at all times, as

$$\begin{aligned} W^{\text{in}}(t) &= e^{-i(K_B + K_T)t}(W_B^{\text{in}} \otimes W_T^{\text{in}})e^{+i(K_B + K_T)t} \\ &= e^{-iK_B t}W_B^{\text{in}}e^{iK_B t} \otimes e^{-iK_T t}W_T^{\text{in}}e^{iK_T t} \\ &= W_B^{\text{in}}(t) \otimes W_T^{\text{in}}(t). \end{aligned} \tag{5.5a}$$

The actual state of the combined system is $W(t)$, which, while it is essentially the same as $W^{\text{in}}(t)$ in the distant past, has a different time development [Equation (1.2)]:

$$W(t) = e^{-iHt}We^{+iHt}. \tag{5.5b}$$

Unlike $W^{\text{in}}(t)$, $W(t)$ is generally not factorable, because the interaction part $V$ of its time-development generator $H = K + V$ acts nontrivially in both $\mathscr{H}_B$ and $\mathscr{H}_T$.

Rather than interrupt the calculations to follow with many explanations, we shall first make some remark about the basis vectors which will be used and their normalizations, and shall then list some needed mathematical relations.

For the basis $\{|E\hat{a}\rangle\}$ of $\mathscr{H} = \mathscr{H}_B \otimes \mathscr{H}_T$ we shall use the direct-product basis consisting of the generalized $K$ eigenvectors

$$|E\hat{a}\rangle = |\mathbf{p}\lambda\rangle \otimes |E_T\eta\rangle. \tag{5.6}$$

Here $|\mathbf{p}\lambda\rangle$ is a generalized eigenvector with polarization $\lambda$ of the momentum operator $\mathbf{P}$ for the beam system, and $|E_T\eta\rangle$ is a (possibly generalized) eigenvector with polarization $\eta$ of the operator $K_T = K_T^{\text{int}}$ for the internal energy of the target system.[8] The generalized eigenvectors $|\mathbf{p}\lambda\rangle$ have the usual $\delta$-function normalization,

$$\langle \mathbf{p}\lambda | \mathbf{p}'\lambda'\rangle = \delta^3(\mathbf{p} - \mathbf{p}')\delta_{\lambda\lambda'} \tag{5.7}$$

while the $|E_T\eta\rangle$'s may be either proper eigenvectors $|E_T^d\eta\rangle$ corresponding to a discrete eigenvalue $E_T^d$ and having normalization

$$\langle E_T^d\eta | E_T^{d'}\eta'\rangle = \delta_{E_T^d E_T^{d'}}\delta_{\eta\eta'}, \tag{5.8a}$$

or generalized eigenvectors $|E_T^c\eta\rangle$ corresponding to the continuous eigenvalue $E_T^c$ and having normalization

$$\langle E_T^c\eta | E_T^{c'}\eta'\rangle = \tilde{\rho}(E_T^c)^{-1}\delta(E_T^c - E_T^{c'})\delta_{\eta\eta'}. \tag{5.8b}$$

Since the $|\mathbf{p}\lambda\rangle$'s are also the generalized eigenvectors of the beam system's energy operator

$$K_B = K_B^{\text{kin}}(\mathbf{P}) + K_B^{\text{int}}(\lambda^{\text{op}}), \tag{5.9}$$

---

[7] Cf. remarks in Section XIII.1 concerning the state of a combination of physical systems.

[8] Though $K_T^{\text{int}}$ is the operator for the internal energy of the target (i.e., has no kinetic-energy term) we shall not exclude the possibility that it also has a continuous spectrum, as may occur, for example, in collision-induced scattering.

the product vectors $|\mathbf{p}\lambda\rangle \otimes |E_T\eta\rangle$ are in fact the generalized eigenvectors of the energy operator $K = K_B + K_T$. In (5.9) $K_B^{\text{kin}}(\mathbf{P})$ is the kinetic energy operator for the projectile, the functional form of which depends on the nature of the projectile; for a massive nonrelativistic projectile

$$K_B^{\text{kin}}(\mathbf{P}) = \mathbf{P}^2/2m, \tag{5.10a}$$

while for a photon

$$K_B^{\text{kin}}(\mathbf{P}) = c(\mathbf{P}^2)^{1/2} = cP, \tag{5.10b}$$

where $c$ denotes the speed of light. $K_B^{\text{int}}(\eta^{\text{op}})$ is the operator for the internal energy of the projectile. The eigenvalues of $K$ on the basis vector (5.6) are therefore $E = E_B + E_T$, where

$$E_B = \frac{\mathbf{p}^2}{2m} + E_B^{\text{int}}(\lambda), \qquad E_B = \frac{\mathbf{p}^2}{2m}, \quad \text{or} \quad E_B = pc \tag{5.11}$$

for a nonrelativistic projectile with internal energy levels $E_B^{\text{int}}(\lambda)$, a structure-less[9] nonrelativistic projectile, or a photon, respectively.

For the sake of definiteness we shall assume that the projectile is non-relativistic and has no internal energy levels, so $E_B = \mathbf{p}^2/2m$. For the other cases in (5.11) or for the massive relativistic case $E_B = (p^2c^2 + m^2c^4)^{1/2}$, minor modifications will be necessary. We shall also use the abbreviated notation

$$\sum_{E_T} = \sum_{E_T^d} + \int \rho(E_T^c)\, dE_T^c \tag{5.12a}$$

and

$$\langle E_T\eta | E_T'\eta\rangle = \delta_{E_T E_T'}\delta_{\eta\eta'} \tag{5.12b}$$

for Equations (5.8a) and (5.8b). We may then express the product basis vectors $|\mathbf{p}\lambda\rangle \otimes |E_T\eta\rangle$ in terms of the total energy $E$ and the spherical coordinates of $\mathbf{p}$:

$$|E\hat{a}\rangle = |EE_T\Omega\lambda\eta\rangle = |\mathbf{p}\lambda\rangle \otimes |E_T\eta\rangle$$
$$= |E_B\Omega\lambda\rangle \otimes |E_T\eta\rangle, \tag{5.13}$$

where

$$E = E_B + E_T = \frac{\mathbf{p}^2}{2m} + E_T, \tag{5.14a}$$

$$\mathbf{p} = p\mathbf{\Omega} = p(\sin\theta\cos\phi, \sin\theta\sin\phi, \cos\theta), \tag{5.14b}$$

---

[9] More precisely, a projectile with only one energy level, which still may have different polarization states.

and

$$\Omega = (\theta, \phi). \tag{5.14c}$$

Since

$$\sum_{\hat{a}} \int \rho_{\hat{a}}(E)dE = \sum_{E_T \lambda \eta} \int d^3 p = \sum_{E_T \lambda \eta} \int p^2 \sin\theta \, dp \, d\theta \, d\phi$$

$$= \sum_{E_T \lambda \eta} \int m \sqrt{2m(E - E_T)} \sin\theta \, d\theta \, d\phi \, dE, \tag{5.15}$$

the basis vectors (5.13) must have normalization

$$\langle E\hat{a} | E'\hat{a}' \rangle = \langle EE_T \Omega \lambda \eta | E'E'_T \Omega' \lambda' \eta' \rangle$$

$$= \frac{1}{m\sqrt{2m(E - E_T)}} \delta(E - E') \delta_{E_T E'_T} \delta^2(\Omega - \Omega') \delta_{\lambda\lambda'} \delta_{\eta\eta'} \tag{5.16}$$

where

$$\delta^2(\Omega - \Omega') = \delta(\cos\theta - \cos\theta')\delta(\phi - \phi') \tag{5.17}$$

in terms of the quantum numbers $E$, $E_T$, $\Omega$, $\lambda$, and $\eta$, i.e., the weight function (measure) $\rho_{\hat{a}}(E)$ of (2.9) is fixed by our normalization choices (5.7) and (5.8) to be

$$\rho_{E_T}(E) = \rho(E - E_T) = m\sqrt{2m(E - E_T)} = mp, \tag{5.18a}$$

and $\delta_{\hat{a}\hat{a}'}$ of (2.9) is

$$\delta_{E_T E'_T} \delta^2(\Omega - \Omega')\delta_{\lambda\lambda'}\delta_{\eta\eta'}. \tag{5.18b}$$

The mathematical relations that we need in the calculations are the identities[10]

$$\int_{-\infty}^{+\infty} dt \, e^{-ist} = 2\pi\delta(s) \tag{5.19}$$

and

$$\frac{1}{s - i0} - \frac{1}{s + i0} = 2\pi i\delta(s) \tag{5.20}$$

for the generalized function $\delta(s)$.

We shall also need the relationship

$$\langle a^+ | W(t) | a'^+ \rangle = \langle a | W^{\text{in}}(t) | a' \rangle, \tag{5.21}$$

which has an immediate derivation (Appendix XV.A) from the developments of Section XV.3 below. Equation (5.21) is plausible for the following reason: $W(t)$ is identical with $W^{\text{in}}(t)$ in the distant past before the interaction $V$ becomes effective, and $|a^+\rangle$ is the same as $|a\rangle$ in the absence of $V$, so (5.21) holds at some time in the distant past. But the time developments of

---

[10] Gel'fand and Shilov (Vol. 1, pp. 168, 94).

$W(t)$ and $W^{in}(t)$ are generated by $H$ and by $K$, respectively, which have the same effect when applied to $|a^+\rangle$ and $|a\rangle$, respectively; consequently, (5.21) holds for all times $t$, not just in the distant past.

We now start the derivation of explicit formulas for the cross section.[11] Inserting (2.20) into (5.3), one obtains

$$-i\sum_{b}\sum_{aa'} \int_{-\infty}^{+\infty} dt\, e^{-i(E_a - E_{a'})t}\langle b|V|a^+\rangle\langle a'^+|V|b\rangle\langle a^+|W|a'^+\rangle$$

$$\times \left(\frac{1}{E_{a'} - E_b - i0} - \frac{1}{E_a - E_b + i0}\right)$$

$$= \int_{-\infty}^{+\infty} dt\, v_0 \sigma \sum_{\lambda} \langle \mathbf{x}\lambda|W_B^{in}(t)|\mathbf{x}\lambda\rangle.$$

The right-hand side may be rewritten by inserting the basis vectors $|\mathbf{p}\lambda\rangle$, using the three-dimensional version

$$\langle \mathbf{x}\lambda|\mathbf{p}\lambda'\rangle = (2\pi)^{-3/2} e^{i\mathbf{p}\cdot\mathbf{x}} \delta_{\lambda\lambda'} \tag{5.23}$$

of (II.8.57), the time development (5.5b) of $W_B^{in}(t)$, and the identity (5.19):

$$\text{RHS of (5.22)} = i\int_{-\infty}^{+\infty} dt\, v_0 \delta \sum_{\lambda} \int d^3p\, d^3p' \langle \mathbf{x}\lambda|\mathbf{p}\lambda\rangle$$

$$\times \langle \mathbf{p}\lambda|e^{-iK_B}W_B^{in}e^{+iK_B}|\mathbf{p}'\lambda\rangle\langle \mathbf{p}'\lambda|\mathbf{x}\lambda\rangle$$

$$= \int_{-\infty}^{+\infty} dt\, v_0 \sigma \sum_{\lambda} \int d^3p\, d^3p' (2\pi)^{-3} e^{i(\mathbf{p}-\mathbf{p}')\cdot\mathbf{x}} e^{-i(E_B - E'_B)t}$$

$$\times \langle \mathbf{p}\lambda|W_B^{in}|\mathbf{p}'\lambda\rangle$$

$$= (2\pi)^{-2} v_0 \sigma \sum_{\lambda\lambda'} \int d^3p\, d^3p'\, e^{i(\mathbf{p}-\mathbf{p}')\cdot\mathbf{x}} \delta(E_B - E'_B)$$

$$\times \langle \mathbf{p}\lambda|W_B^{in}|\mathbf{p}'\lambda'\rangle \delta_{\lambda\lambda'}$$

$$= (2\pi)^{-2} v_0 \sigma \sum_{\lambda\lambda'} \int \rho(E_B)\, dE_B\, \rho(E'_B)\, dE'_B\, d\Omega\, d\Omega'$$

$$\times e^{i(\mathbf{p}-\mathbf{p}')\cdot\mathbf{x}} \delta(E_B - E'_B)\langle E_B\Omega\lambda|W_B^{in}|E'_B\Omega'\lambda'\rangle \delta_{\lambda\lambda'}, \tag{5.24}$$

where

$$\mathbf{p} = p\mathbf{\Omega} = \sqrt{2mE_B}\,\mathbf{\Omega} \quad \text{and} \quad \mathbf{p}' = p'\mathbf{\Omega}' = \sqrt{2mE'_B}\,\mathbf{\Omega}'.$$

The calculation of the left-hand side of (5.22) goes as follows: We use (5.19) to do the $t$-integration, (5.20) to replace the factor in parentheses, and (5.21) to express things in terms of the state $W^{in}(t)$, which may be broken up by

---

[11] The result of this tedious calculation is Equation (5.38).

use of (5.4). The basis system $\{|a\rangle\}$ used in the calculation is that consisting of the generalized eigenvectors $|EE_T\Omega\lambda\eta\rangle = |E_B\Omega\lambda\rangle \otimes |E_T\eta\rangle$ with $E = E_a$. So

$$
\begin{aligned}
\text{LHS of} \atop (5.22) \quad &= -i\sum_b \sum_{aa'} 2\pi\delta(E_a - E_{a'})\langle b|V|a^+\rangle\langle a'^+|V|b\rangle \\
&\quad \times \langle a|W^{\text{in}}|a'\rangle\, 2\pi i\delta(E_a - E_b) \\
&= (2\pi)^2 \sum_b \sum_{\substack{E_T\lambda\eta \\ E'_T\lambda'\eta'}} \int \rho(E - E_T)\, dE\, \rho(E' - E'_T)\, dE'\, d\Omega\, d\Omega' \\
&\quad \times \delta(E - E')\delta(E - E_b)\langle b|V|EE_T\Omega\lambda\eta^+\rangle\langle E'E'_T\Omega'\lambda'\eta'^+|V|b\rangle \\
&\quad \times \langle EE_T\Omega\lambda\eta|W^{\text{in}}|E'E'_T\Omega'\lambda'\eta'\rangle \\
&= (2\pi)^2 \sum_b \sum_{\substack{E_T\lambda\eta \\ E'_T\lambda'\eta'}} \int \rho(E_B)\, dE_B\, \rho(E'_B)\, dE'_B\, d\Omega\, d\Omega' \\
&\quad \times (\delta E_B + E_T - E'_B - E'_T)\delta(E_B + E_T - E_b)\langle b|V|EE_T\Omega\lambda\eta^+\rangle \\
&\quad \times \langle E'E'_T\Omega'\lambda'\eta'^+|V|b\rangle\langle E_B\Omega\lambda|W_B^{\text{in}}|E'_b\Omega'\lambda'\rangle\langle E_T\eta|W_T^{\text{in}}|E'_T\lambda'\rangle
\end{aligned}
$$

$$(5.25)$$

where

$$E = E_B + E_T \quad \text{and} \quad E' = E'_B + E'_T.$$

From (5.24) and (5.25) we then have

$$
\begin{aligned}
0 &= (2\pi)^2(\text{RHS of (5.22)} - \text{LHS of (5.22)}) \\
&= \sum_{\lambda\lambda'} \int \rho(E_B)\, dE_B\, \rho(E'_B)\, dE'_B)\, d\Omega\, d\Omega'\, \langle E_B\Omega\lambda|W_B^{\text{in}}|E'_B\Omega'\lambda'\rangle \\
&\quad \times \left\{ v_0\sigma\delta(E_B - E'_B)e^{i(\mathbf{p}-\mathbf{p}')\cdot\mathbf{x}}\delta_{\lambda\lambda'} \right. \\
&\quad - (2\pi)^4 \sum_b \sum_{\substack{E_T\eta \\ E'_T\eta'}} \delta(E_B + E_T - E'_B - E'_T)\delta(E_B + E_T - E_b) \\
&\quad \left. \times \langle b|V|EE_T\Omega\lambda\eta^+\rangle\langle E'E'_T\Omega'\lambda'\eta'^+|V|b\rangle\langle E_T\eta|W_T^{\text{in}}|E'_T\eta'\rangle \right\}.
\end{aligned}
$$

$$(5.26)$$

We shall continue the calculation under the additional assumption that the beam is completely unpolarized,

$$\langle E_B\Omega\lambda|W_B^{\text{in}}|E'_B\Omega'\lambda'\rangle = \frac{1}{g}\delta_{\lambda\lambda'}\langle E_B\Omega\|W_B^{\text{in}}\|E'_B\Omega'\rangle, \tag{5.27}$$

where $g$ is the number of different polarization states $\lambda$. Were the beam completely polarized with some definite value $\lambda_0$ for $\lambda$, one would have to use

$$\langle E_B\Omega\lambda|W_B^{\text{in}}|E'_B\Omega'\lambda'\rangle = \delta_{\lambda\lambda'}\delta_{\lambda\lambda_0}\langle E_B\Omega\|W_B^{\text{in}}\|E'_B\Omega'\rangle \tag{5.28}$$

instead of (5.27) and would have to replace $\sum_\lambda 1/g$ by $\sum_\lambda \delta_{\lambda\lambda_0}$ in the calculations below.

In the typical scattering experiment, the incident beam is prepared so that it has a well-defined (momentum) direction $\Omega_0$. This condition is expressed by[12]

$$\int d\Omega\, d\Omega'\, \langle E_B\Omega\| W_B^{\mathrm{in}}\| E_B'\Omega'\rangle F(\Omega, \Omega') = \langle E_B\|| W_B^{\mathrm{in}}||| E_B'\rangle F(\Omega_0, \Omega_0) \quad (5.29a)$$

for any smooth function $F(\Omega, \Omega')$. The doubly reduced matrix element

$$\langle E_B\|| W_B^{\mathrm{in}}||| E_B'\rangle = \int d\Omega\, d\Omega' \langle E_B\Omega\| W_B^{\mathrm{in}}\| E_B'\Omega'\rangle$$

$$= \int d\Omega\, \langle E_B\Omega\| W_B^{\mathrm{in}}\| E_B'\Omega\rangle \quad (5.29b)$$

of the state $W_B^{\mathrm{in}}$ with well-defined direction $\Omega_0$ describes the energy distribution in the beam, which will be discussed in more detail below.

The normalization of $W_B^{\mathrm{in}}$ requires that

$$1 = \sum_\lambda \int \rho(E_B)\, dE_B\, d\Omega\, \langle E_B\Omega\lambda| W_B^{\mathrm{in}}| E_B\Omega\lambda\rangle$$

$$= \int \rho(E_B)\, dE_B\, d\Omega \langle E_B\Omega\| W_B^{\mathrm{in}}\| E_B\Omega\rangle$$

$$= \int \rho(E_B)\, dE_B \langle E_B\|| W_B^{\mathrm{in}}||| E_B\rangle. \quad (5.29c)$$

For an unpolarized beam (5.27) with well-defined direction (5.29a), Equation (5.26) becomes

$$0 = \int \rho(E_B)\, dE_B\, \rho(E_B')\, dE_B'\, \langle E_B\|| W_B^{\mathrm{in}}||| E_B'\rangle \Big\{ v_0\sigma\delta(E_B - E_B')e^{i(p-p')\Omega_0\cdot x}$$

$$- \frac{(2\pi)^4}{g} \sum_b \sum_\lambda \sum_{\substack{E_T\eta \\ E_T'\eta'}} \delta(E_B + E_T - E_B' - E_T')\delta(E_B + E_T - E_b)$$

$$\times \langle b| V| EE_T\Omega_0 \lambda\eta^+\rangle \langle E'E_T'\Omega_0 \lambda\eta'^+ | V| b\rangle \langle E_T\eta| W_T^{\mathrm{in}}| E_T'\eta'\rangle \Big\}.$$

---

[12] One can convince oneself that (5.29a) is the continuous-spectrum analogue of (5.28) by calculating for an arbitrary $F_{\lambda\lambda'}$

$$\sum_{\lambda\lambda'} \langle\cdots\lambda| W_B^{\mathrm{in}}|\cdots\lambda'\rangle F_{\lambda\lambda'} = \sum_{\lambda\lambda'} \delta_{\lambda\lambda'}\, \delta_{\lambda\lambda_0}\langle\cdots\| W_B^{\mathrm{in}}\|\cdots\rangle F_{\lambda\lambda'} = \langle\cdots\| W_B^{\mathrm{in}}\|\cdots\rangle F_{\lambda_0\lambda_0}.$$

Intuitively one may want to write as the direct analogue of (5.28)

$$\langle E_B\Omega\| W_B^{\mathrm{in}}\| E_B'\Omega'\rangle = \delta(\Omega - \Omega')\delta(\Omega - \Omega_0)\langle E_B\|| W_B^{\mathrm{in}}||| E_B'\rangle,$$

which, however, would be mathematically incorrect. Equation (5.29a) is an idealization; according to the discussions in Section II.10 every beam must have a finite spread in momentum. Cf. also the discussion following (5.35) below.

As $p = \sqrt{2mE_B}$, the exponential under the integral contributes only unity because of the $\delta(E_B - E_B')$. Hence

$$0 = \int \rho(E_B)\, dE_B\, \rho(E_B')\, dE_B'\, \langle E_B |\|W_B^{\mathrm{in}}\|\| E_B' \rangle \{v_0\sigma\delta(E_B - E_B')$$

$$- \frac{(2\pi)^4}{g} \sum_b \sum_\lambda \sum_{\substack{E_T\eta \\ E_T'\eta'}} \delta(E_B + E_T - E_B' - E_T')\delta(E_B + E_T - E_b)$$

$$\times \langle b|V|EE_T\Omega_0\lambda\eta^+\rangle\langle E'E_T'\Omega_0\lambda\eta'^+|V|b\rangle\langle E_T\eta|W_T^{\mathrm{in}}|E_T'\eta'\rangle\}. \quad (5.30)$$

Recall that we have left open the possibility that the target has continuous as well as discrete energy levels $E_T$; although we have written (5.30) as if the $E_T$ were discrete, it is also valid in the case (5.12) where there is a continuous part to the spectrum of $K_T$. In most experiments (although not for processes like collision-induced scattering) the state of the target system is in a mixture of discrete energy eigenstates (stationary states) before interacting with the beam:

$$[K, W_T^{\mathrm{in}}] = [K_T, W_T^{\mathrm{in}}] = 0. \quad (5.31)$$

Thus

$$W_T^{\mathrm{in}} = \sum_{E_T^d \eta\eta'} \langle \eta\|W_T^{\mathrm{in}}(E_T^d)\|\eta'\rangle |E_T^d\eta\rangle\langle E_T^d\eta'| \quad (5.31)$$

or

$$\langle E_T^d\eta|W_T^{\mathrm{in}}|E_T^{d\prime}\eta'\rangle = \delta_{E_T^d E_T^{d\prime}}\langle \eta\|W_T^{\mathrm{in}}(E_T^d)\|\eta'\rangle. \quad (5.32)$$

Then (5.30) becomes

$$0 = \int \rho(E_B)\, dE_B\, \rho(E_B')\, dE_B'\, \langle E_B |\|W_B^{\mathrm{in}}\|\| E_B' \rangle$$

$$\times \left\{ v_0\sigma\delta(E_B - E_B') - \frac{(2\pi)^4}{g} \sum_b \sum_\lambda \sum_{E_T^d\eta\eta'} \delta(E_B - E_B')\delta(E_B + E_T^d - E_b) \right.$$

$$\left. \times \langle b|V|EE_T^d\Omega_0\lambda\eta^+\rangle\langle E'E_T^d\Omega_0\lambda\eta'^+|V|b\rangle\langle \eta\|W_T^{\mathrm{in}}(E_T^d)\|\eta'\rangle \right\}$$

$$= \int \rho(E_B)\, dE_B\, \rho(E_B)\langle E_B |\|W_B^{\mathrm{in}}\|\| E_B \rangle$$

$$\times \left\{ v_0\sigma - \frac{(2\pi)^4}{g} \sum_b \sum_\lambda \sum_{E_T^d\eta\eta'} \delta(E_B + E_T^d - E_b) \right.$$

$$\left. \times \langle b|V|EE_T^d\Omega_0\eta\lambda^+\rangle\langle EE_T^d\Omega_0\lambda\eta'^+|V|b\rangle\langle \eta\|W_T^{\mathrm{in}}(E_T^D)\|\eta'\rangle \right\}, \quad (5.33)$$

where $E = E_B + E_T^d$ and $E' = E_B' + E_T^d$.

By taking the expectation value of the beam energy $K_B$ in the beam state $W_B^{\text{in}}$ of (5.27) and (5.29),

$$\text{Tr}(K_B W_B^{\text{in}}) = \sum_\lambda \int \rho(E_B) \, dE_B \, d\Omega \, \langle E_B \Omega \lambda | W_B^{\text{in}} K_B | E_B \Omega \lambda \rangle$$

$$= \int \rho(E_B) \, dE_B \, d\Omega \, E_B \langle E_B \Omega \| W_B^{\text{in}} \| E_B \Omega \rangle$$

$$= \int \rho(E_B) \, dE_B \, E_B \langle E_B \| | W_B^{\text{in}} \| | E_B \rangle, \tag{5.34}$$

we may identify

$$F(E_B - E_{B0}) = \rho(E_B) \langle E_B \| | W_B^{\text{in}} \| | E_B \rangle \tag{5.35}$$

as the probability density for obtaining the value $E_B$ when $K_B$ is measured. In writing (5.35) we have used a notation suitable for the assumption that, as is usually the case, the experimental setup is such as to produce a beam whose energy is peaked around some value $E_{B0}$ under the control of the experimenter. Then $F(E_B - E_{B0})$ is one of the functions of Section II.10 that in the unphysical limit case of sharp beam energy will go into $\delta(E_B - E_{B0})$. However, in the general case, the right-hand side of (5.35) could be any well-behaved function of $E_B$. If the resolution of the apparatus is good enough, i.e., if the beam is emitted with a small energy spread $E_{B0} - \Delta E_B < E_B < E_{B0} + \Delta E_B$, then $F(E_B - E_{B0})$ will tend to act like $\delta(E_B - E_{B0})$. More precisely, $F(E_B - E_{B0})$ will act like $\delta(E_B - E_{B0})$ in an integral

$$\int dE_B \, F(E_B - E_{B0}) g(E_B)$$

if the function $g(E_B)$ varies slowly over the range $E_{B0} - \Delta E_B < E_B < E_{B0} + \Delta E_B$. In (5.33) this will be the case if the matrix elements $\langle b | V | E E_T^d \Omega_0 \lambda \eta \rangle$, where $E = E_B + E_T^d$, vary slowly as functions of $E_B$.

This is not always true. Near some values of $E_{B0}$ the matrix elements may change very quickly; such a value $E_{B0}$ is called a *resonance*.

We shall discuss the effect of the finite resolution below and continue the calculation here under the assumption that the beam is ideally mono-chromatic, i.e., that

$$F(E_B - E_{B0}) = \rho(E_B) \langle E_B \| | W_B^{\text{in}} \| | E_B \rangle \to \delta(E_B - E_{B0}). \tag{5.36}$$

In this case the $E_B$ integration in (5.33) can be performed to get

$$0 = \rho(E_{B0}) \left\{ v_0 \sigma - \frac{(2\pi)^4}{g} \sum_b \sum_\lambda \sum_{E_T^d \eta \eta'} \delta(E_{B0} + E_T^d - E_b) \langle b | V | E E_T^d \Omega_0 \lambda \rho^+ \rangle \right.$$

$$\times \left. \langle E E_T^d \Omega_0 \lambda \eta'^+ | V | b \rangle \langle \eta \| W_T^{\text{in}}(E_T^d) \| \eta' \rangle \right\}, \tag{5.37}$$

where $E = E_{B0} + E_T^d$. Solving for $\sigma$ and reverting to rectangular coordinates $\mathbf{p}_0 = \sqrt{2mE_{B0}}\,\mathbf{\Omega}_0$, we at long last have our basic cross-section formula:

$$\sigma(\Lambda \to \mathbf{p}_0) = \frac{(2\pi)^4\hbar^2}{v_0} \sum_b \overline{\sum_\lambda} \sum_{E_T^d \cdot \eta\eta'} \langle b | V | \mathbf{p}_0 \lambda E_T^d \eta^+ \rangle \langle \mathbf{p}_0 \lambda E_T^d \eta'^+ | V | b \rangle$$

$$\times \langle \eta \| W_T^{\text{in}}(E_T^d) \| \eta' \rangle \delta(E_{B0} + E_T^d - E_b), \tag{5.38}$$

where $E_{B0} = p_0^2/2m$, $v_0 = p_0/m$, and where we have introduced the notation

$$\overline{\sum_\lambda} = \frac{1}{g} \sum_\lambda$$

for the averaging over the polarizations in the initial state. We have also restored the $\hbar$'s in order to express this important result in the usual units.

Equation (5.38) is the cross section for the scattering of an unpolarized beam with a well-defined momentum $\mathbf{p}_0$ off a fixed target in a mixture $W_T^{\text{in}}$ of discrete energy eigenstates and into any final configuration whose quantum numbers $b = (E_b, \hat{b})$ appear in the summation $\sum_b$. [To get the differential or partial cross section $d\sigma(b \leftarrow \mathbf{p}_0)$ for scattering into the "state" $\Lambda_b = |b\rangle\langle b|$, one merely omits the summation over $b$ and replaces $\sigma(\Lambda \leftarrow \mathbf{p}_0)$ by $d\sigma(b \leftarrow \mathbf{p}_0)$.]

Suppose that the energy spread $\Delta E_B$ is not negligible, as will be the case if $\langle b | V | EE_T^d \Omega_0 \lambda\eta^+ \rangle$ varies considerably when $E_B = E - E_T^d$ varies within the interval $\Delta E_B$ (this might happen in the neighborhood of a resonance unless $\Delta E_B$ is much smaller than the width of the resonance, as will be discussed in Chapter XVIII). Then the substitution (5.36) is not possible, and consequently one cannot calculate $\sigma(\Lambda \leftarrow \mathbf{p}_0)$ as in (5.37). The quantity one can calculate then from (5.33) is[13]

$$\int \rho(E_B)\, dE_B F(E_B - E_{B0}) v \sigma(E_B) = (\rho v \sigma) * F, \tag{5.39a}$$

where $v = v(E_B) = p(E_B)/m = \sqrt{2mE_B}/m$.

This, according to (5.33), is then equal to

$$(\rho v \sigma) * F = (2\pi)^4 \int \rho(E_B)\, dE_B\, F(E_B - E_{B0}) \sum_b \overline{\sum_\lambda} \sum_{E_T^d \cdot \eta\eta'} \langle b | V | \mathbf{p}\lambda E_T^d \eta^+ \rangle$$

$$\times \langle \mathbf{p}\lambda E_T^d \eta'^+ | V | b \rangle \langle \eta \| W_T^{\text{in}}(E_T^d) \| \eta' \rangle \delta(E_B + E_T^d - E_b). \tag{5.39b}$$

As $\rho(E_B)$ and $v(E_B)$ are slowly varying functions of $E_B$ compared to the rapidly varying functions $F(E_B - E_{B0})$ and $|\langle b | V | EE_T^d \Omega_0 \lambda\eta^+ \rangle|^2$, one can take them out of the integrals on the left-hand side of (5.39a) and on the right-hand side of (5.39b) and replace them by their average values, $\rho(E_{B0})$ and

---

[13] If the spread in velocity is not negligible, then the incident probability per unit area is $\int dz\, \langle \mathbf{x} | W_B^{\text{in}}(t) | \mathbf{x} \rangle = \int dt\, \langle \mathbf{x} | \mathscr{V} W_B^{\text{in}}(t) | \mathbf{x} \rangle$ instead of $\int dt\, v_0 \langle \mathbf{x} | W_B^{\text{in}}(t) | \mathbf{x} \rangle$, where $\mathscr{V}$ is the velocity operator. This results in a replacement of $v_0$ by $v(E_B)$ in (5.33).

$v(E_{B0})$. Then one obtains the cross section formula for coarse-resolution experiments:

$$\sigma * F = \int dE_B \, F(E_B - E_{B0})\sigma(E_B) = (2\pi)^4\hbar^2 \frac{1}{v(E_{B0})} \int dE_B \, F(E_B - E_{B0})$$

$$\times \sum_b \overline{\sum_\lambda} \sum_{E_T^d \eta\eta'} \delta(E_B + E_T^d - E_b)\langle b|V|\mathbf{p}\lambda E_\lambda^d \eta^+\rangle$$

$$\times \langle \mathbf{p}\lambda E_T^d \eta'^+|V|b\rangle\langle\eta\|W_T^{\mathrm{in}}(E_T^d)\|\eta'\rangle. \tag{5.40}$$

We continue now with the expression (5.38) for the idealized energy resolution, but will return to the case of a limited energy resolution in Section XVIII.8.

The cross section $\sigma(\Lambda \leftarrow \mathbf{p}_0)$ is often expressed in terms of the $T$-*matrix*. If the interaction Hamiltonian $V$ is given, the $T$-matrix may be defined by

$$\langle E\hat{a}|T|E\hat{a}'\rangle \equiv \langle E\hat{a}|V|E\hat{a}'^+\rangle. \tag{5.41}$$

This does not fully define a transition operator $T$, because not all matrix elements $\langle E\hat{a}|T|E'\hat{a}'\rangle$ of $T$ are defined by (5.41), but only those matrix elements "on the energy shell" $E = E'$. The matrix elements $\langle E\hat{a}|T|E'\hat{a}'\rangle$ that are "off the energy shell," i.e., those for which $E \neq E'$, may be defined in various ways; and one gets differing transition operators depending on how these off-energy-shell matrix elements are taken. One such operator is $T^+$, defined by

$$\langle E\hat{a}|T^+|E'\hat{a}'\rangle \equiv \langle E\hat{a}|V|E'\hat{a}'^+\rangle. \tag{5.42}$$

Obviously the restriction of (5.34) to the energy shell gives the quantities $\langle E\hat{a}|T|E\hat{a}'\rangle$ of (5.41). Another transition operator $T^-$ is defined by

$$\langle E\hat{a}|T^-|E'\hat{a}'\rangle \equiv \langle E\hat{a}|V|E'\hat{a}'^-\rangle. \tag{5.43}$$

Although not obviously, this too agrees with (5.41) when restricted to the energy shell. Any useful definition of a transition operator $T^\epsilon$ must agree with (5.41) when on the energy shell:

$$\langle E\hat{a}|T^\epsilon|E\hat{a}\rangle = \langle E\hat{a}|T|E\hat{a}'\rangle \tag{5.44}$$

Because of the presence of the factor $\delta(E_{B0} + E_T - E_b)$ (which expresses energy conservation) in (5.38), it is only the on-energy-shell $T$-matrix elements that will contribute, a situation one always encounters when dealing with physically observable quantities. Since all transition operators agree on the energy shell, it makes no difference which $T^\epsilon$ is used.

In the $S$-matrix approach mentioned in the previous section, it is not the existence of an interaction Hamiltonian V that is assumed, but rather that of the $T$-matrix $\langle E\hat{a}|T|E\hat{a}'\rangle$, which is considered the fundamental quantity. The expression of the cross section formula in terms of the $T$-matrix,

$$\sigma(\Lambda \leftarrow \mathbf{p}_0) = \frac{(2\pi)^4\hbar^2}{v_0} \sum_b \overline{\sum_\lambda} \sum_{E_T^d \eta\eta'} \delta(E_{B0} + E_T^d - E_b)$$

$$\times \langle b|T|\mathbf{p}_0\lambda E_T^d\eta\rangle\langle\mathbf{p}_0\lambda E_T^d\eta'|T|b\rangle\langle\eta\|W_T^{\mathrm{in}}(E_T^d)\|\eta'\rangle \tag{5.38'}$$

may therefore be considered valid in either approach.[14] If a $V$ exists, then $\langle E\hat{b}|T|E\hat{a}'\rangle$ is defined in terms of (5.41) and

$$\langle E\hat{b}|T|E\hat{a}'\rangle = \sum_{\hat{a}} \langle E\hat{b}|E\hat{a}\rangle\langle E\hat{a}|T|E\hat{a}'\rangle; \qquad (5.45)$$

otherwise it is the $T$ matrix $\langle E\hat{a}|T|E\hat{a}'\rangle$ that is fundamental and which cannot be further determined.

In order to calculate the cross section from (5.38) or from (5.38'), one has to know both $V$ and $|\mathbf{p}_0 \lambda E_T \eta^+\rangle$ or the matrix $\langle E\hat{b}|T|\mathbf{p}_0 \lambda E_T^d \lambda\rangle$, respectively. Even if $V$ is considered to be known one still requires knowledge of the $|\mathbf{p}_0 \lambda E_T^d \eta^+\rangle = |E_0 E_T^d \Omega_0 \lambda \eta^+\rangle \equiv |E_0 \hat{a}^+\rangle$ $(E_0 = p_0^2/2m + E_T^d)$, which are solutions of the Lippman–Schwinger integral equation. In principle the $|E_0 \hat{a}^+\rangle$ may be obtained by an interation process similar to the one described in Section VIII.1. The approximation that one obtains for the lowest order of the iteration process, i.e., the approximation

$$\langle b|T^+|E_0 \hat{a}\rangle = \langle b|V|E_0 \hat{a}^+\rangle \approx \langle b|V|E_0 \hat{a}\rangle, \qquad (5.46)$$

is called the *Born approximation*, and is usually sufficient in the case of a high initial beam energy $E_{B0}$ and a weak interaction $V$.

Equation (5.38) [or (5.38')] gives the cross section for the rather general situation in which the target state is left unspecified except for the requirement that it be a mixture of discrete energy eigenstates. We shall now give the cross section for more specific target states. We first assume that $W_T$ is diagonal in some of the quantum numbers $\tilde{\eta}$ of $\eta = (\tilde{\eta}, \bar{\eta})$ and that no measurement has been made with respect to the other quantum numbers $\bar{\eta}$. For example, if the target consists of hydrogen or alkali atoms, the internal quantum numbers $\eta$ are $\eta = (n, j, j_3, \pi)$, and $E_T = E_T(nj\pi)$ is a function of $n$, $j$, and $\pi$. Usually the atom is in a specified mixture of energy and angular-momentum and parity eigenstates, while the $z$-component of angular momentum has not been measured. Then

$$\langle njj_3\pi\| W_T^{\text{in}}(E_T)\|nj'j_3'\pi'\rangle = \frac{1}{2j+1}\,\delta_{\pi\pi'}\delta_{j_3j_3'}\delta_{jj'}\,W_T^{\text{in}}(E_T(nj,\pi))\delta_{nn'}.$$

Similarly, in the general situation

$$\langle\bar{\eta}\tilde{\eta}\| W_T^{\text{in}}(E_T^d)\|\tilde{\eta}'\bar{\eta}'\rangle = \frac{1}{g}\,\delta_{\bar{\eta}\bar{\eta}'}\,\delta_{\tilde{\eta}\tilde{\eta}'}\,W_T^{\text{in}}(E_T^d, \tilde{\eta}), \qquad (5.47)$$

---

[14] Strictly speaking, the matrix elements of an operator are the numbers that result from placing the operator between two vectors from the *same* basis. Since the $\hat{b}$ and $\hat{a}$ may refer to eigenvalues of different operators, $\langle E\hat{b}|T|E\hat{a}\rangle$ need not be a matrix element, but it is a linear combination of matrix elements. For convenience, however, we shall hereafter refer to $\langle E\hat{b}|T|E\hat{a}\rangle$ as a $T$-matrix element.

where $g$ is the number of values $\bar{\eta}$ can take on. The cross section obtained in the case (5.47) is

$$\sigma(\Lambda \leftarrow \mathbf{p}_0, \tilde{\eta}) = \frac{(2\pi)^4 \hbar^2 m}{p_0} \sum_b \overline{\sum_\lambda} \sum_{E_T^d \bar{\eta}} \overline{\sum_{\bar{\eta}}} |\langle b|T|\mathbf{p}_0 \lambda E_T^d \tilde{\eta}\bar{\eta}\rangle|^2$$

$$\times W_T^{\text{in}}(E_T^d, \tilde{\eta})\delta(E_{B0} + E_T^d - E_b), \tag{5.48}$$

where we use the notation $\overline{\sum}_\lambda = (1/g)\sum_\lambda$ and $\overline{\sum}_\eta = (1/g)\sum_\eta$ (averaging).

The weight $W(E_T^d \tilde{\eta})$ of each energy level may depend upon all the quantum numbers $\tilde{\eta} E_T^d = (\eta_1 \eta_2 \cdots \eta_\kappa, E_T^d)$ or only upon a subset of them, e.g., $\eta_1, E_T^d$. Very often it depends only upon $E_T^d$: $W(E_T^d \tilde{\eta}) = W(E_T^d)$. For example, if the target system is in thermal equilibrium, then $W_T$ is given by (II.4.50) and $W(E_T^d)$ is given by a Gibbs distribution

$$W(E_n) = \frac{e^{-E_n/kT}}{\displaystyle\sum_{\tilde{\eta} E_n} e^{-E_n/kT}} = \frac{e^{-E_n/kT}}{\displaystyle\sum_{E_n} (\dim \mathcal{H}_n)e^{-E_n/kT}}$$

where $\dim \mathcal{H}_n$ is the dimension of the energy eigenspace with eigenvalue $E_n$.

A very special but rather common situation is the case in which the target is in a pure energy eigenstate, e.g., the ground state of an atom specified by the quantum numbers $(E_T^0, \eta_0)$:

$$\langle \eta \| W_T^{\text{in}}(E_T^d) \| \eta' \rangle = \delta_{\eta\eta_0}\delta_{\eta_0\eta'}\delta_{E_T^d E_T^0}, \tag{5.49}$$

or

$$W_T^{\text{in}} = |E_T^0 \eta_0\rangle\langle E_T^0 \eta_0|. \tag{5.49'}$$

The cross section (5.38) for scattering of an unpolarized beam off such targets then becomes

$$\sigma(\Lambda \leftarrow \mathbf{p}_0 E_T^0 \eta_0) = \frac{(2\pi)^4 \hbar^2 m}{p_0} \sum_b \overline{\sum_\lambda} |\langle b|V|\mathbf{p}_0 \lambda E_T^0 \eta_0^+\rangle|^2 \delta(E_0 - E_b)$$

$$= \frac{(2\pi)^4 \hbar^2 m}{p_0} \sum_b \overline{\sum_\lambda} |\langle b|T|\mathbf{p}_0 \lambda E_T^0 \eta_0\rangle|^2 \delta(E_0 - E_b), \tag{5.50}$$

where $E_0 = E_{B0} + E_T^0$. In the Born approximation and for structureless projectiles, this reduces to the well-known formula

$$\sigma(\Lambda \leftarrow \mathbf{p}_0 E_T^0 \eta) = \frac{(2\pi)^4 \hbar^2 m}{p_0} \sum_b |\langle b|V|\mathbf{p}_0 E_T^0 \eta_0\rangle|^2 \delta(E_{B0} + E_T^0 - E_b). \tag{5.51}$$

Usually, the transition matrix is independent of the polarization $\lambda$. Then the averaging over the polarizations in (5.50) is trivial, and the quantum numbers $\lambda$ could just as well be omitted. For observables that do not depend upon the polarization, an unpolarized beam can be treated like a beam of structureless projectiles; e.g., for an unpolarized electron beam one can neglect the spin of the electron.

The most common experiments are those in which the differential cross section for scattering into a particular solid angle $\Delta\Omega_{D0}$ is measured by placing a counter at a particular angle $\Omega_{D0}$ to the incident direction. $\Lambda\mathscr{H}$ is then the subspace of states with momentum pointing in any of the directions $\Omega_{D0} \pm \Delta\Omega_{D0}$. If the detector detects only those states within a certain energy range, then $\Lambda\mathscr{H}$ is further restricted.

So far we have not specified the basis system $|b\rangle = |E\hat{b}\rangle$ in the space of final states. In order to obtain the differential cross section for scattering into a particular direction specified by the angles $\Omega = (\theta, \phi)$, one conveniently chooses a basis system of generalized eigenvectors labeled by $\Omega$ or by the momentum $\mathbf{k}$ of the detected (scattered) particle. Thus we choose

$$|b\rangle = |E\hat{b}\rangle = |\mathbf{k}\xi\rangle \otimes |\epsilon\zeta\rangle = |E_D\Omega_D\xi\rangle \otimes |\epsilon\zeta\rangle$$
$$= |E_D\epsilon\Omega_D\xi\zeta\rangle. \tag{5.52}$$

where $E = E_D + \epsilon$ and $\mathbf{k} = k\Omega$. $\xi$ are the internal quantum numbers of the detected particle (polarizations) and $\epsilon$, $\zeta$ are the internal energy and other internal quantum numbers of the target. Thus our description is general enough to include the case where the detected particle may be of a different kind than that of the incoming beam, and the particle left behind (post-collision target) may be different from those the target originally consisted of. For example, a photon $\gamma$ might be incident on an atom $A$ and ionize it, leaving behind an ion $A^+$ with an electron coming off: $\gamma + A \rightarrow e + A^+$. A simpler example for the reader to keep in mind is the elastic or inelastic scattering of an electron or photon beam by an atom: $e + A \rightarrow e' + A^*$.

If the particles detected are nonrelativistic, then

$$E_D = k^2/2m_D + E_D^{\text{int}}(\xi), \tag{5.53a}$$

while if the particles detected are photons,

$$E_D = kc. \tag{5.53b}$$

We shall impose the same normalizations as in (5.7) and (5.8) on the $|\mathbf{k}\xi\rangle$ and $|\epsilon\zeta\rangle$:

$$\langle \mathbf{k}\xi | \mathbf{k}'\xi'\rangle = \delta^3(\mathbf{k} - \mathbf{k}')\delta_{\xi\xi'} \tag{5.54}$$

and

$$\langle \epsilon\zeta | \epsilon'\zeta'\rangle = \delta_{\epsilon\epsilon'}\delta_{\zeta\zeta'}. \tag{5.55}$$

Then in the case (5.53a) we have by derivations similar to those of (5.15) and (5.16) that

$$\sum_b = \sum_b \int \bar{\rho}_{\hat{b}}(E)\, dE = \sum_{\zeta\epsilon\xi} \int m_D\sqrt{2m_D(E - E_D^{\text{int}} - \varepsilon)}\, dE\, d\Omega_D$$
$$= \sum_{\xi\epsilon\zeta} \bar{\rho}_{\xi\epsilon}(E)\, dE\, d\Omega_D \tag{5.56}$$

and

$$\langle E\epsilon\Omega_D\xi\zeta | E'\epsilon'\Omega_D'\xi'\zeta'\rangle$$
$$= m_D^{-1}(2m_D(E - E_D^{\text{int}}(\xi) - \epsilon))^{-1/2}\delta(E - E')\delta_{\epsilon\epsilon'}\delta^2(\Omega_D - \Omega_D')\delta_{\xi\xi'}\delta_{\zeta\zeta'} \tag{5.57}$$

Equation (5.50) is then written in detail as

$$\sigma(\Lambda \leftarrow \mathbf{p}_0 E_T^0 \eta_0) = \frac{(2\pi)^4 \hbar^2 m_B m_D}{p_0} \sum_{\lambda} \sum_{\xi'\epsilon'\zeta'} \int dE' d\Omega'_D k'(E', \epsilon', \xi')$$
$$\times |\langle E'\epsilon'\Omega'_D \xi'\zeta' | T | \mathbf{p}_0 \lambda E_T^0 \eta_0 \rangle|^2 \delta(E_0 - E'), \quad (5.58)$$

where

$$E^0 = p_0^2/2m_B + E_T^0 \quad \text{(initial energy)}$$
$$k' = \sqrt{2m_D(E' - E_D^{\text{int}}(\xi') - \epsilon')} \quad \text{(momentum of the detected particle)}$$

The ranges of the summation over $\xi'$, $\epsilon'$, $\zeta'$ and the integration over $E'$, $\Omega'_D$ depend on the nature of the detection apparatus specified by $\Lambda$. From (5.58) we see that the differential cross section per unit solid angle for scattering in the direction $\Omega_D$ is

$$\frac{d\sigma}{d\Omega_D}(\Omega_D \leftarrow \mathbf{p}_0 E_T^0 \eta_0) = \frac{(2\pi)^4 \hbar^2 m_B m_D}{p_0} \sum_{\lambda} \sum_{\xi'\epsilon'\zeta'} \int dE' k'(E', \epsilon', \xi')$$
$$\times |\langle E'\epsilon'\Omega_D \xi'\zeta' | T | \mathbf{p}_0 \lambda E_T^0 \eta_0 \rangle|^2 \delta(E_0 - E'). \quad (5.59)$$

We shall now assume that the projectile and the detected particle are the same ($m_B = m_D = m$) and that it is structureless or that the polarization quantum numbers $\lambda$ and $\xi$ can be ignored.[15] $\eta_0$ and $\zeta$ are then eigenvalues of the same set of operators $\eta^{\text{op}}$. The internal energies $E_T^0$ or $\epsilon'$, which are eigenvalues of the same operator $K_T$, are assumed to be already determined by the internal quantum numbers, i.e., $K_T = K_T(\eta^{\text{op}})$ as is the usual convention for targets with discrete energy spectrum.[16] Then the expression for (5.59) simplifies to

$$\frac{d\sigma}{d\Omega}(\Omega \leftarrow \mathbf{p}_0 \eta_0) = \frac{(2\pi)^4 \hbar^2 m^2}{p_0} \sum_{\zeta'} \int dE' k'(E', \zeta') \delta(E_0 - E')$$
$$\times |\langle E'\Omega, \zeta' | T | \mathbf{p}_0 \eta_0 \rangle|^2, \quad (5.60)$$

where

$$k' = \sqrt{2m(E' - E_T(\zeta'))} \quad \text{and} \quad E_0 = p_0^2/2m + E_T(\eta_0).$$

If not only the initial total energy is fixed (by fixing $\eta_0$ and the beam momentum $p_0$) but also the final total energy (by choosing a detector that registers only a particular momentum $k$), and if also the final internal quantum numbers are fixed (by triggering the detector only if the final internal quantum numbers have the value $\zeta$), then the differential cross section is obtained from (5.60) as

$$\frac{d\sigma}{d\Omega}(\Omega E\zeta \leftarrow \mathbf{p}_0 \eta_0) = \frac{(2\pi)^4 \hbar^2 m^2 k}{p_0} |\langle E\Omega\zeta | T | \mathbf{p}_0 \eta_0 \rangle|^2, \quad (5.61)$$

[15] One could also imagine that the polarization $\lambda$ is incorporated in the quantum numbers $\eta$.
[16] If the target is the Hydrogen atom, then $\eta = (n, j, j_3, \pi)$ and $E_T = E_T = E_T(n)$ given by (VI.5.12).

where

$$k = \sqrt{2m(E - E_T(\zeta))},$$

which must be equal to

$$\sqrt{2m\left(\frac{p_0^2}{2m} + E_T(\eta_0) - E_T(\zeta)\right)}$$

because of the energy-conservation $\delta$-function $\delta(E_0 - E)$. For the case of elastic scattering, i.e., when the internal quantum numbers are not changed ($\eta_0 = \zeta$), one obtains the well-known expression

$$\frac{d\sigma}{d\Omega}(\Omega E_0 \eta_0 \leftarrow \mathbf{p}_0 \eta_0) = (2\pi)^4 \hbar^2 m^2 |\langle E_0 \Omega \eta_0 | T | \mathbf{p}_0 \eta_0 \rangle|^2. \qquad (5.62)$$

The generalized eigenvectors in (5.62) [and also in the preceding expressions (5.61) and (5.60)] are normalized by (5.54) and (5.7) according to

$$\langle E\Omega\eta | E'\Omega'\eta' \rangle = \langle \mathbf{k}\eta | \mathbf{k}'\eta' \rangle = \delta_{\eta\eta'} \delta^3(\mathbf{k} - \mathbf{k}').$$

If one uses generalized eigenvectors with a different normalization, then this expression changes correspondingly.

For spherically symmetric interaction, the $T$-matrix and (therefore the differential cross section) will not depend upon the angle $\phi$ around the incident direction $\mathbf{p}_0/p_0$, but only upon the angle $\theta$ between the incident direction and the final direction $\Omega = \mathbf{k}/k$ (cf. Figure 3.1). (A derivation of this statement will be given in Chapter XVI.) $\theta$ is called the scattering angle. In this case one often uses instead of the $T$-matrix element the elastic-scattering amplitude $T(p_0, \theta)$ defined by

$$T(p_0, \theta) = -4\pi^2 m \langle E_0 \Omega, \eta_0 | T | \mathbf{p}_0 \eta_0 \rangle. \qquad (5.63)$$

The differential cross section (5.62) is then written as

$$\frac{d\sigma}{d\Omega}(E_0, \theta) = |T(p_0, \theta)|^2. \qquad (5.64)$$

## Problems

1.  The quantity $\mathbf{q} = \mathbf{k} - \mathbf{p}_0$, where $\mathbf{p}$ is the projectile momentum before scattering and $\mathbf{k}$ is the projectile momentum after scattering, is called momentum transfer. (It represents the momentum that is transferred to the projectile by the target.)

    (a) Show that the $T$-matrix $\langle \mathbf{k}, \eta_0 | T | \mathbf{p}_0, \eta_0 \rangle$ in the Born approximation is a function of the momentum transfer only and does not depend upon the momenta $\mathbf{p}_0$ and $\mathbf{k}$ separately if the interaction Hamiltonian $V$ is a function of the projectile position.

    (b) Express the magnitude of the momentum transfer in terms of *scattering momentum and scattering angle* (for nonrelativistic particles), and show that in the Born approximation the forward $T$-matrix element $\langle \mathbf{p}_0, \eta_0 | T | \mathbf{p}_0, \eta_0 \rangle$ is independent of scattering energy if the above interaction Hamiltonian is spherically symmetric: $[V, L] = 0$.

(c)   Show that at high energies (when the Born approximation is usually good), the scattering falls off as the scattering angle $\theta$ increases from 0 to $\pi$.

2.   Obtain the expression for the differential cross section (5.62) in terms of the $T$-matrix

$$(E\Omega|T|E\Omega')$$

defined with the generalized eigenvectors $|E\Omega\rangle$, which are normalized according to

$$(E\Omega|E'\Omega') = \delta(E - E')\delta(\Omega - \Omega'),$$

where the solid angle $\delta$-function $\delta(\Omega - \Omega')$ is defined by

$$\int d\Omega\, f(\Omega)\delta(\Omega - \Omega_0) = f(\Omega_0)$$

with $d\Omega = \sin\theta\, d\theta\, d\phi$.

3.   The Yukawa interaction was introduced to describe nuclear interactions and led to the prediction of the meson. It is described by the potential

$$V(r) = ge^{-\mu r}/r$$

and can also serve as a simple model for the screened Coulomb field of an atom.

(a)   Calculate the scattering amplitude $T$ and the differential cross section in Born approximation.

(b)   The second-order term in the Born series of the scattering amplitude is given by [cf. (XV.3.38)]

$$T^{(2)}(\mathbf{k} \leftarrow \mathbf{p}_0) = -4\pi^2 m\langle \mathbf{k}\eta_0|T|\mathbf{p}_0\eta_0\rangle$$

$$= -4\pi^2 m \int d^3p' \,\frac{\langle \mathbf{k}|V|\mathbf{p}'\rangle\langle \mathbf{p}'|V|\mathbf{p}_0\rangle}{E_{p_0} - E_{p'} + i0}$$

Calculate the forward-scattering amplitude, i.e., the scattering amplitude for $\mathbf{p}_0 = \mathbf{k}$, up to the second order of the Born series, and find the values of the parameters $m$, $\mu$, and $g$, and the energies for which the necessary condition for the validity of the Born approximation, $|T^{(2)}| \ll |T^{(1)}|$, is fulfilled.

# Formal Scattering Theory and Other Theoretical Considerations

Notions used in Chapter XIV for the derivation of the cross-section formula will be further discussed in this chapter to provide a deeper understanding of the material. In Section XV.1 the Lippman–Schwinger equation is discussed again. Sections XV.2 and 3 are on formal scattering theory. In the Appendix a derivation of (XIV.2.9b), (XIV.2.17), and (XIV.5.21) is given, using the material developed in this chapter.

## XV.1 The Lippman–Schwinger Equation

The Lippman–Schwinger equation

$$|a^{\pm}\rangle = |a\rangle + \frac{1}{E_a - K \pm i0} V|a^{\pm}\rangle \qquad (1.1\pm)$$

was introduced in Section VIII.2. To each generalized eigenvector $|a\rangle$ of $K$ with eigenvalue $E_a$ the Lippman–Schwinger equation relates a generalized eigenvector $|a^+\rangle$ or $|a^-\rangle$ of $H = K + V$ with the same eigenvalue $E_a$. (In Section VIII.2 it was assumed that the continuous parts of the spectra of $H$ and of $K$ were identical and that $E_a$ was an element of these continuous spectra.) The generalized eigenvectors $|a\rangle$ of $K$ and $|a^{\pm}\rangle$ of $H$ are labeled by the energy $E_a$ together with some additional labels $\hat{a} = (a_1, a_2, \ldots, a_k)$:

$$|a\rangle = |E_a\hat{a}\rangle, \qquad (1.2a)$$

$$|a^{\pm}\rangle = |E_a\hat{a}^{\pm}\rangle. \qquad (1.2b)$$

In Section VIII.1 it was assumed that the $\hat{a}$ were additional quantum numbers for the $K$ basis, i.e., that

$$\{K, A_1, A_2, \ldots, A_k\} \tag{1.3}$$

was a complete system of commuting observables (c.s.c.o.). Suppose that

$$\{H, A_1, A_2, \ldots, A_k\} \tag{1.4}$$

is also a c.s.c.o. Both the generalized eigenvectors $|a\rangle$ of (1.3) and the proper and generalized eigenvectors $|a_n\rangle$ and $|a^+\rangle$ of (1.4) are a basic of the space of physical states.

The assumption that (1.4) is a c.s.c.o. was not used in the derivation of the Lippman–Schwinger equation. Suppose that

$$\{K, B_1, B_2, \ldots, B_k\} \tag{1.5}$$

is a c.s.c.o. but that $\{H, B_1, B_2, \ldots, B_k\}$ is not, i.e., that

$$[B_i, H] = [B_i, V] \neq 0 \tag{1.6}$$

for some or all of the $i$'s ($i = 1, 2, \ldots, k$). By repeating the arguments of Section VIII.2, the basis vectors

$$|b\rangle = |E_b\hat{b}\rangle \tag{1.7}$$

of $K$ may be related by way of the Lippman–Schwinger equation to the generalized eigenvectors

$$|b^\pm\rangle \equiv |E_b(\hat{b})^\pm\rangle = |E_b\hat{b}\rangle + \frac{1}{E_b - K \pm i0} V |E_b(\hat{b})^\pm\rangle \tag{1.8}$$

of $H$. However, the vectors $|E_b(\hat{b})\rangle$ are generally not eigenvectors of the operators $B_i$, because of (1.6). In this case the $b_i$'s are defined not by a c.s.c.o. containing $H$, but only through the Lippman–Schwinger equation and the c.s.c.o. containing $K$. [The parentheses around the label $\hat{b}$ are a temporary notation used to indicate that $\hat{b}$ has a different status—that of a label only— than that of the label $E_b$, which indicates that $|E_b(\hat{b})^\pm\rangle$ is an eigenvector of $H$ corresponding to the eigenvalue $E_b$. Later we shall not make this distinction.]

As an example of these possibilities consider the scattering of structureless particles off a fixed target. Let $\mathbf{Q}$ and $\mathbf{P}$ be the (canonically conjugate) position and momentum operators for the projectile, and assume the interaction of the projectile with the fixed target is given by a spherically symmetric potential $V = V(Q)$ $[Q = (\mathbf{Q}^2)^{1/2}]$; the Hamiltonian for the system is then $H = K + V$, where $K = \mathbf{P}^2/2m$ is the kinetic-energy operator for the projectile. Since $[L_i, V] = [L_i, K] = 0$, it follows that

$$\{K, \mathbf{L}^2, L_3\} \tag{1.9a}$$

*and*

$$\{H, \mathbf{L}^2, L_3\} \tag{1.9b}$$

are both c.s.c.o.'s and that we have the case (1.4). The $|Ell_3\rangle$ and the $|Ell_3^\pm\rangle$ are corresponding generalized eigenvectors. $\{|Ell_3\rangle\}$ is a basis for the entire space of physical states, but $\{|Ell_3^\pm\rangle\}$ is a basis for the "subspace of scattering states" only. The projectiles in scattering experiments are usually prepared as collimated beams going in a specific direction, and it is therefore more convenient to use for the $K$-basis the generalized eigenvectors

$$|b\rangle = |\mathbf{p}\rangle = |Ep_1p_2\rangle = |E\theta\phi\rangle \qquad (1.10)$$

of the momentum operators $P_i$. The c.s.c.o. of which the vectors (1.10) are eigenvectors is

$$\{P_1, P_2, P_3\} \quad \text{or} \quad \{K, P_1, P_2\} \quad \text{or} \quad \{K, \text{directions of } \mathbf{P}\}; \quad (1.11)$$

we then have the case (1.5) with (1.6), because

$$[P_i, H] = [P_i, V] \neq 0.$$

The label $\mathbf{p}$ for the $H$ eigenvectors

$$|(b)^\pm\rangle = |(\mathbf{p})^\pm\rangle \qquad (1.12)$$

does not then mean that the $|(\mathbf{p})^\pm\rangle$ are eigenvectors of the $P_i$ but only means that the $|(\mathbf{p})^\pm\rangle$ are connected to the momentum eigenvectors $|\mathbf{p}\rangle$ by the Lippman–Schwinger equation,

$$|(\mathbf{p})^\pm\rangle = |\mathbf{p}\rangle + \frac{1}{E - K \pm i0} V|(\mathbf{p})^\pm\rangle \qquad (E = \mathbf{p}^2/2m). \qquad (1.13)$$

The assumption that (1.3) is a c.s.c.o. for a particular physical system can only be justified by physical motivation. It is equivalent to the assumption that the set

$$\{|E\hat{a}\rangle : (E, \hat{a}) \in \text{spectrum } (K, A_1, \ldots, A_k)\} \qquad (1.14)$$

of generalized eigenvectors of (1.3) forms a basis for the space of physical states $\mathcal{H}$ of the physical system. Although the set

$$\{|E\hat{a}\rangle_H : (E, \hat{a}) \in \text{spectrum } (H, A_1, \ldots, A_k)\} \qquad (1.15)$$

of generalized eigenvectors of (1.4) is also a basis for $\mathcal{H}$, the sets

$$\{|E\hat{a}^+\rangle : (E, \hat{a}) \in \text{spectrum } (K, A_1, \ldots, A_k)\} \qquad (1.15+)$$

and

$$\{|E\hat{a}^-\rangle : (E, \hat{a}) \in \text{spectrum } (K, A_1, \ldots, A_k)\} \qquad (1.16-)$$

of generalized eigenvectors of (1.4), which are obtained through the Lippman–Schwinger equations $(1.1+)$ and $(1.1-)$, are not generally bases for $\mathcal{H}$, nor are the sets

$$\{|E(\hat{a})^\pm\rangle : (E, \hat{a}) \in \text{spectrum } (K, A_1, \ldots, A_k)\} \qquad (1.17\pm)$$

in the case that (1.4) is not a c.s.c.o. Usually the operator $K$ has only a continuous spectrum, but $H$ has in addition to this (same) continuous spectrum a discrete spectrum $\{E_n\}$ as well. The proper eigenvectors $|E_n\hat{a}\rangle$ with energy eigenvalues $E_n$ in the discrete spectrum of $H$ describe pure physical states corresponding to the bound states of the projectile-target system, and span a *subspace of bound states* $\mathcal{H}_{bnd}$. The generalized eigenvector $|E\hat{a}^+\rangle$ or $|E\hat{a}^-\rangle$ of $H$ with energy eigenvalues $E$ in the continuous spectrum span a *subspace of scattering states* $\mathcal{H}_{scat}$, which is the orthogonal complement[1] of $\mathcal{H}_{bnd}$,

$$\mathcal{H} = \mathcal{H}_{bnd} \oplus \mathcal{H}_{scat}. \tag{1.18}$$

The subspaces $\mathcal{H}_+$ and $\mathcal{H}_-$ that are spanned by the sets $(1.15+)$ and $(1.15-)$ need not necessarily agree with $\mathcal{H}_{scat}$ or with each other. Under very mild conditions on the inveraction $V$, which are generally fulfilled by physical scattering systems, the subspaces $\mathcal{H}_+$, $\mathcal{H}_-$, and $\mathcal{H}_{scat}$ agree:

$$\mathcal{H}_+ = \mathcal{H}_- = \mathcal{H}_{scat} \tag{1.19}$$

The system for which $(1.19)$ is fulfilled is said to be *asymptotically complete*.

Under very general assumptions we thus have the following situation: The space of physical states $\mathcal{H}$ is spanned by the set of $\{|E\hat{a}\rangle\}$ of generalized eigenvectors of $K$, and is the direct sum $\mathcal{H} = \mathcal{H}_{bnd} \oplus \mathcal{H}_{scat}$ of the space $\mathcal{H}_{bnd}$ spanned by the set $\{|E_n\hat{a}\rangle\}$ of proper eigenvectors of $K$ and the space $\mathcal{H}_{scat}$ spanned by either of the sets $\{|E\hat{a}^+\rangle\}$ or $\{|E\hat{a}^-\}$ of generalized eigenvectors of $H$ related to the $|E\hat{a}\rangle$ by the Lippman–Schwinger equation $(1.1\pm)$.

The Lippman–Schwinger equation is an equation for the $|a^\pm\rangle$. It can be iterated as described in Section VIII.1. It can also be solved formally. For any two invertable operators $A$ and $B$ the following easily established relation holds:

$$\frac{1}{A} - \frac{1}{B} = \frac{1}{A}(B-A)\frac{1}{B} = \frac{1}{B}(B-A)\frac{1}{A}. \tag{1.20}$$

A special case of this is

$$\frac{1}{E-H\pm i0} - \frac{1}{E-K\pm i0} = \frac{1}{E-H\pm i0}V\frac{1}{E-K\pm i0}. \tag{1.21\pm}$$

By applying this to $V|a^\pm\rangle$ and then using $(1.1\pm)$ twice, we obtain the "solution"

$$|a^\pm\rangle = |a\rangle + \frac{1}{E_a - H \pm i0}V|a\rangle. \tag{1.22\pm}$$

The operator analogue of Equation (XIV.5.9) is

$$\frac{1}{E-H+i0} - \frac{1}{E-H-i0} = -2\pi i\delta(E-H), \tag{1.23}$$

---

[1] This mathematical result is not difficult to accept if the discrete and continuous spectra of $H$ are disjoint. It may happen, however, that some of the discrete eigenvalues coincide with values of the continuous spectrum (e.g., the doubly excited states of the He atom; see Sections XI.2 and XI.3). To such eigenvalues correspond both proper eigenvectors in $\mathcal{H}_{bnd}$ and generalized eigenvectors in the set of generalized eigenvectors that span $\mathcal{H}_{scat}$. Such proper and generalized eigenvectors are orthogonal, even though they belong to the same eigenvalue of $H$ and the same eigenvalues $\hat{a} = (a_1, a_2, \ldots, a_k)$ of $\{A_1, A_2, \ldots, A_k\}$.

which together with $(1.22+)$ and $(1.22-)$ allows us to relate the $H$ eigenvectors $|a^+\rangle$ and $|a^-\rangle$ to each other:

$$|a^+\rangle - |a^-\rangle = -2\pi i\delta(E_a - H)V|a\rangle. \tag{1.24}$$

The formal solution $(1.22\pm)$ is not of much practical use, but it does serve as a way of introducing the $S$-matrix, which we shall meet in Section XV.3.

## XV.2  In-States and Out-States

In this section we discuss the significance of the labels $+$ and $-$ that appear in the generalized eigenvectors $|E\hat{a}^+\rangle$ and $|E\hat{a}^-\rangle$ of the exact Hamiltonian $H = K + V$. In these discussions we shall consider pure physical states as described by state vectors; the results are then easily carried over to statistical mixtures as described by statistical operators.

Let us first describe the fictitious case where there is no interaction between beam and target. This means that the beam passes through the target without being affected at all. In the simplest case of potential scattering this means that the potential is turned off. The state vector $\Phi(t)$ that describes this fictitious situation develops in time according to the free Hamiltonian $K$:

$$\Phi(t) = e^{-iKt}\Phi \qquad (\Phi = \Phi(0)). \tag{2.1+}$$

It is called the *free state vector*, and the state it describes is called the *free state*. To simplify the discussion we shall assume that $\Phi$ is an eigenvector corresponding to the eigenvalues $\hat{a} = (a_1, a_2, \ldots, a_k)$ of those operators $A_1$, $A_2, \ldots, A_k$ that together with $K$ form a c.s.c.o. (We assume the eigenvalues of $A_1, A_2, \ldots, A_k$ to be discrete.) Sometimes we will write $\Phi(\hat{a}, t)$ instead of $\Phi(t)$ in order to emphasize $\Phi(t)$'s preparation as an eigenstate of the $A_i$'s. The probability distribution for obtaining the energy eigenvalue $E$ when $K$ is measured in the state $\Phi$ is

$$\langle E\hat{a}|\Phi\rangle\langle\Phi|E\hat{a}\rangle = |\langle E\hat{a}|\Phi\rangle|^2. \tag{2.2+}$$

Let us assume that the energy distribution is described by the function $\phi(E)$, which is related to $\langle E\hat{a}|\Phi\rangle$ by

$$\phi(E) = 2\pi\rho_a(E)\langle E\hat{a}|\Phi\rangle, \tag{2.3+}$$

where $\rho_a(E)$ is the normalization function for the generalized eigenvectors of $K$ and the generalized eigenvectors of $H$ as given by (XIV.2.9a) and (XIV.2.9b). That is, we assume that the state $\Phi(t)$ has been prepared at some time such that at $t = 0$, and therefore at any other time $t$ it has the probability distribution:

$$\begin{aligned}\langle E\hat{a}|\Phi(t)\rangle\langle\Phi(t)|E\hat{a}\rangle &= \langle E\hat{a}|\Phi\rangle\langle\Phi|E\hat{a}\rangle \\ &= (2\pi\rho_{\hat{a}}(E))^{-2}|\phi(E)|^2\end{aligned} \tag{2.4+}$$

for the eigenvalue $E$ of the energy operator $K$. $\Phi$ may then be written in terms of the generalized eigenvectors $|E\hat{a}\rangle$ of $K$ as

$$\Phi = \Phi(0) = \int \rho_a(E)\, dE\, \langle E\hat{a}|\Phi\rangle|E\hat{a}\rangle = \frac{1}{2\pi}\int dE\, \phi(E)|E\hat{a}\rangle. \quad (2.5+)$$

The free state vector at any other time $t$ is given by

$$\Phi(t) = e^{-iKt}\Phi = \frac{1}{2\pi}\int dE\, e^{-iEt}\phi(E)|E\hat{a}\rangle. \quad (2.6+)$$

Let us now consider the state vector

$$\Phi^+ = \Phi^+(0) = \frac{1}{2\pi}\int dE\, \phi(E)|E\hat{a}^+\rangle = \int \rho_a(E)\, dE\, \langle E\hat{a}|\Phi\rangle|E\hat{a}^+\rangle, \quad (2.7+)$$

where $|E\hat{a}^+\rangle$ are the generalized eigenvectors of $H$ that are connected with the generalized eigenvectors $|E\hat{a}\rangle$ of $K$ by $(1.1+)$. The $H$ energy distribution of $\Phi^+$ is the expectation value of the "operator" $\sum_{a'}|E\hat{a}'^+\rangle\langle E\hat{a}'^+|$:

$$\langle\Phi^+|\left(\sum_{a'}|E\hat{a}'^+\rangle\langle E\hat{a}'^+|\right)|\Phi^+\rangle = \sum_{a'}|\langle E\hat{a}'^+|\phi^+\rangle|^2 = |\langle E\hat{a}|\Phi\rangle|^2. \quad (2.8+)$$

[Here we have used both the expansion $(2.7+)$ of $\Phi^+$ and the normalization (XIV.2.9b) of the $|E\hat{a}^+\rangle$'s.] Thus, the probability distribution for obtaining the value $E$ when $H$ is measured in the state $\Phi^+$ is the same as the probability distribution for obtaining the value $E$ when the free energy $K$ is measured in the (fictitious) state $\Phi$ [or $\Phi(t)$]. For any other time $t$, earlier or later, the state that develops to or from the state under the influence of the interaction is given by

$$\Phi^+(t) = e^{-iHt}\Phi^+ = \frac{1}{2\pi}\int dE\, e^{-iEt}|E\hat{a}^+\rangle\phi(E). \quad (2.9+)$$

$\Phi^+(t)$ is called the *exact state vector*. $\Phi^+(t)$ and $\Phi(t)$ are states with the same quantum numbers $\hat{a}$ and the same energy distribution $\phi(E)$; their difference is that $\Phi^+(t)$ develops according to the exact Hamiltonian and describes the development of an actual state, while $\Phi(t)$ develops according to the free Hamiltonian and describes the development of the same state as if there were no interaction.

The connection between $\Phi^+(t)$ and $\Phi(t)$ is obtained if one inserts the Lippman–Schwinger equation $(1.1+)$ into the integral of $(2.9+)$:

$$\Phi^+(t) = \frac{1}{2\pi}\int dE\, e^{-iEt}\phi(E)|E\hat{a}\rangle + \frac{1}{2\pi}\int dE\, e^{-iEt}\phi(E)\frac{1}{E-K+i0}V|E\hat{a}^+\rangle$$

$$= \Phi(t) + \frac{1}{2\pi}\int dE\, e^{-iEt}\phi(E)\frac{1}{E-K+i0}V|E\hat{a}^+\rangle \quad (2.10+)$$

The first term in $(2.10+)$ is given by $(2.6+)$; the calculation of the second term is a purely mathematical task and is given below, after $(2.20-)$. Using the result of that calculation one obtains

$$\Phi^+(t) = \Phi(t) + \int_{-\infty}^{\infty} dt' \, G_0^+(t - t')V\phi^+(t') \tag{2.11+}$$

where

$$G_0^+(t) = -i\theta(+t)e^{-iKt} = \begin{cases} -ie^{iKt} & \text{if } t > 0, \\ 0 & \text{if } t < 0. \end{cases} \tag{2.12+}$$

$\theta(t)$ is the unit step function

$$\theta(t) = \begin{cases} 1 & \text{if } t > 0, \\ 0 & \text{if } t < 0. \end{cases} \tag{2.13}$$

Suppose we now take the limit $t \to -\infty$ in (2.11+). As the limit is taken, we eventually get $t < t'$ for any value of $t'$, so that the integrand vanishes. Thus

$$\Phi^+(t) \to \Phi(t) \quad \text{as } t \to -\infty. \tag{2.14+}$$

This means that if in the remote past when the interaction was not effective a state had been prepared so as to have the energy distribution $\phi(E)$ (and the quantum numbers $\hat{a}$), then this state would develop so that at time $t$ it is given by $\Phi^+(t)$ of (2.9+). Thus $\Phi^+(t)$ describes a state that develops from a state $\Phi(t)$ prepared in the remote past when the interaction $V$ was not effective. The prepared state is called an *in-state*:

$$\Phi(t) = \Phi^{\text{in}}(t)[\equiv \Phi^{\text{in}}(\hat{a}, t)] \tag{2.15+}$$

Describing the behavior of $\Phi^+(t)$ in the distant future when the interaction $V$ ceases to be effective is another free state, the *out-state*:

$$\Phi^{\text{out}}(t)[\equiv \Phi^{\text{out}}(\hat{a}, t)] = e^{-iKt}\Phi^{\text{out}} \quad [\Phi^{\text{out}} = \Phi^{\text{out}}(0)]. \tag{2.16+}$$

In the present situation, where the exact state $\Phi^+(t)$ is prescribed by its behavior in the distant past as $\Phi^{\text{in}}(t)$, the outstate is unknown and un-controlled, except through our knowledge of $V$ and our control over $\Phi^{\text{in}}(t)$. As will be discussed in the next section, the connection between the in-state $\Phi^{\text{in}}(t)$ and the out-state $\Phi^{\text{out}}(t) = S\Phi^{\text{in}}(t)$ is the scattering operator $S$.

We shall now repeat the same arguments used above but using the general-ized eigenvectors $|E\hat{b}^-\rangle$ of $H$ that appear in the Lippman–Schwinger equation

$$|E\hat{b}^-\rangle = |E\hat{b}\rangle + \frac{1}{E - K - i0} V|E\hat{b}^-\rangle \tag{2.17-}$$

instead of the $|E\hat{a}^+\rangle$ appearing in (1.1+). Here $\hat{b} = (b_1, b_2, \ldots, b_k)$ are the eigenvalues of a set of operators $B_1, B_2, \ldots, B_k$, which are possibly different from the operators $A_1, A_2, \ldots, A_k$, but which together with $K$ form a c.s.c.o.

(The spectra of $B_1, B_2, \ldots, B_k$ are assumed for simplicity to be discrete.) We thus consider the free state

$$\Psi(t)[\equiv \Psi(\hat{b}, t)] = e^{-iKt}\Psi = \frac{1}{2\pi} \int dE\, e^{-iEt}\psi(E)|E\hat{b}\rangle, \qquad (2.6-)$$

where the function $\psi(E)$ describes the energy distribution of $\Psi(t)$ as it will be prepared at some time. $\psi(E)$ is connected with the probability distribution for the energy operator $K$ by

$$\langle E\hat{b}|\Psi(t)\rangle\langle\Psi(t)|E\hat{b}\rangle = (2\pi\tilde{\rho}_{\hat{b}}(E))^{-2}|\psi(E)|^2. \qquad (2.4-)$$

The function $\tilde{\rho}_{\hat{b}}(E)$ is a weight function for the normalization of the $|E\hat{b}\rangle$'s, just as $\rho_{\hat{a}}(E)$ was a weight function for the normalization (XIV.2.9a) of the $|E\hat{a}\rangle$'s. From its expansion (2.6−), $\Psi(t)$ is obviously an eigenvector of $B_1, B_2, \ldots, B_k$ corresponding to the eigenvalues $\hat{b}$.

The exact state vector $\Psi^-(t)$ corresponding to $\Psi(t)$ is defined to be

$$\Psi^-(t) = e^{-iHt}\Psi^- = \frac{1}{2\pi} \int dE\, e^{-iEt}\psi(E)|E\hat{b}^-\rangle, \qquad (2.9-)$$

where $|E\hat{b}^-\rangle$ are the generalized eigenvectors of $H$ that are connected with the generalized eigenvectors $|E\hat{b}\rangle$ of $K$ by (2.17−). Again $\Psi^-(t)$ and $\Psi(t)$ are states with the same quantum numbers $\hat{b}$ and the same energy distribution $\psi(E)$, but $\Psi^-(t)$ develops according to the exact Hamiltonian $H$, while $\Psi(t)$ develops according to the free Hamiltonian $K$. To obtain the connection between $\Psi^-(t)$ and $\Psi(t)$ we insert (2.17−) into (2.9−) and obtain

$$\Psi^-(t) = \frac{1}{2\pi} \int dE\, e^{-iEt}\psi(E)|E\hat{b}\rangle + \frac{1}{2\pi} \int dE\, e^{-iEt}\psi(E) \frac{1}{E - K - i0} V|E\hat{b}^-\rangle$$

$$= \Psi(t) + \frac{1}{2\pi} \int dE\, e^{-iEt}\psi(E) \frac{1}{E - K - i0} V|E\hat{b}^-\rangle. \qquad (2.10-)$$

The result of a calculation similar to the calculation to be given for the second term in (2.10+) gives

$$\Psi^-(t) = \Psi(t) + \int_{-\infty}^{\infty} dt'\, G_0^-(t - t')V\psi^-(t'), \qquad (2.11-)$$

where

$$G_0^-(t) = +i\theta(-t)e^{-iKt} = \begin{cases} +ie^{-iKt} & \text{if } t < 0, \\ 0 & \text{if } t > 0. \end{cases} \qquad (2.12-)$$

If we now take the limit $t \to +\infty$ in (2.11−), we obtain

$$\Psi^-(t) \to \Psi(t) \quad \text{as } t \to +\infty \qquad (2.14-)$$

This means that if in the distant future a state is measured that has the energy distribution $\psi(E)$ and quantum numbers $\hat{b}$, then this state was given at time $t$ by $\Psi^-(t)$ of (2.9−). Thus $\Psi^-(t)$ describes a state that will develop into a known state in the distant future when the interaction $V$ is no longer effective. This state will be called an *out-state*:

$$\Psi(t) = \Psi^{\text{out}}(t)[\equiv \Psi^{\text{out}}(\hat{b}, t)] \qquad (2.15-)$$

Describing the behavior of $\Psi^-(t)$ in the remote past before the interaction was effective is another free state, the *in-state*:

$$\Psi^{\text{in}}(t)[\equiv \Psi^{\text{in}}(\hat{b}, t)] = e^{-iKt}\Psi^{\text{in}} \qquad [\Psi^{\text{in}} \equiv \Psi^{\text{in}}(0)]. \qquad (2.16-)$$

Now the exact state $\Psi^-(t)$ is prescribed by what its behavior will be in the distant future as $\Psi^{\text{out}}(\hat{b}, t)$, and it is the in-state that is unknown and uncontrolled, except through our knowledge of $V$ and our control over $\Psi^{\text{out}}(t)$. The connection between the in- and out-states is again $\Psi^{\text{out}}(t) = S\Psi^{\text{in}}(t)$— or, since it is our control over $\Psi^{\text{out}}(t)$ that should be emphasized, $\Psi^{\text{in}}(t) = S^{-1}\Psi^{\text{out}}(t) = S^{\dagger}\Psi^{\text{out}}(t)$.

We have thus found the meaning of the labels $+$ and $-$: $\Phi^+(\hat{a}, t)$ describes a state that in the remote past, before the interaction $V$ became effective, was prepared as $\Phi^{\text{in}}(\hat{a}, t)$ with well-defined quantum numbers $\hat{a}$ and a certain energy distribution $\phi(E)$. In the distant future it will again become a free state, an out-state $\Phi^{\text{out}}(t)$; however, this state is not simple, since it is determined not only by the preparation but also by the scattering process. $\Psi^-(\hat{b}, t)$ describes a state that in the distant future, after the interaction $V$ has ceased, will be given by the free state $\Psi_b^{\text{out}}(t)$ with a simple energy distribution $\psi(E)$ and well-determined values $\hat{b}$ for the other quantum numbers. In the remote past $\Psi^-(\hat{b}, t)$ was also a free state $\Psi^{\text{in}}(t)$, but its properties must have been more complicated. Since in scattering experiments it is the behavior of the system in the distant past over which we exercise control, it is the $+$ states $\Phi^+(\hat{a}, t)$ that are natural to use when describing scattering experiments.

Our results may be summarized as integral equations for the exact states $\Phi^+(\hat{a}, t)$ and $\Psi^-(\hat{b}, t)$ in terms of the controlled free states $\Phi^{\text{in}}(\hat{a}, t)$ and $\Psi^{\text{out}}(\hat{b}, t)$:

$$\Phi^+(\hat{a}, t) = \Phi^{\text{in}}(\hat{a}, t) + \int_{-\infty}^{\infty} dt' \, G_0^+(t - t')V\Phi^+(\hat{a}, t'), \qquad (2.18+)$$

$$\Psi^-(\hat{b}, t) = \Psi^{\text{out}}(\hat{b}, t) + \int_{-\infty}^{\infty} dt' \, G_0^-(t - t')V\Psi^-(\hat{b}, t'). \qquad (2.18-)$$

(There are also corresponding integral equations for $\Phi^+(\hat{a}, t)$ in terms of $\Phi^{\text{out}}(\hat{a}, t)$ and for $\Psi^-(\hat{b}, t)$ in terms of $\Psi^{\text{in}}(\hat{b}, t)$, but in view of the uncontrolled nature of $\Phi^{\text{out}}(\hat{a}, t)$ and $\Psi^{\text{in}}(\hat{b}, t)$, such equations have little meaning.)

The preceding results are easily extended by linearity to the case of a mixture. The exact state that was described in the distant past by the statistical operator

$$W^{\text{in}}(t) = \sum_{mn} W_{mn}|\Phi^{\text{in}}(\hat{a}_m, t)\rangle\langle\Phi^{\text{in}}(\hat{a}_n, t)| \qquad (2.19+)$$

is

$$W^+(t) = \sum_{mn} W_{mn} |\Phi_m^+(\hat{a}_m, t)\rangle \langle \Phi_n^+(\hat{a}_n, t)|, \qquad (2.20+)$$

and the exact state that will be described in the distant future by

$$\tilde{W}^{\text{out}}(t) = \sum_{mn} W_{mn} |\Psi_m^{\text{out}}(\hat{b}_m, t)\rangle \langle \Psi_n^{\text{out}}(\hat{b}_n, t)| \qquad (2.19-)$$

is

$$\tilde{W}^-(t) = \sum_{mn} W_{mn} |\Psi_m^-(\hat{b}_m, t)\rangle \langle \Psi_n^-(\hat{b}_n, t)|. \qquad (2.20-)$$

We shall now take up the purely mathematical task of showing that Equations $(2.11\pm)$ follow from $(2.10\pm)$, using results from distribution theory, in particular results concerning the Fourier transforms of generalized functions.

[The *Fourier transform* $\tilde{g} = F[g]$ of a function $g(t)$ is defined by

$$\tilde{g}(E) = F_E(g(t)) = \int_{-\infty}^{\infty} dt \, e^{iEt} g(t). \qquad (2.21)$$

The *inverse Fourier transformation* $F^{-1}$ is given by

$$F_t^{-1}[\tilde{g}(E)] = \frac{1}{2\pi} \int_{-\infty}^{\infty} dE \, e^{-iEt} \tilde{g}(E) \qquad (2.22)$$

Strictly speaking, (2.21) is the definition of a well-behaved function. The Fourier transform of a generalized function $f$, defined by the linear functional $f(\Psi) = (f, \Psi)$, is defined by $(F[f], F[\Psi]) = 2\pi(f, \Psi)$. But as the rules for Fourier transforms are retained for generalized functions, when interpreted properly one may ignore this mathematical precision and use (2.21) and (2.22) also for generalized functions.[2] In particular, one always has

$$g = F^{-1}[\tilde{g}] = F^{-1}[F[g]] = F[F^{-1}[g]]. \qquad (2.23)$$

Translation in the space of Fourier-transformed functions is obviously given by

$$\tilde{g}(E - E') = F_E[e^{-iE't} g(t)]. \qquad (2.24)$$

If we apply (2.24) to the Fourier transforms

$$F_E[\theta[\pm t]] = \pm i \frac{1}{E \pm i0} \qquad (2.25\pm)$$

---

[2] Gel'fand and Shilov (1964, Vol. 2, Chapter III; Vol. 1, Chapter II).

of the generalized function $\theta(+t)$ and $\theta(-t)$, we get

$$\frac{1}{E - E' \pm i0} = F_E[\mp ie^{-iE't}\theta(\pm t)]. \qquad (2.26\pm)$$

The *convolution* $f * g$ of two functions $f$ and $g$ is defined by

$$(f * g)(t) = \int_{-\infty}^{\infty} dt' \, f(t - t')g(t') \qquad (2.27)$$

and has the particularly simple Fourier transform[3]

$$F[f * g] = F[f] \cdot F[g], \qquad (2.28)$$

or equivalently,

$$F^{-1}[\tilde{f} \cdot \tilde{g}] = f * g. \qquad (2.29)$$

With the above mathematical facts we shall show the equivalence of the second term in $(2.10+)$ and $(2.11+)$. If we take the scalar product of $\Phi^+(t)$ as given by $(2.10+)$ with an arbitrary chosen $K$ basis vector $|a'\rangle = |E'\hat{a}'\rangle$ [more precisely, if we consider the value of the functional $|\Phi^\pm(t)\rangle \in \Phi^{\times\times} = \Phi$ at the point $\langle a'| \in \Phi^\times$], we obtain

$$\langle a'|\Phi^+(t)\rangle = \langle a'|\Phi(t)\rangle + \frac{1}{2\pi} \int dE \, e^{-iEt}\phi(E)\frac{1}{E - E' \pm i0}$$
$$\times \langle a'|V|E\hat{a}^+\rangle. \qquad (2.30+)$$

Equation $(2.9+)$ implies

$$\langle a'|V|\Phi^+(t)\rangle = \frac{1}{2\pi} \int dE e^{-iEt}\phi(E)\langle a'|V|E\hat{a}^+\rangle$$
$$= F_t^{-1}[\phi(E)\langle a'|V|E\hat{a}^+\rangle], \qquad (2.31+)$$

or equivalently,

$$\phi(E)\langle a'|V|E\hat{a}^+\rangle = F_E[\langle a'|V|\Phi^+(t)\rangle]. \qquad (2.32+)$$

The second term on the right-hand side of $(2.30+)$ is now recognized as the inverse Fourier transform of the product of two Fourier transforms [cf. $(2.26+)$ and $(2.32+)$]. By using $(2.29)$ we may rewrite $(2.30+)$ as

$$\langle a'|\Phi^+(t)\rangle = \langle a'|\Phi(t)\rangle + (-i\theta(+t)e^{-iE't}) * (\langle a'|V|\Phi^+(t)\rangle)$$
$$= \langle a'|\Phi(t)\rangle - i \int_{-\infty}^{\infty} dt' \, \theta(+(t - t'))e^{-iE'(t-t')}$$
$$\times \langle a'|V|\Phi^+(t')\rangle. \qquad (2.33+)$$

---

[3] This statement has to be qualified somewhat in the case of generalized functions. See Gel'fand and Shilov (1964, Vol. 2, Chapter III).

Since $\lceil a'\rangle$ was an arbitrary $K$ basis vector, we have shown

$$|\Phi^+(t)\rangle = |\Phi(t)\rangle + \int_{-\infty}^{\infty} dt'\, G_0^+(t - t')V|\Phi^+(t')\rangle, \quad (2.11+)$$

as we set out to do. The equivalence of the second terms in (2.10−) and (2.11−) may be shown by a similar calculation.

The operators

$$G_0^{\pm}(t) = \mp i\theta(\pm t)e^{-iKt}$$

are seen by (2.26±) to be the inverse Fourier transforms of the resolvents

$$\tilde{G}_0^{\pm}(E) = (E - K \pm i0)^{-1} \quad (2.34)$$

of the operator $K$. The operator $G_0^+(t)$ is known as the *retarded Green's function*, and $G_0^-(t)$ is known as the *advanced Green's function*. (In physics the distinction between a function $g(t)$ and its Fourier transform $\tilde{g}(E) = F_E[g(t)]$ is not always reflected by the terminology, and so the resolvents $\tilde{G}_0^{\pm}(E)$ are sometimes also called "Green's functions," as mentioned in Section VIII.1. Physicists frequently blur the notation as well, and write $G_0^{\pm}(E)$ for $\tilde{G}_0^{\pm}(E)$; the argument of the function is then used to tell whether the function or its Fourier transform is meant.)⟧

The operators $G_0^{\pm}(t)$ are both *free* Green's functions, as opposed to the *exact* Green's functions given by

$$G^+(t) = -i\theta(+t)e^{-iHt} = \begin{cases} 0 & \text{if } t < 0, \\ -ie^{-iHt} & \text{if } t > 0, \end{cases} \quad (2.35+)$$

$$G^-(t) = +i\theta(-t)e^{-iHt} = \begin{cases} +ie^{-iHt} & \text{if } t < 0, \\ 0 & \text{if } t > 0. \end{cases} \quad (2.35-)$$

These latter Green's functions may be used to express formal solutions of (2.18+) and (2.18−):

$$\Phi^+(\hat{a}, t) = \Phi^{in}(\hat{a}, t) + \int_{-\infty}^{+\infty} dt'\, G^+(t - t')V\Phi^{in}(\hat{a}, t'), \quad (2.36+)$$

$$\Psi^-(\hat{b}, t) = \Psi^{out}(\hat{b}, t) + \int_{-\infty}^{+\infty} dt'\, G^-(t - t')V\Psi^{out}(\hat{b}, t'). \quad (2.36-)$$

The derivation of these solutions, which is left as an exercise, parallels the derivations of (2.11±) [or equivalently of (2.18±)] except that where the derivations of the latter use the Lippman–Schwinger equation (1.1+) and (2.17−), the derivations of the former use the formal solution (1.22+) and the formal solution for the $|E\hat{b}^-\rangle$'s corresponding to (1.22).

One final remark should be made: None of our arguments have depended on the assumptions that $\Phi^{in}(t)$ is an eigenvector $\Phi^{in}(\hat{a}, t)$ of the operators

$A_1, A_2, \ldots, A_k$ and that $\Psi^{\text{out}}$ is an eigenvector $\Psi^{\text{out}}(\hat{b}, t)$ of the operators $B_1, B_2, \ldots, B_k$. All equations for pure states (as opposed to statistical mixtures) are therefore equally valid, by linearity, if we remove the restrictions that $\Phi^{\text{in}}(t)$ and $\Psi^{\text{out}}(t)$ must be eigenvectors of the operators $A_1, A_2, \ldots, A_k$ and of the operators $B_1, B_2, \ldots, B_k$, respectively.

## XV.3 The S-Operator and the Møller Wave Operators

In a scattering experiment a state is prepared before the projectile and target start interacting with each other, and a state is detected after they stop interacting with each other. Thus in scattering experiments in-states are transformed into out-states. The operator that describes this transformation is called the *scattering operator* or *S-operator*. For every physical system undergoing collisions we postulate the existence of a unitary operator $S$ that transforms in-states into out-states. The S-operator is so immediately related to the directly observable quantities (like cross sections) that knowledge of it leads at once to the prediction of these quantities.

As remarked in Section XIV.4, the concept of the S-operator makes sense even if a generator $H$ of time development is not defined and the time-development axiom (V in Chapter XII) does not hold. If $H$ is not defined, then neither are its generalized eigenvectors $|a^{\pm}\rangle$ nor the exact states $\Phi^+(t)$ and $\Psi^-(t)$. However, the states $\Phi^{\text{in}}$ and $\Psi^{\text{out}}$, which describe the prepared and detected states, and the operator

$$S : \Phi^{\text{in}} \to \Phi^{\text{out}} = S\Phi^{\text{in}}, \tag{3.1}$$

which transforms the prepared (and hence controlled) in-state $\Phi^{\text{in}}$ into the uncontrolled out-state $\Phi^{\text{out}}$, are still valid concepts. If the time development axiom does hold (and there has been no evidence to the contrary in non-relativistic quantum physics), then the S-operator and its matrix, the S-matrix

$$(\Psi^{\text{out}}, S\Phi^{\text{in}}) \tag{3.2}$$

may be expressed in terms of more fundamental and less directly observable quantities.

We shall now do this by giving a precise definition of $S$ in terms of previously introduced quantities. The modulus squared of the matrix element

$$(\Psi^-(\hat{b}, t), \Phi^+(\hat{a}, t)) \tag{3.3}$$

gives the probability of finding the state $\Psi^-(\hat{b}, t)$, which is observed after the interaction $V$ has ceased to be effective as the state $\Psi^{\text{out}}(\hat{b}, t)$, if the state of the system is $\Phi^+(\hat{a}, t)$, which was prepared before $V$ became effective as the state $\Phi^{\text{in}}(\hat{a}, t)$. That is to say, (3.3) describes the probability for a transition from an initial configuration, described by the quantum numbers $\hat{a}$ and the energy distribution $\phi(E)$, into a final configuration, described by the quantum numbers $\hat{b}$ and the energy distribution $\psi(E)$. We first develop an expression for (3.3) in terms of the in-state $\Phi^{in}$ and the out-state $\Psi^{out}$. To do this we shall use the formal solutions (2.36+) and (2.36−),

$$\Phi^+(t) = \Phi^{in}(t) + \int_{-\infty}^{+\infty} dt'\, G^+(t - t')V\Phi^{in}(t') \qquad (3.4+)$$

and

$$\Psi^-(t) = \Psi^{out}(t) + \int_{-\infty}^{+\infty} dt'\, G^-(t - t')V\Psi^{out}(t'), \qquad (3.4-)$$

of (2.18+) and (2.18−). Equations (2.6±) and (2.15±) imply

$$\Phi^{in}(t') = e^{-iK(t'-t)}\Phi^{in}(t) \qquad (3.5+)$$

and

$$\Psi^{out}(t') = e^{-iK(t'-t)}\Psi^{out}(t). \qquad (3.5-)$$

These expressions may be substituted into the integrands of (3.4+) and (3.4−) to get

$$\Phi^+(t) = \Omega^+\Phi^{in}(t) \qquad (3.6+)$$

and

$$\Psi^-(t) = \Omega^-\Psi^{out}(t), \qquad (3.6-)$$

where we have defined the *Møller wave operators*

$$\Omega^\pm \equiv I + \int_{-\infty}^{+\infty} dt'\, G^\pm(t - t')Ve^{-iK(t'-t)}$$

$$= I \mp i \int_{-\infty}^{+\infty} dt'\, G^\pm(t - t')VG_0^\mp(t' - t)$$

$$= I \mp i \int_{-\infty}^{+\infty} dt''\, G^\pm(-t'')VG_0^\pm(t''). \qquad (3.7\pm)$$

An immediate consequence of (3.6±) is that the mixtures $W^+(t)$ of (2.20+) and $\tilde{W}^-(t)$ of (2.20−) are related to the mixtures $W^{in}(t)$ of (2.19+) and $\tilde{W}^{out}(t)$ of (2.19−) by

$$W^+(t) = \Omega^+ W^{in}(t)\Omega^{+\dagger} \qquad (3.8+)$$

and by

$$\tilde{W}^-(t) = \Omega^- \tilde{W}^{out}(t)\Omega^{-\dagger}. \qquad (3.8-)$$

Use of (3.6±) in (3.3) gives

$$(\Psi^-(\hat{b}, t), \Phi^+(\hat{a}, t)) = (\Psi^{out}(\hat{b}, t), \Omega^{-\dagger}\Omega^+\Phi^{in}(\hat{a}, t)) \qquad (3.9)$$

This last equation motivates

$$S \equiv \Omega^{-\dagger}\Omega^+ \qquad (3.10)$$

as the definition of the scattering operator.

The last expression for $\Omega^\pm$ in (3.7±) shows $\Omega^\pm$ to be time-independent, which implies that $S$ is also time-independent. So is the matrix element (3.9). For $\Phi^+(t)$ is given both by

$$\Phi^+(t) = e^{-iHt}\Phi^+ = e^{-iHt}\Omega^+\Phi^{in}(0)$$

and by

$$\Phi^+(t) = \Omega^+\Phi^{in}(t) = \Omega^+ e^{-iKt}\Phi^{in}.$$

Since $\{\Phi^{in}\}$ spans the space of physical states, we conclude that

$$e^{-iHt}\Omega^\pm = \Omega^\pm e^{-iKt}. \tag{3.11±}$$

[The upper equations follow from the stated argument; a similar argument gives (3.11 −).] Equivalently, one has the so-called *intertwining relations*

$$H\Omega^\pm = \Omega^\pm K. \tag{3.12±}$$

It is then easily shown that

$$[S, K] = 0, \tag{3.13}$$

a consequence of which is the time independence of (3.9):

$$(\Psi^-(\hat{b}, t), \Phi^+(a, t)) = (\Psi^{out}(\hat{b}, 0), e^{+iKt}Se^{-iKt}\Phi^{in}(\hat{a}, 0))$$
$$= (\Psi^{out}(\hat{b}, 0), S\Phi^{in}(\hat{a}, 0)). \tag{3.14}$$

Equation (3.13), it should be noted, may be interpreted as a statement of energy conservation between the in-state $\Phi^{in}$ and the out-state $\Psi^{out}$.

If we substitute (2.6+) and (2.9+) into the left-hand side and right-hand side of (3.6+), respectively, we obtain

$$\frac{1}{2\pi}\int dE\, e^{-iEt}\phi(E)|E\hat{a}^+\rangle = \frac{1}{2\pi}\int dE\, e^{-iEt}\phi(E)\Omega^+|E\hat{a}\rangle.$$

Since $\phi(E)$ may be any well-behaved function, it follows that

$$\Omega^+|E\hat{a}\rangle = |E\hat{a}^+\rangle. \tag{3.15+}$$

Equations (2.6−), (2.9−), and (3.6−) may be used in a similar manner to conclude

$$\Omega^-|E\hat{a}\rangle = |E\hat{a}^-\rangle. \tag{3.15−}$$

Since we have assumed the system is asymptotically complete [Equation (1.19)], we may conclude that the Møller operators $\Omega^+$ and $\Omega^-$ map the space of physical states $\mathscr{H}$ onto the space of scattering states $\mathscr{H}_{scat}$:

$$\Omega^+\mathscr{H} = \Omega^-\mathscr{H} = \mathscr{H}_{scat}. \tag{3.16}$$

In other words, the domain of definition of both $\Omega^+$ and $\Omega^-$ is the entire space $\mathscr{H}$, but the range of $\Omega^+$ and $\Omega^-$ is the (proper if bound states exist) subspace $\mathscr{H}_{scat}$.

[An operator $A$ is said to be *isometric* if it preserves the norm of vectors, i.e., if

$$(\psi, \psi) = (A\psi, A\psi) = (\psi, A^\dagger A\psi) \tag{3.17}$$

for all vectors $\psi$. An equivalent definition is that $A$ satisfies the condition

$$A^\dagger A = I. \tag{3.18}$$

If $A$ satisfies the stronger condition

$$A^\dagger A = I \quad \text{and} \quad AA^\dagger = I, \tag{3.19}$$

then $A$ is *unitary*. A unitary operator $A$ is defined on all of $\mathscr{H}$ and has all of $\mathscr{H}$ as its range, i.e., maps $\mathscr{H}$ onto $\mathscr{H}$ in a one-to-one fashion. A unitary operator is necessarily isometric, but not vice versa.⟧

Because $\Omega^\pm \mathscr{H} = \mathscr{H}_{\text{scat}}$ is usually not all of $\mathscr{H}$, we should not expect the Møller operators to be unitary; they are, however, isometric, a fact we shall now show.

By inserting

$$\Phi^+(t') = e^{-iH(t'-t)}\Phi^+(t)$$

[which follows from (2.9+)] into (2.18+), we obtain

$$\Phi^{\text{in}}(t) = \Phi^+(t) - \int_{-\infty}^{+\infty} dt'\, G_0^+(t-t')Ve^{-iH(t'-t)}\Phi^+(t)$$

$$= \left(I + i\int_{-\infty}^{+\infty} dt'\, G_0^+(t-t')VG^-(t'-t)\right)\Phi^+(t). \tag{3.20}$$

$$= \left(I + i\int_{-\infty}^{\infty} dt''\, G_0^+(-t'')VG^-(+t'')\right)\Phi^+(t).$$

The second line follows from the definition (2.35−) of $G^-(t)$ and the fact that $G_0^+(t-t')$ is nonzero only for $t-t' > 0$. From the definitions (2.12±) and (2.35±) of the Green's functions and from the Hermiticity of $K$ and $H$ it easily follows that

$$G_0^\pm(-t)^\dagger = G_0^\mp(t) \tag{3.21a$\pm$}$$

and

$$G^\pm(t)^\dagger = G^\mp(-t). \tag{3.21b$\pm$}$$

Upon use of (3.21±) and the definition (3.7+) of $\Omega^+$, Equation (3.20) becomes

$$\Phi^{\text{in}}(t) = \left(I - i\int_{-\infty}^{\infty} dt''\, G^+(-t'')VG_0^-(t'')\right)^\dagger \Phi^+(t) = \Omega^{+\dagger}\Phi^+(t). \tag{3.22+}$$

But $\Phi^+(t) = \Omega^+\Phi^{\text{in}}(t)$, so

$$\Phi^{\text{in}}(t) = \Omega^{+\dagger}\Omega^+\Phi^{\text{in}}(t).$$

We have not made use of any property of $\Phi^{in}(t)$ except that it can be expanded in terms of the generalized eigenvectors $|E\hat{a}\rangle$ of $K$, so $\Phi^{in}(t)$ is an arbitrary element of $\mathcal{H}$. Consequently,

$$\Omega^{+\dagger}\Omega^{+} = I, \qquad (3.23+)$$

i.e., $\Omega^{+}$ is isometric. A similar derivation gives

$$\Omega^{-\dagger}\Omega^{-} = I. \qquad (3.23-)$$

The failure of $\Omega^{\pm}$ to be unitary is now easily shown: By use of $(3.15\pm)$ and (XIV.2.11) one obtains

$$\Omega^{\pm}\Omega^{\pm\dagger} = \left(\sum_{a}\Omega^{\pm}|a\rangle\langle a|\right)\left(\sum_{a'}\Omega^{\pm}|a'\rangle\langle a'|\right)^{\dagger}$$

$$= \sum_{aa'}|a^{\pm}\rangle\langle a|a'\rangle\langle a'^{\pm}| = \sum_{a}|a^{\pm}\rangle\langle a^{\pm}| \qquad (3.24\pm)$$

$$= \Pi_{scat} = I - \Pi_{bnd}$$

where

$$\Pi_{scat} = \sum_{a}|a^{\pm}\rangle\langle a^{\pm}| \qquad (3.25a)$$

and

$$\Pi_{bnd} = \sum_{a}|\alpha_{n}\rangle\langle\alpha_{n}| \qquad (3.25b)$$

are the projectors onto the subspaces of scattering states and bound states, $\mathcal{H}_{scat}$ and $\mathcal{H}_{bnd}$, respectively. Unless the *unitary deficiency* $\Pi_{bnd}$ is zero, i.e., unless there are no bound states $|\alpha_{n}\rangle$, the operators $\Omega^{\pm}$ can only be isometric and not unitary.

The adjoint Møller operators $\Omega^{+\dagger}$ map $\mathcal{H}_{scat}$ back onto $\mathcal{H}$; more precisely, they annihilate the $\mathcal{H}_{bnd}$ part of $\mathcal{H} = \mathcal{H}_{bnd} \oplus \mathcal{H}_{scat}$ and map the rest, $\mathcal{H}_{scat}$, back onto $\mathcal{H}$:

$$\Omega^{\pm\dagger}|\alpha_{n}\rangle = \left(\sum_{a'}\Omega^{\pm}|a'\rangle\langle a'|\right)^{\dagger}|\alpha_{n}\rangle = \sum_{a}|a'\rangle\langle a'^{\pm}|\alpha_{n}\rangle = 0, \quad (3.26\pm)$$

or equivalently,

$$\Omega^{\pm\dagger}\Pi_{bnd} = 0; \qquad (3.26'\pm)$$

and

$$\Omega^{\pm\dagger}|a^{\pm}\rangle = \Omega^{\pm\dagger}\Omega^{\pm}|a\rangle = I|a\rangle = |a\rangle. \qquad (3.27\pm)$$

We are now in a position to show that $S$ is unitary:

$$SS^{\dagger} = (\Omega^{-\dagger}\Omega^{+})(\Omega^{+\dagger}\Omega^{-}) = \Omega^{-\dagger}(I - \Pi_{bnd})\Omega^{-} = I \qquad (3.28a)$$

and

$$S^{\dagger}S = (\Omega^{+\dagger}\Omega^{-})(\Omega^{-\dagger}\Omega^{+}) = \Omega^{+\dagger}(I - \Pi_{bnd})\Omega^{+} = I. \qquad (3.28b)$$

In writing Equation (3.28) we have used the definition (3.10) of $S$, the iso-metricity (3.23±) of $\Omega^{\pm}$, and Equations (3.24±) and (3.26±); in order for Equation (3.28) to make sense—indeed, for the definition (3.10) of $S$ to make sense—we had to assume domain $\Omega^{\pm\dagger} = $ range $\Omega^{\mp\dagger} = \Omega^{\mp}\mathcal{H}$, or, in other words, asymptotic completeness in the form (3.16). In this way we have proven that the $S$-operator defined by (3.10) and (3.7±) in terms of $V$ and $K$ is unitary.

We define the $S$-*matrix* to be the matrix $\{S_{aa'}\}$ of the scattering operator with respect to the $K$ basis:

$$S_{aa'} \equiv \langle a|S|a'\rangle = \langle a|\Omega^{-\dagger}\Omega^{+}|a'\rangle = \langle a^{-}|a'^{+}\rangle. \qquad (3.10')$$

By using the relation (1.24) we may write $S_{aa'}$ either as

$$\begin{aligned} S_{aa'} &= [\langle a^{+}| - 2\pi i\langle a|V\delta(E_a - H)]|a'^{+}\rangle \\ &= \langle a^{+}|a'^{+}\rangle - 2\pi i\delta(E_a - E_{a'})\langle a|V|a'^{+}\rangle \end{aligned} \qquad (3.29)$$

or as

$$\begin{aligned} S_{aa'} &= \langle a^{-}|(|a'^{-}\rangle - 2\pi i\delta(E_{a'} - H)V|a'\rangle \\ &= \langle a^{-}|a'^{-}\rangle - 2\pi i\delta(E_{a'} - E_a)\langle a^{-}|V|a'\rangle. \end{aligned} \qquad (3.30)$$

If we use (XIV.2.9b) (which we shall finally prove at the end of this section), we see that the first terms on the right-hand sides of (3.29) and (3.30) are equal. Thus

$$a|V|a'^{+}\rangle = \langle a^{-}|V|a'\rangle \quad \text{when } E_a = E_{a'} = E,$$

i.e.,

$$\langle E\hat{a}|V|E\hat{a}'^{+}\rangle = \langle E\hat{a}^{-}|V|E\hat{a}'\rangle. \qquad (3.31)$$

In Section XIV.5 we introduced the on-the-energy-shell $T$-matrix

$$T_{aa'}(E) \equiv \langle E\hat{a}|V|E\hat{a}'^{+}\rangle \qquad (3.32)$$

(which does not define an operator) and its off-the-energy-shell extension

$$T^{+}_{aa'} = \langle a|T^{+}|a'\rangle \equiv \langle a|V|a'^{+}\rangle = \langle E_a\hat{a}|V|E_{a'}\hat{a}'^{+}\rangle, \qquad (3.33)$$

which defines an operator $T^{+}$. Equation (3.31) suggests a different extension of the $T$-matrix, namely

$$T^{-}_{aa'} = \langle a|T^{-}|a'\rangle \equiv \langle a^{-}|V|a'\rangle = \langle E_a\hat{a}^{-}|V|E_{a'}\hat{a}'\rangle, \qquad (3.34)$$

which defines the operator $T^{-}$. $\{T^{+}_{aa'}\}$ and $\{T^{-}_{aa'}\}$ are extension of $\{T_{aa'}(E)\}$ in the sense that

$$T^{+}_{aa'} = T^{-}_{aa'} = T_{aa'}(E) \quad \text{when } E_a = E_{a'} = E. \qquad (3.35)$$

The $S$-matrix element $S_{aa'}$ may then be expressed as

$$\begin{aligned} S_{aa'} &= \langle a|a'\rangle - 2\pi i\delta(E_a - E_{a'})T_{aa'}(E_a) \\ &= \rho(E_a)^{-1}\delta_{aa'}\,\delta(E_a - E_{a'}) - 2\pi i\delta(E_a - E_{a'})T_{aa'}(E_a). \end{aligned} \qquad (3.36)$$

Note that it is the on-the-energy-shell *T*-matrix elements that contribute to the *S*-matrix; matrix elements of $S$ between generalized eigenvectors $|E_a\hat{a}\rangle$ and $|E_{a'}\hat{a}'\rangle$ vanish for $E_a \neq E_{a'}$.

An integral equation for the $T^+$-matrix may be obtained by inserting the Lippman–Schwinger equation (1.1+) into (3.33):

$$T^+_{aa'} = \langle a| V \left( |a'\rangle + \frac{1}{E_{a'} - K + i0} V |a'^+\rangle \right)$$

$$= \langle a|V|a'\rangle + \sum_{a''} \frac{\langle a|V|a''\rangle\langle a''|V|a'^+\rangle}{E_{a'} - E_{a''} + i0} \tag{3.37}$$

$$= \langle a|V|a'\rangle + \sum_{a''} \frac{\langle a|V|a''\rangle T^+_{a''a}}{E_{a'} - E_{a''} + i0}.$$

Equation (3.37) may be iterated by repeated substitution of the right-hand side into itself:

$$T^+_{aa''} = \langle a|V|a'\rangle + \sum_{a''} \frac{\langle a|V|a''\rangle\langle a''|V|a'\rangle}{E_{a'} - E_{a''} + i0} + \cdots. \tag{3.38}$$

The *Born approximation*

$$T^+_{aa''} \approx \langle a|V|a'\rangle = \langle E_a\hat{a}|V|E_{a'}\hat{a}'\rangle, \tag{3.39}$$

in which only the first term of the iterative solution (3.38) is kept, is usually adequate in the case of high energies $E_a = E_{a'} = E$ and a weak interaction $V$.

From the unitarity of the operator $S$ it follows that the *S*-matrix is also unitary:

$$\sum_{a''} S_{aa''} S^*_{a'a''} = \sum_{a''} S^*_{a''a} S_{a''a} = \delta_{aa'}, \tag{3.40}$$

or in more detail,

$$\sum_{a''} \int dE_{\hat{a}''} \rho_{\hat{a}''}(E_{a''}) \langle E_a\hat{a}|S|E_{a''}\hat{a}''\rangle \langle E_{a'}\hat{a}'|S|E_{a''}\hat{a}''\rangle^*$$

$$= \sum_{a''} \int dE_{a''} \rho_{\hat{a}''}(E_{a''}) \langle E_{a''}\hat{a}''|S|E_a\hat{a}\rangle^* \langle E_{a''}\hat{a}''|S|E_{a'}\hat{a}'\rangle \tag{3.40'}$$

$$= \rho(E_a)^{-1} \delta_{\hat{a}\hat{a}'} \delta(E_a - E_{a'}).$$

By inserting (3.40') into (3.36) we may express the unitarity of $\{S_{aa'}\}$ in terms of the matrix elements $T_{\hat{a}\hat{a}'}(E)$:

$$T_{\hat{a}\hat{a}'}(E) - T^*_{\hat{a}'\hat{a}}(E) = -2\pi i \sum_{\hat{a}''} \rho_{\hat{a}''}(E) T_{\hat{a}\hat{a}''}(E) T^*_{\hat{a}'\hat{a}''}(E). \tag{3.41}$$

The unitarity condition as expressed by (3.41) is sometimes called the *generalized optical theorem*.

In summary, we have obtained both the *S* matrix and the *T* matrix by assuming the axiom V of continuous time development. We have expressed the $S$-matrix and $T$-matrix in terms of quantities, like $V$ and $|a^+\rangle$, that depend on the existence of the time-development generator $H$ [cf. (3.32) and (3.36)]. The generalized eigenvectors $|E\hat{a}^+\rangle$ of $H$ and the corresponding states $|\Phi^+(\hat{a}, t)\rangle$ describe the situation at the time of collision; they are therefore quantities that are never accessible to experimental determination. In contrast, the generalized eigenvectors $|E\hat{a}\rangle$ of $K$ and the corresponding states $|\Phi^{in}(\hat{a}, t)\rangle$ and $|\Phi^{out}(\hat{a}, t)\rangle$ are experimentally accessible by means of the initial preparation and final observation of a collision experiment. They, together with the operator that describes the transition between them, are indispensable in the theoretical description of a collision experiment; whereas experimentally nonaccessible quantities like $|\Phi^+(\hat{a}, t)\rangle$ are theoretical tools and rest on additional theoretical assumptions (e.g., the axiom V).

The theoretical description which is based only on the experimentally accessible quantities is called *S-matrix theory*. The $S$-matrix and the on-energy-shell $T$-matrix, which are connected by way of (3.36), are then the fundamental physical quantities; furthermore, the only assumption entering into (3.36) is the conservation of energy between the in-state and the out-state ($[S, K] = 0$). Unitarity, which implies the condition (3.41) on the $T$-matrix, is a separate fundamental assumption in $S$-matrix theory and is justified by the requirement of conservation of overall probability. Another fundamental assumption in $S$-matrix theory is the analyticity of the $T$-matrix as a function of energy and other physical parameters. This is argued to be connected with causality. (We shall discuss this connection in Section XVIII.4.) In the conventional formulation based on the assumption of a continuous time development, properties such as unitarity and analyticity (in a certain domain) are derived from the properties of $H$ and of $V$.

Symmetry considerations place the most significant conditions on the $S$-matrix. In the conventional formulation symmetries are formulated as transformation properties of $V$ and $H$ under various symmetry transformations, and these transformation properties then lead to derived symmetry properties of the $S$-matrix. (We shall discuss the consequences of rotational symmetry in Chapter XVI.) In $S$-matrix theory the symmetries are formulated directly in terms of symmetries of the $S$-matrix or in terms of transformation properties of the $T$-operator.

If $V$ and $H$ are known, then the conventional approach to the description of collision processes is more "fundamental," although based on more assumptions. After Equation (3.37) has been solved for the $T$-matrix, one may predict experimental results. As mentioned in previous chapters, in physics we most often have the reverse situation: One wants to conjecture from experimental data (e.g., cross sections) the properties of the fundamental observables and then use these properties to make predictions about other experimental quantities. In this situation a formulation of the properties in terms of the $T$-operator is just as valid as a formulation in terms of $V$, and is often more practical.

## XV.A  Appendix

With the help of the isometricity $(3.23\pm)$ of the Møller operators, $\Omega^{\pm\dagger}\Omega^{\pm} = I$, it is now trivial to prove two equalities that were used in Chapter XIV. (The proof of isometricity did not depend upon these equalities.) Equation (XIV.2.9b) follows from $(3.15\pm)$ and isometricity:

$$\langle a^{\pm}|a'^{\pm}\rangle = (\langle a|\Omega^{\pm\dagger})(\Omega^{\pm}|a'\rangle)$$
$$= \langle a|(\Omega^{\pm\dagger}\Omega^{\pm})|a'\rangle = \langle a|a'\rangle. \tag{A.1}$$

To prove (XIV.5.21) we use $(3.15+)$ and $(3.8+)$ together with isometricity:

$$\langle a^{+}|W^{+}(t)|a'^{+}\rangle = (\langle a|\Omega^{\pm\dagger})(\Omega^{+}W^{\text{in}}(t)\Omega^{\pm\dagger})(\Omega^{+}|a\rangle)$$
$$= \langle a|(\Omega^{\pm\dagger}\Omega^{+})W^{\text{in}}(t)|(\Omega^{\pm\dagger}\Omega^{+})|a'\rangle \tag{A.2+}$$
$$= \langle a|W^{\text{in}}(t)|a'\rangle;$$

in a similar fashion one can prove

$$\langle b^{-}|\tilde{W}^{-}(t)|b'^{-}\rangle = \langle b|\tilde{W}^{\text{out}}(t)|b'\rangle. \tag{A.2-}$$

Lastly we show that the term II given by (XIV.2.17) vanishes. The key to doing so is Equation $(A.2+)$. If we substitute (XIV.2.9b) and $(A.2+)$ at $t = 0$ into (XIV.2.17), then

$$\text{II} = \sum_{b}\sum_{aa'} e^{-i(E_a - E_{a'})t}\langle b|V|a\rangle\rho(E_b)^{-1}\delta_{\hat{a}'\hat{b}}\delta(E_{a'} - E_b)\langle a|W^{\text{in}}|a'\rangle$$

$$= \int\rho(E_b)\,dE_b\sum_{\hat{b}}\int\rho(E_a)\,dE_a\,\rho(E_{a'})\,dE_{a'}\sum_{\hat{a}\hat{a}'} e^{-i(E_a - E_{a'})t}$$
$$\times \langle E_b\hat{b}|V|E_a\hat{a}\rangle\rho(E_b)^{-1}\delta_{\hat{a}'\hat{b}}\delta(E_{a'} - E_b)\langle E_a\hat{a}|W^{\text{in}}|E_{a'}\hat{a}'\rangle$$

$$= \int\rho(E_b)\,dE_b\sum_{\hat{b}}\int\rho(E_a)\,dE_a\sum_{\hat{a}} e^{-i(E_a - E_b)t}\langle E_b\hat{b}|V|E_a\hat{a}\rangle$$
$$\times \langle E_a\hat{a}|W^{\text{in}}|E_b\hat{b}\rangle$$

$$= \sum_{b}\sum_{a} e^{-i(E_a - E_b)t}\langle b|V|a\rangle\langle a|W^{\text{in}}|b\rangle. \tag{A.3}$$

But the summation $\sum_{b}$ is only over those values of $b$ for which $|b\rangle \in \Lambda\mathcal{H}$, and by hypothesis the initial state $W^{\text{in}}$ is so prepared that $|A\rangle \notin \Lambda\mathcal{H}$ for all states $|A\rangle$ that give a nonzero contribution $(w_A \neq 0)$ to

$$W^{\text{in}} = \sum_{A} w_A|A\rangle\langle A|.$$

Therefore

$$W^{\text{in}}|b\rangle = 0$$

for all $b$ in the summation, so that

$$\text{II} = 0. \tag{A.4}$$

The contribution of II both to the transition rate $d\langle\Lambda\rangle_t/dt$ of (XIV.2.5) and to the transition probability $\langle\Lambda\rangle_{\infty}$ of (XIV.5.2) is then, of course, zero.

# CHAPTER XVI

# Elastic and Inelastic Scattering for Spherically Symmetric Interactions

In this chapter we derive general results for spherically symmetric situations, i.e., for situations when in addition to the free Hamiltonian $K$, the interaction Hamiltonian $V$ and the transition operator $T$ also commute with the operator of angular momentum. For the sake of simplicity we consider the spin-zero case. In Section XVI.1, the partial-wave expansion is discussed and the partial cross sections are derived. In Section XVI.2, the phase shift is introduced as a consequence of the unitarity of the $S$-matrix. In Section XVI.3, a graphical representation of the partial-wave amplitude, the Argand diagram, is introduced.

## XVI.1  Partial-Wave Expansion

Many scattering problems are spherically symmetric. This means that both $H = K + V$ and $V$ commute with the (orbital) angular-momentum operators $L_i$ or, if the scattering is described by a transition operator $T$, that $T$ (and hence the scattering operator $S$) commutes with the $L_i$'s:

$$[V, L_i] = 0, \qquad [T, L_i] = 0. \tag{1.1}$$

Then

$$\{K, \mathbf{L}^2, L_3, \eta^{\mathrm{op}}\} \tag{1.2a}$$

and

$$\{H, \mathbf{L}^2, L_3, \kappa^{\mathrm{op}}\}, \tag{1.2b}$$

where $\eta^{op}$ and $\kappa^{op}$ stand for the internal observables (their eigenvalues $\eta$ and $\kappa$ being the internal quantum numbers), are both c.s.c.o.'s, and one may choose collections of generalized eigenvectors

$$|E_a l\, l_3 \eta\rangle \qquad (1.3a)$$

and

$$|E_\alpha l\, l_3 \eta^+\rangle \qquad (1.3b)$$

of these systems as bases.

If $[V, \eta^{op}] = 0$ and therefore $[H, \eta^{op}] = 0$, then the collection of operators $\kappa^{op}$ in (1.2b) can be chosen to be $\eta^{op}$, and the generalized eigenvectors (1.3b) are eigenvectors of the c.s.c.o. (1.2b). In general $[V, \eta^{op}] \neq 0$, and then (1.3b) are not eigenvectors of the c.s.c.o. (1.2b) and the meaning of the label H is defined through the Lippman–Schwinger equation as explained in Section XV.1.

Still another c.s.c.o. consists of the projectile momentum $\mathbf{P}$ (or the momentum $\mathbf{P}$ in the center-of-mass system when the target is not fixed) and the operators for the internal quantum numbers $\eta^{op}$.

$$\{\mathbf{P}_1, \mathbf{P}_2, \mathbf{P}_3, \eta^{op}\}. \qquad (1.4)$$

The corresponding basis consists of the generalized momentum eigenvectors

$$|\mathbf{p}\,\eta\rangle. \qquad (1.5)$$

The additional quantum numbers $\eta = (\eta_1 \eta_2 \cdots \eta_k)$ may refer to the internal structure of the target, of the projectile, or of both. For example, if the target is a hydrogen atom, then H will contain the quantum numbers $n, j, j_3$, and $\pi$ that characterize the state of the electron in the hydrogen atom (cf. Chapters VI and IX). According to (1.2), $[L_i, \eta^{op}] = 0$; thus we exclude the possibility that $\eta$ includes the quantum numbers of an internal angular momentum. Then the total angular momentum is equal to the orbital angular momentum $L_i$, and we avoid the complication of coupling the internal angular momentum with the orbital angular momentum to get the total angular momentum. Most processes encountered in experiment involve at least one internal angular momentum; however, to understand the principal problems of this chapter it is easier if one omits at the beginning the inessential complications introduced by the coupling of angular momenta.

The simplest case we can discuss is the case of a structureless projectile incident on a structureless target. Whether the target and/or projectile may be considered to be structureless depends on the energies involved. If, for example, the target is a helium atom, then it may be considered structureless for projectile energies less than the energy of the first excited state, and the quantum numbers $\eta$ are unnecessary. If the projectile energy is comparable in size to the difference between the energy levels, then the target's structure must be considered, and $\eta$ will contain the quantum numbers $n, j, j_3, \pi$ (singlet) or $n, j, j_3, \pi$ (triplet) of the helium atom.

We would like to express the matrix element

$$\langle \mathbf{p}_b \eta_b | T | \mathbf{p}_A \eta_A\rangle \qquad (1.6)$$

of the transition operator $T$ with respect to the momentum basis (1.5) in terms of the matrix elements of $T$ with respect to the angular-momentum basis (1.3a). To do this we expand $|\mathbf{p}\,\eta\rangle$ in terms of the angular-momentum basis,

$$|\mathbf{p}\,\eta\rangle = \sum_{l l_3} \int \rho_\eta(E)\, dE\, |E\, l\, l_3\, \eta\rangle \langle E\, l\, l_3\, \eta | \mathbf{p}\,\eta\rangle. \qquad (1.7)$$

The determination of the transition coefficients $\langle \mathbf{p}\,\eta | E\, l\, l_3\, \eta\rangle = \langle E\, l\, l_3\, \eta | \mathbf{p}\,\eta\rangle^*$ is a problem similar to the determination of the wave functions $\langle x | n\, l\, m\rangle$ of the hydrogen atom, which was discussed in Section VII.3. Instead of (VII.3.5) we have the analogue

$$\langle \mathbf{p}\,\eta | L_i | \psi\rangle = -\frac{1}{i}\, \epsilon_{ijk} P_j\, \frac{\partial}{\partial p_k}\, \langle p\,\theta_p\,\phi_p\,\eta | \psi\rangle, \qquad (1.8)$$

from which follow the analogue of (VII.3.6$_3$) and (VII.3.6′). Here $(p\,\theta_p\,\phi_p)$ are the spherical coordinates of the momentum vector $\mathbf{p}$. Instead of the "radial equation" (VII.3.7) we have

$$\langle p\,\theta_p\,\phi_p\,\eta | \left(\frac{P^2}{2M} + K^{\text{int}}\right) | E\, l\, l_3\, \eta\rangle = E\langle p\,\theta_p\,\phi_p\,\eta | E\, l\, l_3\, \eta\rangle, \qquad (1.9)$$

which is a consequence of the (nonrelativisitic) relation (cf. XIV.5.31)

$$K = \frac{P^2}{2M(\eta^{\text{op}})} + K^{\text{int}}(\eta^{\text{op}}) \qquad (1.10)$$

between the energy operator $K$, the momentum $\mathbf{P}$, and the internal energy $K^{\text{int}}$. Although the point has not been raised previously, the mass of the particle may depend upon the internal quantum numbers $\eta$ of the system; this dependence allows us to consider cases where the outgoing particle is different from the incoming particle. From (1.9) and the analogue of (VII.3.6′) and (VII.3.6$_3$) it follows that the transition coefficients have the form [cf. (VII.3.11)]

$$\langle p\,\theta_p\,\phi_p\,\eta | E\, l\, l_3\, \eta\rangle = f_\eta(p)\delta\left(E - \frac{p^2}{2m_\eta} - E_\eta^{\text{int}}\right) Y_{l l_3}(\theta_p, \phi_p) \qquad (1.11)$$

The function $f_\eta(p)$ is determined (up to a phase factor) by the normalization (XIV.2.9a) of the generalized $K$-eigenvectors $|E\, l\, l_3\, \eta\rangle$, for

$$\rho_\eta(E)^{-1}\delta(E - E')\delta_{l l'}\,\delta_{l_3 l_3'} = \langle E\, l\, l_3\, \eta | E'\, l'\, l_3'\, \eta\rangle$$

$$= \int d^3p \langle E\, l\, l_3\, \eta | \mathbf{p}\,\eta\rangle \langle \mathbf{p}\,\eta | E'\, l'\, l_3'\, \eta\rangle$$

$$= \int p^2 \sin\theta_p\, dp\, d\theta_p\, d\phi_p\, f_\eta^*(p)\, \delta\left(E - \frac{p^2}{2m_\eta} - E_\eta^{\text{int}}\right)$$

$$\times Y_{l l_3}^*(\theta_p, \phi_p) f_\eta(p)\delta\left(E' - \frac{p^2}{2m_\eta} - E_\eta^{\text{int}}\right) Y_{l' l_3'}(\theta_p, \phi_p)$$

$$= m_\eta \rho_\eta(E)|f_\eta(p_\eta(E))|^2 \delta(E - E')\delta_{l l'}\,\delta_{l_3 l_3'}. \qquad (1.12)$$

In the last step we have used the fact that if $g(x)$ has only simple zeros $x_i$, then

$$\delta(g(x)) = \sum_i \frac{\delta(x - x_i)}{|g'(x_i)|}; \qquad (1.13)$$

the quantity $p_\eta(E)$ that appears on the right-hand side of (1.12) is therefore the positive zero of the equation

$$E - \frac{p^2}{2m_\eta} - E_\eta^{\text{int}} = 0,$$

i.e. is

$$p_\eta(E) = \sqrt{2m_\eta(E - E_\eta^{\text{int}})}. \qquad (1.14)$$

Equation (1.12), together with

$$\rho_\eta(E) = m_\eta p_\eta(E) \qquad (1.15)$$

of (XIV.5.18a), gives

$$|f_\eta(p_\eta(E))| = \rho_\eta(E)^{-1}, \qquad (1.16)$$

so that with an appropriate choice of phase for the generalized eigenvectors $|\mathbf{p}\, \eta\rangle$, the transition coefficients (1.11) are given by

$$\langle p\, \theta_p \phi_p \eta | E\, l\, l_3 \eta \rangle = (\rho_\eta(E))^{-1} \delta\left( E - \frac{p^2}{2m_\eta} - E_\eta^{\text{int}} \right) Y_{ll_3}(\theta_p, \phi_p). \qquad (1.17)$$

Equation (1.1) tells us that $T$ is a scalar operator; its matrix elements with respect to the angular-momentum basis are then [using the Wigner–Eckart theorem (V.3.6)]

$$\langle E\, l\, l_3\, \eta_b | T | E'\, l'\, l_3'\, \eta_A \rangle = \langle l'\, l_3'\, 0\, 0 | l\, l_3 \rangle \langle E\, l\, \eta_b \| T \| E'\, l'\, \eta_A \rangle$$

$$= \delta_{ll'} \delta_{l_3 l_3'} \langle E\, l\, \eta_b \| T \| E'\, l\, \eta_A \rangle. \qquad (1.18)$$

Use of (1.7), (1.17), and (1.18) in (1.6) gives

$$\langle \mathbf{p}_b \eta_b | T | \mathbf{p}_A \eta_A \rangle = \sum_{ll_3} \int dE\, dE'\, \delta\left( E - \frac{\mathbf{p}_b^2}{2m_b} - E_b^{\text{int}} \right) \delta\left( E' - \frac{\mathbf{p}_A^2}{2m_A} - E_A^{\text{int}} \right)$$

$$\times \langle E\, l\, \eta_b \| T \| E'\, l\, \eta_A \rangle Y_{ll_3}(\Omega_b) Y_{ll_3}^*(\Omega_A), \qquad (1.19)$$

where $\Omega_A = \mathbf{p}_A / |\mathbf{p}_A| = (\theta_A, \phi_A)$ is the direction of the incident particle and $\Omega_b = \mathbf{p}_b / |\mathbf{p}_b| = (\theta_b, \phi_b)$ is the direction of the detected particle. If we perform the summation $\sum_{l_3}$ by using the addition theorem (VII.3.22)

$$\sum_{l_3 = -l}^{+l} Y_{ll_3}(\Omega_b) Y_{ll_3}^*(\Omega_A) = \frac{2l + 1}{4\pi} P_l(\Omega_b \cdot \Omega_A) \qquad (1.20)$$

for spherical harmonics, then (1.19) becomes

$$\langle \mathbf{p}_b\, \eta_b | T | \mathbf{p}_A\, \eta_A \rangle = \frac{1}{4\pi} \sum_l (2l + 1) P_l(\cos\theta) \langle E_b\, l\, \eta_b \| T \| E_A\, l\, \eta_A \rangle, \qquad (1.21)$$

360 Elastic and Inelastic Scattering for Spherically Symmetric Interactions

where $\theta$ is the angle between $\mathbf{\Omega}_A$ and $\mathbf{\Omega}_b$ (cf. Figure XIV.3.1), and where[1]

$$E_A = E(p_A, \eta_A) = \frac{p_A^2}{2m_A} + E_{\eta_A}^{\text{int}} \tag{1.22a}$$

$$E_b = E(p_b, \eta_b) = \frac{p_b^2}{2m_b} + E_{\eta_b}^{\text{int}}. \tag{1.22b}$$

The reduced matrix element $\langle E_b\, l\, \eta_b \| T^+ \| E_A\, l\, \eta_A \rangle$ depends upon the energy and the internal quantum numbers $\eta_b$ and $\eta_A$, but does not depend upon the direction. As one would have expected for a spherically symmetric problem in which the initial conditions distinguish only the direction $\mathbf{\Omega}_A$, the $T$-matrix element (1.21) does not depend upon the angle $\phi$ around the direction $\mathbf{\Omega}_A$.

Equation (1.21) proves the statement preceding and justifying the definition (XIV.5.63). Written in terms of the scattering amplitude, (1.21) becomes, for $\eta_b = \eta_A$ and $E_b = E_A$,

$$T(p_A, \theta) = \sum_l (2l + 1) P_l(\cos \theta)(-\pi m)\langle E_A\, l\, \eta_A \| T \| E_A\, l\, \eta_A \rangle \tag{1.21'}$$

In the expressions for the cross section [e.g., (XIV.5.38), (XIV.5.50), and (XIV.5.60)] and in any other physical quantity, the factor $\delta(E_b - E_A)$ appears as a consequence of energy conservation [cf. (XV.3.13)]. Consequently only the reduced matrix elements

$$\langle \eta_b \| T_l(E_A) \| \eta_A \rangle \equiv \langle E_b = E_A\, l\, \eta_b \| T \| E_A\, l\, \eta_A \rangle \tag{1.23}$$

on the energy shell $E_b = E_A$ have physical meaning. Energy conservation for the nonrelativistic case is given by

$$p_A^2/2m_A + E_{\eta_A}^{\text{int}} = E_A = E_b = p_b^2/2m_b + E_{\eta_b}^{\text{int}}. \tag{1.24}$$

In nonrelativistic collision experiments one usually has $m_b = m_A$, but does not have $E_{\eta_A}^{\text{int}} = E_{\eta_b}^{\text{int}}$ except in the case of elastic scattering. In the following we shall examine in detail the case in which the detected particle is the same as the incident particle, so that $m_b = m_A$; first, however, we give the expressions for the cross section in the general (spinless) case.

The differential cross section per unit solid angle for scattering from an initial state with momentum $\mathbf{p}_A$ and internal quantum numbers $\eta_A$ into a final state with internal quantum numbers $\eta_b$ is [cf. (XIV.5.60) or (XIV.5.61')] at the end of Section XIV.4]

$$\frac{d\sigma^{\eta_b}}{\partial\Omega_b} = (2\pi)^4 \frac{m_A m_b}{p_A} \int p_b(E_b)\, dE_b |\langle E_b \theta_b \phi_b \eta_b | T | p_A \theta_A \phi_A \eta_A \rangle|^2$$

$$\times\, \delta(E_b - E_A). \tag{1.25}$$

---

[1] Note the change in notation: $E_A$ here denotes the total energy, whereas in Section XIV.5 $E_A$ denoted the projectile (kinetic) energy.

By using (1.21) we may rewrite $d\sigma^{\eta_b}/d\Omega_b$ in terms of the reduced matrix elements (1.23):

$$\frac{d\sigma^{\eta_b}}{d\Omega_b} = \pi^2 \frac{m_A m_b}{p_A} p_b(E_A)|\sum_l (2l+1)P_l(\cos\theta)\langle\eta_b\|T_l(E_A)\|\eta_A\rangle|^2. \quad (1.26)$$

The dependence of the momentum $p_b = p_b(E_b, \eta_b)$ of the detected particle on $E_b (= E_A)$ and $\eta_b$ is

$$p_b(E_A) = \sqrt{2m_b(E_A - E_b^{\text{int}})}, \quad (1.27)$$

as determined by (1.24) or (1.14).

The differential cross section given by (1.26) may be expressed as a series in the Legendre polynomials $P_l$, but the result is complicated (this will be done in Section XVII.8, where it is applied to the phase shift analysis). We therefore give here only the expression for the total cross section, which may be expressed simply in terms of the reduced matrix elements. By using the orthogonality property

$$\int_{-1}^{+1} d\xi \, P_l(\xi)P_{l'}(\xi) = \frac{2}{2l+1}\delta_{ll'} \quad (1.28)$$

of the Legendre polynomials one obtains

$$\sigma^{\eta_b} = \int d\Omega_b \frac{d\sigma^{\eta_b}}{d\Omega_b}$$

$$= 4\pi^3 \frac{m_A m_b}{p_A} p_b(E_A) \sum_l (2l+1)|\langle\eta_b\|T_l(E_A)\|\eta_A\rangle|^2 \quad (1.29)$$

for the total cross section. This suggests that we define the $l$th *partial cross section*

$$\sigma_l^{\eta_b} \equiv 4\pi^3 \frac{m_A m_b}{p_A} p_b(E_b)(2l+1)|\langle\eta_b|T_l(E_A)\|\eta_A\rangle|^2 \quad (1.30)$$

and that we write the total cross section as

$$\sigma^{\eta_b} = \sum_l \sigma_l^{\eta_b}. \quad (1.31)$$

$\sigma_l^{\eta_b}$ and $\sigma^{\eta_b}$ in (1.30) and (1.31) are the partial and total cross sections, respectively, for scattering from a state with internal quantum numbers $\eta = \eta_A$ into a state with internal quantum numbers $\eta = \eta_b$. Often the detector registers not just those states with a well-defined value $\eta_b$ of $\eta$ but all states with values of $\eta$ within some range. This happens, for example, if $\eta$ are the internal quantum numbers of the target and the detector registers scattered projectiles with any possible energy. The projectile may excite the target into any internal state $\eta$ by losing the corresponding amount of kinetic energy, and the detector cannot distinguish between elastically scattered projectiles and inelastically scattered projectiles. In such a case the $l$th partial cross section is given by

$$\sigma_l = \sum_{\eta_b} \sigma_l^{\eta_b} = 4\pi^3 \frac{m_A m_b}{p_A} (2l + 1) \sum_{\eta_b} p_b(E_A) |\langle \eta_b \| T_l(E_A) \| \eta_A \rangle|^2, \quad (1.32)$$

and the total cross section is given by

$$\sigma = \sum_{\eta_b} \sigma^{\eta_b} = \sum_l \sigma_l = \sum_{\eta_b l} \sigma_l^{\eta_b}. \quad (1.33)$$

Instead of the reduced matrix elements $\langle \eta_b \| T_l(E_A) \| \eta_A \rangle$ it is convenient to use the "partial-wave amplitudes." They are commonly used in nonrelativistic scattering theory (particularly in the case of elastic scattering), and formulas written in terms of them can be immediately taken over to the relativistic theory. One proceeds from the assumption that the initial state is fixed by the internal quantum numbers $\eta_A$, which usually describes the state of lowest energy. From the reduced matrix elements one defines the *partial-wave amplitudes* by[2]

$$T_{l\eta_A}^{\eta_b}(E_A) = T_l^{\eta_b}(p_A) \equiv -\pi \sqrt{m_A m_b} \, \langle \eta_b \| T_l(E_A) \| \eta_A \rangle, \quad (1.34)$$

where $p_A$ is given in terms of $E_A$ and $\eta_A$ by (1.22a) and in terms of $p_b$, $E_{\eta_b}^{int}$, and $E_{\eta_A}^{int}$ by (1.24). No additional quantum numbers $\eta_b$ are required to describe elastic scattering, for which $\eta_b = \eta_A$, and one simply writes[3]

$$T_l(p_A) \equiv T_{l\eta_A}^{\eta_A}(p_A) = -\pi \sqrt{m_A m_A} \, \langle \eta_A \| T_l(E_A) \| \eta_A \rangle. \quad (1.35)$$

The partial-wave amplitude $T_l^{\eta_b}(p_A)$ for the inelastic case $(\eta_b \neq \eta_A)$ is often called the *partial-wave reaction amplitude*, in distinction to the *partial-wave amplitude* $T_l(p_A)$ for elastic scattering. In terms of the partial-wave amplitudes the differential cross section (1.26) is written

$$\frac{d\sigma^{\eta_b}}{d\Omega_b} = \frac{p_b(E_A)}{p_A} \left| \sum_l (2l + 1) P_l(\cos \theta) T_l^{\eta_b}(p_A) \right|^2, \quad (1.36)$$

while the $l$th partial cross section (1.30) is written

$$\sigma_l^{\eta_b} = 4\pi \frac{p_b(E_A)}{p_A} (2l + 1) |T_l^{\eta_b}(p_A)|^2. \quad (1.37)$$

The $l$th partial cross section (1.32) for elastic scattering is then

$$\sigma_l(\text{elastic}) = \sigma_l^{\eta_A} = 4\pi(2l + 1) |T_l(p_A)|^2, \quad (1.38)$$

and the $l$th partial cross section for inelastic scattering is

$$\sigma_l(\text{inelastic}) = \sum_{\eta_b \neq \eta_A} \sigma_l^{\eta_b} = \frac{4\pi}{p_A} (2l + 1) \sum_{\eta_b \neq \eta_A} p_b(E_A) |T_l^{\eta_b}(p_A)|^2. \quad (1.39)$$

---

[2] See footnote 3 below.
[3] We shall suppress the lable $\eta_A$ designating the internal properties of the intial state, whenever this does not obfuscate the notation.

The final momentum $p_b(E_A)$ that appears in the above equations may be expressed in terms of the initial momentum and the difference between the initial and final internal energies by

$$p_b(E_A)^2/2m_b = p_A^2/2m_A - (E_b^{int} - E_A^{int}). \qquad (1.40)$$

The $S$-matrix is given in terms of the (on-the-energy-shell) $T$-matrix by[4]

$$\langle E_b\, l\, l_3\, \eta_b | S | E_A\, l'\, l_3'\, \eta_A \rangle = \langle E_b\, l\, l_3\, \eta_b | E_A\, l'\, l'\, \eta_A \rangle$$
$$- 2\pi i \delta(E_b - E_A) \langle E_A\, l\, l_3\, \eta_b | T | E_A\, l'\, l_3'\, \eta_A \rangle. \qquad (1.41)$$

Spherical symmetry, which was initially stated by (1.1), is equivalently stated by

$$[S, L_i] = 0. \qquad (1.42)$$

$S$ is thus a scalar operator and by the Wigner–Eckart theorem has matrix elements of the form

$$\langle E_b\, l\, l_3\, \eta_b | S | E_A\, l' l_3'\, \eta_A \rangle = \delta_{ll'} \delta_{l_3 l_3'} \langle E_b\, l\, \eta_b \| S \| E_A\, l\, \eta_A \rangle. \qquad (1.43)$$

From energy conservation [Equation (XV.3.13)],

$$[S, K] = 0, \qquad (1.44)$$

it follows that

$$\langle E_b\, l\, \eta_b \| S \| E_A\, l\, \eta_A \rangle = \delta(E_b - E_A)\rho_b(E_A)^{-1/2} \rho_A(E_A)^{-1/2} \langle \eta_b \| S_l(E_A) \| \eta_A \rangle, \qquad (1.45)$$

where $\rho_\eta(E)$ is given by (1.14) and (1.15). The factoring off of the weight functions $\rho_b^{-1/2}$ and $\rho_A^{-1/2}$ is a convention made in the definition of $\langle \eta_b \| S_l(E_A) \| \eta_A \rangle$ [but not in definition of $\langle \eta_b \| T_l(E_A) \| \eta_A \rangle$] in order to simplify later expressions (e.g., the unitarity condition in terms of the $S$- and $T$-matrices, which then can be carried over directly to the relativistic case). After substitution of (1.18) and (1.23) and of (1.43) and (1.45), the connection (1.41) between the reduced matrix elements of $S$ and $T$ becomes

$$\langle \eta_b \| S_l(E_A) \| \eta_A \rangle = \delta_{\eta_b \eta_A} - 2\pi i \sqrt{\rho_b(E_A)\rho_A(E_A)} \langle \eta_b \| T_l(E_A) \| \eta_A \rangle. \qquad (1.46)$$

In terms of the partial-wave amplitudes this connection is given by

$$S_l(E_A) \equiv \langle \eta_A \| S_l(E_A) \| \eta_A \rangle = 1 + 2ip_A\, T_l(p_A) \qquad (1.47)$$

for elastic scattering and by

$$\langle \eta_b \| S_l(S_A) \| \eta_A \rangle = 2i\sqrt{\rho_b(E_A)p_A}\, T_l^{\eta_b}(p_A) \qquad (\eta_b \neq \eta_A) \qquad (1.48)$$

for inelastic reaction processes.

----

[4] The $S$- and $T$-operators were discussed in more detail in Section XV.3, and (1.41) is identical with (XV.3.36). If Chapter XV was skipped in the first reading, the reader may take (1.41) as the definition of the $S$-matrix.

## XVI.2  Unitarity and Phase Shifts

In the previous section we expressed the cross section in terms of the transition matrix and also in terms of the partial-wave amplitudes [Equations (1.25) and (1.36)]. The only condition used was the assumption (1.1) of spherical symmetry. This allowed us to write the cross section in terms of some known functions of the scattering angle and the partial-wave amplitudes. For a given scattering problem the partial-wave amplitudes can be determined once the potential is known, in the same way that the scattering amplitude can be determined from the potential, by solving the integral equation (XV.3.37). There is, however a general condition that permits us to obtain some general properties of the scattering amplitude. This is the unitarity of the $S$-operator, which is an expression of the conservation of overall probability and which is a consequence (for the case that time development is described by a Hamiltonian, as specified by the axiom V) of the Hermiticity of the Hamiltonian.

The unitarity of the scattering operator,

$$S^\dagger S = I \qquad \text{(and } SS^\dagger = I\text{),} \tag{2.1}$$

is written in the angular-momentum basis as

$$\sum_{l l_3 \eta_b} \int \rho_b(E_b)\, dE_b \langle \tilde{E}\, \tilde{l}\, \tilde{l}_3\, \tilde{\eta}| S^\dagger | E_b\, l\, l_3\, \eta_b\rangle \langle E_b\, l\, l_3\, \eta_b| S | E_A\, l'\, l'_3\, \eta_A\rangle$$

$$= \langle \tilde{E}\, \tilde{l}\, \tilde{l}_3\, \tilde{\eta}| E_A\, l'\, l'_3\, \eta_A\rangle = \rho_A(E_A)^{-1}\delta(\tilde{E} - E_A)\delta_{\tilde{l}l'}\,\delta_{\tilde{l}_3 l'_3}\delta_{\tilde{\eta}\eta_A}. \tag{2.2}$$

Using (1.43) and (1.45), this leads to the unitarity condition for the reduced $S$-matrix elements:

$$\sum_{\eta_b} \langle \eta_b \| S_l(E_A) \| \tilde{\eta}\rangle^* \langle \eta_b \| S_l(E_A) \| \eta_A\rangle = \delta_{\tilde{\eta}\eta_A}. \tag{2.3}$$

For the special case $\tilde{\eta} = \eta_A$ Equation (2.3) may be written as

$$|S_l(E_A)|^2 + \sum_{\eta_b \neq \eta_A} |\langle \eta_b \| S_l(E_A) \| \eta_A\rangle|^2 = 1, \tag{2.4}$$

where $\sum_{\eta_b}$ extends over all possible values of $\eta_b$ except $\eta_b = \eta_A$. Inserting (1.47) and (1.48) into (2.4) leads to the unitarity condition expressed in terms of the partial-wave amplitudes:

$$\mathrm{Im}\, T_l(p_A) = \sum_{\eta_b} \rho_b(E_A)| T_l^{\eta_b}(p_A)|^2$$

$$= p_A| T_l(p_A)|^2 + \sum_{\eta_b \neq \eta_A} \rho_b(E_A)| T_l^{\eta_b}(p_A)|^2. \tag{2.5}$$

Equation (2.5) is the partial-wave amplitude form of the "generalized optical theorem" (XV.3.41). The first term on the right-hand side corresponds to elastic scattering; the second term corresponds to inelastic reaction processes.

For a particular collision experiment the values of $\eta_b$ over which the sum in (2.3) and (2.5) extends may be limited to those permitted by the initial energy. If the kinetic energy of the projectile is lower than the energy necessary to excite the target, i.e., if

$$p_A^2/2m_A < E_{\eta_b}^{int} - E_{\eta_A}^{int} \qquad (\eta_b \neq \eta_A)$$

(for an "infinitely heavy" target that remains always at rest), then the projectile can only be scattered elastically, and all $\langle \eta_b \| S_l(E_A) \| \eta_A \rangle$ vanish for $\eta_b \neq \eta_A$. One says that only the "elastic channel" is open. For instance, if in $e - H$ scattering the kinetic energy $p^2/2m$ of the incident electron is lower than the energy difference between the ground state and the first excited state, $E(n = 1) - E(n = 2)$, then only elastic scattering

$$e + H \to e + H$$

is energetically possible, so only the elastic channel is open. As the kinetic energy of the incident beam is increased, one reaches a point above which it is energetically possible to raise the target into the first excited state $\eta_1$ with the projectile losing a corresponding amount of kinetic energy. This point is the *threshold* of an inelastic process. The threshold momentum for an infinitely heavy target is given by

$$p_{thresh}^2/2m_A = E_{\eta_1}^{int} - E_A^{int}.$$

Above this value for $p_A$, both $\langle \eta_A \| S_l(E_A) \| \eta_A \rangle$ and $\langle \eta_1 \| S_l(E_A) \| \eta_A \rangle$ are different from zero. One says that an "inelastic channel" has opened up. As the incident energy is further increased, other inelastic processes may become possible, and an increasing number of reduced $S$-matrix elements in (2.4) become nonzero.

There can be different kinds of inelastic channels. The simplest are the excitation channels, in which the internal state of the projectile (target) remains unchanged whereas the target (projectile) performs a transition from the initial state into an excited state. $\eta_A$ and $\eta_b$ then refer to the different internal quantum numbers of the target (projectile). (If the projectile is also a system with internal structure and its internal quantum numbers are contained in the set $\eta$, then these quantum numbers will remain fixed.) In the example of e-H scattering the excitation channels are the final states of the process

$$e + H(n = 1) \to e + H^*(n \geq 2).$$

The threshold energy for this process is $|E(n = 1) - E(n = 2)|$. If the incident energy is increased further, the target may ionize, i.e., the target may be raised to the energy at which it breaks up into two particles; one then says that the "ionization channel" has opened up. For e-H scattering the ionization channel is the final state of the process

$$e + H(n = 1) \to e + e + H^+$$

and the ionization threshold is $|E(n = 1) - E(n \to \infty)| = 13.53$ eV.

The internal quantum numbers $\eta_b$, or at least a subset of them, now also extend over a continuous set of values. In these cases it is often more practical to use for the postcollision system a different set of quantum numbers than for the precollision system. This becomes even more necessary for more complicated collision processes like rearrangement collision, in which the projectile is absorbed by the target and another particle is emitted. A simple example of such a process in atomic physics is

$$\gamma + \text{He} \rightarrow \text{He}^+ + e.$$

For more complicated precollision systems the number of possible scattering channels increases—e.g., in the collision of He atoms by protons, where the following channels are possible:

$$
\begin{aligned}
\text{H}^+ + \text{He} \rightarrow \text{H}^+ + \text{He} & \quad \text{(elastic channel)} \\
\rightarrow \text{H}^+ + \text{He}^* & \quad \text{(excitation channels)} \\
\rightarrow \text{H}^+ + \text{He}^+ + e & \quad \text{(ionization channel)} \\
\rightarrow \text{H}^+ + \text{He}^{+*} + e & \quad \text{(ionization-excitation channels)} \\
\rightarrow \text{H} + \text{He}^+ & \quad \text{(rearrangement channel)} \\
\rightarrow \text{H}^* + \text{He}^+ & \quad \text{(rearrangement-excitation channels)} \\
\vdots &
\end{aligned}
$$

Many-channel processes occur also in nuclear physics and in particle physics, and it is in these areas that many-channel problems become particularly important. For instance in pion-proton ($\pi$-$p$) scattering one has (among others) the following channels:

$$
\begin{aligned}
\pi^- + p \rightarrow \pi^- + p & \quad \text{(elastic channel)} \\
\rightarrow \pi^- + \pi^0 + p & \\
\rightarrow \eta + n & \\
\rightarrow K^0 + \Lambda & \\
\rightarrow \pi^- + p^* &
\end{aligned}
$$

($\eta$ and $K^0$ denote mesons heavier than $\pi$; $n$ and $\Lambda$ denote the neutron and the "strange" baryon $\Lambda$, which is heavier than the proton and the neutron; $p^*$ denotes various excited states of the proton.)

   All scattering channels (i.e., the elastic channel and all inelastic channels) are simultaneously open if the initial energy is high enough. These scattering channels are various final states, differing in the internal quantum numbers $\eta_b$, and the sums in (2.4) and (2.5) must be extended over all channels that are open, i.e., energetically possible. The interaction can, of course, be such that an open channel, say $\eta_{\bar{b}}$, does not contribute much to a particular partial-wave amplitude, i.e., that $|\langle\eta_{\bar{b}}\|S_l(E_A)\|\eta_A\rangle|$ is small or zero even for energies above the threshold for the channel $\eta_{\bar{b}}$. These are detailed properties of the interaction ("the dynamics"), which require further information.

   We initially assume that all inelastic channels are closed. Equation (2.4) then becomes

$$|S_l(E_A)|^2 = 1, \tag{2.6}$$

For elastic scattering it thus follows that $S_l(E_A)$ is a function of $E_A$ (or, equivalently, of $p_A$) of modulus 1 and consequently may be written

$$S_l(E_A) = e^{2i\delta_l(E_A)}; \qquad (2.7)$$

$\delta_l(E_A)$ is a real-valued function defined up to an integral multiple of $\pi$ by (2.7). $\delta_l(E_A)$ is called the *scattering phase shift*, for reasons we will meet later.

The elastic partial-wave amplitude that appears in (1.47) may also be expressed in terms of the phase shift,

$$T_l(p_A) = \frac{S_l(p_A) - 1}{2ip_A} = \frac{1}{2ip_A}(e^{2i\delta_l(p_A)} - 1)$$

$$= \frac{1}{p_A} e^{i\delta_l(p_A)} \sin\delta_l(p_A) = \frac{1}{p_A}\frac{\tan\delta_l(p_A)}{1 - i\tan\delta_l(p_A)} \qquad (2.8)$$

$$= \frac{1}{p_A}\frac{1}{\cot\delta_l(p_A) - i},$$

which in turn allows us to express the *l*th partial elastic cross section of (1.38) in terms of the phase shift:

$$\sigma_l(\text{elastic}) = \frac{4\pi}{p_A^2}(2l + 1)\sin^2\delta_l(p_A). \qquad (2.9)$$

We would like to extend the phase-shift formalism to include the inelastic or reaction channels. If the sum over the reaction channels in (2.4) is nonzero (i.e., if $E_A$ is sufficiently high that some of the reaction channels are open), then $S_l(E_A)$ can no longer be written as in (2.7) with real $\delta_l(E_A)$; but it may be parametrized in the form[5]

$$S_l(E_A) = \eta_l(E_A)e^{2i\delta_l(E_A)} \qquad (2.10)$$

with real $\delta_l(E_A)$ and with real $\eta_l(E_A)$ satisfying

$$0 \le \eta_l(E_A) \le 1. \qquad (2.11)$$

That $\eta_l$ satisfies (2.11) may be seen by inserting (2.10) into (2.4),

$$\eta_l(E_A)^2 = 1 - \sum_{\eta_b \ne \eta_A} |\langle\eta_b\|S_l(E_A)\|\eta_A\rangle|^2, \qquad (2.12)$$

and observing that

$$\sum_{\eta_b \ne \eta_A} |\langle\eta_b\|S_l(E_A)\|\eta_A\rangle|^2$$

[5] The $\eta_l$ that appears in (2.10) is not to be confused with our use of $\eta$ to denote the internal quantum numbers. We use it here because this notation of $\delta_l$ and $\eta_l$ is often used in particle physics; in other areas of physics the notation may be different—e.g., in atomic physics $\eta_l$ is often the symbol for the phase shift denoted by $\delta_l$ here.

cannot be larger than 1. Equation (2.12) can be written in terms of the partial-wave reaction amplitudes by using (1.48):

$$\eta_l(E_A)^2 = 1 - 4p_A \sum_{\eta_b \neq \eta_A} p_b(E_A) |T_l^{\eta_b}(p_A)|^2. \qquad (2.13)$$

The partial-wave amplitude for the elastic channel may be expressed in terms of $\delta_l(p_A)$ and $\eta_l(p_A)$ as [cf. (1.47)]

$$T_l(p_A) = \frac{S_l(p_A) - 1}{2ip_A} = \frac{\eta_l(p_A)e^{2i\delta_l(p_A)} - 1}{2ip_A}$$

$$= \frac{\eta_l(p_A)\sin 2\delta_l(p_A)}{2p_A} + i\frac{1 - \eta_l(p_A)\cos 2\delta_l(p_A)}{2p_A}. \qquad (2.14)$$

From (1.38) and (2.14) it then follows that the $l$th partial elastic cross section is

$$\sigma_l(\text{elastic}) = \frac{\pi}{p_A^2}(2l + 1)(1 + \eta_l(p_A)^2 - 2\eta_l(p_A)\cos 2\delta_l(p_A)), \qquad (2.15)$$

while from (1.39) and (2.13) we have

$$\sigma_l(\text{inelastic}) = \frac{\pi}{p_A^2}(2l + 1)(1 - \eta_l(p_A)^2) \qquad (2.16)$$

for the $l$th partial inelastic cross section. The total $l$th partial cross section is the sum of these:

$$\sigma_l = \sigma_l(\text{total}) = \frac{2\pi}{p_A^2}(2l + 1)(1 - \eta_l(p_A)\cos 2\delta_l(p_A)). \qquad (2.17)$$

For $\eta_l = 1$ both (2.15) and (2.17) are identical with (2.9), which gives $\sigma_l$ in the case that all inelastic channels are closed, and (2.16) vanishes. From this and from (2.10) and (2.11) one sees that $\eta_l$ is a quantitative characterization of the amount of inelasticity involved in the scattering; $\eta_l$ is therefore called the *inelasticity coefficient*.

The total $l$th partial cross section may also be expressed in terms of the $l$th elastic partial-wave amplitude by adding (1.38) and (1.39) and using the generalized optical theorem (2.5) to get

$$\sigma_l = \frac{4\pi}{p_A}(2l + 1)\,\text{Im}\,T_l(p_A). \qquad (2.18)$$

If we sum (2.18) over all $l$ and use the imaginary part at $\theta = 0$ of the *elastic-scattering amplitude*

$$T(p_A, \theta) \equiv \sum_l (2l + 1)P_l(\cos\theta)T_l(p_A), \qquad (2.19)$$

we then obtain

$$\sigma = \sum_l \sigma_l = \frac{4\pi}{p_A} \sum_l (2l + 1)\,\text{Im}\,T_l(p_A) = \frac{4\pi}{p_A}\,\text{Im}\,T(p_A, \theta = 0). \qquad (2.20)$$

[Also used was the property $P_l(1) = 1$.] Equation (2.20), which connects the total cross section and the imaginary part of the forward-scattering amplitude, is usually called the *optical theorem*.

## XVI.3 Argand Diagrams

In order to familiarize ourselves with the properties of the scattering amplitude and its behavior under certain conditions we consider a graphical representation in which $p_A T_l(p_A)$ is regarded as a vector in the complex plane. These graphical representations are called *Argand diagrams* and play a particularly important role in the description and detection of resonances.

We start from the expression (2.14) for the elastic partial wave amplitude, which is written in the form

$$p_A T_l(p_A) - \frac{i}{2} = -\frac{i}{2} \eta_l(p_A) e^{2i\delta_l(p_A)}. \tag{3.1}$$

For the case of purely elastic scattering (e.g., in the region of low momentum where no inelastic channels are open) $\eta_l(p_A) = 1$, so $p_A T_l(p_A) - i/2$ is a complex number that lies on a circle of radius $1/2$. If $\delta_l(p_A)$ varies between $0 \leq \delta_l \leq \pi$ (modulo $\pi$) when $p_A$ is varied over the range in which only elastic scattering occurs, then all points of this circle are covered. In general, however, $\delta_l(p_A)$ will not vary over all of the interval $0 \leq \delta_l \leq \pi$, and hence $p_A T_l(p_A) - i/2$ will not vary over the full circle as $p_A$ is varied over that range in which only elastic scattering is possible.

The values of

$$p_A T_l(p_A) = \frac{i}{2} - \frac{i}{2} e^{2i\delta_l(p_A)} \qquad [\eta_l(p_A) = 1] \tag{3.2}$$

lie on a circle with center at $i/2$ and radius $1/2$. This is in agreement with the optical theorem (2.18), according to which Im $T_l(p_A) \geq 0$. The circle around $i/2$ with radius $1/2$ is called the *unitary circle*.

If there is absorption of energy by inelastic processes, then

$$p_A T_l(p_A) = \frac{i}{2} - \frac{i}{2} \eta_l(p_A) e^{2i\delta_l(p_A)} \qquad [\eta_l(p_A) < 1], \tag{3.3}$$

and the value of $p_A T_l(p_A)$ moves on a path inside the unitary circle as $p_A$ is varied. Figure 3.1 shows the graphical representation (Argand diagram) of $p_A T_l(p_A)$. One calls the circle the "unitary circle" because, as a consequence of the unitary condition, which is expressed by (3.3), it follows that the path of $p_A T_l(p_A)$ must lie inside this circle. The exact path followed depends upon the particular interaction. The most important case of these diagrams is the

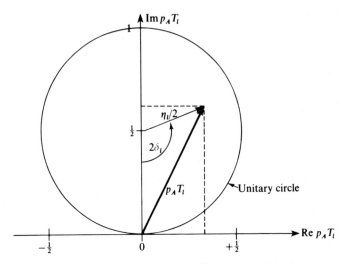

Figure 3.1   The elastic-scattering amplitude $p_A T_l(p_A)$ in the complex plane. (Argand diagram).

case in which the interaction between projectile and target is resonant. We will consider these cases below in Section XVIII.8.

Argand diagrams may also be drawn for the partial-wave amplitude $T_l^{\eta_b}(p_A)$ of an inelastic channel ($\eta_b \neq \eta_A$). One expresses the partial-wave amplitude $T_l^{\eta_b}(p_A)$ [or the reduced $S$-matrix element $\langle \eta_b \| S_l(E_A) \| \eta_A \rangle$ that corresponds to it by (1.48)] in a form analogous to (2.10):[6]

$$\sqrt{p_b(E_A)p_A}\, T_l^{\eta_b}(p_A) = \frac{1}{2i} \langle \eta_b \| S_l(E_A) \| \eta_A \rangle = \frac{1}{2i}\, \eta_l^b(p_A)e^{2i\delta_l^b(p_A)}. \quad (3.4)$$

Unitarity of $S$ ((2.3) or (2.4)) imposes the restrictions

$$0 \leq \eta_l^b(p_A) \leq 1 \quad (3.5a)$$

and

$$\eta_l(p_A)^2 + \sum_{\eta_b \neq \eta_A} \eta_l^b(p_A)^2 = 1. \quad (3.5b)$$

In the Argand diagram, $\sqrt{p_b(E_A)p_A}\, T_l^{\eta_b}(p_A)$ then moves as a function of $p_A$ on a path that lies inside the circle with center at the origin and radius $1/2$ (cf. Figure 3.2). (The path would lie on the circle if there were neither elastic scattering nor inelastic scattering into any channel other than the one with quantum numbers $\eta_b$.)

---

[6] $\eta_b, \eta_A$ denote the internal quantum numbers (channels). $\eta_l^b$ denotes the inelasticity coefficient for channel $\eta_b$ as in Equation (3.4).

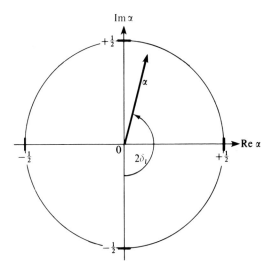

Figure 3.2   The amplitude $\alpha = \sqrt{p_b(E_A)p_A}\,T_l^{\eta b}(p_A)$ for scattering from channel $\eta_A$ to an inelastic channel $\eta_b$. In this case the center of the circle is at the origin and the amplitude is restricted to $\|\alpha\| \leq \frac{1}{2}$.

## Problem

1. (a)  Show that the Born approximation violates unitarity.
   (b)  Where in the Argand diagram for elastic scattering can be Born approxima-
        tion be considered a good approximation?

CHAPTER XVII

# Free and Exact Radial Wave Functions

The present chapter treats the spatial properties of the scattering process as described by the wave function. After an introduction in Section XVII.2 the differential equation for the radial wave function (the Schrödinger equation) is derived. Section XVII.3 presents the solutions of the free radial wave equation, and lists some of their properties. The properties of the exact radial wave functions, in particular their asymptotic forms and their connection to the phase shifts and the S-matrix, are discussed in Section XVII.4. In Section XVII.5 the connection between bound states and poles of the S-matrix on the positive imaginary axis is established. Some material about functions of a complex variable, which is needed in this chapter and in Chapter XVIII, is reviewed in a mathematical appendix.

## XVII.1  Introduction

In the preceding sections we discussed those properties of the scattering amplitude and the cross section that were consequences of very general assumptions like unitarity and spherical symmetry. We derived relations between the phase shifts, the inelasticity coefficients, the partial-wave amplitudes, and the $l$th partial $S$-matrix elements. The cross section was expressed in terms of these quantities. Nowhere in our discussion did we consider the spatial properties of the scattering process; in particular, we did not have to make use of position-dependent quantities such as the probability wave function.

The probability wave function, however, serves to provide a pictorial representation of the scattering process. In fact, it is in scattering theory that the wave function has a direct relation to observation, since it represents the probability for position measurements that can actually be performed. (For stationary states such measurements are not performed in practice, as has been discussed in the first half of the book; the wave function then is merely an auxiliary mathematical quantity.)

In the present chapter we introduce the wave function. As in the previous chapters, we consider two different points of view: Firstly, we assume time development of the system as given by a Hamiltonian $H = K + V$. This will provide a "deeper" foundation for the relations between the more directly accessible quantities. We then see, however, that these relations have a more general validity.

In addition to providing us with the wave function, this chapter will also establish the connection between our more general coordinate-free formulation of scattering theory and the usual formulation in terms of solutions of the Schrödinger wave equation. This latter formulation is directly obtained if the internal quantum numbers $\eta$ are ignored.

Though the Schrödinger differential equation constitutes a powerful tool in many cases (described by an interaction potential), it will turn out that the integral equation for the wave function (which is, in fact, just the Lippman–Schwinger equation in coordinate form) is of more immediate use.

## XVII.2 The Radial Wave Equation

We start with the Lippman–Schwinger equation in the angular momentum basis

$$|Ell_3\eta^+\rangle = |Ell_3\eta\rangle + \frac{1}{E - K + i0} V|Ell_3\eta^+\rangle.$$

It is assumed that the position operator for projectile relative to target, $\mathbf{Q}$, commutes with the internal observables $\eta^{op}$ of the system:

$$[Q_i, \eta^{op}] = 0.$$

By taking the scalar product of the Lippman–Schwinger equation with the generalized position eigenvector

$$|\mathbf{x}\,\eta'\rangle \equiv |x_1 x_2 x_3 \eta'\rangle \equiv |r\theta\phi\eta'\rangle,$$

we then get

$$\langle r\theta\phi\eta'|Ell_3\eta^+\rangle = \langle r\theta\phi\eta'|Ell_3\eta\rangle + \langle r\theta\phi\eta'|\frac{1}{E - K + i0} V|Ell_3\eta^+\rangle, \quad (2.1)$$

where $\langle r\theta\phi\eta'|Ell_3\eta\rangle$ and $\langle r\theta\phi\eta'|Ell_3\eta^+\rangle$ are called the *free* and *exact wave functions*, respectively.

After scattering, the projectile's position $\mathbf{x}$ and momentum $\mathbf{p}$ have the same direction (see Figure XIV.3.1):

$$\mathbf{x}/|\mathbf{x}| = \mathbf{p}/|\mathbf{p}|. \quad (2.2)$$

Thus $(\theta, \phi)$ are the angular coordinates of $\mathbf{x}$ as well as of $\mathbf{p}$. On account of spherical symmetry one can write, using the same arguments that led to Equation (VII.3.11),

$$\langle r\theta\phi\eta' | Ell_3\eta\rangle = \langle\theta\phi|ll_3\rangle\langle r\eta'|E\eta\rangle_l$$
$$= Y_{ll_3}(\theta\phi)\delta_{\eta\eta'}\langle r|E\rangle_l^\eta, \qquad (2.3a)$$

$$\langle r\theta\phi\eta' | Ell_3\eta^+\rangle = \langle\theta\phi|ll_3\rangle\langle r\eta'|E\eta^+\rangle_l$$
$$= Y_{ll_3}(\theta\phi)\rangle r\eta'|E\eta^+\rangle_l. \qquad (2.3b)$$

The reduced matrix elements $\sqrt{\pi/2}\langle r|E\rangle_l^\eta$ and $\sqrt{\pi/2}\langle r\eta'|E\eta^+\rangle_l$ are called the *free* and *exact radial wave functions* respectively.[1]

Using the Wigner–Eckart theorem for the scalar operator $V$, one obtains

$$\langle r\theta\phi\eta' | V | Ell_3\eta^+\rangle = \sum_{\tilde{E}\tilde{l}\tilde{l}_3\tilde{\eta}} \langle r\theta\phi\eta' | \tilde{E}\tilde{l}_3\tilde{\eta}^+\rangle\langle \tilde{E}\tilde{l}_3\tilde{\eta}^+ | V | Ell_3\eta^+\rangle$$

$$= \sum_{\tilde{E}\tilde{l}\tilde{l}_3\tilde{\eta}} \langle r\theta\phi\eta' | \tilde{E}\tilde{l}_3\tilde{\eta}^+\rangle \delta_{l\tilde{l}}\delta_{l_3\tilde{l}_3}\langle \tilde{E}\tilde{\eta}^+ \| V \| E\eta^+\rangle_l$$

$$= \sum_{\tilde{E}\tilde{\eta}} Y_{ll_3}(\theta, \phi)\langle r\eta' | \tilde{E}\tilde{\eta}^+\rangle_l\langle \tilde{E}\tilde{\eta}^+ \| V \| E\eta^+\rangle_l$$

where $\langle \tilde{E}\tilde{\eta}^+ \| V \| E\eta^+\rangle_l$ is the reduced matrix element of $V$. Thus

$$\langle r\theta\phi\eta' | V | Ell_3\eta^+\rangle = Y_{ll_3}(\theta, \phi)\langle r\eta' \| V \| E\eta^+\rangle_l, \qquad (2.3c')$$

where we have defined the reduced transition matrix element $\langle r\eta' \| V \| E\eta^+\rangle_l$ in terms of the reduced matrix elements of $V$ by

$$\langle r\eta' \| V \| E\eta^+\rangle_l = \sum_{\tilde{E}\tilde{\eta}} \langle r\eta' | \tilde{E}\tilde{\eta}^+\rangle_l\langle \tilde{E}\tilde{\eta}^+ \| V \| E\tilde{\eta}^+\rangle_l.$$

The reduced transition matrix element is a function of $l, r, E, \eta'$, and $\eta$ only. For spherically symmetric problems, $V$ is an operator function of the radius operator $Q \equiv (\mathbf{Q}^2)^{1/2}$. In general it also depends upon the internal observables, i.e., upon the operators that change the internal quantum numbers $\eta$.

We use the fact that $[Q, V] = 0$ to write

$$\langle r\eta' \| V \| \tilde{r}\tilde{\eta}\rangle = \langle r|\tilde{r}\rangle\langle\eta'\|\|V(r)\|\|\tilde{\eta}\rangle,$$

where $\langle r|\tilde{r}\rangle$ is the $\delta$-function with respect to the continuous summation over $\tilde{r}$ and $\langle\eta'\|\|V(r)\|\|\tilde{\eta}\rangle$ is a doubly reduced matrix element depending only on $r, \tilde{\eta}$, and $\eta'$.[2] Therefore

$$\langle r\eta' \| V \| E\eta^+\rangle_l = \sum_{\tilde{r}\tilde{\eta}} \langle r\eta' \| V \| \tilde{r}\tilde{\eta}\rangle\langle\tilde{r}\tilde{\eta} | E\eta^+\rangle_l$$

$$= \sum_{\tilde{r}\tilde{\eta}} \langle r|\tilde{r}\rangle\langle\eta'\|\|V(r)\|\|\tilde{\eta}\rangle\langle\tilde{r}\tilde{\eta} | E\eta^+\rangle_l.$$

[1] $\langle r\eta'|E\eta^+\rangle_l$ cannot, in general, be written as $\langle r|E^+\rangle_l^\eta\delta_{\eta\eta'}$, because $[H, \eta^{op}] = 0$ may not be true. Thus in general $|E\eta^+\rangle$ may not be an eigenvector of $\eta^{op}$. Only if $[H, \eta^{op}] = [V, \eta^{op}] = 0$ is this true. Cf. the remark following Equation (XVI.1.3).

[2] Applying the Wigner–Eckart theorem for the group generated by $Q$ to the scalar operator with respect to this group, $V$.

Performing the summation over $\tilde{r}$ and inserting the result,

$$\langle r\eta' \| V \| E\eta^+ \rangle_l = \sum_{\tilde{\eta}} \langle \eta' \| |V(r)| \| \tilde{\eta} \rangle \langle r\tilde{\eta} | E\eta^+ \rangle_l,$$

into (2.3c'), one obtains

$$\langle r\theta\phi\eta' | V | Ell_3\eta^+ \rangle = Y_{ll_3}(\theta, \phi) \sum_{\tilde{\eta}} \langle \eta' \| |V(r)| \| \tilde{\eta} \rangle \langle r\tilde{\eta} | E\eta^+ \rangle_l. \qquad (2.3c)$$

Using the same arguments, one can also show that

$$\left\langle r\theta\phi\eta' \left| \frac{1}{E - K + i} V \right| Ell_3\eta \right\rangle = Y_{ll_3}(\theta, \phi) \left\langle r\eta' \left| \frac{1}{E - K + i} V \right| E\eta^+ \right\rangle_l. \qquad (2.3d)$$

We shall now use the results (2.3) to obtain the Schrödinger differential equation for the radial wave function. In Section XVII.3 we shall consider the differential equation for the free radial wave function. Then, from the Lippman–Schwinger equation (2.1) for the wave function, we shall obtain, in Section XVII.4, the integral equation for the radial wave function. The advantage of the integral equation over the differential equation is that the former incorporates the boundary conditions; in this particular case the boundary conditions appropriate to the Lippman–Schwinger equation for $|E\hat{a}^+\rangle$ are those for a free incident state, as discussed in Section XV.2.

We assume that the incident beam has projectiles of one kind, whose mass $m$ is not changed by the scattering process. The differential equation is then obtained by taking the transition matrix element of the operator $H = K + V = \mathbf{P}^2/2m + K^{int}(\eta^{op}) + V$, where $\mathbf{P}$ is the projectile momentum operator:

$$E\langle r\theta\phi\eta' | Ell_3\eta^+ \rangle = \langle r\theta\phi\eta' | H | Ell_3\eta^+ \rangle$$

$$= \langle r\theta\phi\eta' | (\mathbf{P}^2/2m + K^{int} + V) | Ell_3\eta^+ \rangle$$

$$= \frac{1}{2m} \left( \frac{-1}{r^2} \frac{d}{dr} \left( r^2 \frac{d}{dr} \right) + \frac{l(l+1)}{r^2} + 2mE^{int} \right)$$

$$\times \langle r\theta\phi\eta' | Ell_3\eta^+ \rangle + \langle r\theta\phi\eta' | V | Ell_3\eta^+ \rangle \qquad (2.4)$$

[The second equality follows from Equation (VII.2.6) and (VII.3.4).]

Using (2.3a, b, c) in (2.4), we obtain a system of coupled differential equations:

$$\sum_{\tilde{\eta}} \left[ \left\{ \frac{1}{r^2} \frac{d}{dr} \left( r^2 \frac{d}{dr} \right) - \frac{l(l+1)}{r^2} + p_{\tilde{\eta}}(E)^2 \right\} \delta_{\eta\tilde{\eta}} \right.$$

$$\left. - 2m \langle \eta' \| |V(r)| \| \tilde{\eta} \rangle \right] \langle r\tilde{\eta} | E\eta^+ \rangle_l = 0, \qquad (2.5)$$

where

$$p_\eta(E) = \sqrt{2m(E - E_\eta^{int})}. \qquad (2.6)$$

In order to solve this system of equations one has to known the reduced matrix elements of the potential $\langle \eta' \|| V(r) \|| \eta \rangle$. These depend, of course, upon the particular process under investigation. For instance, in the many-channel problem of the elastic and inelastic scattering of electrons by atoms these reduced matrix elements can be determined from the Coulomb interaction and the atomic wave function.[3]

We shall not discuss the many-channel problem here any further[4] but restrict ourselves to the case

$$[V, \eta^{op}] = 0. \tag{2.7}$$

Then the problem reduces to a one-channel problem for each value of $\eta$ (i.e., for each set of values of the internal quantum numbers). To perform the reduction, we use the above arguments to write

$$\langle \eta' \|| V(r) \|| \tilde{\eta} \rangle = \delta_{\eta' \eta} V_{\tilde{\eta}}(r). \tag{2.8}$$

Since we now have $[H, \eta^{op}] = 0$, we can also write

$$\langle r\eta' | E\eta^+ \rangle_l = \delta_{\eta' \eta} \langle r | E^+ \rangle_l^\eta. \tag{2.9}$$

Under these conditions (1.5) goes over into a set of uncoupled equations, one for each channel $\eta$:

$$\left[ \frac{1}{r^2} \frac{d}{dr} \left( r^2 \frac{d}{dr} \right) - \frac{l(l+1)}{r^2} - 2mV_\eta(r) + p_\eta(E)^2 \right] \langle r | E^+ \rangle_l^\eta = 0. \tag{2.10}$$

In the following sections we shall discuss the properties of the radial wave functions $\langle r | E^+ \rangle_l^\eta$. Though $V_\eta(r)$ may vary with $\eta$, and $p_\eta^2$ may differ from $2mE = p^2$ by different constants $2mE_\eta^{int}$, the label $\eta$ is inessential, as it does not change in the scattering process. We shall therefore suppress it in most of what follows, reinstating it at the end to emphasize the fact that further quantum numbers may be present.

## XVII.3   The Free Radial Wave Function

The equation for the free radial wave function $\langle r | E \rangle_l$ is obtained from the eigenvalue equation of $K$:

$$E \langle r\theta\phi\eta' | Ell_3\eta \rangle = \langle r\theta\phi\eta' | K | Ell_3\eta \rangle$$

in the same way as (2.5) is obtained from (2.4). It differs from (2.5) only in that the term in $V$ is missing and that since $[K, \eta^{op}] = 0$,

$$\langle r\tilde{\eta} | E\eta \rangle_l = \delta_{\tilde{\eta}\eta} \langle r | E \rangle_l^\eta. \tag{3.1}$$

We thus have

$$\left\{ \frac{1}{r^2} \frac{d}{dr} \left( r^2 \frac{d}{dr} \right) - \frac{l(l+1)}{r^2} + p^2 \right\} \langle r | E \rangle_l^\eta = 0, \tag{3.2}$$

---

[3] For the scattering of electrons by hydrogen atoms this is discussed in Massey et al. (1969, Vol. 1, Section 7.1). A more general case is treated in Smith (1971, Section 2.1).

[4] We shall return to the many-channel problem in Chapter XX.

where $p = p_n(E) \equiv \sqrt{2m(E - E_n^{int})}$. We shall write $\langle r|E\rangle_l^\eta = \langle r|p\rangle_l^\eta$. Equation (3.2) is a well-known equation, which has as its linearly independent solutions the spherical Bessel and Neumann functions. The solution that is regular at $r = 0$, to which $\langle r|p\rangle_l$ is proportional, is the (proper) *spherical Bessel function* (spherical Bessel function of the first kind, or Riccati–Bessel function):[5]

$$j_l(pr) = j_l(z) = \left(\frac{\pi}{2z}\right)^{1/2} J_{l+1/2}(z)$$

$$= (-z)^l \left(\frac{1}{z}\frac{d}{dz}\right)^l \left(\frac{\sin z}{z}\right) \tag{3.3}$$

$$= z^{l+1} \sum_{n=0}^{\infty} \frac{(-z^2/2)^n}{n'(2l + 2n + 1)!!}, \qquad pr = z,$$

and the solution irregular at $r = 0$ is the *spherical Neumann function* (spherical Bessel function of the second kind or Riccati–Neumann function),

$$n_l(z) = -(-1)^l \left(\frac{\pi}{2z}\right)^{1/2} J_{-l-1/2}(z)$$

$$= -(-z)^l \left(\frac{1}{z}\frac{d}{dz}\right)^l \left(\frac{\cos z}{z}\right) \tag{3.4}$$

$$= z^{-l} \sum_{n=0}^{\infty} \frac{(-z^2/2)^n(2l - 2n - 1)!!}{n!}.$$

Here $J_{l+1/2}(z)$ and $J_{-l-1/2}(z)$ are ordinary Bessel functions of the first kind. The *spherical Hankel functions*, defined by

$$h_l(z) = j_l(z) + in_l(z), \tag{3.5a}$$

$$h_l^*(z) = j_l(z) - in_l(z), \tag{3.5b}$$

are then also independent solutions of (3.2). We note for later use that all the above relations also hold for complex values of $z$.

The asymptotic behavior of $j_l(z)$, $n_l(z)$, and $h_l(z)$ for real $z \to \infty$ is

$$j_l(z) \underset{z\to\infty}{\sim} \frac{1}{z} \sin\left(z - \frac{l\pi}{3}\right), \tag{3.6a}$$

$$n_l(z) \underset{z\to\infty}{\sim} -\frac{1}{z} \cos\left(z - \frac{l\pi}{2}\right), \tag{3.6b}$$

$$h_l(z) \underset{z\to\infty}{\sim} \frac{(-i)^{l+1}}{z} e^{iz}. \tag{3.6c}$$

---

[5] The double factorial is defined by

$$n!! \equiv \begin{array}{ll} n(n - 2)\cdots(5)(3)(1) & \text{if } n \text{ is odd,} \\ n(n - 2)\cdots(4)(2) & \text{if } n \text{ is even.} \end{array}$$

For future reference we note the bounds for complex values of $z$:

$$|j_l(z)| \leq \text{const} \left( \frac{|z|}{1 + |z|} \right)^{l+1} e^{|\text{Im } z|}, \tag{3.6d}$$

$$|h_l(z)| \leq \text{const} \left( \frac{|z|}{1 + |z|} \right)^{-l} e^{-\text{Im } z}. \tag{3.6e}$$

Near the origin we have

$$j_l(z) \xrightarrow[z \to 0]{} \frac{z^l}{(2l + 1)!!}, \tag{3.7a}$$

$$n_l(z) \xrightarrow[z \to 0]{} [-(2^l - 1)!!]z^{-l-1}. \tag{3.7b}$$

The normalization of the $j_l$'s is

$$\int_0^\infty r^2 \, dr \, j_l(pr)j_l(p'r) = \frac{\pi}{2p^2} \delta(p - p'), \tag{3.8a}$$

$$\int_0^\infty p^2 \, dp \, j_l(pr)j_l(pr') = \frac{\pi}{2r^2} \delta(r - r'). \tag{3.8b}$$

The spherical Neumann function cannot be delta-function normalized in this way. All spherical Bessel functions $j_l(z)$, $n_l(z)$, $h_l(z)$, and $h_l^*(z)$ fulfill the functional equation

$$\frac{d}{dz} f_l(z) = \frac{1}{2l + 1} (lf_{l-1}(z) - (l + 1)f_{l+1}(z)) \tag{3.9a}$$

and the following equality for the indefinite integral:

$$\int r^2 \, dr \, f_l(pr)f_l^*(p'r) = \frac{r^2}{p^2 - p'^2} [p'f_l(pr)f_{l-1}^*(p'r) - pf_{l-1}(pr)f_l^*(p'r)], \tag{3.9b}$$

of which (3.8) is a consequence.

To find the proportionality factor between $\langle r|p \rangle_l$ and $j_l(pr)$, we calculate the normalization integral of the free radial wave function $\langle r|p \rangle_l$. In spherical coordinates the $\delta$-function normalization of the generalized position eigenvectors reads

$$\langle \mathbf{x}|\mathbf{x}' \rangle = \delta^3(\mathbf{x} - \mathbf{x}') = \frac{1}{r^2 \sin \theta} \delta(r - r')\delta(\theta - \theta')\delta(\phi - \phi'); \tag{3.10}$$

and comparison with

$$\langle x | x' \rangle = \sum_{ll_3} \int \rho(E) \, dE \, \langle r\theta\phi | Ell_3 \rangle \langle Ell_3 | r'\theta'\phi' \rangle$$

$$= \sum_{ll_3} \int \rho(E) \, dE \, Y_{ll_3}(\theta, \phi) \langle r | E \rangle_l Y^*_{ll_3}(\theta', \phi') \langle r' | E \rangle^*_l$$

$$= \frac{1}{\sin \theta} \delta(\theta - \theta')\delta(\phi - \phi') \int \rho(E) \, dE \, \langle r | E \rangle_l \langle r' | E \rangle^*_l,$$

where we have used

$$\sum_{ll_3} Y_{ll_3}(\theta, \phi) Y^*_{ll_3}(\theta', \phi') = \frac{1}{\sin \theta} \delta(\theta - \theta')\delta(\phi - \phi'), \qquad \text{(VII.3.21)}$$

shows that the normalization of the radial wave function is

$$\int \rho(E) \, dE \, \langle r | E \rangle_l \langle r' | E \rangle^*_l = \frac{1}{r^2} \delta(r - r'), \qquad (3.11)$$

or, in terms of $p$,

$$\int p^2 \, dp \langle r | p \rangle_l \langle r' | p \rangle^*_l = \frac{1}{r^2} \delta(r - r'). \qquad (3.11')$$

As $\langle r | p \rangle_l$ is a solution of (3.2) regular at $r = 0$, and therefore proportional to $j_l(pr)$, one obtains by comparison of (3.11') with (3.8b) that

$$\langle r | p \rangle_l = \sqrt{2/\pi} j_l(pr). \qquad (3.12)$$

## XVII.4 The Exact Radial Wave Function

The precise properties of the exact radial wave functions $\langle r | E^+ \rangle_l$ depend upon the interaction Hamiltonian $V$, through the interaction potentials $V_\eta(r) = \langle x\eta | V | x\eta \rangle$. However, provided $V$ fulfills some very general conditions, one can make some general statements about $\langle r | E^+ \rangle_l$ without knowing the detailed form of $V$. In our cavalier treatment of the mathematical aspects these conditions have never been precisely stated, but their content is roughly as follows:

1. The interaction is not effective for large projectile-target separation, i.e., $V(r) \to 0$ as $r \to \infty$. Which precise assumption one has to make for the rate of falling off $V$ depends on what one wants to achieve: Usually one assumes that $V(r)$ falls off at least as fast as $1/r^3$. For the derivation of some properties one has to assume that $V(r)$ decreases faster than any power of $1/r$, and often one assumes a finite range for $V$.
2. The interaction does not change too abruptly as the projectile approaches the target: $V(r)$ is a continuous function of $r$ except at a finite number of finite discontinuities.

3.  There exists a lowest energy value (since the binding energy must be
    finite) or in other words the spectrum of $H$ has a lower bound. In terms
    of the potential this means that $V(r)$ is not too singular at the origin
    $r = 0$ (less singular than $r^{-3/2}$ if $V(r) < 0$).

Almost all physical potentials fulfill these conditions—potentials for
electrons scattering off an atom, potentials for atom-atom scattering, the
Yukawa potential, the square-well potential, etc. The Coulomb potential
does not fulfill condition 1 and therefore requires a separate mathematical
treatment. In practice, however, this is not of much consequence, because
electromagnetic screening is always present: One never really has the exact
Coulomb potential $1/r$, but a screened potential $\psi(r)/r$, with $\psi(r) = 0$ for
large $r$.[6] For the following discussion we assume that the above three conditions are fulfilled in whatever precise form we need.
   If

$$V(r) > 0 \quad \text{for all } r, \tag{4.1}$$

then the potential is said to be *repulsive*, while if

$$V(r) < 0 \quad \text{for all } r, \tag{4.2}$$

then it is said to be *attractive*.[7] If $V(r) < 0$ in some interval $r_1 < r < r_2$, then
$V(r)$ is said to be attractive in that interval. If $V(r)$ is attractive in some
interval, there may be a certain number of solutions of the differential equation (2.10) for a certain number of discrete, negative values of $p_n^2/2m = E_n$.
These solutions correspond to bound states of the projectile-target system
with binding energies $|E_n|$, as we shall discuss in the next section. They must
fall off sufficiently rapidly as $r \to \infty$, in addition to being regular at $r = 0$.
   The exact radial wave function $\langle r|E^+\rangle_l = \langle r|p^+\rangle_l$ for $p^2 \geq 0$ is the
solution of the differential equation (2.10) with certain boundary conditions:
In addition to being regular (i.e., $\langle r|p^+\rangle < \infty$ at $r = 0$), one would want, as
$r \to \infty$, $\langle r|p^+\rangle_l$ to represent the $l$th spherical component of an incident
plane wave and an outgoing scattered wave, and that for $V \to 0$, $\langle r|p^+\rangle_l \to$
$\langle r|p\rangle_l = \sqrt{2/\pi}j_l(pr)$, the acceptable solution of the free wave equation.
   Instead of solving (2.10) under these boundary conditions (see Problem
XVII.1), we shall obtain the exact radial wave function $\langle r|p^+\rangle_l$ from the
Lippman–Schwinger equation (2.1), which has the boundary conditions
already built in. Inserting (2.3a), (2.3b), (2.3d), (2.8), and (2.9) into (2.1), one
obtains the integral equation for the radial wave function $\langle r|E^+\rangle_l$:

$$\langle r|E^+\rangle_l$$
$$= \langle r|E\rangle_l + \int \rho(E')\, dE'\, r'^2\, dr'\, \langle r|E'\rangle_l \frac{1}{E - E' + i0} \langle r'|E'\rangle_l^* V(r') \langle r'|E^+\rangle_l. \tag{4.3}$$

---

[6] Note also that in general any assumed $V(r)$ is a mathematical idealization of the actual
situation.

[7] The forces between projectile and target are then also called repulsive and attractive,
respectively.

Now define the (*l*th-partial-wave) Green's function by

$$G_l^p(r;r') = \frac{1}{2m} \int_0^\infty \rho(E')\, dE'\, \frac{\langle r|E'\rangle_l \langle r'|E'\rangle_l^*}{E - E' + i0} \tag{4.4}$$

$$= \int_0^\infty p'^2\, dp'\, \frac{\langle r|p'\rangle_l \langle r'|p'\rangle_l^*}{p^2 - p'^2 + i0} \tag{4.5}$$

$$= \frac{2}{\pi} \int_0^\infty p'^2\, dp'\, \frac{j_l(p'r)j_l(p'r')}{p^2 - p'^2 + i0} \tag{4.6}$$

$$= \frac{1}{\pi} \int_{-\infty}^{+\infty} p'^2\, dp'\, \frac{j_l(p'r)j_l(p'r')}{p^2 - p'^2 + i0}, \tag{4.7}$$

where we have used the fact that the integrand is even. The integral equation (4.3) then becomes

$$\langle r|p^+\rangle_l = \langle r|p\rangle_l + 2m \int_0^\infty r'^2\, dr'\, G_l^p(r;r')V(r')\langle r'|p^+\rangle_l$$

$$= \sqrt{2/\pi}j_l(pr) + 2m \int_0^\infty r'^2\, dr'\, G_l^p(r;r')V(r')\langle r'|p^+\rangle_l. \tag{4.8}$$

Note that if the potential has a finite range, then the integral is over a finite interval.

The Green's function is evaluated using Equation (3.5) to write

$$G_l^p(r;r') = \frac{1}{2\pi} \int_{-\infty}^{+\infty} dq\, q^2\, \frac{j_l(qr)h_l(qr')}{p^2 - q^2 + i\epsilon} + \frac{1}{2\pi} \int_{-\infty}^{+\infty} dk\, k^2\, \frac{j_l(kr)h_l^*(kr')}{p^2 - k^2 + i\epsilon}. \tag{4.9}$$

Suppose $r' > r$. Then because of the asymptotic behavior (3.6e) of $h_l$, and (3.6d), the product $j_l(qr)h_l(qr')$ decreases exponentially in the upper half complex $q$-plane, and one may extend the integration path of the first integral to include a large semicircle in the upper half plane (cf. Figure 4.1). From Equations (3.3), (3.4), and (3.5) it is seen that

$$j_l(kr) = (-1)^l j_l(-kr) \quad \text{and} \quad h_l^*(kr') = (-1)^l h_l(-kr')$$

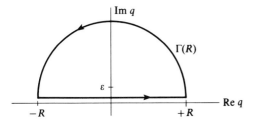

Figure 4.1   The integration path $\Gamma(R)$ used to evaluate the first integral in Equation (4.9). As $R \to \infty$, the contribution from the semicircle vanishes, leaving only the integral along the Re $q$ axis.

and hence with the substitution $q = -k$, the second integral of (3.9) is the same as the first integral. We may now rewrite (3.9) as

$$G_l^p(r; r') = \lim_{R \to \infty} \frac{-2}{2\pi} \oint_{\Gamma(R)} dq \frac{q^2 j_l(qr) h_l(qr')}{(p+q)(q-p-i\epsilon')} \qquad \left(\epsilon' = \frac{\epsilon}{p+\epsilon}\right),$$

where $\Gamma(R)$ is the closed contour indicated by Figure 4.1. The spherical Bessel functions are entire functions, and the function $q^2 j_l(qr) h_l(qr')/(p+q)$ is analytic inside $\Gamma(R)$, so by the Cauchy residue theorem[8] we obtain

$$G_l^p(r; r') = -2i \frac{p^2}{p+p} j_l(pr) h_l(pr')$$

$$= -ipj_l(pr) h_l(pr') \qquad (r' > r). \tag{4.10a}$$

A similar procedure, but with $r$ and $r'$ interchanged, gives the Green's function for $r' < r$:

$$G_l^p(r; r') = -ipj_l(pr') h_l(pr) \qquad (r' < r). \tag{4.10b}$$

Equations (4.10) are often combined as

$$G_l^p(r; r') = ipj_l(pr_<) h_l(pr_>) \tag{4.11}$$

where

$$r_< = \min(r, r') \tag{4.12a}$$

and

$$r_> = \max(r, r'). \tag{4.12b}$$

We already know the asymptotic behavior of the free radial wave function $\langle r|p \rangle_l = \sqrt{2/\pi} j_l(pr)$ for $r \to \infty$ [cf. (3.6a)]. Let us now examine the asymptotic behavior of the exact radial wave function $\langle r|p^+ \rangle_l$. Using (3.6a), (3.6c), and (4.10b) in (4.8), we obtain

$$\sqrt{\frac{\pi}{2}} \langle r|p^+ \rangle_l \underset{r \to \infty}{\sim} \frac{1}{pr} \sin\left(pr - \frac{l\pi}{2}\right)$$

$$+ 2m \sqrt{\frac{\pi}{2}} \int_0^\infty r'^2 \, dr' \left[ -ipj_l(pr') \frac{(-i)^{l+1}}{pr} e^{ipr} \right] V(r') \langle r'|p^+ \rangle_l$$

$$= \frac{-1}{2ipr} \left\{ e^{-i(pr - l\pi/2)} \right.$$

$$\left. - e^{i(pr - l\pi/2)} \left[ 1 - (2ip)(\sqrt{2\pi m}) \int_0^\infty r'^2 \, dr' \, j_l(pr') V(r') \langle r'|p^+ \rangle_l \right] \right\}$$

$$\tag{4.13}$$

---

[8] For a review of important results in the theory of functions of a complex variable, see Appendix XVII.A.

The first term represents an incoming partial wave, whereas the second term represents an outgoing one. (We shall discuss this further at the end of this section.) The outgoing wave is modified with respect to the incoming one by the factor in square brackets. This factor may be related to the elastic scattering phase shift defined by Equation (XVI.2.7) for every value of the internal quantum number $\eta$. Note that having chosen the quantum numbers $\eta$ such that (2.7) holds, $\langle \eta_b \| S_l(E) \| \eta_A \rangle$ vanishes for $\eta_b \neq \eta_A$. Thus we have here only elastic scattering for every value of $\eta$.

Recall, now, that the $l$th partial-wave amplitude defined by (XVI.1.35), (XVI.1.18), and (XVI.1.23) is

$$T_l(p) = -\pi m \langle Ell_3 | T | Ell_3 \rangle = -\pi m \langle Ell_3 | V | Ell_3^+ \rangle, \qquad (4.14)$$

where $E = E(p) = p/2m + E^{\text{int}}$, the label $\eta$ having been suppressed. In terms of the radial wave function this is written

$$T_l(p) = -\pi m \int r^2 \sin\theta \, dr \, d\theta \, d\phi \langle Ell_3 | r\theta\phi \rangle \langle r\theta\phi | V | Ell_3^+ \rangle$$

$$= -\pi m \left[ \int \sin\theta \, d\theta \, d\phi \; Y_{ll_3}^*(\theta, \phi) Y_{ll_3}(\theta, \phi) \right]$$

$$\times \left[ \int r^2 \, dr \, \langle r | E \rangle_l^* V(r) \langle r | E^+ \rangle_l \right]$$

$$= -\pi m \int_0^\infty r^2 \, dr \, \sqrt{\frac{2}{\pi}} j_l(pr) V(r) \langle r | p^+ \rangle_l, \qquad (4.15)$$

where Equations (2.3a), (2.3c), (2.8), (2.9), (VII.3.16), and (3.12) have been used. Using (XVI.1.47) and XVI.2.7), it follows from (4.15) that

$$S_l(p) = e^{2i\delta_l(p)} = 1 + 2ip T_l(p)$$

$$= 1 - (2ip)(\sqrt{2\pi}m) \int_0^\infty r^2 \, dr \, j_l(pr) V(r) \langle r | p^+ \rangle_l. \qquad (4.16)$$

Although Equations (4.15) and (4.16) are important in themselves, expressing the $l$th partial-wave amplitude and the phase shift in terms of the potential and the exact radial wave function, they are of little practical use. This is because the differential equation (2.10) and the integral equation (4.8) are not easily solved for $\langle r | p^+ \rangle_l$, nor is $\langle r | p^+ \rangle_l$ as directly related to the experimental data as are $\delta_l(p)$ and $T_l(p)$.

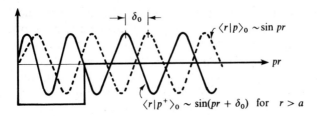

Figure 4.2   The $l = 0$ free and exact radial wave functions for an attractive square well.

We make use of (4.15) and (4.16) to rewrite (4.13) in various useful forms:

$$\sqrt{\pi/2}\langle r|p^+\rangle_l \underset{r\to\infty}{\sim} \frac{1}{pr}\sin\left(pr - \frac{l\pi}{2}\right) + pT_l(p)\frac{e^{i(pr - l\pi/2)}}{pr} \qquad (4.17a)$$

$$\underset{r\to\infty}{\sim} \frac{1}{2ipr}\{e^{i(pr - l\pi/2)}e^{2i\delta_l(p)} - e^{-i(pr - l\pi/2)}\} \qquad (4.17b)$$

$$\underset{r\to\infty}{\sim} \frac{e^{i\delta_l(p)}}{pr}\sin\left(pr - \frac{l\pi}{2} + \delta_l(p)\right) \qquad (4.17c)$$

$$\underset{r\to\infty}{\sim} \frac{e^{-il\pi/2}}{2ipr}\{e^{ipr}S_l(p) - (-1)^l e^{-ipr}\}. \qquad (4.17d)$$

Comparison of (4.17c) with the asymptotic form of the free wave function obtained from (4.12) and (3.6a),

$$\sqrt{\frac{\pi}{2}}\langle r|p\rangle_l = j_l(pr) \underset{r\to\infty}{\sim} \frac{1}{pr}\sin\left(pr - \frac{l\pi}{2}\right) \qquad (4.18)$$

explains why $\delta_l(p)$ is called the "phase shift": The interaction shifts the phase of the asymptotic exact radial wave function $\langle r|p^+\rangle_l$ by $\delta_l(p)$ as compared with the asymptotic free radial wave function $\langle r|p\rangle_l$. This property also serves as an alternative definition of $\delta_l$ in place of (XVI.2.7). The phase shift for the $l = 0$ wave in a square-well potential is shown in Figure 4.2.

For an attractive potential (like the square well in Figure 4.2) the wave function is "pulled" into the interaction region and the phase shift is positive. The exact wave function emerges from the interaction region with its phase in advance of the free wave function. For a repulsive potential the wave function is "pushed" out of the interaction region and the phase shift is negative.

That this connection between the sign of the phase shift and the sign of the potential is generally fulfilled is best established in the small-phase-shift approximation:

$$e^{i\delta_l(p)}\sin\delta_l(p) \approx \delta_l(p).$$

From (4.15), using (XVI.2.8), one then obtains

$$\delta_l(p) \approx -\pi m p \int_0^\infty r^2 \, dr \, \sqrt{2/\pi} j_l(pr) V(r) \langle r | p^+ \rangle_l,$$

which in the Born approximation is given by

$$\delta_l(p) \approx -2mp \int_0^\infty r^2 \, dr \, \{j_l(pr)\}^2 V(r).$$

To fully justify the above statement that (4.13) and (4.17) represent an incoming and an outgoing partial wave, we have to consider the time-dependent wave function of a state $\phi^+(t)$ that develops in time from a prepared in-state.

The exact radial wave function $\langle r | p^+ \rangle_l$ is the $p$th component of the radial wave function of an angular-momentum eigenstate $\phi_{ll_3}^+(t)$ that develops in time according to the equation

$$\phi_{ll_3}^+(t) = \frac{1}{2\pi} \int dE \; \tilde{\phi}(E) | E l l_3^+ \rangle e^{-iEt}, \tag{4.19}$$

where $\tilde{\phi}(E) \equiv \tilde{\phi}(p)$, $E = p^2/2m$, is the energy (momentum) distribution of this state in the remote past.[9] The wave function of this state is

$$\langle r\theta\phi | \phi_{ll_3}^+(t) \rangle = Y_{ll_3}(\theta\phi) \langle r | \phi^+(t) \rangle_l, \tag{4.20}$$

where the radial wave function $\langle r | \phi^+(t) \rangle_l$ of the state $\phi^+(t)$ is given by

$$\langle r | \phi^+(t) \rangle_l = \frac{1}{2\pi} \int_0^\infty dE \; \tilde{\phi}(E) \langle r | E^+ \rangle_l e^{-iEt}. \tag{4.21}$$

Multiplying (4.17d) by $\sqrt{2/\pi}(1/2\pi)\tilde{\phi}(E)e^{-iEt}$ and integrating over $E$ then gives

$$\langle r | \phi^+(t) \rangle_l \underset{r\to\infty}{\sim} \frac{e^{-i(\pi/2)(l-1)}}{2\pi^2 m} \int_0^\infty dp \; \tilde{\phi}(p) e^{-iEt} \frac{1}{r} (e^{-ipr}(-1)^l - S_l(p)e^{ipr})$$

$$\sim \frac{e^{-i(\pi/2)(l-1)}}{2\pi^2 m} \left[ \int_0^\infty dp \; \tilde{\phi}(p) \frac{e^{-i(pr+Et)}(-1)^l}{r} \right.$$

$$\left. - \int_0^\infty dp \; \tilde{\phi}(p) S_l(p) \frac{e^{i(pr-Et)}}{r} \right]. \tag{4.22}$$

Thus the radial wave function of the angular-momentum eigenstate $\phi_{ll_3}^+(t)$ consists asymptotically of an incoming wave with the original momentum distribution and an outgoing wave in which the original momentum distribution has been modified by $S_l(p)$ due to the interaction. Equations (4.13) and (4.17) merely state this result in the time-independent form for

---

[9] That this $\phi^+(t)$, connected with the basis vectors $|Ell_3^+\rangle$, is a state that develops from a prepared in-state has been shown in Section XV.2 and is here of no further consequence.

the $p$th component of the radial wave function, i.e., for an (unphysical) exact momentum eigenstate.

The asymptotic forms of the radial wave function (4.22) and (4.17) have been derived from the Lippman–Schwinger equation under the assumption that the interaction is described by an interaction Hamiltonian $V$. Then $S_l(p)$ is given in terms of $V(r)$ by (4.16). Even if the time-development axiom is not assumed, there exists a unitary operator $S$ that transforms states before the interaction into states after the interaction. If this interaction is of finite range, then outside the interaction region the wave function should still be a superposition of incoming and outgoing waves. Thus the radial wave function for large distances $r$ should still be given by (4.22), where $S_l(p)$, the $l$th $S$-matrix element, is now the fundamental quantity. Consequently, the $p$th component of the radial wave function, $\langle r | p^+ \rangle_l$, should have the asymptotic form (4.17).

Thus in the framework of a Hamiltonian time development, the asymptotic forms (4.17) and (4.22) are derived, while in the framework of $S$-matrix theory the asymptotic forms (4.17) and (4.22) are assumed and are justified as a consequence of the superposition principle for the wave function outside the interaction region.

## XVII.5  Poles and Bound States[10]

In scattering processes, the variable $p$ that we have been considering in the preceding section is the magnitude of the incident momentum, and therefore a real positive quantity. Accordingly, the variable $E = p^2/2m$ representing the kinetic energy (or $E - E_n^{\text{int}} = p^2/2m$ if the internal energy has to be taken into account) is a real positive quantity. We know, however, that other values of $E$ also have physical significance. For example, the solutions of (2.10) for negative values of $p^2 = 2mE$ are the bound-state radial wave functions, and the discrete negative energy values $E_n$, for which (2.10) has normalizable solutions $\langle r | E_n \rangle_l$, are the energy levels of the projectile-target bound system (as has been discussed in Section VII.3).

One should then also be able to extend the integral equation (4.8) [with $G_l^p$ given by (4.11)], which follows from (2.10), to negative-energy solutions. Such solutions can be obtained by the following line of reasoning: Replace $p$ by the complex variable $z$. For $\operatorname{Im} z \geq 0$, the arguments that lead to the Green's function (4.11) are still valid. (For $\operatorname{Im} z < 0$, similar arguments, wherein one closes the contour in the lower half of the complex $z$-plane, hold, and one gets analogous expressions with $h_l$ replaced by $h_l^*$.)

---

[10] For this section, as well as for some sections of Chapter XVIII, we shall require some basic results from the theory of functions of a complex variable. These results are outlined in a mathematical appendix to this chapter, Appendix XVII.A. The reader who is unfamiliar with complex variable theory, or who merely wishes to refresh his memory, may consult the appendix before reading this section.

The fact that (4.11) holds in this general case does not necessarily mean that (4.8) can also be continued to Im $z > 0$. Let us assume, however, that the potential $V(r)$ is such that it can be so continued, and let us assume that $H$ has a bound state with energy $E = -\alpha^2/2m$. Then, at the point $z = i\alpha$, the radial wave function $\langle r|i\alpha^+\rangle_l$ (being, as a wave function of a bound state, an element of the space of infinitely differentiable, rapidly decreasing functions that is a realization of the Schwartz space) must asymptotically decrease faster than any power of $r^{-1}$, e.g., it must decrease exponentially. Now according to (3.6d), $j_l(i\alpha r)$ diverges exponentially. The second term in the RHS of (4.8) is bounded as $r \to \infty$, because $h_l(i\alpha r)$ decreases exponentially according to (3.6e). Thus $\langle r|i\alpha^+\rangle_l$ can decrease for $r \to \infty$ only if the term $\langle r|p\rangle_l = \sqrt{2/\pi}j_l(pr)$, originating according to (4.13) or (4.17a) from the incident state, drops out of Equation (4.8). This can be achieved by considering, instead of (4.8), the equation obtained by multiplying Equation (4.8) by $(z - i\alpha)$:

$$(z - i\alpha)\langle r|z^+\rangle_l = (z - i\alpha)\langle r|z\rangle_l$$
$$+ 2m\alpha \int_0^\infty r'^2 \, dr' \, j_l(i\alpha r_<)h_l(i\alpha r_>)V(r')(z - i\alpha)\langle r'|z^+\rangle_l.$$

$$(5.1)$$

Equation (5.1) is identically fulfilled at $z = i\alpha$ unless $\langle r|z^+\rangle_l$ has a pole at $z = i\alpha$, i.e., in the neighborhood of $i\alpha$ is of the form[11]

$$\langle r|z^+\rangle_l = \frac{1}{z - i\alpha} \langle r|E = -\alpha^2/2m\rangle_l$$

$$+ \text{(function of } z \text{ analytic around } z = i\alpha).  \qquad (5.2)$$

Then (5.1) goes over into

$$\langle r|E = -\alpha^2/2m\rangle_l = 2m\alpha \int_0^\infty r'^2 \, dr' \, j(i\alpha r_<)h_l(i\alpha r_>)V(r')\langle r'|E = -\alpha^2/2m\rangle_l,$$

$$(5.3)$$

which is the homogeneous integral equation for the discrete number of normalizable solutions of the Schrödinger equation (2.10) with eigenvalues $E = -\alpha^2/2m$. Thus we see that a bound-state solution is obtained if we assume that $\langle r|z^+\rangle$ is of the form (5.2), i.e., that it has a pole on the positive imaginary momentum axis at the point $z = i\alpha = i\sqrt{2m|E|}$, where $E$ is the bound-state energy eigenvalue for angular momentum $l$. The bound-state radial wave function $\langle r|E = -\alpha^2/2m\rangle_l$ is then the residue at this pole.

From (4.15) and (4.16) one then concludes that the $l$th partial-wave amplitude and the $l$th $S$-matrix element have poles on the positive imaginary

---

[11] The above arguments will also hold for a pole of higher order, but one can prove that these poles are simple (which we shall not, however, do here).

momentum axis at the values $p = i\sqrt{2m|E_n^l|}$, where $E_n^l$ $(n = 1, 2, \ldots)$ are the discrete eigenvalues of the Hamiltonian $H$. In a theory with Hamiltonian time development, these discrete eigenvalues correspond to the energy levels of the projectile-target bound states of angular momentum $l$.

Though the above arguments[12] show that a bound state of the Hamiltonian with energy $E_n^l$ corresponds to a pole of the $l$th partial $S$-matrix element $S_l(p)$ on the positive imaginary axis, the reverse need not hold: $T_l(p)$ and $S_l(p)$ may have singularities that are not connected with the bound states.[13]

Nevertheless, the hypothesis that bound states of angular momentum $l$ correspond to poles of $S_l(p)$ on the positive imaginary axis (and vice versa) has become quite generally accepted. Especially in those cases where a Hamiltonian time development is not assumed to exist (relativistic $S$-matrix theory) and there is no operator whose spectral properties characterize the bound states, one usually assumes a one-to-one correspondence between bound states and poles on the positive imaginary axis.

## XVII.6  Survey of Some General Properties of Scattering Amplitudes and Phase Shifts

The precise properties of the scattering amplitudes and the phase shifts depend upon the interaction. If the interaction is described by a potential, one can calculate phase shifts and partial-wave scattering amplitudes as a function of scattering energy, as has been done in some of the problems. There are, however, general properties of these functions that do not depend upon the particular form of the interaction, and we want to list them here.

The Born approximation for the partial-wave amplitude, obtained from (4.15) by inserting there $\langle r | p^+ \rangle \approx \langle r | p \rangle = \sqrt{2/\pi} j_l(pr)$ namely

$$T_l^{(\text{Born})}(p) = -\pi m \frac{2}{\pi} \int_0^\infty r^2 \, dr \, j_l(pr) V(r) j_l(pr), \tag{6.1}$$

is believed to be good for high energies and weak potentials, because in the higher-order terms the potential and the spherical Bessel functions, which fulfill (3.6a), appear in higher powers.

The asymptotic behavior of the $l$th partial $S$-matrix element is then

$$S_l(p) \to 1 \quad \text{as } p \to \infty. \tag{6.2}$$

This follows immediately from (4.16) using the Born approximation (6.1) and the asymptotic behavior (3.6a) of $j_l(pr)$. An intuitive argument for (6.2)

---

[12] For potential scattering, the connection between simple poles and bound-state eigenvalues of $H$ can be made more precise, Cf. Taylor (1972, Chapter 12).

[13] These "redundant" poles may occur if the asymptotic expressions in the complex $p$-plane differ from (4.17). For interactions that are cut off at a finite distance and also for potentials that fall off at infinity faster than any exponential (e.g. $e^{-\mu r}$), "redundant" poles are absent.

is that with increasing energy of the projectile the effect of a given interaction will become less important.

The phase shift $\delta_l(p)$, therefore, tends to a multiple of $\pi$ as $p \to \infty$. As the phase shift is defined by (XVI.2.7), only within a multiple of $\pi$ one can remove this modulo $\pi$ ambiguity by defining

$$\delta_l(p) \to 0 \quad \text{as } p \to \infty. \tag{6.3}$$

Requiring then that $\delta_l(p)$ be a continuous function of $p$ (which is possible because $S_l(p)$ is continuous) make $\delta_l(p)$ unique.

Clearly $S_l(p) \to 1$ also if the interaction goes to zero. Then with the convention (6.3) it follows that also

$$\delta_l(p) \to 0 \quad \text{for interaction going to zero } [V(r) \to 0]. \tag{6.4}$$

For a given potential and energy

$$T_l \to 0 \quad \text{and} \quad S_l \to 1 \quad \text{as } l \to \infty. \tag{6.5}$$

Intuitively, this can be understood from (2.10) by regarding the term $l(l + 1)/2mr^2$ as a repulsive centrifugal potential. The larger $l$ is, the more repulsive is this centrifugal barrier and the less effective does the actual potential $V(r)$ become. Let $R$ be the range of the interaction. Then for values $l$ such that $l(l + 1)/2mR^2$ is much larger than the kinetic energy $E = p^2/2m$ the projectile is unlikely to penetrate into the range of the interaction. Thus for

$$l(l + 1) \gg p^2 R^2, \quad \text{or } l \gg pR,$$

the partial-wave amplitude $T_l(p)$ will be negligible:

$$T_l(p) \approx 0, \quad S_l(p) \approx 1 \quad \text{for } l \gg pR. \tag{6.6}$$

It follows from (6.6) that for the phase shift

$$\delta_l \approx n\pi, \quad n \text{ integer} \quad \text{for } l \gg pR. \tag{6.7}$$

Equations (6.6) and (6.7) are important properties of which one makes use when one expresses the differential cross section in terms of the partial-wave amplitudes as in (XVI.1.36). For a given value of $E = p^2/2m$ the infinite sum in (XVI.1.36) can, because of (6.6), be approximated by a finite sum.

To obtain the phase-shift behavior at low energies, we turn to (4.15). For small values of $p$, $\langle r | p^+ \rangle_l$ and $j_l(pr)$ have the same $p$-dependence, given by (3.7a), since according to (4.11),

$$G_l^p(r, r') \to -ip \frac{p^l r_<^l}{(2l + 1)!!} \left( \frac{p^l r_>^l}{(2l + 1)!!} - i(2l + 1)!! p^{-l-1} r_<^{-l-1} \right)$$

$$G_l^p(r, r') \to -\frac{r_<^l r_>^{-l-1}}{2l + 1} \quad \text{for } p \to 0 \tag{6.8}$$

and $\langle r|p^+\rangle_l$ and $j_l(pr)$ are connected by (4.8). Consequently, it follows from (4.15), for potentials that vanish for $r$ greater than some $R$, that

$$T_l(p) \rightarrow -\pi m \left( \int_0^R r^2 \, dr \, \sqrt{\frac{2}{\pi}} \, r^l V(r) r^l \, \frac{1}{(2l+1)!!} \, \frac{\langle r|p^+\rangle_l}{j_l(pr)} \right) p^{2l}$$

$$T_l(p) \rightarrow -a_l p^{2l} \quad \text{for } p \rightarrow 0 \tag{6.9}$$

where $a_l$ is a constant given by the integral in brackets, which does not depend upon $p$, because $\langle r|p^+\rangle_l/j_l(pr) \rightarrow$ (function of $r$) for $p \rightarrow 0$.[14]

The constants $a_l$ are called the *scattering lengths*. Only the s-wave scattering length $a_0$ has the dimension of a length. (6.9) shows that for $p = 0$ all partial-wave amplitudes except the s-wave vanish. For the s-wave

$$T_0(p) \rightarrow -a_0 \quad \text{as } p \rightarrow 0, \tag{6.10}$$

and consequently by (XVI.1.37) and (XVI.1.39),

$$\frac{d\sigma}{d\Omega} \rightarrow a_0^2, \quad \sigma \rightarrow 4\pi a_0^2, \quad \text{as } p \rightarrow 0. \tag{6.11}$$

Using (XVI.2.8) it follows further that

$$\delta_l(p) - n\pi \rightarrow -a_l p^{2l+1} \quad \text{as } p \rightarrow 0, \tag{6.12}$$

and consequently also

$$\tan \delta_l(p) \rightarrow -a_l p^{2l+1} \quad \text{as } p \rightarrow 0. \tag{6.13}$$

Equation (6.13) shows that

$$p^{2l+1} \cot \delta_l(p) \rightarrow -\frac{1}{a_l}$$

or their inverses are the appropriate functions to consider near $p = 0$ (as long as the scattering length $a_l \neq 0$ or $a_l \neq \infty$). The power series for this quantity near $p = 0$ is:

$$p^{2l+1} \cot \delta_l(p) = -\frac{1}{a_l} + \frac{r_l}{2} p^2 + O(p^4). \tag{6.14}$$

That only even powers appear in the power-series expansion of $p^{2l+1} \cot \delta_l(p)$ follows from $\delta_l(-p) = -\delta_l(p)$, which is established in Equation (XVIII.5.7) below.

For $l = 0$, (6.14) is a very useful approximation called the *effective-range approximation*:

$$p \cot \delta_l(p) = -\frac{1}{a_0} + \frac{r_0}{2} p^2. \tag{6.15}$$

---

[14] If the potential does not have a finite range, some modifications are necessary, but one can show that for $V(r)$ decreasing faster than any power of $1/r$ the result (6.9) still holds.

The constant $r_0$ is called the effective range of the potential; one can show that $r_0$ is roughly proportional to the range of the potential. Equation (6.15) is a good approximation for energies small compared to the potential energy and has been one of the most important parameterizations of low-energy scattering data, in particular for neutron-proton scattering.

Without giving a derivation, we mention that the sign of $a_0$ is related to whether or not an $S$-wave bound state is present. If the scattering length $a_0 < 0$, no bound state is possible, and for $a_0 > 0$, a bound state of target and projectile is formed.

## XVII.A Mathematical Appendix

In this appendix we review some basic *results* in the theory of analytic functions of a complex variable. For proofs of these results, and for more on the theory of analytic functions, the reader is referred to, e.g., Smirnov (1964, Vol. II, Part 2).

### a. Definition of Analyticity

The function $f(z) = u + iv$ of the complex variable $z = x + iy$ is said to be analytic in some region $R$ of the $z$-plane when

1.  $df/dz$ is continuous and independent of the direction of $dz$ in the region $R$. (The following conditions are equivalent.)
2.  $\partial u/\partial x = \partial v/\partial y \ \partial u/\partial y = -\partial v/\partial x$ (Cauchy–Riemann conditions). These derivatives are also continuous in $R$.
3.  $\oint_C f(z)\, dz = 0$ for any closed contour $C$ lying within $R$.

If $f(z)$ is analytic in $R$, then so is $d^n f/dz^n$ for all $n = 0, 1, 2, \dots$.

### b. Cauchy Integral Formula

If $f(z)$ is analytic within some simply connected region $R$, then

$$\oint \frac{f(z)\, dz}{z - a} = 2\pi i f(a), \tag{A.1a}$$

$$\oint \frac{f(z)\, dz}{(z - a)^{n+1}} = \frac{2\pi i}{n!} f^{(n)}(a) \equiv \frac{2\pi i}{n!} \frac{d^n f(a)}{da^n} \tag{A.1b}$$

for contours entirely within $R$ and enclosing the point $a$. If $a$ is excluded, the integrals vanish. (Integration is assumed to be counterclockwise.) Thus the values of $f(z)$ within $R$ are completely determined by the values on the boundary.

## c. Taylor Series

If $f(z)$ is analytic within and on a circle $C$ centered at $z = z_0$, then one can expand $f(z)$ about $z_0$:

$$f(z) = \sum_{n=0}^{\infty} a_n(z - z_0)^n, \tag{A.2a}$$

where

$$a_n = \frac{1}{n!} f^{(n)}(z_0). \tag{A.2b}$$

This Taylor series is uniformly convergent as long as the radius of $C$ is less than the "radius of convergence," which is the distance from $z_0$ to the nearest singularity of $f(z)$, i.e., the nearest point where $f$ is nonanalytic. Conversely, any power series represents an analytic function within its radius of convergence.

## d. Laurent Series

If $f(z)$ is analytic within and on the annular region between two concentric circles $C_1$ and $C_2$ centered at $z = z_0$, then for any point $z$ within the annular region one has the uniformly convergent power series

$$f(z) = \sum_{n=-\infty}^{\infty} a_n(z - z_0)^n, \tag{A.3a}$$

where

$$a_n = \frac{1}{2\pi i} \oint_C \frac{f(z)\, dz}{(z - z_0)^{n+1}}, \tag{A.3b}$$

$C$ being any contour within the annulus.

## e. Singularities

Points of nonanalyticity of $f$ are called singularities of $f$.

1.  A single-valued function $f$ can have *isolated singularities* of the following kinds:

    (a)  A singularity at $z = a$ is called a *pole of order n* if as $z \to a$

$$f(z) \to \frac{g(z)}{(z - a)^n} \qquad (n = 1, 2, \ldots), \tag{A.4}$$

   where $g(z)$ is analytic at $z = a$, and $g(a) \neq 0$. A pole of order 1 is called a *simple pole*.

(b) If $f(z)(z - a)^n$ diverges as $z \to a$ for all finite $n$, then $a$ is called an essential singularity of $f$.

In the Laurant series for $f$ about a pole of order $n$, $-n$ is the lowest power one encounters. The Laurent series about an essential singularity contains infinitely many terms of negative power.

2. A multivalued function $f$ (e.g., $f = z^{1/2}$) has *branch points*, which always occur in pairs. At a branch point $z = a$, the function $f(a + \varepsilon e^{i\phi})$, where $\varepsilon$ is such that $a + \varepsilon e^{i\phi}$ does not include another branch point of $f$ for all $\phi$, is not periodic in $\phi$ with period $2\pi$.

A function that is analytic in a region of the complex plane, except for a finite number of poles in that region, is said to be *meromorphic* in that region. If the function is meromorphic in the entire $z$-plane, except at the point $\infty$, it is called a *meromorphic function* of $z$. If the function is analytic in the entire $z$-plane, except at $\infty$, it is called an *entire* function.

## f. Cauchy Residue Theorem

The coefficient $a_{-1}$ in the Laurent series is called the *residue* of $f(z)$ at $z = z_0$. If $f(z)$ is analytic within and on a closed contour $C$, except for a finite number of poles and essential singularities, then

$$\oint_C f(z)\, dz = 2\pi i \sum (\text{residues in } C) \qquad (A.5)$$

## g. Analytic Continuation

Let two functions be analytic in a region $R$, and have the same values (a) in some subregion of $R$, or (b) on a line segment in $R$, or (c) at a denumerably infinite number of points having a limit point within $R$. Then they are identical throughout $R$. However, if one knows only that they differ at most by an amount $\epsilon$ in (a), (b), or (c), then, no matter how small $|\epsilon|$ is, the two functions may differ vastly within the rest of $R$.

The facts stated above illustrate the justification and the limitations of *analytic continuation*: If $f_1(z)$ is defined on a line segment $\Gamma$, and $f_2(z)$ is analytic in a region $R$ containing $\Gamma$, and if $f_1(z) = f_2(z)$ on $\Gamma$, then $f_2(z)$ is a unique analytic continuation of $f_1(z)$ onto $R$.

In the same spirit one can also do the following: Let $f_1(z)$ be given within a circle $R_1$, e.g., by a power series about the center $z_1$. Let $f_2(z)$ be the power-series expansion of $f_1(z)$ about a point $z_2$ near the periphery of $R_1$, the circle of convergence being $R_2$, which extends beyond $R_1$. Then the values of $f_2(z)$ in $R_2$ are uniquely determined by the values of $f_1(z)$ in $R_1$, and $f_2(z)$ is the analytic continuation of $f_1(z)$ into $R_2$. One can repeat this process

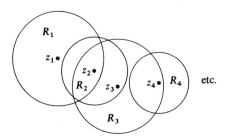

Figure A.1   Extension of a function defined in $R_1$ into the region $R_2 \cup R_3 \cup R_4 \cup \cdots$ by analytic continuation.

several times, obtaining a unique analytic continuation $F(z)$ of $f_1(z)$ into the region $R_1 \cup R_2 \cup R_3 \cup \ldots$ (see Figure A.1). A function cannot be analytically continued into a region containing a singularity of that function.

After repeating the process of analytic continuation several times, the $n$th circle of convergence may overlap the first one. Then the values of $f_n(z)$ in the overlap region may or may not coincide with those of $f_1(z)$. In the latter case the function is multivalued, and the region of encirclement (see Figure A.2) contains a simple branch point of the function.

We have seen above how the values of an analytic function in a large region of the complex plane are determined by its values in an arbitrary small region of analyticity. However, the analytic continuation does not depend on these initial values in a continuous manner. Therefore, it is impossible to express $F(z)$ in $R$ explicitly in terms of the values in a certain small region. Though $F(z)$ is uniquely prescribed by its values in the subregion, the problem of constructing $F(z)$ in $R$ from its values in the small subregion is not a correctly formulated one.

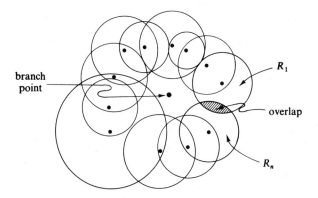

Figure A.2   Analytic continuation of a multivalued function around a branch point.

## h. Schwarz Reflection Principle

If $f(z)$ is analytic in a region $R$ that includes a segment of the real axis and if $f(z)$ is real on this segment, then $f(z)$ can be continued onto the region $R^* = \{z^* | z \in R\}$ and satisfies $f(z) = [f(z^*)]^*$ for all $z \in R \cup R^*$.

## i. Multivalued Functions

The function $w(z)$ can be considered as a map of the $z$-plane onto the $w$-plane. Consider the example $w(z) = z^2$, Figure A.3. It is clear from the figure that the upper half of the $z$-plane, including the positive real axis and excluding the negative real axis, is mapped onto the entire $w$-plane. So also is the lower half of the $z$-plane, including the negative real axis and excluding the positive real axis. Thus one complete circuit of the $z$-plane (e.g., *efgpqr*) involves two complete circuits of the $w$-plane (e.g., *e'f'g'p'q'r'*) provided the former circuit encloses the origin.

The inverse of a function such as $w(z)$ is a *multivalued function*. To continue with the above example, consider the inverse "function"

$$f(z) = z^{1/2}.$$

Writing $z$ and $f$ in polar form,

$$z = \rho e^{i\phi}, \qquad f = r e^{i\theta},$$

we have

$$r = \rho^{1/2}, \qquad \theta = \phi/2.$$

In making one complete circuit around the origin of the $z$-plane, $0 \leq \phi < 2\pi$, one has

$$f_1 = \rho^{1/2} e^{i\phi/2},$$

and the part of the $f$-plane so covered is the upper half including the $+\mathrm{Re}\, f$ axis and excluding the $-\mathrm{Re}\, f$ axis. Going around the $z$-plane once more

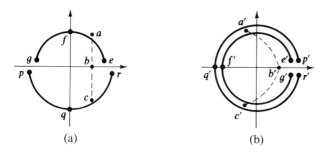

(a)                              (b)

Figure A.3   The function $w(z) = z^2$ (a) $z$-plane; (b) $w$-plane.

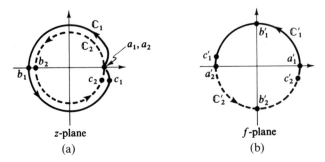

Figure A.4   The function $f(z) = z^{1/2}$ (a) $z$-plane; (b) $f$-plane.

$(2\pi \leq \phi < 4\pi)$, the point with polar angle $\phi$ now has polar angle $\phi + 2\pi$; consequently

$$f_2 = \rho^{1/2}e^{i(\phi + 2\pi)/2} = -\rho^{1/2}e^{i\phi/2}.$$

Making a third circuit would give back the value $f_1$ for $f$. Clearly $f$ is a double-valued function of $z$ (Figure A.4).

We have seen that it requires two circuits around the origin of the $z$-plane to encircle the origin of the $f$-plane, thus giving rise to the double-valued-ness of $f$. On the other hand, starting with some point $z_0$ ($\neq 0$) in the $z$-plane with well-defined argument[15] $\phi$ and following a closed contour $C$, one traces out a closed contour $C'$ in the $f$-plane (Figure A.5). Thus, as long as one does not encircle the origin one can treat $f$ as a single-valued function.

The two values of $z^{1/2}$ that occur for each value of $z$ form two independent sets, called *branches* of $z^{1/2}$ (e.g., upper and lower $z^{1/2}$-plane). On crossing the positive real $z$-axis one goes over from one branch of $z^{1/2}$ into the other. The positive real axis is called a *branch line*. If one analytically continues one branch of the function $z^{1/2}$ along a circle that enclosed the origin, one ends up with the other branch. The region of encirclement then contains a singularity of $z^{1/2}$. Since the circle can be made arbitrarily small, it is clear that the singularity is $z = 0$. This is called a *branch point* of the function $z^{1/2}$.

At a branch point, the multivalued function $f(z)$ has the same value for all branches of $z$. Branch points always occur in pairs, with a branch line joining a pair of branch points. In our example $z = \infty$ is the other branch point. Any curve joining these two branch points can serve as a branch line—the choice is a matter of definition and usually dictated by convenience.

## j. Riemann Surfaces

The theory of analytic functions outlined above for single-valued functions can be extended to a wide class of multivalued functions using a geometrical construct called a *Riemann surface*.

---

[15] More precisely, defined up to integral multiples of $4\pi$.

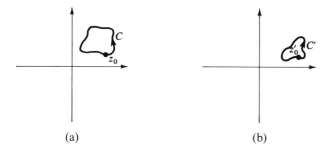

(a)                                         (b)

Figure A.5   Closed contour in one branch of the function $f(z) = z^{1/2}$
(a) $z$-plane; (b) $f$-plane.

We shall describe the Riemann surface for the function $f(z) = z^{1/2}$. We take two $z$-planes, call them $R_1$ and $R_2$, and define the branch line to be the positive real axis. We "cut" $R_1$ along the branch line (a *branch cut*), or actually "a little below" it; we do the same for $R_2$.

We now join the lower lip of $R_1$ to the upper lip of $R_2$, and the lower lip of $R_1$ to the upper lip of $R_2$. The surface thus obtained is the Riemann surface for the function $z^{1/2}$ (Figure A.6).

We now *define* $\phi$ on $R_1$ to have the range $0 \le \phi < 2\pi$, and $\phi$ on $R_2$ to have the range $2\pi \le \phi < 4\pi$. Thus on the entire Riemann surface $\phi$ takes on the values $0 \le \phi < 4\pi$. Starting from $\phi = 0$ in $R_1$ and encircling the branch point, one reaches the branch cut. Crossing the branch cut takes us onto the upper region, $R_2$. Crossing the branch cut once more after encircling the origin, one returns to $R_1$. For $z$ in $R_1$ $[R_2]$ one has $f(z) = f_1(z)$ $[f_2(z)]$, i.e., each *Riemann sheet* of the composite Riemann surface corresponds to a single-valued branch of the composite multivalued function.

What we have achieved is the following: From a sequence of single-valued functions [branches of the multivalued $f(z)$] defined on *one* complex $z$-plane, we have obtained *one* continuous single-valued function defined on the Riemann surface. $f(z)$ is now analytic over the entire Riemann surface except for the branch points, which are now to be treated as (isolated) singularities.

A branch point is said to be of order $n$ if the (multivalued) function is returned to its original value after encircling the branch point at least $n + 1$ times. If this cannot be done for finite $n$, the branch point is said to be of

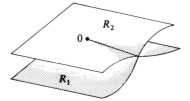

Figure A.6   Riemann surface for $f(z) = z^{1/2}$.

infinite order. Thus for the function $z^{1/n}$, the branch points $z = 0$ (and $z = \infty)^{16}$ are of order $n - 1$. The Riemann surface has $n$ sheets.

An example of an infinite-valued function is the function

$$f(z) = \ln z$$
$$= \ln \rho + i(\phi + 2\pi n); \qquad (A.6)$$

each encirclement of the branch point $z = 0$ increases the value of $\ln z$ by $2\pi i$. Accordingly, $z = 0$ is a branch point of infinite order, and the Riemann surface has an infinity of sheets.

## Problems

1.  The square-well potential is given by

$$V(r) = \begin{cases} -V_0 & \text{if } r < a. \\ 0 & \text{if } r \geq a. \end{cases}$$

(a)   Calculate the Born approximation of the scattering amplitude.

(b)   Show that the exact expression for the $l$th partial-wave amplitude is given by

$$T_l(p) = -\frac{1}{p} \frac{pj_l(ka)j_l'(pa) - kj_l'(ka)j_l(pa)}{kj_l'(ka)h_l(pa) - pj_l(ka)h_l'(pa)},$$

where $k = \sqrt{p^2 + 2mV_0}$ (momentum inside the well) and $j_l'(z) = dj_l/dz$.

(c)   Show that at low energies the $s$-wave, $l = 0$, dominates so that $T(p, \theta) \approx T_0(p)$.

(d)   Compare the correct low-energy amplitude with the Born approximation and show that at low energies the Born approximation is good only for a shallow well.

2.   Calculate the partial cross section $\sigma_{l=0}$ for scattering by a square-well potential in the limit of zero scattering energy.

---

[16] $f(z)$ is said to have a singularity at $z = \infty$ if $f(1/z)$ has a singularity at $z = 0$.

# CHAPTER XVIII

# Resonance Phenomena

Resonance phenomena constitute some of the most interesting and striking features of scattering experiments. This chapter discusses in detail the connection between quasistationary states and resonance phenomena, and culminates in the derivation of the Breit–Wigner formula. In Section XVIII.2 the concept of "time delay" is introduced and its relation to the phase shift derived. Various formulations of causality are given in Section XVIII.3. In Section XVIII.4 the causality condition is used to derive certain analyticity properties of the $S$-matrix. These properties are discussed further in Section XVIII.5. In Section XVIII.6, the central section of this chapter, the connection between quasistationary states, defined by a large time delay, and resonances, defined by characteristic structures in the cross section, is derived. Section XVIII.7 describes the observable effects of virtual states. Section XVIII.8 discusses the effect resonances have on the Argand diagram. The actual appearance of resonances in experimental data when the effects of the resonant phase shift, the nonresonant background, and the limited resolution of the apparatus are taken into account is discussed in Section XVIII.9.

## XVIII.1 Introduction

In our study of the discrete energy spectra of atoms and molecules we treated the excited states as if they were infinitely long-lived. When we discussed in Section XI.4 the helium atom and its levels, which have the same energy value as the physical system consisting of an $He^+$ ion and an electron, we ignored the presence of the $He^+$-$e^-$ system and used an approximate description in which the helium atom was considered to be an isolated stationary system, even though the helium-atom system can never really be isolated from the $He^+$-$e^-$ system with which it interacts. In this approximate description of excited states the transition processes could not be described. In the more accurate description, which does not ignore transitions, excited states have a finite lifetime. This lifetime is fairly long for the ordinary discrete energy levels, but is not so long for the discrete energy levels that lie in the continuous spectrum of the He-($He^+$-$e^-$) system; the lifetimes of both sorts of states are still, however, much longer than the time taken by the transition processes.

The helium energy-loss experiment, whose result is depicted in Figure XI.4.1, is an inelastic scattering process in which a long-lived intermediate state He* is produced, which subsequently decays either into a helium ion and electron or into a helium atom and photon:

$$e^- + He \rightarrow e^{-\prime} + He^*$$

$$\begin{array}{l} \phantom{e^- + He \rightarrow e^{-\prime} + } \llcorner \rightarrow He^+ + e^- \\ \phantom{e^- + He \rightarrow e^{-\prime} + } \llcorner \rightarrow He + \gamma \end{array} \qquad (1.1)$$

The lifetime of a singly excited He*(S), which has an energy level below the first ionization threshold, is orders of magnitude larger than the lifetime of the doubly excited He*(D).[1] This is due to the fact that the interaction $V_{II}$ that causes the transition

$$He^*(D) \rightarrow He^+ + e^-$$

is much stronger than the interaction $V_I$ that causes the transitions

$$He^*(D) \rightarrow He + \gamma$$

and

$$He^*(S) \rightarrow He + \gamma.$$

[The transition $He^*(S) \rightarrow He^+ + e$ is energetically impossible.] In the approximation of Section XI.3, in which both $V_I$ and $V_{II}$ are ignored, the energy operator is given by

$$H = K + V,$$

where $K = K_{electron} + K_{He}$ with $K_{He}$ given by (XI.1.2), and $V$ describes the interaction between He and $e^-$. Consequently both He*(S) and He*(D) have infinite lifetimes. If $V_I$ is ignored but the stronger interaction $V_{II}$ is not, then He*(S) lives infinitely long, but He*(D) has a finite lifetime, which is determined by $V_{II}$. If neither $V_I$ nor $V_{II}$ is ignored, then both He*(S) and He*(D) have finite lifetimes.

---

[1] The width of the bumps in Figure XI.3.1 is mainly due to the resolution of the apparatus—caused, for example, by the energy spread in the incident electron beam—and not a measure of the lifetime.

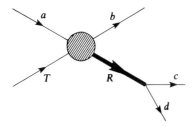

Figure 1.1   Schematic diagram of a production process.

Systems that have finite lifetimes are called *quasistationary* or *metastable states*, in distinction to the *stationary* or *stable states*, which are "infinitely" long lived. Below in Section XVIII.6 we will justify the name *resonances* for the quasistationary states. When to consider a state as a resonance and when to consider it as stable is a matter of the accuracy required of the description; the distinction between (relatively) stable states [particles like He*(*S*)] and unstable states [resonances like He*(*D*)] is in principle quantitative rather than qualitative.

There are two main types of experiments in which resonances can be obtained: *formation experiments* and *production experiments*. The process (1.1) is an example of a production experiment, which experiments are characterized by processes of the form

$$a + T \rightarrow b + R \rightarrow b + c + d, \qquad (1.2)$$

and which are depicted by diagrams like that of Figure 1.1. Production experiments always involve inelastic scattering processes. Formation experiments, on the other hand, consist of simpler processes

$$a + T \rightarrow R \rightarrow a' + T' \qquad (1.3)$$

and are illustrated by diagrams like that of Figure 1.2.

A beautiful example of a resonance obtained in a formation experiment in atomic physics is the Schulz resonance in the elastic scattering process

$$e^- + \text{He} \rightarrow \text{He}^- \rightarrow \text{He} + e^-. \qquad (1.4)$$

If one measures the elastic scattering cross section as a function of the scattering energy (kinetic energy of the projectile $e^-$ with respect to the target He), one observes that at the energy $E = E_R = 19.31$ eV something unusual happens. At this value the cross section changes violently, as shown in the data of various elastic-collision experiments [cf. Figures 1.3, 1.4, and 1.5,

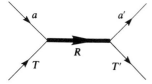

Figure 1.2   Schematic diagram of a formation process.

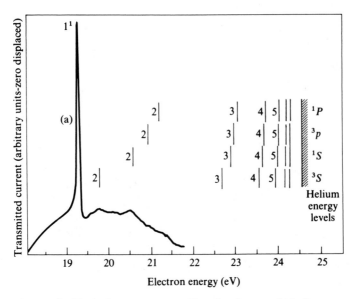

Figure 1.3   Typical current-energy plots for electrons in helium as observed by C. E. Kuyatt, J. A. Simpson, and S. R. Mielczarek. [From *Physical Review* 138, A385 (1965), with permission.]

which show the observed cross sections (Figures 1.4a, 1.4b, 1.5) as well as the intensity of the transmitted electron current, i.e., the intensity of the current of electrons that are not scattered (Figures 1.3, 1.4b, 1.4d)]. We shall show that the effect shown in these figures can be explained from the hypothesis that He and $e^-$ form a long-lived compound that belongs to one of the energy levels of the $He^-$ ion, as indicated by Equation (1.4).

A *resonance* is usually defined as a sharp structure in the cross section together with a rapidly increasing phase shift going through $\pi/2$. In the remainder of this chapter we shall see that in general the formation of a quasistationary state is connected with these effects, and shall study the characteristic properties of the $S$-matrix and phase shifts for these phenomena. We shall in this chapter restrict ourselves to the case where only one channel is open. This means that we are in an energy range below the first inelastic threshold, where only the elastic channel

$$a + T \rightarrow R \rightarrow a + T \tag{1.5}$$

is open, so that the only reduced $S$-matrix element different from zero is

$$S_l(E_A) = \langle \eta_A \| S_l(E_A) \| \eta_A \rangle = e^{i\delta_l(E_A)}$$

(Recall that $\eta_A$ are the quantum numbers that determine the internal state of the $a$-$T$ system; we omit the label $\eta_A$ on $S_l$ and $\delta_l$.)

Resonances in a multichannel system will be discussed in Chapter XX; it will be shown there that with a few changes multichannel resonances may be treated in much the same manner as the single-channel resonance case studied in this chapter.

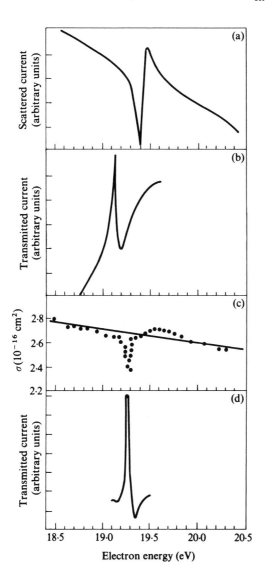

Figure 1.4   (a) Variation of intensity of scattering of electrons at 72° by helium atoms with electron energy, as observed by Schulz, *Phys. Rev. Lett.* **13**, 583 (1964). (b) Variation with electron energy of the transmission of electrons through helium as observed by Simpson, *Phys. Rev. Lett.* **11**, 158 (1964). (c) Variation with electron energy of the total cross section for elastic scattering of electrons by helium atoms, as observed by Golden and Bandel, *Phys. Rev.* **138**, A14 (1965). ●, Experimental points. (d) Variation with electron energy of the transmission of electrons through helium as observed by Golden and Nakano, *Phys. Rev.* **144**, 71 (1966).

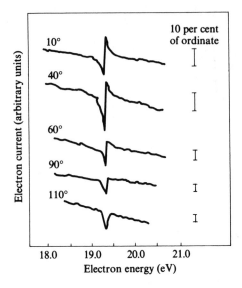

Figure 1.5   Variation with incident electron energy of the intensity of
electrons elastically scattered in helium at different angles of scattering.
The intensity variations may be judged from the lines shown on the
right-hand side, which indicate 10 percent of the total intensity near
the resonance. From D. Andrick and H. Ehrhardt, *Z. Phys.* **192**, 99
(1966).

## XVIII.2  Time Delay and Phase Shifts

The lifetime of a quasistationary state formed in a scattering experiment of
the form $(1.3)^2$ is roughly measured by the time by which the projectile is
delayed in the scattering region as a consequence of the interaction. Let $\mathscr{R}$,
the interaction region around the target, be characterized by a radius $R$,
which is chosen large enough such that there is no interaction outside of $\mathscr{R}$.
This is possible because the interactions encountered in practice do have a
finite range. We describe the state of the projectile-target system in the
Schrödinger picture by the statistical operator $W(t)$. We also imagine a
fictitious noninteracting system described by the operator $W^{in}(t)$. If $H =
K + V$ is the total-energy operator, then the time development of $W(t)$ is
generated by $H$, while that of $W^{in}(t)$ is generated by $K$ [cf. Equation
(XIV.5.5)]. Since we shall consider only elastic scattering, with the target
remaining in the pure state $W_T = |\eta_A\rangle\langle\eta_A|$, we can ignore the factor $W_T$ in
$W^{in} = W_B^{in} \times W_T$ and consider $W$ as referring to the projectile alone.
$W(t)$ then describes the state as it develops when the target is present in the
center of $\mathscr{R}$, while $W^{in}(t)$ describes the state when there is no target in $\mathscr{R}$.

---

[2] To keep the discussion simple we assume that projectile mass remains unchanged, i.e.,
$m_{a'} = m_a = m$.

The probability that the projectile is in the region $\mathcal{R}$ at the time $t$ is given by

$$I(t) = \int_{\mathcal{R}} d^3x \, \langle \mathbf{x} | W(t) | \mathbf{x} \rangle \qquad (2.1)$$

(cf. Section II.10 and II.11), where $|\mathbf{x}\rangle$ are the generalized eigenvectors of the projectile position operator with respect to the target. The total time the projectile spends in $\mathcal{R}$ is then

$$T = \int_{-\infty}^{+\infty} dt \, I(t) = \int_{-\infty}^{\infty} dt \int_{\mathcal{R}} d^3x \, \langle \mathbf{x} | W(t) | \mathbf{x} \rangle. \qquad (2.2)$$

If there were no interaction between projectile and target, the time spent in $\mathcal{R}$ would be

$$T^{\text{in}} = \int_{-\infty}^{+\infty} dt \int_{\mathcal{R}} d^3x \, \langle \mathbf{x} | W^{\text{in}}(t) | \mathbf{x} \rangle. \qquad (2.3)$$

The time that the projectile is delayed by the interaction with the target is therefore

$$t_D^{(\mathcal{R})} = T - T^{\text{in}} = \int_{-\infty}^{+\infty} dt \int_{\mathcal{R}} d^3x \, (\langle \mathbf{x} | W(t) | \mathbf{x} \rangle - \langle \mathbf{x} | W^{\text{in}}(t) | \mathbf{x} \rangle). \qquad (2.4)$$

This "time delay"[2a] may be positive or negative. If a quasistationary state is formed it will be positive and large.

We now investigate the relationship of the time delay $t_D^{(\mathcal{R})}$ with the phase shifts and the $S$-matrix. We insert complete systems of $K$ and $H$ eigenvectors into Equation (2.4) and use Equations (XIV.5.5) and (XIV.5.21) (which was proved in Appendix XV.A) to obtain

$$
\begin{aligned}
t_D^{(\mathcal{R})} &= \int_{-\infty}^{\infty} dt \int_{\mathcal{R}} d^3x \sum_{aa'} \{ \langle \mathbf{x} | a^+ \rangle \langle a^+ | e^{-itH} W e^{itH} | a'^+ \rangle \langle a'^+ | \mathbf{x} \rangle \\
&\quad - \langle \mathbf{x} | a \rangle \langle a | e^{-itK} W^{\text{in}} e^{itK} | a' \rangle \langle a' | \mathbf{x} \rangle \} \\
&= \int_{-\infty}^{\infty} dt \sum_{aa'} e^{-it(E_a - E_{a'})} \langle a | W^{\text{in}} | a' \rangle \\
&\quad \times \int_{\mathcal{R}} d^3x \{ \langle \mathbf{x} | a^+ \rangle \langle a'^+ | \mathbf{x} \rangle - \langle \mathbf{x} | a \rangle \langle a' | \mathbf{x} \rangle \}. \qquad (2.5)
\end{aligned}
$$

$\mathcal{R}$ is the region within which the interaction takes place. The wave function $\langle \mathbf{x} | a^+ \rangle$ in this region depends strongly upon the particular nature of the interaction. On the other hand, the wavefunction has a very general form outside the region $\mathcal{R}$, where the projectile moves freely, the effect of the interaction being expressed by a phase shift relative to the asymptotic form of the free wave function $\langle \mathbf{x} | a \rangle$. Since we want to obtain statements about the time delay that are very generally true, we attempt to replace the integral over $\mathcal{R}$

---

[2a] Time delay has been introduced previously in a similar way by F. T. Smith: Phys. Rev. *fff*, 349 (1960). Time delay functions computed for particular potentials are given in R. J. LeRoy, R. B. Bernstein: J. Chem. Phys. **54**, 5114 (1971).

by an integral outside of $\mathscr{R}$. To this end, we first show that the time delay over all space $(R \to \infty)$ vanishes. Denote all space by $\infty$, and the region outside $\mathscr{R}$ by $(\infty - \mathscr{R})$. Then

$$\int_{\infty} d^3x \, \{\langle x|a^+\rangle\langle a'^+|x\rangle - \langle x|a\rangle\langle a'|x\rangle\}$$

$$= \int_{\infty} d^3x \, \{\langle a'^+|x\rangle\langle x|a^+\rangle - \langle a'|x\rangle\langle x|a\rangle\} \qquad (2.6)$$

$$= \langle a'^+|a^+\rangle - \langle a'|a\rangle,$$

where we have used the relation

$$\int_{\infty} d^3x \, |x\rangle\langle x| = I.$$

According to Equation (XIV.2.9a,b), the generalized eigenvectors of $H$ and $K$ have the same normalization. (This is proved in Appendix XV.A.) Thus the right-hand side of (2.6) and consequently $t_D^{(\infty)}$ vanish. The time delay over region $\mathscr{R}$ can then be written as the time delay over $(\infty - \mathscr{R})$:

$$t_D^{(\mathscr{R})} = \int_{-\infty}^{\infty} dt \sum_{aa'} e^{-it(E_a - E_{a'})}\langle a|W^{in}|a'\rangle$$

$$\times \int_{(\infty - \mathscr{R})} d^3x \, \{\langle x|a\rangle\langle a'|x\rangle - \langle x|a^+\rangle\langle a'^+|x\rangle\}. \qquad (2.7)$$

Assuming spherical symmetry, i.e., $[H, L_i] = [K, L_i] = 0$, we may choose angular-momentum bases $\{|Ell_3\eta^+\rangle\}$ and $\{|Ell_3\eta\rangle\}$ for $\{|a^+\rangle\}$ and $\{|a\rangle\}$.[3] Then the time delay is given by

$$t_D^{(\mathscr{R})} = \int_{-\infty}^{+\infty} dt \sum_{ll_3} \int \rho(E) \, dE \, \rho(E') \, dE' \, e^{-it(E - E')}\langle Ell_3|W^{in}|E'll_3\rangle$$

$$\times \int_R^{\infty} r^2 \, dr \, \{\langle r|E\rangle_l\langle r|E'\rangle_l^* - \langle r|E^+\rangle_l\langle r|E'^+\rangle_l^*\}, \qquad (2.8)$$

where $\rho$ is the normalization function of (XIV.2.9a), and where the relations

$$\langle r\theta\phi|Ell_3\rangle = Y_{ll_3}(\theta, \phi)\langle r|E\rangle_l, \qquad (2.9a)$$

$$\langle r\theta\phi|Ell_3^+\rangle = Y_{ll_3}(\theta, \phi)\langle r|E^+\rangle_l, \qquad (2.9b)$$

$$\int d\Omega \, Y_{ll_3}(\Omega)Y_{l'l_3'}(\Omega) = \delta_{ll'}\delta_{l_3l_3'} \qquad (2.9c)$$

have been used. We can write (2.8) in the form

$$t_D^{(\mathscr{R})} = \int_{-\infty}^{\infty} dt \sum_{ll_3} \int \rho(R) \, dE \, \rho(E') \, dE' \, e^{-it(E - E')}\langle Ell_3|W^{in}|E'll_3\rangle J, \qquad (2.10)$$

---

[3] Here $\eta$ can be any set of additional quantum numbers, and we shall usually suppress the label $\eta$.

where

$$J \equiv \int_R^\infty r^2 \, dr \, \{\langle r | E \rangle_l \langle r | E' \rangle_l^* - \langle r | E^+ \rangle_l \langle r | E'^+ \rangle_l^* \}. \qquad (2.11)$$

In Equation (2.11), which holds for all $R$ larger than the radius of the inter-action region $\mathscr{R}$, we choose $R$ to lie in the asymptotic region. The integral $J$ can then be evaluated using the asymptotic forms (XVII.4.17) and (XVII.4.18) of the exact and free radial wave functions.[3a] Though derived under the as-sumption of a Hamiltonian time development, these asymptotic forms exist independently of $H$, depending as they do only on the $S$-matrix through the phase shift $\delta_l$. We have, therefore,

$$J = \frac{2}{\pi} \int_R^\infty r^2 \, dr \, \frac{1}{r^2 p p'} \left\{ \sin\left( pr - \frac{l\pi}{2} \right) \sin\left( p'r - \frac{l\pi}{2} \right) \right.$$

$$\left. - e^{i(\delta_l(p) - \delta_l(p'))} \sin\left( pr - \frac{l\pi}{2} + \delta_l(p) \right) \sin\left( p'r - \frac{l\pi}{2} + \delta_l(p') \right) \right\}. \qquad (2.12)$$

Because the $t$-integration in (2.10) gives a factor $\delta(E - E')$, one only needs to evaluate $J$ in the limit $E' \to E$, i.e., $p' \to p$. The evaluation of this integral is an exercise in distribution theory, with the result

$$J(p \to p') = \frac{1}{\pi} \frac{1}{p^2} \left\{ \frac{d\delta_l(p)}{dp} - (-1)^l \frac{1}{p} \cos(2pR + \delta_l(p)) \sin \delta_l(p) \right\}. \qquad (2.13)$$

[The calculation that gives (2.13) proceeds in the following way. We rewrite the integrand in (2.12) as follows:

$$\pi p p' J = \frac{1}{2} \int_{-\infty}^{+\infty} dr \, \theta(r - R) \{ e^{i(p - p')r}[1 - e^{2i(\delta_l(p) - \delta_l(p'))}]$$

$$+ (-1)^l [e^{i(p + p')r}(e^{2i\delta_l(p)} - 1) + e^{-i(p + p')r}(e^{-2i\delta_l(p')} - 1)] \}. \qquad (2.14)$$

The integral over the first term in the integrand is the Fourier transform of the $\theta$-function and may be calculated using Equation (XV.2.25):

$$\text{first term} = \frac{1}{2} \int_{-\infty}^{+\infty} dr \theta(r - R) e^{i(p - p')r}[1 - e^{2i(\delta'(p) - \delta_l(p'))}]$$

$$= \tfrac{1}{2}[1 - e^{2i(\delta_l(p) - \delta_l(p'))}] e^{i(p - p')R} \int_{-\infty}^\infty dr \theta(r - R) e^{i(p - p')(r - R)}$$

$$= \tfrac{1}{2}[1 - e^{2i(\delta_l(p) - \delta_l(p'))}] e^{i(p - p')R} \frac{i}{p - p' + i0}. \qquad (2.15)$$

---

[3a] Only for $l = 0$ are the exact and free radial wave functions given by the trigonometric function such that for $l = 0$ (2.12) holds for any value of $R$ outside the *interaction* region and one can choose for $R$ the smallest possible value, the effective radius of the scatterer. The larger the value of $l$ the larger one has to choose the value of $R$ in order to be in the asymptotic region.

Using the relation[4] (VII.2.5):

$$\frac{1}{p - p' \pm i0} = \frac{1}{p - p'} \mp i\pi\delta(p - p'), \qquad (2.16)$$

we rewrite (2.15), noting that the contribution of the term containing $\delta(p - p')$ vanishes, since

$$\delta(p - p')[1 - e^{2i(\delta_l(p) - \delta_l(p'))}] = 0.$$

We are left with

$$\text{first term} = \tfrac{1}{2}[1 - e^{2i(\delta_l(p) - \delta_l(p'))}]e^{i(p - p')R}\,\frac{i}{p - p'}. \qquad (2.17)$$

Now let $p \to p'$. Expanding the exponential in the square brackets, we get

$$\text{first term} \to \tfrac{1}{2}[-2i(\delta_l(p) - \delta_l(p'))]e^{i(p - p')}\,\frac{i}{p - p'} \to \frac{d\delta_l(p)}{dp}. \qquad (2.18)$$

The integrals over the second and third terms in (2.14) are obtained similarly, with the result

$$\text{second and third terms} = (-1)^l\left[ i\,\frac{e^{i(p + p')R}}{p + p' + i0} \cdot \tfrac{1}{2}(e^{+2i\delta_l(p')} - 1)\right.$$

$$\left. - i\,\frac{e^{-i(p + p')R}}{p + p' - i0} \cdot \tfrac{1}{2}(e^{-2i\delta_l(p')} - 1)\right]. $$

$$(2.19)$$

Here there is no problem with taking the limit $p' \to p$, and after elementary simplification one obtains the second term on the right-hand side of (2.13). ]

If we insert $J$ of (2.13) into (2.10), perform the integration over $t$ [letting the bound of the integral go to infinity and using (XIV.5.19)], and then integrate over $E'$, we obtain

$$t_D^{(\mathcal{R})} = 2\sum_{ll_3} \int \rho^2(E)\, dE\, \langle Ell_3 | W^{\text{in}} | Ell_3 \rangle$$

$$\times \frac{1}{p^2}\left[\frac{d\delta_l(p)}{dp} - (-1)^l \frac{1}{p}\cos(2pR + \delta_l(p))\sin\delta_l(p)\right]. \qquad (2.20)$$

Since the expression in the square brackets does not depend upon $l_3$, we can sum over that component of the angular momentum. Denoting

$$\sum_{l_3} \langle Ell_3 | W^{\text{in}} | Ell_3 \rangle = W_l(E) = W_l(p), \qquad (2.21)$$

---

[4] Gel'fand and Shilov (1964, Vol. 1).

which because of the normalization of $W^{in}$ fulfills

$$\sum_{l_3 l} \int \rho(E)\, dE \, \langle Ell_3 | W^{in} | Ell_3 \rangle = \sum_l \int \rho(E)\, dE \, W_l(E) = 1, \qquad (2.22)$$

one can write (2.20) as

$$t_D^{(\mathcal{R})} = \sum_l \int \rho(E)\, dE \, W_l(E) t_D^{(\mathcal{R})l}(E), \qquad (2.23)$$

where for a nonrelativistic particle with mass $m$, the quantities $p$, $E$ and $\rho$ are related by [cf. (XIV.5.11) and (XIV.5.18a]

$$E = p^2/2m, \qquad \rho(E) = mp(E), \qquad (2.24)$$

and where we have defined

$$t_D^{(\mathcal{R})l}(E) = 2\rho(E) \frac{1}{p^2} \left[ \frac{d\delta_l(p)}{dp} - \frac{(-1)^l}{p} \cos(2pR + \delta_l(p)) \sin \delta_l(p) \right]$$

$$= 2m \frac{1}{p} \left[ \frac{d\delta_l(p)}{dp} - \frac{(-1)^l}{p} \cdot \tfrac{1}{2}(\sin 2(pR + \delta_l) - \sin 2pR) \right]. \quad (2.25)$$

The second term on the right-hand side of (2.25) is of limited variation and depends upon the radius $R$ of the region $\mathcal{R}$, which can be arbitrarily chosen, the only restriction being that it be greater than the range of the interaction.[4a] We therefore average over $R$ to obtain a quantity that does not depend upon the irrelevant exact choice of the radius $R$ and is characteristic of the scattering process only. This average of $t_D^{(\mathcal{R})l}$ is

$$t_D^l(E) = 2\rho(E) \frac{1}{p^2} \frac{d\delta_l(p)}{dp} = 2m \frac{1}{p} \frac{d\delta_l(p)}{dp} = 2 \frac{d\delta_l(E)}{dE}. \qquad (2.26)$$

The average time delay, which for short we again just call the time delay, is thus

$$t_D = \sum_l \int \rho(E)\, dE \, W_l(E) t_D^l(E). \qquad (2.27)$$

The incident beam $W^{in}$ is usually prepared, not with a definite value of the orbital angular momentum $l$, but with a momentum pointing in a well-defined direction. $W_l(E)$ then describes the weight of the $l$th angular-momentum component in the mixture $W^{in}$. However, if one arranges that the incident "beam" has a well defined angular momentum $l_A$, then one has

$$\rho(E)W_l(E) = \delta_{ll_A} F_l(E - E_A), \qquad (2.28)$$

where $F_{l_A}(E - E_A)$ fulfills, because of (2.22), the condition

$$\int dE \, F_{l_A}(E - E_A) = 1$$

---

[4a] Note that for $l \neq 0$ $R$ in the above derivation has to be chosen in the asymptotic region.

and describes the energy distribution in the "beam" [cf. XIV.5.35)]. The incident beam is usually not monoenergetic. However, if it can be considered monoenergetic, then [cf. XIV.5.36)]

$$F_{l_A}(E - E_A) = \delta(E - E_A). \tag{2.29}$$

Inserting (2.29) and (2.28) into (2.27), one obtains for the time delay of this particular "state" with angular momentum $l_A$ and energy $E_A$

$$t_D = t_D^{l_A}(E_A). \tag{2.30}$$

Thus the quantity (2.25) or (2.26) is the time delay occurring in the scattering of a state with angular momentum $l$ and energy $E$. And the time delay in the state $W^{\text{in}}$ (2.27) is the weighted average of the time delays $t_D^l(E)$ over all angular-momentum and energy values.

Restoring the dependence upon the additional quantum numbers $\eta$, the time delay of a "state" with quantum numbers $\eta l E$ is given by:

$$t_D^{\eta l}(E) = 2 \frac{d\delta_l^\eta(E)}{dE} = \frac{2m}{p} \frac{d\delta_l^\eta(p)}{dp}. \tag{2.31}$$

A monoenergetic beam describes a steady state for which time delay really has no observable meaning. For a measurement of the time delay one needs a beam of pulses of finite duration and, therefore, of nonzero energy spread.[5] Then the quantity $t_D$ of (2.27) and also the time delay for a particular angular-momentum value,

$$t_D{}^{\eta l} = \int dE \, F(E - E_A) t_D^l(E) = 2 \int dE \, F(E - E_A) \frac{d\delta_l^\eta(E)}{dE}, \tag{2.32}$$

can—at least in a *Gedanken* experiment—be measured as the difference in time that a pulse of projectiles needs to pass through a region $\mathcal{R}$ with and without the target present. Scattering experiments are, however, usually not carried out in a way that allows the measurement of such a delay, and the significance of (2.31) or (2.26) lies in its theoretical consequences, which we shall discuss now.

If during the scattering process for a particular set of values $(\eta l E) = (\eta_R l_R E_R)$ a quasistationary state is formed, that means, if the projectile is temporarily captured by the target, then for this set of values the delay time $t_D^{\eta l}(E)$ must be large. Thus the condition for a quasistationary state is

$$\frac{d\delta_l^\eta(E)}{dE} \text{ has a sharp positive maximum at } (\eta_R l_R E_R), \tag{2.33}$$

which in particular means that

$$\left. \frac{d^2 \delta_{l_R}^{\eta_R}(E)}{dE^2} \right|_{E=E_R} = 0, \quad \left| \frac{d^3 \delta_l(E)}{dE^3} \right| \text{ is large.} \tag{2.34}$$

---

[5] Cf. the discussion in Section II.11.

The main contribution in the expression (2.27) for the delay time then comes from the term with these particular values $(l_E E_R)$.

It may, of course, happen that for a particular scattering system all $\delta_l^\eta(E)$ are such functions of $E$ that the $d\delta_l^\eta(E)/dE$ do not have any maxima, and therefore no quasistationary state exists. However, more often than not at least some of the $d\delta_l^\eta(E)/dE$ do have such isolated maxima, and we shall investigate the consequences of such an occurrence below, in particular in Section XVIII.6.

## XVIII.3 Causality Conditions

The time $T = T^{(\mathscr{R})}$ which a projectile spends within a certain region $\mathscr{R}$ of radius $R$ around the target must always be positive. Thus the condition

$$T^{(\mathscr{R})} \geq 0$$

is obviously fulfilled. This condition is, through the definition of $T$ in (2.2), connected with the *von Kampen causality condition*,

$$\int_{\mathscr{R}} d^3x \, \langle \mathbf{x} | W(t) | \mathbf{x} \rangle \geq 0 \quad \text{at any time } t, \tag{3.1}$$

which is also obviously fulfilled as it represents the probability to find the projectile in the region $\mathscr{R}$. The strongest form of the condition (3.1) is obtained if one chooses for $R$ its smallest possible value, i.e., the effective radius of the scatterer.

The time $T^{(\mathscr{R})}$ that the projectile spends in the interaction region can be obtained according to (2.4) from

$$T^{(\mathscr{R})} = t_D^{(\mathscr{R})} + T^{\text{in}(\mathscr{R})}, \tag{3.2}$$

where $t_D^{(\mathscr{R})}$ is given by (2.23) and (2.25) and $T^{\text{in}(\mathscr{R})}$ is given by

$$T^{\text{in}(\mathscr{R})} = \int_{-\infty}^{+\infty} dt \sum_{l l_3} \int \rho(E) \, dE \, \rho(E') \, dE' \, e^{-it(E-E')} \langle E l l_3 | W^{\text{in}} | E' l l_3 \rangle (-J^{\text{in}}) \tag{3.3}$$

with $J^{\text{in}}$ given by

$$(-J^{\text{in}}) = \int_0^R r^2 \, dr \, \langle r | E \rangle_l \langle r | E' \rangle_l^* = \frac{2}{\pi} \int_0^R r^2 \, dr \, j_l(pr) j_l(p'r). \tag{3.4}$$

Equation (3.3) with (3.4) is obtained by the same arguments that led from (2.4) to (2.10) with (2.11), except that in (3.4) we did not have to replace the integral over the interaction region $\mathscr{R}$ by an integral over $(\infty - \mathscr{R})$.

The integral in (3.4) is calculated using (XVII.3.9b) and (XVII.3.7a), the result being

$$(-J^{in}) = \frac{2}{\pi} \frac{R^2}{(p+p')(p-p')} (p'j_l(pR)j_{l-1}(p'R) - pj_{l+1}(pR)j_l(p'R))$$

$$\approx \frac{2}{\pi} \frac{R^2}{(p+p')(p-p')} \left( \frac{1}{pR_2} \sin\left(pR - \frac{\pi l}{2}\right) \sin\left(p'R - \frac{\pi(l-1)}{2}\right)\right.$$

$$\left. - \frac{1}{p'R^2} \sin\left(pR - \frac{\pi(l+1)}{2}\right) \sin\left(p'R - \frac{\pi l}{2}\right)\right). \qquad (3.5)$$

For the second equality in (3.5) the asymptotic form (XVII.3.6a) was used, which is justified for $R$ in the asymptotic region or in case $l = 0$ for $R$ just outside the interaction region. Using well-known relations between trigonometric functions, this becomes

$$-J^{in} = \frac{1}{\pi} \frac{1}{pp'} \frac{1}{p-p'} \sin(p-p')R - \frac{(-1)^l}{p+p'} \sin(p+p')R \qquad (3.6)$$

and for $p' \to p$

$$-J^{in} = \frac{1}{\pi p^2} \left( R - \frac{(-1)^l}{2p} \sin 2pR \right). \qquad (3.7)$$

This expression is inserted into (3.3) after one has performed in (3.3) the integration over $t$ using $\int_{-\infty}^{+\infty} e^{itx} dt = 2\pi\delta(x)$ and over $E'$ which leads to

$$T^{in(\mathcal{R})} = \sum_{ll_3} \int \rho^2(E) \, dE \, \langle Ell_3 | W^{in} | Ell_3 \rangle \left( \frac{2R}{p^2} - \frac{(-1)^l}{p^3} \sin 2pR \right). \qquad (3.8)$$

With the definition (2.21) and (2.24) one obtains

$$T^{in(\mathcal{R})} = \sum_{l} \int \rho(E) \, dE \, W_l(E) T^{in(\mathcal{R})l}(E) \qquad (3.9)$$

where

$$T^{in(\mathcal{R})l}(E) = m\left( \frac{2R}{p(E)} - \frac{(-1)^l}{p(E)^2} \sin^2 p(E)R \right). \qquad (3.10)$$

Equations (3.9) and (3.10) have been written in complete analogy to (2.23) and (2.25). The quantity $T^{in(\mathcal{R})l}(E)$ is thus the time which a "state" with angular momentum $l$ and energy $E$ spends in the region $\mathcal{R}$ if no interaction takes place. The time $T^{(\mathcal{R})}$ is in analogy to (3.9) and (2.23) also written as

$$T^{(\mathcal{R})} = \sum_{l} \int \rho(E) \, dE \, W_l(E) T^{(\mathcal{R})l}(E), \qquad (3.11)$$

where according to (3.2)

$$T^{(\mathcal{R})l}(E) = t_D^{(\mathcal{R})l}(E) + T^{\text{in}(\mathcal{R})l}(E). \tag{3.12}$$

Inserting (2.25) and (3.10) in (3.12) and using some elementary relations between the trigonometric functions, one obtains

$$T^{(\mathcal{R})l}(E) = 2m \frac{1}{p} \left( \frac{d\delta_l}{dp} + R \right) - (-1)^l \frac{m}{p^2} \sin 2(pR + \delta_l(p)) \tag{3.13}$$

$T^{(\mathcal{R})l}(E)$ is the time that the "part" of the projectile with energy $E$ and angular momentum $l$ spends in the interaction region.

The condition (3.1) must be fulfilled for any arbitrary state of the projectile, i.e., for any set of positive functions $W_l(E)$. Consequently we conclude from (3.11) and (3.1) that

$$T^{(\mathcal{R})l}(E) \geq 0 \quad \text{for all } l \text{ and } E. \tag{3.14}$$

This condition (3.14) can be rewritten using (3.13):

$$\frac{d\delta_l}{dp} \geq -R + (-1)^l \frac{\sin 2(\delta_l + pR)}{2p} \geq -\left( R + \frac{1}{2p} \right). \tag{3.15}$$

In the form (3.15) the condition is called *Wigner's causality inequality*. The smaller $R$ the stronger is the condition (3.15); but only for $l = 0$ can one choose in our above derivation for $R$ the effective radius of the scatterer.

Wigner's causal inequality states that the phase shift cannot decrease faster than at a certain rate. Thus if the phase changes rapidly, then it must be increasing. We shall make use of this fact below.

The condition (3.1) is so obviously fulfilled that one may wonder why the conditions (3.1) and (3.15) were given the name *causality conditions*. The connection with causality becomes most apparent if one rewrites (3.15) into yet another form using (3.12) with (3.10):

$$t_D^{(\mathcal{R})l}(E) \geq -T^{\text{in}(\mathcal{R})l}(E) = -\left[ \frac{m \cdot 2R}{p} - (-1)^l \frac{m}{p} \frac{\sin 2pR}{p} \right]. \tag{3.16}$$

The first term on the right-hand side and on the right-hand side of (3.10) is the free time of flight that a classical projectile with velocity $m/p$ would need to traverse the region of diameter $2R$. The second term represents the quantum effects; it is negligible for sufficiently fast particles, $pR \gg 1$, and arises from the fact that quantum particles cannot be localized within dimensions smaller than a de Broglie wavelength.

If the region of diameter $2R$ is the interaction region in which the projectile interacts with the target, then—depending upon the kind of interaction—the projectile may be delayed ($t_D > 0$) or hastened ($t_D < 0$) in it. It may be delayed an arbitrarily long time, and we shall discuss the phenomena connected with it in the subsequent sections. However, it cannot be advanced an arbitrarily large time, because according to causality outgoing projectiles cannot appear before the incoming projectiles have reached the scattering region. Classically, the maximal time advance occurs if the projectile has hardly entered the interaction region when it is repelled at the surface, and leaves the interaction region $\mathscr{R}$ at a time $m2R/p$ earlier than a particle that passes through such a region without interaction—i.e., its time delay is $-m2R/p$. A more negative time delay is not possible, because then the outgoing projectile would have to leave the surface of the interaction region before the incoming projectile had reached it. For quantum systems these arguments have to be corrected by the term $(m/p)(\sin 2pR)/p$, and then lead to (3.16) as a mathematical formulation of the above statement of causality.

## XVIII.4  Causality and Analyticity

In Chapter XVI we considered the $l$th partial $T$-matrix elements $\langle\eta_p\|T_l(E)\|\eta_A\rangle$ and the $l$th partial $S$-matrix elements $\langle\eta_b\|S_l(E)\|\eta_A\rangle$; in particular we considered the $l$th partial $T$- and $S$- matrix elements for elastic scattering, $T_l(E) = T_l(p)$ and $S_l(E) = S_l(p)$, where the momentum $p$ and energy $E$ of the projectile of mass $m$ are connected by $E = p^2/2m$. If the interaction is described by a Hamiltonian $H = K + V$, then $T$ and $S$ are connected to the interaction Hamiltonian $V$ by expressions like (XIV.5.41), but as mentioned above (e.g., in Section XIV.4), the notions of $S$- and $T$-matrix make sense also if there exists no Hamiltonian time development. Then the $S$- or $T$-matrix is the ultimate basis for the description of all the information required to compute observable quantities, such as cross sections [(XVI.1.26), (XVI.1.29), (XVI.1.30)]. The properties of the $S$-matrix then follow from general physical principles believed to be satisfied by any inter-action. One kind of general physical principles are symmetry principles, and we have, in fact, already made use of such a symmetry principle, namely the rotational symmetry (XVI.1.1), when we derived that the $S$-matrix elements are independent of $l_3$. The general physical principle that we shall utilize now is the causality condition discussed in the preceding section. The causality condition leads to analytic properties of the $S$-matrix elements when the energy and momentum of the projectile are extended to complex values.[6] When the interaction is described in terms of a potential function $V(r)$, then this kind of analyticity of the $S$-matrix can be derived from the properties of the potential function.[7] Following the spirit of the presentation in this book, we shall not start from a potential function, but establish some analyticity of $S_l(p)$ as a consequence of causality. We shall not give an extensive discussion of this subject, especially as not all questions on this subject have been answered yet, and content ourselves in this section with the formulation of the analyticity properties of $S_l(p)$ that will be needed in the subsequent sections in connection with resonance phenomena.

---

[6] One obtains dispersion relations when these analytic properties are expressed in terms of integral relations between different matrix elements for real values of the variables.

[7] Taylor (1972, Chapter 12).

The probability of finding the projectile at an arbitrary time $t$ anywhere in space is unity:

$$1 = \int_{-\infty}^{+\infty} d^3x \, \langle \mathbf{x} | W(t) | \mathbf{x} \rangle = \int_{\mathscr{R}} d^3x \, \langle \mathbf{x} | W(t) | \mathbf{x} \rangle + \int_{(\infty - \mathscr{R})} d^3x \, \langle \mathbf{x} | W(t) | \mathbf{x} \rangle.$$

$$(4.1)$$

Consequently the van Kampen causality condition (3.1) can be reformulated to read

$$\int_{(\infty - \mathscr{R})} d^3x \, \langle \mathbf{x} | W(t) | \mathbf{x} \rangle \le 1 \quad \text{at any time } t. \tag{4.2}$$

Inserting a complete system of basis vectors $|Ell_3^+\rangle$ and using (XIV.5.21) and (2.9a,b,c), one obtains for (4.2) [by the same calculation that led from (2.4) to (2.7) except for the change from the integral over $\mathscr{R}$ to the integral over $(\infty - \mathscr{R})$],

$$\sum_{ll_3} \int \int \rho(E) \, dE \, \rho(E') \, dE' \, e^{-it(E-E')} \langle Ell_3 | W^{\text{in}} | E'll_3 \rangle$$

$$\times \int_{r=\mathscr{R}}^{r=\infty} \langle r | E^+ \rangle_l \langle E'^+ | r \rangle_l r^2 \, dr \le 1. \tag{4.3}$$

In the region $(\infty - \mathscr{R})$, where no interaction takes place and the projectiles move freely, the wave function $\langle r | E^+ \rangle_l = \langle r | p^+ \rangle_l$ has the general asymptotic form (XVII.4.17), in which the effect of the interaction is expressed in terms of the $S$-matrix $S_l(p) = e^{2i\delta_l(p)}$:

$$\langle r | p^+ \rangle \sim \sqrt{\frac{2}{\pi}} \frac{1}{2ipr} (e^{i(pr - l\pi/2)} S_l(p) - e^{-i(pr - l\pi/2)}). \tag{4.4}$$

Inserting this into the integral over $r$ in (4.3), one obtains for $R$ from the asymptotic region

$$\text{integral} = \int_{-\infty}^{+\infty} r^2 \, dr \, \theta(r - R) \langle r | p^+ \rangle_l \langle r | p'^+ \rangle_l^*$$

$$= \frac{2}{\pi} \frac{1}{4pp'} e^{-i(p-p')R} \left\{ S_{lR}(p) S_{lR}^*(p') \int_{-\infty}^{+\infty} dr \, \theta(r - R) e^{i(p-p')(r-R)} \right.$$

$$+ \int_{-\infty}^{+\infty} dr\theta \, (r - R) e^{-i(p-p')(r-R)}$$

$$- e^{-il\pi} S_{lR}(p) \int_{-\infty}^{+\infty} dr \, \theta(r - R) e^{i(p+p')(r-R)}$$

$$- e^{il\pi} S_{lR}^*(p') \int_{-\infty}^{+\infty} dr \, \theta(r - R) e^{-i(p+p')(r-R)} \left. \right\},$$

where we have defined[8]

$$S_{lR}(p) = e^{2ipR}S_l(p). \tag{4.5}$$

The Fourier transform of the $\theta$-function has been calculated in (XV.2.25+); using that result one obtains

$$\text{integral} = \frac{2}{\pi} \frac{e^{-i(p-p')R}}{4pp'} \left\{ S_{lR}(p)S_{lR}^*(p) \frac{i}{p-p'+i0} + \frac{i}{p'-p+i0} \right.$$

$$\left. + (-1)^l \left( S_{lR}^*(p') \frac{i}{p+p'-i0} - S_{lR}(p) \frac{i}{p+p'+i0} \right) \right\}.$$

Using the relation between the distributions, $(2.16) = (VIII.2.5)$, this can be rewritten

$$\text{integral} = \frac{1}{2\pi pp'} e^{-i(p-p')R} \left\{ (S_{lR}(p)S_{lR}^*(p') - 1) \frac{i}{p-p'} \right.$$

$$+ (-1)^l (S_{lR}^*(p') - S_{lR}(p)) \frac{i}{p+p'}$$

$$+ \pi\delta(p-p')(S_{lR}(p)S_{lR}^*(p') + 1)$$

$$\left. + \pi(-1)^l\delta(p+p')(S_{lR}^*(p') + S_{lR}(p)) \right\}$$

Inserting this integral into (4.3), one obtains, using $\rho\, dE = p^2\, dp$,

$$\sum_{l l_3} \int_0^\infty \int_0^\infty p^2\, dp\, p'^2\, dp'\, e^{-it(E-E')} \langle p l l_3 | W^{in} | p' l l_3 \rangle \frac{1}{2\pi pp'} e^{-i(p-p')R}$$

$$\times \left\{ (1 - S_{lR}(p)S_{lR}^*(p')) \frac{1}{i(p-p')} - (-1)^l (S_{lR}^*(p') - S_{lR}(p)) \frac{1}{i(p+p')} \right.$$

$$\left. \pi\delta(p-p')(S_{lR}(p)S_{lR}^*(p') + 1) - \pi(-1)^l\delta(p+p')(S_{lR}^*(p') + S_{lR}(p)) \right\} \le 1. \tag{4.6}$$

Performing the integral over $p$ and $p'$ in the last term gives a zero contribution because of $\delta(p+p')$. After integrating the next to the last term over $p'$ and

---

[8] The physical meaning of the exponential factor $e^{-2ipR}$ in

$$S_l(p) = e^{2i\delta_l(p)} = e^{2i(-pR)}S_{lR}(p)$$

is that it represents the phase advance corresponding to a path difference from the surface of the interaction region to the center and back. According to the description of Figure XVII.4.1, a repulsive potential gives a negative phase shift. An impenetrable sphere of radius $R$ has a phase shift of $-pR$, because the wave is reflected at the surface of the sphere, as compared with the one going through the center. Correspondingly an outgoing signal can appear up to $2R/v = 2Rm/p$ earlier than would have been possible in the absence of the scatterer. $S_{lR}(p)$ then describes the deviation of the effect of the scatterer from the effect of an impenetrable sphere with the radius of the interaction region.

using the unitary condition $S_{lR}(p)S^*_{lR}(p) = 1$, one obtains the contribution

$$\sum_{ll_3} \int dp \, p^2 \langle pll_3 | W^{in} | pll_3 \rangle = \tfrac{1}{2} \cdot 2 = 1.$$

For $t = 0$ the above inequality (4.6) therefore becomes

$$\sum_{ll_3} \int_0^\infty \int_0^\infty p^2 \, dp \, p'^2 \, dp' \, \langle pll_3 | W^{in} | p'll_3 \rangle \frac{1}{2\pi pp'} e^{-i(p-p')R}$$

$$\times \, (S_{lR}(p)S^*_{lR}(p') - 1) \frac{1}{i(p - p')}$$

$$\geq \sum_{ll_3} \int_0^\infty \int_0^\infty p^2 \, dp \, p'^2 \, dp' \langle pll_3 | W^{in} | p'll_3 \rangle \frac{1}{2\pi pp'} e^{-i(p-p')R}(-1)^l$$

$$\times \, (S^*_{lR}(p') - S_{lR}(p)) \frac{1}{i(p + p')}. \tag{4.7}$$

As this has to be fulfilled for any state $W^{in}$, it follows that a corresponding inequality must hold for any term in the sum $\sum_{ll_3}$ separably and for any physically permitted momentum distribution function.

One can write (4.7) in a different form by introducing for $W^{in}$ the momentum wave function $A^{ll_3}_R(p)$ [in analogy to the $f(p - p')$ of (II.11.1p)] by

$$A^{ll_3}_R(p)A^{ll_3}_R(p') = pe^{-ipR}\langle l_3 lp | W^{in} | p'll_3 \rangle e^{ip'R}p'. \tag{4.8}$$

Thus one obtains from (4.7)

$$\int_0^\infty \int_0^\infty dp \, dp' \, A^{ll_3}_R(p)A^{ll_3}_R(p')(S_{lR}(p)S^*_{lR}(p') - 1) \frac{1}{i(p - p')}$$

$$\geq \int_0^\infty \int_0^\infty dp \, dp' \, (-1)^l A^{ll_3}_R(p)A^{ll_3*}_R(p')(S^*_{lR}(p') - S_{lR}(p)) \frac{1}{i(p + p')}. \tag{4.9}$$

for every value of $l$ and $l_3$ and for any well-behaved function (i.e., for any element of the Schwartz space), $A^{ll_3}_R(p)$.

From the form (4.9) of the van Kampen causal inequality it follows, by a purely technical proof that we shall not reproduce here:[9]

> $S_l(p)$ has an analytic continuation without singularities in the first quadrant $0 \leq \arg p \leq \pi/2 - \delta \, (\delta > 0)$ of the complex $p$ plane. (4.10)

In fact, one can prove more about the analyticity property of $S_l(p)$ than just this, and in Section XVIII.5 below we shall summarize some results. It is, however, only (4.10) that we shall need for our derivation of the Breit–Wigner

[9] N. G. van Kampen, *Phys. Rev.* **91**, 1267 (1953), Section II [for odd $l$ all statements are for $-S_l(p)$].

formula in Section XVIII.6. We wish to emphasize that (4.10) holds for any value that one chooses for $R$, as long as it is finite.

## XVIII.5 Brief Description of the Analyticity Properties of the S-Matrix

Though a derivation of the following result on the analyticity properties of $S_l(p)$ exceeds the scope of this book, and we shall also not make explicit use of these properties in what follows, the knowledge of these properties will greatly enhance the picture of the $S$-matrix. We shall, therefore, state these analyticity properties here with a few connecting remarks.[10] These properties are independent of the specific form of the interaction and follow from very general principles—essentially from the causality condition in the form (4.9).

In order to obtain the properties of $S_l(p)$ beyond the first quadrant, we have to continue $S_l(p)$ across the imaginary axis and into the lower half plane. In order to do the first we have to know the properties of $S_l(p)$ on the positive imaginary axis. We know already from the discussions in Section XVII.5 that bound states correspond to poles on the positive imaginary axis, so that we cannot expect $S_l(p)$ to be analytic there. Figure 5.1 shows these poles and zeros, which are, of course, also possible on the real axis.

As a further consequence of the causality condition (4.9) one can prove that

$$\mathrm{Im}\, S_{lR}(p) \leq (-1)^l \frac{\mathrm{Re}\, p}{\mathrm{Im}\, p}, \quad \mathrm{Im}\, S_{lR}(p) \leq 1 \quad \text{in } 0 \leq \arg p \leq \pi/2 \quad (5.1)$$

and

$$|S_{lR}(p)| \text{ is bounded in the first quadrant.}[11] \tag{5.2}$$

This is not sufficient to obtain information about the nature of the singularities on the imaginary axis, except to conclude that if these are poles, they cannot be of higher than the first order.

---

[10] For more on this subject see Nussenzveig (1972).

[11] This is a very weak condition for large values of $R$, because it allows $S_l(p)$ to vary rapidly on the real axis also in the absence of any poles or zeros. But in the above derivation of the causality condition (4.9) $R$ has to be chosen from the asymptotic region because (4.4) is the asymptotic form for the radial wave function outside the interaction region. The smaller the value of $l$, the smaller can one choose the value of $r$ for which the asymptotic form (4.4) is valid. But only for the case $l = 0$ is (4.4) identical with the radial wave function outside the interaction region. Therefore, only for $l = 0$ can one choose for $R$ the smallest possible value namely the effective radius of the scatterer. In order to obtain the strongest result, i.e. the above statements for $R$ equal to the effective radius of the scatterer, also for $l > 0$, one has to take instead of the asymptotic form (4.4) the exact expression for the radial wave function for $r$ outside the interaction region:

$$\langle r|p^+\rangle_l = \sqrt{\tfrac{1}{2}\pi}(S_l(p)h_l(pr) + h_l^*(pr)) \tag{4.4'}$$

where $h_l(pr)$ and $h_l^*(pr)$ are the spherical Hankel functions (XVII.2.5). The mathematical proof that (5.1) and (5.2) follow also if one takes (4.4') instead of (4.4) has not yet been given.

In order to conclude anything more, one has to make an assumption in addition to the causality condition. If the interaction is described by a Hamiltonian $H$, this assumption amounts to the requirement that $H$ be a semibounded operator, i.e., there is a lowest energy eigenvalue $B$ and the binding energy cannot be infinite. In the case where one does not have a Hamiltonian time development and the bound states are described by singularities on the imaginary $p$-axis, one would assume this to translate into the statement that there are no singularities of $S_l(p)$ on the imaginary $p$-axis above the value $iK$, where $K = +\sqrt{2mB}$.

It has in fact been shown[12] that from a precise formulation of the finiteness of the binding energy it follows that

$$\text{Im } S_{lR}(p) \geq -(-1)^l \frac{\text{Re } p}{\text{Im } p} \quad \text{in the first quadrant,} \tag{5.3}$$

which together with (5.1) gives

$$\text{Im } S_{lR}(p) \to 0 \quad \text{for Re } p \to 0_+, \tag{5.4}$$

so that

$$S_l(p) \text{ is real on the positive imaginary axis except between } 0 \text{ and } iK. \tag{5.5}$$

For scattering states the physical values of the momentum $p$ must be positive; however, $p$ occurs quadratically in the Schrödinger equation and also in the Lippman–Schwinger equation. Consequently the equations are invariant with respect to the change of sign of $p$, and therefore, if $p$ is replaced by $-p$ in the radial wave function, the resulting function must again be a solution of these equations. Replacing $p$ by $-p$ in (XVII.4.17d) gives

$$\sqrt{\frac{\pi}{2}} \langle r | -p^+ \rangle \sim \frac{e^{-i(\pi/2)(l+1)}}{\sqrt{2\pi}} (-1)^l S_l(-p) \frac{1}{pr} ((-1)^l e^{-ipr} - S_l^{-1}(-p) e^{ipr}).$$

$$\tag{5.6}$$

As $\langle r | p^+ \rangle$ and $\langle r | -p^+ \rangle$ describe the same physical content, the right-hand side of (XVII.4.17d) and (5.6) must both describe an incoming spherical wave and an outgoing spherical wave modified by the scattering matrix. Therefore one requires that for negative values of $p$ the $S$ matrix element $S_l$ be given by

$$S_l^{-1}(-p) = S_l(p) \tag{5.7a}$$

or

$$S_l(-p) = S_l^{-1}(p). \tag{5.7b}$$

Using unitarity,

$$S_l(p) S_l^*(p) = 1, \tag{5.8}$$

---

[12] N. G. van Kampen, *Physica* (Utrecht) **20**, 115 (1954).

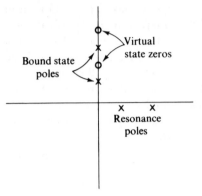

Figure 5.1   Some poles (indicated by x) and zeros (indicated by 0) of the S-matrix element in the p-plane.

one obtains

$$S_l(-p) = S_l^*(p) \quad \text{for real } p.  \tag{5.7c}$$

This is called the *symmetry relation* for the *l*th partial S-matrix element.

The *l*th partial S-matrix element is usually considered as a function of $E = p^2/2m$, $S_l(E)$. In the mapping from $p$ to $E$ the complex $p$-plane (see Figure 5.1) is mapped onto a two-sheeted Riemann surface [cf. XVII.Ai and XVII.Aj] with the branch cut from 0 to $\infty$. The upper half $p$-plane Im $p > 0$ corresponds to the first sheet; the first quadrant of the $p$-plane corresponds to the upper half of the first sheet. The physical meaningful values of $p$ in scattering states, $p$ real, $p \geq 0$, correspond to the upper rim of the first sheet. The first sheet is called the "physical" sheet (Figure 5.2). If one continues through the cut one comes to the second sheet, called the "unphysical" sheet, which corresponds to the lower half of the $p$-plane. $S_l(E)$ is a function on the two-sheeted Riemann surface. The bound-state

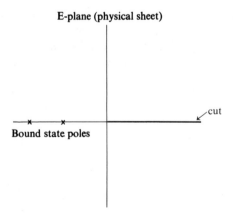

Figure 5.2   Poles and cut in the first ("physical") sheet of the energy plane.

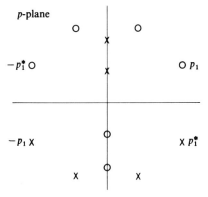

p-plane

Figure 5.3   Relation between zeros (indicated by ○) and poles.

poles of $S_l(p)$ on the positive imaginary $p$-axis correspond to poles on the negative real $E$-axis. Statement (4.10) then means that

> $S_l(E)$ is analytic in the upper half plane of the physical sheet.   (5.9)

The relation (5.7c) for $S_l$ considered as function of $E$ is then stated as

$$S_l(E - i\epsilon) = S_l^*(E + i\epsilon), \qquad E > 0. \tag{5.10}$$

From the Schwarz reflection principle (XVII.Ah) it then follows that $S_l(E)$ can be continued into the lower half energy plane and is given there by

$$S_l(E) = S_l^*(E^*) \tag{5.11}$$

Since the lower half energy plane corresponds to the second quadrant of the momentum plane and $E^*$ corresponds to $-p^*$, (5.11) leads to

$$S_l(p) = S_l^*(-p^*) \tag{5.12}$$

as an extension of the symmetry relation (5.7c). Thus, at points symmetrically placed with respect to the imaginary axis, $S_l(p)$ takes on complex conjugate values. As it has no singularity in the first quadrant, it has also no singularity in the second quadrant. If $S_l(p)$ has a zero at $p_1$, as shown in Figure 5.3, then it will also have a zero at $-p_1^*$.

One can now continue $S_l(p)$ into the lower half $p$-plane (Figure 5.3), by extending the relation (5.7a) to complex values of $p$. For every value in the second quadrant (5.7a) defines a value of $S_l(p)$ in the fourth quadrant, except at zeros in the second quadrant which produce poles in the fourth quadrant. In the same way (5.7a) defines a function analytic in the third quadrant except for poles coming from zeros in the first quadrant. Thus, $S_l(p)$ is in general not analytic in the lower half plane and if it has zeros above the positive $p$-axis it will have poles below the $p$-axis, the resonance poles of Figure 5.1 and their counterparts in the third quadrant.

In this way we have seen that $S_l(p)$ is a meromorphic function in the whole $p$-plane that is analytic in the upper half $p$-plane except perhaps on the imaginary axis. The poles in the lower half plane occur in pairs, symmetrically distributed with respect to the negative imaginary axis, except for those on

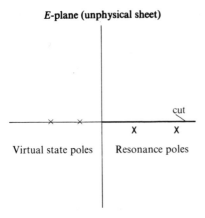

Figure 5.4   Positions of possible resonance poles, virtual-state poles and cuts (capture state poles not shown) in the second ("unphysical") sheet of the energy-plane.

the negative imaginary axis, corresponding to zeros on the positive imaginary axis. [The number of poles can be infinite, but they cannot have an accumulation point for finite values of $p$, as an analytic function (in the upper half plane) cannot have a finite accumulation point of zeros.] The singularities on the imaginary axis can be shown to lie within the interval $(-iK, +iK)$, where $B = K^2/2m$ is the maximal binding energy. Nothing can be derived from causality about the behavior of $S_l(p)$ in the neighborhood of this interval. It is, however, believed that for an interaction that is effective only in a region of finite size, one has poles, which must then be simple.[13]

If we translate these properties of $S_l(p)$ into statements about $S_l(E)$ in the energy plane, we can summarize our results, which are principally a consequence of the causality condition, in the following way (Figures 5.2 and 5.4): $S_l(E)$ is a meromorphic function on the two-sheeted Riemann surface with a branch point at $E = 0$ and a cut from 0 to $\infty$. The physical values of $E$ in collision processes lie on the upper edge of the cut on the "physical sheet." Bound-state poles lie on the negative real axis of the physical sheet, and except for these bound-state poles $S_l(E)$ is an analytic function on the physical sheet. Further poles (of any order) may lie on the second, "unphysical" sheet, coming from possible zeros on the first sheet. Poles at various locations on the "unphysical" sheet have various physical interpretations. Poles on the negative real axis of the unphysical sheet [coming from zeros of $S_l(p)$ on the positive imaginary axis] are called virtual-state poles. A virtual state is one that would be bound if the interaction were more attractive and a

---

[13] If the interaction is described by a potential $V(r)$ that decreases faster than any exponential for $r \to \infty$, then $S_l(p)$ is meromorphic in the whole $p$-plane and analytic in the upper half plane except for a finite number of simple poles. On the other hand, for a Yukawa-type potential $V(r) = \int_m^\infty \sigma(\mu)e^{-\mu r}/r \, d\mu$, $m > 0$, one has, in addition to possible poles of $S_l(p)$, also cuts along the imaginary axis from $im/2$ to $i\infty$ and from $-im/2$ to $-i\infty$.

virtual state close to threshold causes a large cross section at low energy; we will discuss such states in Section XVIII.7. Poles of $S_l(E)$ on the unphysical sheet, if they are close to the positive real axis, are of particular importance. They are called resonance poles or Siegert poles, and we shall study them in the remainder of this book. Each pole in the second sheet below the real axis has a counterpart in the second sheet above the real axis (capture states).

## XVIII.6  Resonance Scattering—Breit–Wigner Formula for Elastic Scattering

Experimentally, resonances are usually associated with a sharp variation of the cross section as a function of energy. If the elastic or inelastic cross sections exhibit sharp maxima or minima, one says that a resonance has occurred. From (XVI.2.9) for the elastic $l$th partial cross section one sees that a maximum occurs either at the energies for which $\delta_l(E) = \pi/2$ (modulo $\pi$) or for which $\delta_l(E)$ has a maximum. The latter possibility, however, cannot lead to a sharp maximum, because according to the causality condition (3.15), the phase shift can never decrease with a steep slope. Thus a sharp maximum in the $l$th partial cross section occurs at those value of $E$ for which $\delta_l(E) = \pi/2$ (modulo $\pi$); and the more rapidly $\delta_l(E)$ increases by $\pi$, the sharper is the maximum. Similarly, a minimum for the elastic $l$th partial cross section occurs for $\delta_l(E) = \pi$ (modulo $\pi$). The connection between the change in cross sections and the change in phase shifts for these two cases and for some intermediate cases is shown in Figure 6.1. Note that as a consequence of the causality condition (3.15), $\delta_l(E)$ must be mainly increasing; if it decreases, this decrease can be only very slowly. In every case a sharp structure in the cross section $\sigma_l(E)$ is connected with a sharp increase of the phase shift $\delta_l(E)$ by $\pi$.

We therefore want to take as the preliminary definition of a resonance of angular momentum $l$ at the energy $E_0$ that

$\delta_l(E)$ increases rapidly by approximately $\pi$ when $E$ passes through $E_0$.   (6.1)

If there is a sharp increase of one $\delta_l$ by $\pi$ around $E_0$, and if all other phase shifts are constant or slowly varying around $E_0$, then rapid variations will also show up in the total cross section $\sigma = \sum \sigma_l$ or the differential cross section at certain angles. However, not all structures in the cross section should be ascribed to resonances. Resonances can usually be distinguished from nonresonant phenomena in the cross sections by the fact that resonances appear only in the single partial wave, while nonresonant phenomena are the result of cooperative contributions from many partial waves.

We now want to explain the association of resonances and quasistationary states that we have already alluded to in the introduction (Section XVIII.1). According to our considerations in Section XVIII.2, a quasistationary state occurs at those values of $E = E_0$ for which $d\delta_l(E)/dE$ has a sharp

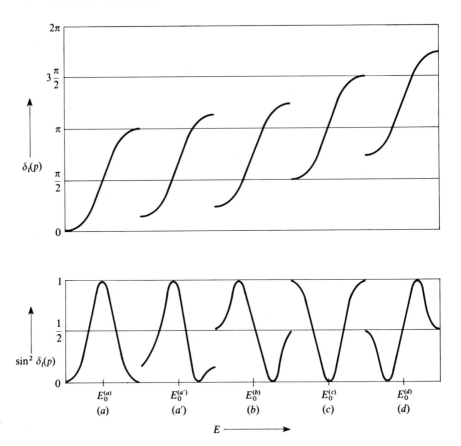

Figure 6.1    Phase shifts and resonance profiles.

*maximum.* We therefore take as the preliminary definition of a quasistationary state of angular momentum $l$ at the energy $E_0$ that

$$\frac{d^2\delta_l(E)}{dE^2}\bigg|_{E=E_0} = 0, \quad \frac{d^3\delta_l(E)}{dE^3}\bigg|_{E=E_0} < 0; \quad \frac{1}{E_0^2}\bigg|\frac{d\delta_l}{dE}\bigg|_{E_0} \ll \bigg|\frac{d^3\delta_l(E)}{dE^3}\bigg|_{E_0} \quad (6.2)$$

Whereas the first two conditions of (6.2) state that $E_0$ is a maximum of $d\delta_l(E)/dE$, the last condition of (6.2) states that this maximum is to be very sharp. The first part of the conditions (6.2) is already incorporated in the drawings of the phase shifts in Figure 6.1, where the point $E_0$ around which $\delta_l(E)$ changes rapidly is a point of inflection. However, it is not clear that the two phenomena characterized by (6.1) and (6.2) are related. We shall show in this section that as a consequence of very general physical assumptions (causality), these two phenomena are indeed connected: that a quasistationary state leads to a resonance and that a resonance leads to a quasistationary state. We will see that (6.2) leads to a pole in the $l$th partial $S$-matrix below and near to the positive real $E$-axis. And we shall obtain from (6.2) a representation of $S_l(E)$, valid in the neighborhood of $E_0$, from which (6.1) follows immediately.[14] The notion of resonance poles was introduced in Section 5, when we described the analyticity property of the S-matrix without, however, giving a complete derivation. We concluded there that there may exist some resonance poles. Here we show that they appear as a consequence of the existence of a quasistationary state, (6.2), and provide therewith a physical explanation of their existence.

To start our derivation we introduce the function

$$f(E) = \left( \frac{d\delta_l(E)}{dE} \right)^{-1}, \tag{6.3}$$

which as a consequence of (6.2) has a sharp minimum at $E = E_0$, so that $f' = f'(E)_{E=E_0} = 0$ and $f'' = f''(E)_{E=E_0}$ is large. More precisely, a simple calculation shows that

$$f'' = - \left( \frac{d\delta_l}{dE} (E_0) \right)^{-2} \frac{d^3\delta_l}{dE^3} (E_0), \tag{6.4}$$

and consequently,

$$\sqrt{\frac{2f}{E_0^2 f''}} = \sqrt{- \frac{2}{E_0^2} \frac{d\delta_l}{dE} (E_0) \bigg/ \frac{d^3\delta_l}{dE^3} (E_0)} \ll 1 \tag{6.5}$$

from the last two inequalities (6.2).

We want to expand $f(E)$ in a Taylor series around $E = E_0$:

$$f(E) = f + \frac{f''}{2} (E - E_0)^2 + F(E - E_0), \tag{6.6}$$

where the "remainder" term is

$$F(E - E_0) = \frac{f'''}{3!} (E - E_0)^3 + \cdots + \frac{f^{(n)}}{n!} (E - E_0)^n + \cdots \tag{6.7}$$

The $n$th derivative of $f$ at the value $E = E_0$, $f^{(n)} = d^n f(E)/dE^n|_{E=E_0}$ can easily be calculated to be

$$f^{(n)} = - \frac{\delta^{(n+1)}}{(\delta^{(1)})^2} \quad \text{where} \quad \delta^n = \frac{d^n \delta_l}{dE^n} \bigg|_{E=E_0} \tag{6.8}$$

The Taylor series, therefore, converges if for all $n$ larger than a certain value $N$

$$\delta^{(n+1)} < \delta^{(n)} \frac{n}{(E - E_0)}, \tag{6.9}$$

certainly not too stringent a condition.

We shall first assume that $F(E - E_0)$ is negligibly small $[F(E - E_0) = 0]$ in a certain energy range around $E_0$; this would be the case if $\delta_l^{(4)}$ and the higher derivatives of the phase shifts at $E_0$ could be neglected. In a second step we will discuss the corrections that arise from a small $F(E - E_0)$.

---

[14] This derivation is based on Goldberger and Watson (1964, Section 8.5).

With $F(E - E_0) = 0$ it follows from (6.3) and (6.6) that

$$\delta_l(E) = \int \frac{dE}{f + f''/2(E - E_0)^2}$$

$$= \sqrt{\frac{2}{ff''}} \arctan \left( \frac{E - E_0}{\sqrt{2f/f''}} \right) + \gamma_l'$$

$$= \sqrt{\frac{2}{ff''}} \operatorname{arccot} \frac{E - E_0}{\sqrt{2f/f''}} + \gamma_l \qquad (6.10')$$

$$\delta_l(E) = \sqrt{\frac{2}{ff''}} \arctan \left( \frac{\sqrt{2f/f''}}{E - E_0} \right) + \gamma_l, \qquad (6.10)$$

where $\gamma_l'$ and $\gamma_l$ are arbitrary integration constants, $\gamma_l - \gamma_l' = \pi/2$. Equation (6.10) suggests the introduction of the two new parameters:

$$r = \sqrt{\frac{2}{ff''}} = \sqrt{-\frac{2(\delta_l^{(1)})^3}{\delta_l^{(3)}}} \qquad (6.11)$$

and

$$\Gamma = 2 \sqrt{\frac{2f}{f''}} = \sqrt{-\frac{2\delta_l^{(1)}}{\delta_l^{(3)}}}. \qquad (6.12)$$

$r$ is dimensionless, and $\Gamma$ has the dimension of energy. As a consequence of (6.5) (which followed from the requirement that there exist a quasistationary state at $E_0$), $\Gamma$ fulfills the condition

$$\Gamma \ll 2E_0. \qquad (6.13)$$

Inserting (6.11) and (6.12) into (6.10), one obtains for the $l$th partial $S$-matrix element

$$S_l(E) = e^{i2\delta_l(E)} = e^{i2r \arctan((\Gamma/2)/E_0 - E)} e^{i2\gamma_l} \qquad (6.14)$$

The phase shift (6.10) is written as a sum of two terms:

$$\delta_l(E) = \delta_l^{(R)}(E) + \gamma_l, \qquad (6.15)$$

where $\delta_l^{(R)}(E)$ is a rapidly varying function due to (6.13):

$$\delta_l^{(R)}(E) = r \arctan \frac{\Gamma/2}{E_0 - E}, \qquad (6.16)$$

whose value changes by almost $r\pi$ when $E$ passes through $E_0$. $\gamma_l$ is a constant under the assumption $F(E - E_0) = 0$.

If $F(E - E_0)$ is not zero, then one obtains instead of (6.10'):

$$\delta_l(E) = \int \frac{dE}{f + f''/2(E - E_a)^2 + F(E - E_0)} = \int \frac{dE}{f + f''/2(E - E_0)^2}$$

$$- \int \frac{dE \, F(E - E_0)}{[f + f''/2(E - E_0)^2][f + f''(E - E_0)^2 + F]}. \qquad (6.17)$$

Writing $\delta_l(E)$ again in the form (6.15), we obtain for $\gamma_l$

$$\gamma_l = \gamma_l(E) = -\int \frac{dE\, F(E - E_0)}{[f + f''/2(E - E_0)^2][f + f''/2(E - E_0)^2 + F]}, \quad (6.18)$$

which for small $F(E - E_0)$ is a slowly varying function of $E$.

Thus the phase shift in the neighborhood of a quasistationary state defined by the requirements (6.2) is the sum of a term $\delta_l^{(R)}(E)$, which changes rapidly by almost $\pi r$, and a slowly varying term $\delta_l(E)$:

$$\delta_l(E) = \delta_l^{(R)}(E) + \gamma_l(E), \quad (6.19)$$

e.g., as depicted in Figure 6.2. The quantity $\gamma_l(E)$ is called the background phase shift, or also the potential part[15] of the phase shift, as opposed to $\delta_l^{(R)}(E)$, which is called the resonant part. Thus, if $r = 1$, we have derived the condition (6.1) as a consequence of the condition (6.2). In particular, if the background phase shift is zero, we have the situation depicted in Figure 6.1a.

We should now, therefore, ask the question, what values the parameter $r$ can take. In order to answer this question, we use the analyticity property (4.10).[16]

The function

$$\arctan \frac{\Gamma/2}{E_0 - E} = \frac{i}{2} [\ln(E - E_0 + i\Gamma/2) - \ln(E - E_0 - i\Gamma/2)] \quad (6.20)$$

has a branch point of infinite order at $E = E_0 + i\Gamma/2$ (and also one at $E = E_0 - i\Gamma/2$).[17] If $E$ moves from the real axis around this branch point and returns to the initial point on the real axis, always staying in the domain of analyticity, then $\ln(E - E_0 - i\Gamma/2)$ increases by $2\pi i$, and $\arctan[(\Gamma/2)/(E_0 - E)]$ by $\pi$. Thus moving along this closed contour around the point $E_0 + i\Gamma/2$ results, according to (6.14), in the change

$$S_l(E) \rightarrow S_l(E)e^{2\pi ir}. \quad (6.21)$$

On the other hand, since $S_l(E)$ is analytic in this domain, its value should not change. This is only possible if

$$r = 0, 1, 2, \ldots. \quad (6.22)$$

We shall now discuss the simplest nontrivial case, $r = 1$. From (6.14) one then obtains

$$S_l(E) = \frac{E_0 - E + i\Gamma/2}{E_0 - E - i\Gamma/2} e^{i2\gamma_l} = \left(1 + \frac{i\Gamma}{E_0 - E - i\Gamma/2}\right) e^{i2\gamma_l}$$

$$= e^{i2\delta_l^{(R)}(E)} e^{i2\gamma_l}, \quad (6.23)$$

---

[15] If the interaction is described by a potential, then $\gamma_l$ comes from the long-range component of this potential, whereas $\delta_l^R$ comes from a short-range attractive component.

[16] For the present discussion not all of the analyticity (4.10) is required; one just needs that $S_l(E)$ and, therewith, $T_l(E)$ are analytic in a domain above the real positive energy axis that contains the point $E = E_0 + i\Gamma/2$, a very weak condition indeed.

[17] Cf. Appendix XVIII.A.

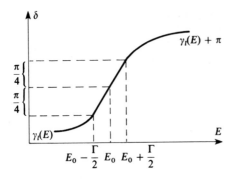

Figure 6.2   Change of phase shift at a quasistationary state.

and from (6.16),

$$\tan \delta_l^R(E) = \frac{\Gamma/2}{E_0 - E}. \tag{6.24}$$

The $S$-matrix element (6.23) is then written

$$S_l(E) = 1 + 2ipT_l(E) = e^{i2\gamma_l}(1 + 2ipT_l^{(R)}(E)), \tag{6.25}$$

where one has defined the resonant partial-wave amplitude $T_l^{(R)}(E)$ in such a way that it agrees with the partial-wave amplitude $T_l(E)$ if the background can be ignored ($\gamma_l = 0$). From comparison of (6.25) and (6.23) one then obtains for the resonant partial-wave amplitude

$$T_l^{(R)}(E) = \frac{1}{p} \frac{\Gamma/2}{(E_0 - E) - i\Gamma/2}. \tag{6.26}$$

Thus, if a quasistationary state with angular momentum $l$ is formed at the energy $E_0$, then the phase shift has the resonant behavior specified by (6.1), and—if the background can be neglected—the partial-wave amplitude is given for the simplest case $r = 1$ by (6.26).

The $l$th partial cross section that one obtains from (6.26), which is therefore the pure resonance cross section (any background being neglected), is then given [using (XVI.1.38)] by

$$\sigma_l^R(E) = \frac{4\pi}{p^2}(2l + 1)\frac{(\Gamma/2)^2}{(E_0 - E)^2 + (\Gamma/2)^2} \tag{6.27}$$

This function of $E$ is shown in Figure 6.3 together with the corresponding resonance phase shift. We see that the resonance cross section has a maximum at $E_0$, and the full width at half maximum of this peak is $\Gamma$. $E_0$ is called the *resonance energy*, and $\Gamma$ is called the *resonance width*.

At resonance energy the cross section takes the value

$$\sigma_l^{(R)} = \frac{4\pi}{p^2}(2l + 1), \tag{6.28}$$

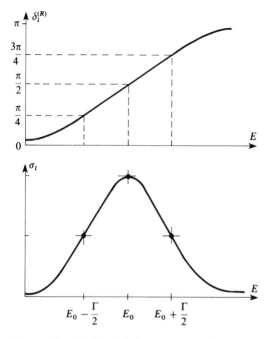

Figure 6.3    The Breit–Wigner cross section and its relation to the resonance phase shift.

which is the maximum value for the elastic partial cross section compatible with unitarity, as seen from (XVI.2.9).

The quantity $\Gamma$ that has emerged in these discussions can be calculated from the phase shifts using (6.12) if these are known, e.g., from the interaction potential in potential scattering. However, as mentioned already in several places, the situation is more often than not the reverse: One knows the cross section from an experiment and notices that it shows, in the neighborhood of a certain value $E_0$, the features shown in the lower part of Figure 6.1. Then one tries to fit this experimental cross section with a formula

$$\sigma_l^{\text{exp}} = 4\pi(2l + 1)|T_l^R(E) + T^{\text{bg}}(E)|^2, \qquad (6.29)$$

where $T_l^R(E)$ is the function (6.26) and $T^{\text{bg}}(E)$ is a slowly varying background term. In this way one determines the values $(E_0, \Gamma)$.

Equation (6.27) is the celebrated Breit–Wigner formula; (6.26) is called the Breit–Wigner amplitude. It hardly ever gives a completely accurate description of the experimental situation, which indeed one would not expect in view of the idealizations involved. It almost never fails to give a useful parametrization whenever resonance phenomena are involved. It is, perhaps, the most frequently used formula of quantum physics. Its fundamental significance comes from its introduction of the parameter $\Gamma$: On the level of accuracy at which excited states are considered stationary, they are characterized by one parameter,[18] their energy value $E_0$; on the level of accuracy at which they are considered *quasistationary states*, they *are characterized by two parameters, their energy and width* $(E_0, \Gamma)$. From Chapter II we know that stationary states are described by eigenvectors of the energy operator with eigenvalue $E_0$; in Chapter XXI we will see that, analogously, quasistationary states are described by generalized eigenvectors of the ⟦essentially selfadjoint⟧ energy operator with eigenvalue $E_0 + i\Gamma/2$.

Like every theoretical description, the description of a quasistationary state by a Breit–Wigner resonance (6.26) is an approximate description. If one wants a more accurate description, one can consider $\Gamma$ not as a parameter but as a function of energy, $\Gamma = \Gamma(E)$, which may depend upon one or more parameters (e.g., the radius of the interaction and the angular momentum $l$). We content ourselves here with the degree of accuracy with which the quasistationary state is characterized by the two parameters $(E_0, \Gamma)$.

We now want to establish the connection between resonances and poles of $S_l$. We have already introduced, in Section XVIII.5, the name resonance pole for a pole of $S_l(E)$ in the second sheet of the energy plane immediately below the positive real axis. As seen, e.g., from (6.23), the Breit–Wigner formula furnishes such a pole in the lower half plane at $E = E_0 - i\Gamma/2$. The smaller the value of $\Gamma$, and consequently the sharper the resonance is, the closer this pole is to the real axis. We have already mentioned that there is no pole on the lower half of the first sheet of the energy plane.[19] Consequently, this pole furnished by the Breit–Wigner formula must be on the second sheet of the plane, and thus there must be a pole of $S_l(p)$ in the fourth quadrant of the $p$-plane close to the real axis.

On the other hand, every simple pole of $S_l(p)$ at the point $p = p_R = k_R - iK_R$ in the fourth quadrant near the real axis $(K_R \ll k_R)$ may, but does not necessarily, lead to the resonance phenomenon. This is easily seen by expanding $S_l(p)$ around this pole in a Laurent series. In a sufficiently small neighborhood of the pole, $S_l(p)$ may be approximated by the principal part of the Laurent series $\sim 1/(p - p_R)$. However, $S_l(p)$ must be unitary $[|S_l(p)| = 1]$, so the factor multiplying $1/(p - p_R)$ must be chosen to be $p - p_R^*$ if the rest of the series is to be expressed as an exponential $e^{i2\gamma_l}$. Thus in the neighborhood of the pole the expansion of $S_l(p)$ compatible with unitarity is

$$S_l(p) = e^{i2\delta_l(p)} \approx e^{i2\gamma_l(p)} \frac{p - p_R^*}{p - p_R} = e^{i2\gamma_l(p)} \frac{p - k_R + iK_R}{p - k_R - iK_R}. \qquad (6.30)$$

Multiply denominator and numerator by $p/m$ and note that

$$p(p - k_R)\frac{1}{m} = \frac{p^2}{m} - \frac{pk_R}{m} = \frac{p^2}{2m} - \frac{k_R^2}{2m} + \frac{(p - k_R)^2}{2m} \approx \frac{p^2}{2m} - \frac{k_R^2}{2m} + \frac{K_R^2}{2m}$$

$$(6.31)$$

---

[18] In addition to their other quantum numbers like $l$, $\eta$.

[19] We did not give a complete proof of the statements in Section XVIII.5, but argued that causality, the finiteness of the binding energies for possible bound states, and the finite range of the interaction are sufficient, though certainly much less is necessary.

for real values of $p$ that differ from $k_R$ by approximately $K_R \ll k_R$, i.e., for values of $p$ that fulfill $|p - k_R| \approx K_R$. Then with

$$E = \frac{p^2}{2m}, \qquad E_R = \frac{1}{2m}(k_R^2 - K_R^2), \tag{6.32}$$

one obtains

$$S_l(E) = e^{i2\gamma_l(E)} \frac{E - E_R - i\Upsilon/2}{E - E_R + i\Upsilon/2} \quad \text{with } \Upsilon = -\frac{2K_R p}{m} = \Upsilon(p). \tag{6.33}$$

With the same accuracy as the approximate equality (6.31),

$$\Upsilon(p) \approx -\frac{2K_R k_R}{m} \equiv \Gamma. \tag{6.34}$$

Then (6.33) is identical with the resonance formula (6.23), and consequently a pole of $S_l(p)$ at $p_R$ describes a resonance.

The above considerations would suggest a one-to-one correspondence between poles of the $S$-matrix immediately below the positive real axis and a resonance phenomenon characterized by (6.1); however, they do not prove such a correspondence, because we have always assumed that the $\gamma_l$ has the nice feature of a background phase shift. If $\gamma_l$ varies rapidly without having the Breit–Wigner form (6.24), then it may compensate the effect of the pole, and the total phase shift $\delta_l = \delta_l^{(R)} + \gamma_l$ may be slowly varying. It can, indeed, be shown[20] that phase shifts $\tilde{\gamma}_l$ can be constructed that approach the function (6.24) as close as one wishes in an interval $E_0 - \Delta \le E \le E_0 + \Delta$, $\Delta \gg \Gamma$, and that therefore describe a resonance phenomenon (6.1), but that are not associated with a singularity of $e^{i2\tilde{\gamma}_l}$. Thus choosing $S_l = e^{i2\tilde{\gamma}_l}$ gives an $S$-matrix with a resonance phenomenon but without a pole, and choosing $\gamma_l = -\tilde{\gamma}_l$ in (6.33) gives an $S$-matrix with a pole but without a resonance phenomenon.

However, if the fourth and higher derivatives of $\delta_l$ are small, then, as we saw in the first part of this section, $\gamma_l$ will be slowly varying. Therefore, one generally assumes the *one-to-one correspondence between poles immediately below the real axis and resonances.*

We also have seen above that a quasistationary state characterized by a sharp maximum in the time delay is a resonance. Further, as from (6.19), (6.24), and (2.26) it follows immediately that

$$\tfrac{1}{2} t_D^l(E) = \frac{d\delta_l(E)}{dE} = \frac{d\delta_l^{(R)}}{dE} + \frac{d\gamma_l}{dE} = \frac{\Gamma/2}{(E_0 - E)^2 + \Gamma^2/4} + \frac{d\gamma_l(E)}{dE}, \tag{6.35}$$

we see that a resonance leads to a sharp maximum in the time delay for the resonating partial wave. Thus *a quasistationary state, a resonance, and a pole of $S_l$ immediately below the real axis are in fact one and the same phenomenon.*

[20] L. Fonda, *Fortschr. Phys.* **20**, 135 (1972).

We shall now discuss the connection between the width $\Gamma$ and the time delay. According to the discussions in Section XVIII.2, (6.35) is the "time delay for a monoenergetic beam," $\Delta E \ll \Gamma$. If a resonance occurs, then the derivative of the background phase shift is negligible, so that the resonant time delay is well approximated by

$$\tfrac{1}{2}t_D^l(E) = \frac{d\delta_l^{(R)}}{dE} = \frac{\Gamma/2}{(E_0 - E)^2 + \Gamma^2/4}. \tag{6.35'}$$

The "time delay for a monoenergetic beam" that has energy $E = E_0$ identical to the resonance energy is therefore

$$t_D^l(E)|_{E=E_0} = \frac{4}{\Gamma}. \tag{6.36}$$

This idealized time delay is not really a physical quantity, because in the case of a monoenergetic beam there is a steady state, as mentioned already in Section XVIII.2.

In the opposite extreme, $\Delta E \gg \Gamma$, the time delay is essentially determined by the experimental energy distribution of the beam, as we see by inserting (6.35) into (2.32):

$$t_D^l = 2 \int dE\, F(E - E_A) \frac{\Gamma/2}{(E_0 - E)^2 + \Gamma^2/4}. \tag{6.37}$$

Noting that for $\Delta E \gg \Gamma$ one can take $\Gamma/2 \to 0$ in the integral, and using

$$\frac{\Gamma/2}{(E_0 - E)^2 + \Gamma^2/4} \to \pi\delta(E_0 - E) \quad \text{for } \Gamma \to 0,$$

we obtain

$$t_D^l = 2\pi F(E_0 - E_A). \tag{6.38}$$

For a Gaussian energy distribution of the incident beam,[21]

$$F(E - E_A) = \frac{1}{\sqrt{2\pi}} \frac{1}{\Delta E} e^{-(E - E_A)^2/2(\Delta E)^2}, \tag{6.39}$$

one obtains at resonance energy

$$t_D^l|_{E_A=E_0} = \sqrt{2\pi} \frac{1}{\Delta E}. \tag{6.40}$$

And for a Lorentzian energy distribution of the incident beam,[21]

$$F(E - E_A) = \frac{1}{\pi} \frac{\Delta E/2}{(E_A - E)^2 + (\Delta E/2)^2}, \tag{6.41}$$

---

[21] The normalization of $F$ has to be chosen so that $F(E - E_A) \to \delta(E - E_A)$ as $\Delta E \to 0$ according to (XIV.5.36). Note that $F$ is the energy distribution of the incident noninteracting beam at $t = 0$, i.e., the beam has been prepared in the remote past in such a way that an energy measurement at $t = 0$ with no interaction present would give the result $F(E - E_A)$ for the probability distribution to measure $E$.

with $\Delta E \gg \Gamma$, one obtains

$$t_D^l \big|_{E_A = E_0} = \frac{1}{\Delta E}. \tag{6.42}$$

In general, if the energy distribution in the incident beam caused by the resolution of the apparatus has a width $\Delta E$ of the same order as the resonance width $\Gamma$, one obtains a time delay that depends on $\Gamma$ and on the experimental energy distribution. A particular situation arises when the experimental energy distribution is Lorentzian [Equation (6.41)] with an energy spread $E$ that is equal to the resonance width $\Delta E = \Gamma$. Then one obtains by inserting (6.41) into (6.37)

$$t_D^l = 2/\Gamma \quad (= 2\hbar/\Gamma \text{ in conventional units of time).} \tag{6.43}$$

Thus we have seen that the time delay has something to do with the inverse of the resonance width. For an idealized monochromatic beam this relation is given by (6.36). But for more realistic situations the effect of the energy resolution of the incident beam will result in other connections between time delay and width—e.g., the one given in (6.43).

So far we have considered for the parameter $r$, which we introduced in (6.10), only the value $r = 1$. However, according to (6.22) it can take any integer value.

The case $r = 1$ seems to be the only case that is realized in quantum-scattering experiments, though some time ago there was thought to exist in particle physics some evidence for the existence of a "dipole," which is a quasistationary state with $r = 2$. We shall discuss this case briefly here. With $r = 2$ we obtain from (6.14), ignoring the background ($\gamma_l = 0$),

$$S_l(E) = \left( \frac{E_0 - E + i\Gamma/2}{E_0 - E - i\Gamma/2} \right)^2 \tag{6.44}$$

and therewith for the $l$th partial cross section

$$\sigma_l^{(r=2)}(E) = \frac{4\pi(2l+1)}{p^2} \frac{\Gamma^2(E - E_0)^2}{[(E - E_0)^2 - \Gamma^2/4]^2 + \Gamma^2(E - E_0)^2}. \tag{6.45}$$

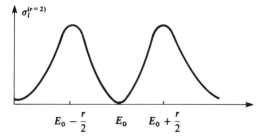

Figure 6.4   Illustration of the energy dependence of the scattering cross section (6.45) for a dipole resonance.

The shape of the cross section for such a dipole resonance is depicted in Figure 6.4.

This is to be compared with Figure 6.1a. When background scattering is present, the cross section will be modified accordingly.

## XVIII.7  The Physical Effect of a Virtual State

Besides the bound-state poles and the resonance poles, we mentioned in Section XVIII.5 the virtual-state poles on the negative imaginary axis of the momentum plane. If a virtual-state pole is sufficiently close to the real axis, it can have an appreciable influence on the cross section at low energies.

Denoting the momentum value of the virtual state pole by $p_V = -i\kappa$, $\kappa > 0$, we can write the $l = 0$ S-matrix element in the neighborhood of the pole as

$$S_0(p) \approx \frac{p - p_V^*}{p - p_V} = \frac{ip + \kappa}{ip - \kappa}. \tag{7.1}$$

This can be arrived at by the same arguments leading to (6.30), the background phase shift having been chosen suitably. The 0th partial cross section is then

$$\sigma_0 = 4\pi |T_0|^2 = \frac{4\pi\kappa^2}{p^2(p^2 + \kappa^2)} \approx \frac{4\pi}{\kappa^2 + p^2}. \tag{7.2}$$

In the limit of zero scattering energy, when the virtual pole is close to the real axis, this goes into

$$\sigma_0 \approx \frac{4\pi}{\kappa^2}, \tag{7.3}$$

which is identical to the total cross section, since according to (XVII.5.6), higher partial waves can be ignored at low energies. Thus we see that for a virtual-state pole close to the real axis the scattering cross section can be extremely large. Also large is the scattering length, given according to (XVII.6.11) by

$$a_0 = -\frac{1}{\kappa}. \tag{7.4}$$

It is easy to see that a bound state with sufficiently low binding energy will also lead to (7.2) and (7.3) for the cross section. Therefore, a virtual state can only be detected in the low-energy scattering cross section if low-energy bound states are absent.

Virtual states have been observed in nature. In the case of proton-neutron scattering there exists a bound state, the deuteron (isospin 0, angular momentum 1). The scattering length is $a_0^{(0,\,1)} = 5.4 \times 10^{-13}$ cm. For angular momentum equal to zero (isospin 1), no bound state is present, but $a_0^{(1,\,0)} = -23.7 \times 10^{-13}$ cm. Thus, there is a virtual state in the singlet (angular momentum 0) p-n system, whose position is closer to the threshold than the triple bound state and which dominates low-energy scattering.

Virtual states also cause long time delay. As we are at low energies, the phase shift can be obtained from (cf. XVII.6.15)

$$p \cot \delta_0(p) = -\frac{1}{a_0}. \tag{7.5}$$

From this we obtain

$$\frac{d\delta_0}{dp} = -\frac{a_0}{1 + p^2 a_0^2} = \frac{1}{\kappa} \frac{1}{1 + p^2/\kappa^2}.$$

Thus the average time delay (2.26),

$$t_D^{(\text{virt})} = 2 \frac{m}{p} \frac{1}{\kappa} \frac{1}{1 + p^2/\kappa^2}, \tag{7.6}$$

can reach very high values. A better illustration of this effect is obtained if we compare the time delay with the time delay of a noninteracting classical particle in a region of radius $R$, which is given by (cf. Section XVIII.3)

$$T^{\text{cl}} = \frac{m \cdot 2R}{p}$$

Thus we see that the ratio of the time delay caused by a virtual state to the classical time delay is

$$\frac{t_D^{\text{virt}}}{T^{\text{cl}}} = \frac{1}{R\kappa} \frac{1}{1 + p^2/\kappa^2} = -\frac{a^0}{R} \frac{1}{1 + p^2 a_0^2}. \tag{7.7}$$

Depending upon the value of $\kappa$ and $R$, this ratio can be very high. E.g., for the $p$-$n$ virtual state, taking $R \approx 2.5 \times 10^{-13}$ cm and $a_0 \approx -24 \times 10^{-13}$ cm, we see that the time delay caused by a virtual state is an order of magnitude larger than the classical time.

A virtual state is thus not simply a fictitious, mathematical state, but a physical state in which projectile and target spend a considerable length of time together.

## XVIII.8 Argand Diagrams for Elastic Resonances and Phase-Shift Analysis[22]

The Argand diagram introduced in Section XVI.3 is a useful tool for the detection and display of resonances and the determination of their parameters $(E_R, \Gamma)$. To understand this we investigate in detail the behavior of the resonant scattering amplitude in the Argand diagram. We first study the idealized case of an elastic resonance without background. For this the scattering amplitude is given by (6.26), which we write in the form

$$p T_l^{(R)}(E) = \frac{1}{\epsilon - i}, \tag{8.1}$$

[22] G. Bialkowski has helped me with the writing of this section.

where we have introduced the abbreviation

$$\epsilon = (E_R - E)\frac{2}{\Gamma} = \cot \delta_l^{(R)}(E). \tag{8.2}$$

Figure 8.1 shows the Argand diagram for Equation (8.1). As we see, this elastic Breit–Wigner amplitude lies on the unitary circle. The top of the circle corresponds to the resonant energy $\epsilon = 0$, at which value the phase shift $\delta_l(E_l)$ is $\pi/2$ (cf. Figure XVI.4.1). With increasing energy $E$, $pT_l^R$ moves in a counterclockwise sense—slowly along the lower part of the circle, and more rapidly as $E$ gets closer to the resonance value $E_R$. This is indicated by the values for $\epsilon$ on the path of $pT_l^R$ in Figure 8.1.

To take the background into account we use (6.25) and obtain

$$pT_l = e^{i\gamma_l}\sin\gamma_l + e^{2i\gamma_l}\,pT^{(R)} = pT_l^{(bg)} + e^{2i\gamma_l}pT_l^{(R)}, \tag{8.3}$$

where we have defined the background amplitude $T^{(bg)}$ in analogy to (XVI.2.8) by

$$T_l^{(bg)} = \frac{1}{p}e^{i\gamma_l}\sin\gamma_l = \frac{1}{2ip}(e^{i2\gamma_l} - 1). \tag{8.4}$$

From (8.3) we see that the effect of the background is to rotate the resonant amplitude by the angle $2\gamma_l$. The quantity $pT_l$ for constant background amplitude is shown in Figure 8.2. The resonance circle starts at the point $B$ and traverses the unitary circle, the only difference with the case of Figure 8.1 being that now the resonance energy is not at the point where the resonance amplitude is pure imaginary, but at the point diametrically opposite $B$. If the background changes with energy, then the point of resonance is not well defined.

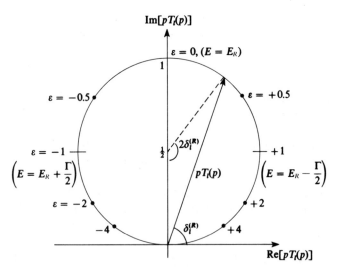

Figure 8.1   Argand diagram for the resonant elastic scattering amplitude: $pT_l(p) = (\epsilon - i)^{-1}$ and $\cot\delta_l^{(R)} = \epsilon$.

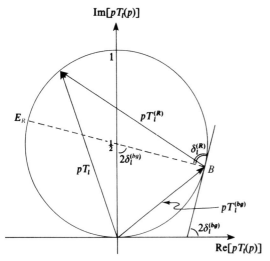

Figure 8.2   Argand diagram for the resonant elastic scattering amplitude in the presence of background: $T_l = T_l^{(R)} + T_l^{(bg)}$.

As we have seen above, the resonance amplitude has a characteristic circular path in the Argand diagram, the resonance energy being the value where the phase shift changes most rapidly. As we shall see in Section XX.3, this feature will remain even when inelasticity is taken into account, only there the circles become smaller and distorted.

The Argand diagram can be used to display the energy and width of a resonance for a particular partial wave. For this purpose one has just to draw $pT_l$ as function of energy in the Argand diagram. If $pT_l$ runs along a circle close to the unitary circle, one knows that this particular partial wave has an elastic resonance, and the energy value at which the amplitude has its most rapid change will be the resonance energy $E_R$. If there is no background, $E_R$ will be the point at which $pT_l$ crosses the imaginary axis. In order to draw $pT_l(E)$ one has to determine it from the experimental data, and this is the main difficulty of phase-shift analysis. In the following we will give a brief description of this procedure.

The data determined in an experiment are the cross sections: the total cross section at various momenta $\sigma(p)$ and the differential cross sections $d\sigma/d\Omega(\theta, p)$ at various angles and momenta (or the differential cross sections $d\sigma/d\cos\theta = 2\pi \, d\sigma/d\Omega$). According to (XVI.1.3) these cross sections are connected to the partial-wave amplitudes by

$$\frac{d\sigma}{d\cos\theta}(p, \cos\theta) = 2\pi \left| \sum_{l=0}^{\infty} (2l+1)P_l(\cos\theta)T_l(p) \right|^2. \tag{8.5}$$

And by the optical theorem (XVI.2.20),

$$\sigma(p) = \frac{4\pi}{p} \sum_{l=0}^{\infty} (2l+1) \, \mathrm{Im} \, T_l(p). \tag{8.6}$$

These sums extend to infinity and would be of little practical use for the determination of the partial-wave amplitude $T_l(p)$. Fortunately, however, according to (XVII.6.6) the higher partial-wave amplitudes for a particular value of the momentum $p$ can be neglected, so that for a particular value of $p$ the sum in (8.5) extends in fact only over a finite number of terms:

$$\frac{d\sigma}{d\cos\theta}(p,\cos\theta) = 2\pi\left|\sum_{l=0}^{L}(2l+1)P_l(\cos\theta)T_l(p)\right|^2, \qquad (8.5')$$

and so does the sum in (8.6). How high a value of $L$ one has to take for a particular value of $p$ depends upon the problem under investigation and has to be determined empirically by the "principle of stability." One does this by choosing some value $L_0$. Then one determines, by the procedure described below, $T_0(p), \ldots, T_{L_0}(p)$; chooses $L_0 + 1$; and determines from the same set of data a new set of amplitudes $\tilde{T}_0(p), \ldots, \tilde{T}_{L_0}(p), \tilde{T}_{L_0+1}(p)$. If $T_l$ and $\tilde{T}_l$ determined in this way differ significantly, $L_0$ was chosen too small for that particular value of energy and one has to try a higher value.

Using the relation for the Legendre polynomials[23]

$$P_l(\cos\theta)P_{l'}(\cos\theta) = \sum_{L=|l-l'|}^{l+l'}\langle l\,0\,l'\,0|L0\rangle^2 P_L(\cos\theta), \qquad (8.7)$$

where $\langle ll'mm'|L, m+m'\rangle$ are the SO(3) Clebsch–Gordan coefficients, one can write (8.5') in the form

$$\frac{d\sigma}{d\cos\theta} = 2\pi\sum_{j=0}^{2L}P_j(\cos\theta)C_j(p). \qquad (8.8)$$

The inversion of (8.8) is

$$C_l(p) = \frac{2l+1}{4\pi}\int_{-1}^{+1}d\cos\theta\,\frac{d\sigma(p,\theta)}{d\cos\theta}P_l(\cos\theta). \qquad (8.8')$$

For the sake of definiteness let us do this for the particular values $L = 1$ and $L = 2$. For $L = 1$, (8.5') becomes

$$\frac{d\sigma}{d\cos\theta} = 2\pi\{|T_0|^2 + 3(T_0\,T_1^* + T_0^*T_1)P_1 + 9|T_1|^2P_1^2\} \qquad (8.9_1)$$

For $L = 2$ it becomes

$$\frac{d\sigma}{d\cos\theta} = 2\pi\{[|T_0|^2 + 3|T_1|^2 + 5|T_2|^2]P_0(\cos\theta)$$

$$+ 6\,\mathrm{Re}(T_0^*T_1 + T_1^*T_2)\,P_1(\cos\theta)$$

$$+ [10\,\mathrm{Re}\,(T_0^*T_2) + 6|T_1|^2 + \tfrac{50}{7}|T_2|^2]P_2(\cos\theta)$$

$$+ 18\,\mathrm{Re}\,(T_1^*T_2)P_3(\cos\theta) + \tfrac{90}{7}|T_2|^2P_4(\cos\theta)\}, \qquad (8.9_2)$$

---

[23] This is a special case of (VII.3.22b).

where the numerical factors are determined from the values of $2l + 1$ and of the Clebsch–Gordan coefficients. Thus for example for $L = 2$,

$$C_0(p) = |T_0|^2 + 3|T_1|^2 + 5|T_2|^2,$$

$$C_1(p) = 6 \, \mathrm{Re}(T_0^* T_1 + T_1^* T_2), \qquad \text{etc.} \qquad (8.10)$$

Now if $d\sigma/d \cos \theta \, (p, \cos \theta)$ has been measured at a particular value of $p$, then one calculates the $C_l(p)$ from (8.8'). These "experimental" $C_l(p)$ one uses in equations like (8.10) to determine the "experimental" values of the partial-wave amplitudes. There are $2L + 1$ equations (8.10). In addition one has the optical theorem

$$\sigma(p) = \frac{4\pi}{p} \sum_{l=0}^{L} (2l + 1) \, \mathrm{Im} \, T_l(p). \qquad (8.6')$$

So one has $2L + 2$ nonlinear equations for the $2L + 2$ real quantities $\mathrm{Im} \, T_l(p)$, $\mathrm{Re} \, T_l(p)$, $l = 0, 1, \dots, L$, or the $2L + 2$ quantities $\delta_l(p)$ and $\eta_l(p)$ of (XVI.2.10).

If the scattering at that particular energy is elastic, then one has only the $L + 1$ unknowns $\delta_l(p)$, and one need not consider the $C_j(p)$ for $j > L$. There are then $L + 1$ equations for $L + 1$ unknowns in (8.10). [The optical theorem (8.6') is then not independent of (8.10).]

One has to go through the above procedure of determining the partial-wave amplitude for every value of $p$ that is plotted in the Argand diagram. For small momenta, small values of $L$ (e.g., $L = 0$) are sufficient, but as the momentum increases, higher values of $L$ have to be chosen. As the equations are nonlinear, the solutions are not unique. Further, the experimental data have errors. The uncertainties that result can be overcome partially by considering $p T_l(p)$ at several values of $p$ and requiring that it changes smoothly with momentum. Sometimes further relations, such as dispersion relations, are used to obtain this smoothness. In this way one obtains a picture of $p T_l$ as a function of $p$.

## XVIII.9 Comparison with the Observed Cross Section: the Effect of Background and Finite Energy Resolution

Due to the effect of the background in the resonating partial cross section and the background of other nonresonating partial cross sections, the experimental cross section near resonance does not usually look as simple as in Figure 6.1a. We first consider in detail the interference between the resonance and the background.

If we insert (8.3) into (XVI.1.38), we obtain for the resonant partial cross section

$$\sigma_l = 4\pi(2l + 1) \left| \frac{1}{p} e^{i\gamma_l} \sin \gamma_l + e^{2i\gamma_l} T_l^R \right|^2. \qquad (9.1)$$

We introduce the Fano shape parameter[24]

$$q = +\cot \gamma_l, \qquad \epsilon = (E_R - E)\frac{2}{\Gamma} = \cot \delta_l^{(R)}(E). \tag{9.2}$$

$q$ is called the profile index or shape profile parameter, and $\epsilon$ is defined as in (8.2). After some straightforward calculation one obtains for (9.1)

$$\sigma_l = 4\pi(2l + 1)\left[\frac{1}{p^2}\sin^2 \gamma_l + |T_l^R|^2\left(\frac{q^2 + 2\epsilon q - 1}{q^2 + 1}\right)\right]. \tag{9.3}$$

This is rewritten as

$$\sigma_l = \sigma_l^{(bg)} + \sigma_l^{(R)}\left(\frac{q^2 + 2\epsilon q - 1}{q^2 + 1}\right), \tag{9.4}$$

where

$$\sigma_l^{(R)} = \frac{4\pi}{p^2}(2l + 1)\frac{1}{\epsilon^2 + 1} \tag{9.5}$$

is the Breit–Wigner cross section of (6.27), and $\sigma_l^{(B)}$ has been defined in analogy to (XVI.2.9) by

$$\sigma_l^{(bg)} = \frac{4\pi}{p^2}(2l + 1)\sin^2 \gamma_l. \tag{9.6}$$

Inserting (9.5) and (9.6) with (9.2) back into (9.4), one can write the resonant partial cross section also as

$$\sigma_l = \frac{4\pi}{p^2}(2l + 1)\frac{(q + \epsilon)^2}{(q^2 + 1)(q^2 + 1)} = \frac{4\pi}{p^2}(2l + 1)\sin^2 \gamma_l \frac{(q + \epsilon)^2}{q^2 + 1}, \tag{9.7}$$

or

$$\sigma_l = \sigma_l^{(bg)}\frac{(q + \epsilon)^2}{\epsilon^2 + 1}. \tag{9.8}$$

The factor $(q + \epsilon)^2/(\epsilon^2 + 1)$, which gives the shape of the resonance, is shown for various values of the shape profile parameter $q$ in Figure 9.1. We see that the shape of the resonance in the resonant partial cross section depends considerably upon the value of the parameter $q$, which is connected with the background phase shift $\gamma_l$. (If the background phase shift depends upon the energy $\epsilon$, which is generally the case, then the shape becomes even more complicated.) For positive values of $q$ we see that for increasing energies $E$, i.e., for increasing values of $-\epsilon$, the cross section increases above background, then falls sharply at the value of the resonance energy $\epsilon = 0$, is zero at $\epsilon = -q$, and increases thereafter.

For negative values of $q$ the cross section decreases with increasing energy, reaches zero at $\epsilon = q$ (i.e., before resonance), and then increases sharply.

---

[24] The original parameters introduced by U. Fano [*Phys. Rev.* **124**, 1866 (1961)] were the negative of the parameters defined by (9.2).

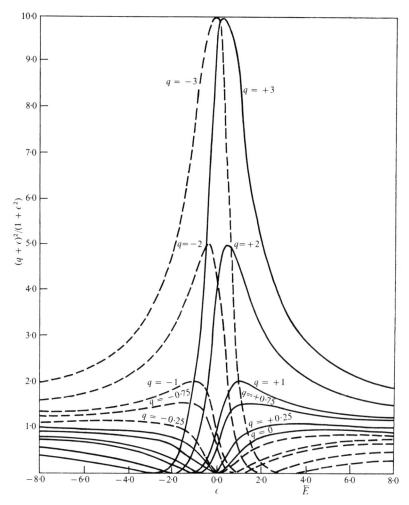

Figure 9.1    The resonance profile function for different values of the profile index $q$. The arrow above $E$ indicates the direction in which $E$ increases. [From Massey & Burhop (1969), vol. 1, with permission.]

We are now in a position to explain the behavior of the experimental data depicted in some of the figures of Section XVIII.1. As we have so far only discussed the effect of the background, with the present results we can only interpret those experiments in which the energy distribution in the beam and the resolution of the detector can be ignored. These are the high-resolution experiments whose results are depicted in Figure 1.3 and Figure 1.4d. These experiments measure the transmitted current, which is proportional to the incident current minus the total cross section. As the total cross section is the sum of the resonant $l$th partial cross section (9.8) plus the slowly varying background cross section from the other partial waves, a dip in the resonant partial cross section should show up as a peak in the transmitted current, and a peak in the resonant partial cross section as a dip in the transmitted current. According to the results of Figure 9.1, the resonance in the $l$th partial cross section should show up in the following way:

For $q > 0$, as a dip around resonance followed by a peak right after resonance.

For $q < 0$, as a peak right before resonance followed by a dip right after resonance.

In the high-resolution experiment (very narrow spread of the electron beam) of Figure 1.4d, and in particular in the very high-resolution experiment of Figure 1.3, one can see that the situation described for $q < 0$ is observed. (If one knew the exact value for $E_R$, one could even infer the value of $q$). Thus our hypothesis of Section XVIII.1, that in the collision process (1.4) a quasistationary state $He^-$ is formed at around 19.31 eV, is justified. We shall see that the other experiments of Figure 1.4 and Figure 1.5 lead to the same conclusion. For these experiments the resolution of the apparatus is coarse, and the quantity measured is not the cross section at a particular energy value, but the total cross section in a whole energy interval.

The total cross section (XVI.1.33) is written as

$$\sigma = \sigma_l + \sum_{l' \neq l} \sigma_{l'} = \sigma_l + \sigma_b, \tag{9.9}$$

where $\sigma_l$ is the resonant partial cross section given by (9.8) (describing the resonance and the background) and it is assumed that there is no resonance of any angular momentum $l'$ other than $l$. The background cross section $\sigma_b$ is then given by

$$\sigma_b = \frac{4\pi}{p^2} \sum_{l' \neq l} (2l' + 1) \sin^2 \gamma_{l'}, \tag{9.10}$$

where $\gamma_{l'}(E)$ are slowly varying background phase shifts.

Inserting (9.10) and (9.8) into (9.9), we obtain for the cross section

$$\sigma = \sigma_l^{(bg)} \frac{(q + \epsilon)^2}{\epsilon^2 + 1} + \sigma_b, \tag{9.11}$$

where $\sigma_l^{(bg)}$ and $\sigma_b$ vary slowly with energy.

At energies far away from the resonance energy, $\epsilon \to \infty$, we obtain from (9.11)

$$\sigma \to \sigma_l^{(bg)} + \sigma_b \equiv \sigma_{(nr)} \tag{9.12}$$

where we called $\sigma_{(nr)}$ the slowly varying non-resonant background. The maximum of the cross section (9.11) occurs at $\epsilon = +1/q$, and at this maximum the cross section is

$$\sigma_{max} = \sigma_l^{(bg)} + \sigma_b + q^2 \sigma_l^{(bg)} \tag{9.13}$$

Because of (9.12) the cross section (9.11) is conveniently written in the form

$$\sigma = \sigma_l^{(bg)} + \sigma_b + \frac{q^2 + 2q\epsilon - 1}{1 + \epsilon^2} \sigma_l^{(bg)} = \sigma_{(nr)} + \frac{q^2 + 2q\epsilon - 1}{1 + \epsilon^2} \sigma_l^{(bg)}, \tag{9.14}$$

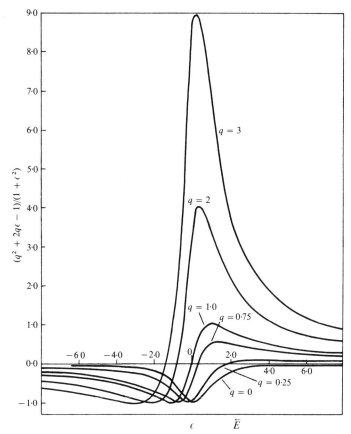

Figure 9.2    The function that determines the form of the resonance effect relative to the smooth background. [Massey Burhop [1969], vol. 1, with permission]

which shows that the resonance occurs above a gradually varying background $\sigma_{(nr)}$ and has the characteristic shape determined by the factor $(q^2 + 2q\epsilon - 1)/(1 + \epsilon^2)$. Thus, if there is background from other partial waves, then in distinction to (9.8), the cross section no longer vanishes at $\epsilon = q$ but has its minimum value there, $\sigma_{min} = \sigma_b$, which lies below the value for the background $\sigma_{(nr)}$. The shape is now determined by the shape factor shown in Figure 9.2.

If $|q| < 1$,

$$\tfrac{1}{4}\pi < \gamma_l < \tfrac{3}{4}\pi,$$

then the resonance effect results in a reduction from the background on one side of the resonance that is greater than the increase above the background on the other side. For $|q| > 1$ the reverse is true.

So far in our discussions we have always assumed that the energy resolution of the experimental apparatus is ideal. This means that the initially prepared beam can be considered as ideally monoenergetic, so that the energy distribution function can be described by the $\delta$-function according to (XIV.5.36):

$$F(E - E_A) = \rho(E)\langle E|||W_B^{in}|||E\rangle = \delta(E - E_A), \qquad (9.15)$$

where $E_A$ is the central energy value in the beam. Such an ideal resolution can be achieved in two ways: either by an ideal preparation apparatus (a high-resolution monochromator), or by an ideal detection apparatus (a high-resolution analyzer in the experiment of Figure II.4.1). Both will lead to the same result because of energy conservation. As already mentioned in Chapter XIV, what constitutes ideal resolution depends upon the effects that are to be observed: The energy spread of the beam $\Delta E$ must be small compared to the width of the structure in the scattering amplitude in order that (9.15) may be used.

The observed quantity is not the cross section $\sigma(E)$ at a definite value $E$, but the energy-averaged cross section

$$\sigma * F = \int dE \, F(E - E_A)\sigma(E), \qquad (9.16)$$

where $F$ is determined by the energy resolution of the detection (or preparation) apparatus, because every realistic detection apparatus detects particles not only with a definite energy $E_A$, but with an energy in a finite interval $\Delta E$ around $E_A$. If $\sigma(E)$ varies slowly as compared to $F(E - E_A)$, then $\sigma * F \approx \sigma(E_A)$, which is equivalent to using (9.15). But if $\Delta E$ is not small compared with the width $\Gamma$ of a resonance, and consequently $\sigma(E)$ given by (6.27) does not vary slowly, then one cannot approximate $\sigma * F$ by $\sigma(E_A)$. The observed quantity is then the quantity given in terms of the $T$-matrix elements by (XIV.5.40). $\sigma(E)$ is just an abbreviation for the expression under the integral besides $F(E - E_A)$ on the right-hand side of (XIV.5.40) and is not an observable quantity. Thus for a resonance of the $l$th partial wave with background in the $l$th partial wave and the other partial waves, $\sigma(E) = \sigma(\epsilon)$ given by (9.14) is not the observable physical quantity if the width $\Gamma$ is of the same order as the energy resolution $\Delta E$. The observable physical quantity is

$$\langle\sigma(E_A)\rangle = \sigma * F = \int dE \, \sigma(E)F(E - E_A)$$

$$= \int dE \, (\sigma_l^{(bg)}(E) + \sigma_b(E))F(E - E_A)$$

$$+ \int \frac{q^2 + 2q\epsilon - 1}{1 + \epsilon^2} \sigma_l^{(bg)}(E)F(E - E_A) \, dE. \qquad (9.17)$$

The precise value of this depends upon the energy distribution [the detector efficiency function $F(E - E_A)$] for the particular experiment. For the sake of definiteness let us assume that the probability of measuring the energy $E$

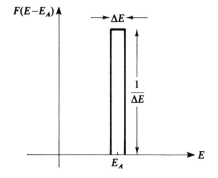

Figure 9.3   Assumed energy distribution for the beam with energy resolution $\Delta E \gg \Gamma$.

is zero except for values in an interval of $\Delta E$ around $E_A$. Then

$$F(E - E_A) = \begin{cases} \dfrac{1}{\Delta E} & \text{for } E_A - \dfrac{\Delta E}{2} \le E \le E_A + \dfrac{\Delta E}{2}, \\ 0 & \text{otherwise}, \end{cases} \qquad (9.18)$$

depicted in Figure 9.3, will describe the apparatus resolution. Equation (9.18) is of course also an unphysical distribution like $\delta(E - E_A)$, and a Gaussian around $E_A$ would often best reflect reality. However, for a beam with energy spread $\Delta E \gg \Gamma$, the assumption (9.18) is of sufficient accuracy. With (9.18) we obtain for the observed cross section (9.17):

$$\langle \sigma(E_A) \rangle = \frac{1}{\Delta E} \int_{E_A - \Delta E}^{E_A + \Delta E} \sigma(E) \, dE = \frac{1}{\Delta E} \int_{E_A - \Delta E}^{E_A + \Delta E} (\sigma_l^{(\text{bg})}(E) + \sigma_b(E)) \, dE$$

$$+ \frac{1}{\Delta E} \int_{E_A - \Delta E}^{E_A + \Delta E} \frac{q^2 + 2q\epsilon - 1}{1 + \epsilon^2} \sigma_l^{(\text{bg})}(E) dE. \qquad (9.19)$$

$\sigma_l^{(\text{bg})}(E)$ and $\sigma_b(E)$ are slowly varying functions of $E$. Therefore

$$\langle \sigma_{\text{nr}}(E_A) \rangle = \frac{1}{\Delta E} \int_{E_A - \Delta E}^{E_A + \Delta E} (\sigma^{(\text{bg})}(E) + \sigma_l(\epsilon)) dE$$

$$\approx \sigma_l^{\text{bg}}(E_A) + \sigma_b(E_A) \approx \sigma_l^{\text{bg}}(E_R) + \sigma_b(E_R)$$

is a slowly varying function of the average beam energy $E_A$ whose value is well approximated by the value it takes when the beam energy $E_A$ is equal the resonance energy $E_R$.

The second integral on the right-hand side of (9.19) is given by

$$\langle \sigma_r(E_A) \rangle = -\frac{\Gamma}{2} \sigma_l^{(\text{bg})}(E_A) \int_{\epsilon_A + 2\Delta E/\Gamma}^{\epsilon_A - 2\Delta E/\Gamma} \frac{q^2 + 2q\epsilon - 1}{1 + \epsilon^2} \, d\epsilon,$$

$$\epsilon_A = (E_R - E_A)\frac{2}{\Gamma}.$$

For $E_A$ far away from $E_R$, i.e., for $\epsilon_A$ far away from zero, this will be close to zero, as can be seen from Figure 9.2, and the averaged cross section will be given there essentially by $\langle \sigma_{nr}(E_A) \rangle$. At the value of the average beam energy $E_A = E_R$, the value of $\langle \sigma_r(E_A) \rangle$ is given by

$$\langle \sigma_r(E_R) \rangle = -\frac{\Gamma}{2} \sigma_l^{(bg)}(E_R) \int_{2\,\Delta E/\Gamma}^{-2\,\Delta E/\Gamma} \frac{q^2 + 2q\epsilon - 1}{1 + \epsilon^2} \, d\epsilon$$

$$\approx \frac{\Gamma}{2} \sigma_l^{(bg)} \int_{-\infty}^{+\infty} \frac{q^2 + 2q\epsilon - 1}{1 + \epsilon^2} \, d\epsilon$$

$$= \frac{\Gamma}{2} \sigma_l^{(bg)} \pi (q^2 - 1).$$

In performing the integration we have taken the limits of the integral over the shape factor as $\pm\infty$, which for $\Delta E \gg \Gamma$ does not cause an appreciable error, as can be seen from Figure 9.2. Inserting this and

$$\sigma_l^{(bg)}(E_R) = \frac{4\pi}{(p(E_R))^2} (2l + 1) \sin^2 \gamma_l(E_R) \tag{9.20}$$

into (9.19), we obtain for the value of the averaged cross section when the beam energy $E_A$ is set at the resonance energy $E_R$

$$\langle \sigma(E_R) \rangle = \langle \sigma_{nr}(E_R) \rangle + \langle \sigma_r(E_R) \rangle$$

$$= \langle \sigma_{nr}(E_R) \rangle + \frac{2\pi^2}{(p(E_R))^2} \Gamma(2l + 1) \cos 2\gamma_l(E_R). \tag{9.21}$$

Thus if a value of $\langle \sigma(E_R) \rangle$ is observed in a cross-section measurement with an apparatus with poor energy resolution, then it may be above or below or even equal to the slowly varying background $\langle \sigma_{nr}(E_A) \rangle$, depending upon the value of the background phase shift in the resonant partial wave, $\gamma_l(E_R) \approx \gamma_l(E_A)$. If $\cos 2\gamma_l > 0$, then *the resonance will show* in the observed cross section $\langle \sigma(E_A) \rangle$ *as a peak above* background $\langle \sigma_{nr}(E_A) \rangle$ at the value $E_A = E_R$. If $\cos 2\gamma_l < 0$, then *the resonance will show as a depression* below background. For $\cos 2\gamma_l \approx 0$ the resonance will not show up at all in the observed cross section.

With these results we can interpret the experimental data in Figures 1.4b, 1.4c, and 1.5, which were taken in an experiment with energy spread of the electron beam, $\Delta E$, larger than the resonance width. Figures 1.4b and 1.4c show an increase of the transmitted current and a decrease of the total elastic cross section around the value $E_A \approx E_R \approx 19.3$ eV. This is what one expects from (9.21) if the background phase shift $\gamma_l$ satisfies $\cos 2\gamma_l(E_R) < 0$, or

$$\tfrac{1}{4}\pi < \gamma_l(E_R) < \tfrac{3}{4}\pi. \tag{9.22a}$$

The same effect is observed in the averaged differential cross section at all angles depicted in Figure 1.5.

From the data of the high-resolution experiments in Figure 1.3 and Figure 1.4d, we concluded above that $q < 0$. Using (9.2), this means that

$$\tfrac{1}{2}\pi < \gamma_l(E_R) < \pi. \tag{9.22b}$$

Thus combining (9.22a) and (9.22b), we conclude as a result of our qualitative considerations that

$$\tfrac{1}{2}\pi < \gamma_l(E_R) < \tfrac{3}{4}\pi. \tag{9.23}$$

This shows that the background phase shift in the resonant partial wave for the process $e + \text{He} \to \text{He}^- \to \text{He} + e$ is not at all small. Taking just $T_l^R$ of (6.26) as the $l$th partial scattering amplitude and $\sigma_l^{(R)}$ of (6.27) as the cross section around resonance energy would certainly be unjustified.

We have now explained the experimental data for the elastic scattering process $e + \text{He} \to e + \text{He}$ as a resonance of narrow width around 19.3 eV interfering with the background phase shift fulfilling (9.23). In order to obtain more information about this $\text{He}^-$ resonance, we turn to the experiment of Figure 1.5. This is a low-resolution experiment with $\Delta E = (50\text{--}100) \times 10^{-3}$ eV in which the intensity of the elastically scattered electrons is measured as a function of average beam energy at various values of the scattering angle, $\theta = 10°, 40°, 60°, 90°, 110°$. Thus Figure 1.5 gives us the averaged elastic differential cross section as function of scattering angle and energy:

$$\left\langle \frac{d\sigma}{d\Omega}(E_A, \theta) \right\rangle = \int dE \, \frac{d\sigma}{d\Omega}(E, \Omega) F(E - E_A),$$

where $F$ is the energy-resolution function of the apparatus in this experiment.

With the experimental data of Figure 1.5, one can perform the phase-shift analysis as described in the second part of Section XVIII.8. Assuming that in the energy range considered the partial-wave amplitudes $T_l$ with $l > 2$ can be ignored, one can use (8.9) for the determination of the phase shifts $\delta_0$, $\delta_1$, and $\delta_2$ at various values of the momentum. As the inelasticity coefficients $\eta_l$ are zero in the present case, where only the elastic channel is open, the $T_l(p)$ of (8.9) are connected with $\delta_l(p)$ by (cf. XVI.2.8)

$$pT_l(p) = e^{i\delta_l} \sin \delta_l. \tag{9.24}$$

The Legendre polynomials that occur in (8.9) are

$$P_0 = 1, \qquad P_1 = \cos\theta, \qquad P_2 = \tfrac{1}{2}(3\cos^2\theta - 1),$$

$$P_3 = \tfrac{1}{2}(5\cos^3\theta - 3\cos\theta), \qquad P_4 = \tfrac{1}{8}(35\cos^4\theta - 30\cos^2\theta + 3). \tag{9.25}$$

One can first ignore the partial-wave amplitudes above $T_1$ and use $(8.9_1)$. At $\theta = 90°$ one has, according to $(8.9_1)$, only the contributions from $T_0$. As according to Figure 1.5 there is still a resonance effect at $\theta = 90°$, one concludes that the resonance occurs in the $S$-wave:

$$l_R = 0. \tag{9.26}$$

One can now apply (9.21) to the cross section $d\sigma/d\Omega(E_A, \theta = 90°)$. Instead of the apparatus resolution function (8.18), one may have to take a different one, depending upon the experiment.[25] Then the expression on the right-hand side of (9.21) will be different, but the principle of determining the resonant phase shift $\delta_0 = \pi/2 + \gamma_0$ is the same. The curve in Figure 1.5 for $\theta = 90°$ represents

$$\left\langle \frac{d\sigma}{d\Omega}(E_A, \theta = 90°) \right\rangle.$$

It shows again the decline below the nonresonant differential cross section characteristic of a resonating partial wave with background phase shift given by (9.22a). From the values of

$$\left\langle \frac{d\sigma}{d\Omega}(E_A, \theta = 90°) \right\rangle$$

at various energies $E_A$ around the resonance energy $E_R$, one can determine the background phase shift $\gamma_0$ and the true resonance width $\Gamma$. The values obtained from this experiment are

$$\Gamma \approx (15\text{--}20) \times 10^{-3}\,\text{eV}, \qquad \gamma_0 \approx 100°. \qquad (9.27)$$

$\gamma_0$ is in agreement with the restrictions (9.23) obtained from the other experiments above.

One may now use the data of Figure 1.5 at $\theta = 60°$. $P_2(\cos 60°)$ is close to zero, and the influence of a possible $T_2$ term, which would be considerable for $P_2(\cos \theta) \neq 0$ according to (8.9$_2$), is minimized. Fitting (8.9$_1$) to the experimental data for $\theta = 60°$ with $\delta_0^R = \pi/2$ and $\gamma_0 = 100°$, one obtains

$$\delta_1 = \gamma_1 \approx 25°. \qquad (9.28)$$

To check that the contribution from the higher partial waves is small, one can now fit the experimental data for the other scattering angles $\theta$ with (8.9$_2$). For example, with $\gamma_0 = 100°$ and $\gamma_1 = 25°$, one obtains from the experimental curve for $\theta = 10°$

$$\delta_2 = \gamma_2 \approx 4°. \qquad (9.29)$$

This is indeed small, and justifies the working hypothesis made at the beginning of our analysis that $T_2$ and all higher partial waves can be ignored.

The He in the collision process

$$e + \text{He} \rightarrow \text{He}^- \rightarrow \text{He} + e \qquad (1.4)$$

---

[25] Most frequently the apparatus resolution function is a Gaussian (6.39). The observed cross section is then according to (9.17) the convolution of a Gaussian and a Lorentzian function, which is known as the Voigt integral or Voigt profile. The Voigt integral cannot be found analytically, and has been computed and tabulated. The Lorentzian width $\Gamma$ and the Gaussian width $\Delta E$ can be obtained from a fit of the experimental data with the Voigt profile. See, e.g., H. G. Kuhn, *Atomic Spectra*, Academic Press, 1969, Section VIID.

is in the ground state $^1S_0$ with $l = 0$, $s = 0$. The $e$ in this process has $s = \frac{1}{2}$, and as we know from (9.26), $l = 0$. Consequently, the quasistationary state He$^-$ has $l = 0$ and $s = \frac{1}{2}$, i.e., it is a $^2S_{1/2}$ state. It is very reminiscent of the $^2S_{1/2}$ ground state in the lithium spectrum (cf. Figure VII.2.1). The only difference is that the charge of the nucleus in lithium is $+3$, whereas that in He$^-$ is $+2$, and the He$^-$ state has a width different from zero and given by (9.27).

## Problems

1.  (a)  Obtain the $S$-wave phase shift and the $l = 0$ $S$-matrix element $S_0(p)$ for the square-well potential.
    (b)  Find the energy values of the bound states of angular momentum 0 as poles of $S_0(p)$. Discuss the relations for the mass, the depth $V_0$, and the radius $R$ of the well that have to be fulfilled in order for there to be (i) no bound state, (ii) one bound state, (iii) two bound states.
    (c)  What are the virtual states of angular momentum zero? Under which conditions do there exist no virtual states for $l = 0$?
    (d)  Consider a very deep well fulfilling $2mV_0R^2 \gg 1$. Show that there are regularly spaced $s$-wave resonances. Determine energy and width of these resonances.
    (e)  Obtain the scattering matrix element $S_0(p)$ at energies close to a resonance energy, and obtain the background phase shifts. Calculate the elastic scattering cross section ($l = 0$), and show that it can be written as the sum of three terms: one from background scattering, one from resonance scattering, and one from the interference between resonance and background. Give a graphical representation of the elastic scattering cross section as a function of energy.

2.  Calculate the time delay caused by a narrow resonance of angular momentum $l_R$ and with resonance parameters $(E_R, \Gamma)$ of a beam with an energy distribution given by

$$\sum_{l_3} \langle l_3 lE | W^{\text{in}} | Ell_3 \rangle = W_l(E) = \frac{1}{\rho(E)} F(E - E_R) \quad \text{for every } l,$$

where

$$F(E - E_R) = \begin{cases} \dfrac{1}{\Delta E} & \text{for } E_R - \dfrac{\Delta E}{2} \le E \le E_R + \dfrac{\Delta E}{2}, \\ 0 & \text{otherwise}, \end{cases}$$

with $\Delta E \gg \Gamma$. Give the result in the approximation of constant background phase shifts.

3.  Calculate the time delay of a beam with a Lorentzian energy distribution (6.41) caused by a resonance with resonance parameters $(E_0 = E_A, \Gamma = \Delta E)$.

4.  Taking for the scattering amplitude an expression of the form

$$\tilde{T}_l = \frac{1}{p} \frac{\Gamma/2}{(E - E_0) - i(\Gamma/2)}$$

will also lead to a Breit–Wigner cross section (6.27). Why can $\tilde{T}_l$ not describe the partial-wave amplitude for resonance scattering?

5. (a) Describe the behavior of the resonance phase shift for a "dipole" resonance (6.44).

 (b) Discuss the cross section near resonance energy as a function of energy for a dipole resonance, taking the effect of a slowly varying background $\gamma_l(E)$ into account. Draw graphs in analogy to Figure 6.1.

6. Compare the case of a dipole resonance with the case of two resonances close to each other that are described by a scattering amplitude

$$T_l(p) = \frac{1}{p} e^{i2\gamma_l^R(E)} \left( \frac{\frac{1}{2}\gamma_R}{E_R - E - i(\Gamma_R/2)} + \frac{\frac{1}{2}\gamma_X e^{i2(\gamma_l^X(E) - \gamma_l^R(E))}}{E_X - E - i(\Gamma_X/2)} \right)$$

where $\gamma_R, \Gamma_R, \gamma_X, \Gamma_X$ are real constants; $E_i \gg \Gamma_i > \gamma_{i'}$ $(i = R, X)$ and $|E_R - E_X|$ are of the same orders as $\Gamma_R$ and $\Gamma_X$; and $\gamma_l^R(E)$, $\gamma_l^X(E)$ are slowly varying functions of $E$. Consider first the special case of zero background shifts $\gamma_l^R = \gamma_l^X = 0$.

7. (a) For a resonance, $pT_l(E)$ moves counterclockwise in the Argand diagram. Show that this is related to causality.

 (b) Discuss when $pT_l(E)$ can move clockwise in the Argand diagram.

8. Consider the quasistationary state with energy $E_0$ and width $\Gamma$ for which the parameter $r$ in

$$S_l(E) = \exp\left( i2r \arctan \frac{\Gamma/2}{E_0 - E} \right)$$

has the value $r = 3$. Obtain the slope of the resonant cross section for such a "tripole" resonance. (Ignore the background.)

9. Explain, using the shape of the function $\sigma_l(E)$ in Figure 6.1 for a background phase shift $\gamma_l = \pi/4$, why the cross section $\langle \sigma(E_A) \rangle$ observed by an apparatus with poor energy resolution $\Delta E \gg \Gamma$ need not show a structure at the energy value $E_A$ equal to the resonance energy $E_R$.

10. Show that the effect of a bound state with low binding energies in elastic scattering is the same as that of a virtual-state pole close to the real axis.

# CHAPTER XIX

# Time Reversal

For the discussion of multichannel resonances in Chapter XX, we require the properties of the $S$-matrix that follow from time-reversal invariance. In Section XIX.1 space-inversion invariance is applied to determine some properties of the $S$-matrix. Time reversal is introduced in Section XIX.2. The main purpose of Section XIX.1 is to introduce Section XIX.3, where the consequences of time-reversal invariance for the properties of the $S$-matrix are examined.

## XIX.1 Space-Inversion Invariance and the Properties of the $S$-Matrix

In Section V.4 we introduced the observable *parity*, represented in the space of physical states by the unitary operator $U_P$ fulfilling the relations (V.4.1)–(V.4.3) with the position, momentum, and angular-momentum observables. We mentioned in Section V.4 that not all energy operators commute with the parity operator—in particular, the interaction Hamiltonian that causes the weak transitions, and consequently the transition operator for the weak interaction, do not commute with $U_P$. The weak interaction violates parity invariance. However, many scattering and decaying systems are parity-invariant, i.e., in addition to

$$[K, U_P] = 0 \tag{1.1}$$

they also fulfill

$$[V, U_P] = 0, \quad \text{or} \quad [T, U_P] = 0, \tag{1.2}$$

or in terms of the $S$-operator

$$S = U_P^\dagger S U_P \tag{1.3}$$

Parity invariance leads to some constraints on the $T$-matrix and $S$-matrix elements. For example, it follows from (V.4.2) and (V.4.5) that

$$U_P |\mathbf{p}\eta\rangle = \pi_\eta |-\mathbf{p}\eta^\pi\rangle, \tag{1.4}$$

where $\pi_\eta$ is $+1$ or $-1$, and for the angular-momentum eigenvectors (XVI.1.3), in analogy to (V.4.39),

$$U_P |Ell_3\eta\rangle = (-1)^l \pi_\eta |Ell_3\eta^\pi\rangle. \tag{1.5}$$

The additional, internal quantum numbers $\eta$ usually do not change under $U_P$; thus $\eta^\pi = \eta$. The internal parity $\pi_\eta$ may be an element of the set of additional quantum numbers $\eta$.

From (1.4) it follows for the $S$- and $T$-matrix elements in the momentum basis, using (1.3) or (1.2), that

$$\langle \mathbf{p}'\eta'|S|\mathbf{p}\eta\rangle = \langle \mathbf{p}'\eta'|U_P^\dagger S U_P|\mathbf{p}\eta\rangle$$
$$= \langle -\mathbf{p}'\eta'|S|-\mathbf{p}\eta\rangle \pi_{\eta'} \pi_\eta, \tag{1.6}$$

or

$$\langle \mathbf{p}'\eta'|T|\mathbf{p}\eta\rangle = \langle -\mathbf{p}'\eta'|T|-\mathbf{p}\eta\rangle \pi_{\eta'} \pi_\eta. \tag{1.7}$$

From (1.5) it follows for the $T$-matrix element in the angular-momentum basis (assuming $T$ is spherically symmetric) that

$$\langle El'l_3'\eta'|T|Ell_3\eta\rangle = \langle El'l_3'\eta'|T|Ell_3\eta\rangle(-1)^{l+l'}\pi_{\eta'}\pi_\eta. \tag{1.8}$$

The conditions (1.7) and (1.8) constitute restrictions on the properties of the $T$-matrix elements.

We shall also restrict ourselves to the case discussed in Chapter XVI, namely, where the internal quantum numbers $\eta$ do not contain internal angular momenta. Then, using (XVI.1.18), one obtains from (1.8)

$$\langle El'l_3'\eta'|T|Ell_3\eta\rangle = \pi_{\eta'}\pi_\eta \langle El l_3'\eta'|T|Ell_3\eta\rangle \delta_{ll'}\delta_{l_3 l_3'} \tag{1.10}$$

or for the partial-wave amplitude defined by (XVI.1.34), (XVI.1.23), (XVI.1.18),

$$T_{ln}^{\eta'}(E) = \pi_{\eta'}\pi_\eta T_{ln}^{\eta'}(E). \tag{1.11}$$

This shows that in the case of elastic scattering of spin-zero particles, parity invariance does not impose any additional restrictions that are not already consequences of rotational invariance. For $\eta \neq \eta'$, (1.11) states that the partial-wave reaction amplitude will be different from zero only if the internal parity of the initial state is equal to the internal parity of the final state $\pi_{\eta'} = \pi_{\eta'}$ or if the internal parity is conserved. Parity invariance (1.2) leads to parity conversation (1.7), (1.8), (1.11), in the same way as rotational invariance (XVI.1.1) led to angular-momentum conservation (XVI.1.18).

## XIX.2  Time Reversal

The main purpose of the preceding discussion on parity invariance is to serve as an introduction to another kind of invariance, which superficially appears to be very similar to parity invariance but is of a completely different kind: the invariance under time reversal. Naively, time reversal is the replacement of $t$ by $t^R = -t$; however, this is not a physical operation like taking the mirror image $x_i \to x_i^R = -x_i$, as we cannot turn time backwards. But we can compare the state of a physical system having a position $x_i$ and a velocity $v_i$ with the state of the physical system having the same position $x_i$ and the velocity $-v_i$. If in a scattering process all positions and velocities undergo this transition at all times:

$$\mathscr{T} : \mathbf{x} \to +\mathbf{x}, \mathscr{T} : \mathbf{v} \to -\mathbf{v}, \quad \text{or } \mathscr{T} : \mathbf{x} \to \mathbf{x}, \mathscr{T} : \mathbf{p} \to -\mathbf{p}, \tag{2.1}$$

then one obtains a process which looks like the original process going backward in time.

The time-reversal operation $\mathscr{T}$ is the velocity reversal (2.1). For a quantum physical system this transformation $\mathscr{T}$ is represented by an operator in the space of physical states. We call this operator the time-reversal operator $A_T$. If $A_T$ represents a usual observable, like those we have considered so far, then $A_T$ should be a linear operator. We shall see instantly that $A_T$ cannot be a linear operator.

In analogy to (V.4.1) and (V.4.2) for the parity operator $U_P$, one would translate (2.1) for the quantum physical observables $Q_i$ and $P_i$ into

$$A_T Q_i A_T^{-1} = Q_i, \tag{2.2}$$

$$A_T P_i A_T^{-1} = -P_i. \tag{2.3}$$

We shall justify this below.

For the orbital-angular-momentum operator $L_i = \epsilon_{ijk} Q_j P_k$ and thence for any angular-momentum operator $J_i$, it follows from (2.2) and (2.3) that

$$A_T J_i A_T^{-1} = -J_i. \tag{2.4}$$

Let us now consider the commutation relation between the momentum operator $P_i$ and the angular-momentum operator $J_i$:

$$[J_i, P_k] = i\epsilon_{ikl} P_l, \tag{2.5}$$

and insert (2.3) and (2.4) into (2.5):

$$A_T [J_i, P_k] A_T^{-1} = -i\epsilon_{ikl} A_T P_l A_T^{-1}. \tag{2.6}$$

If $A_T$ were an ordinary linear operator, then we would obtain by multiplying (2.6) with $A_T^{-1}$ from the left and $A_T$ from the right

$$[J_i, P_k] = -i\epsilon_{ikl} P_l.$$

This would contradict (2.5). Thus if $A_T$ is to reverse the direction of the momenta (2.6), keeping the positions unchanged, then $A_T$ cannot be a linear operator.

〚A one-to-one mapping $A$ of a linear space $\mathscr{H}$ over the complex numbers $\mathbb{C}$ is called a *semilinear* operator if

$$A(\psi + \phi) = A\psi + A\phi \quad \text{for all } \psi, \phi \in \mathscr{H}, \qquad A\phi = 0 \Rightarrow \phi = 0, \tag{2.7a}$$

and

$$A\alpha\phi = \alpha'A\phi\alpha \in \mathbb{C}\alpha' = \alpha, \text{ or } \alpha' = \alpha^*. \tag{2.7b}$$

A semilinear operator is called *antilinear* if $\alpha' = \alpha^*$ (complex conjugate of $\alpha$), i.e., if

$$A\alpha\phi = \alpha^*A\phi. \tag{2.7c}$$

A semilinear operator is called *linear* if $\alpha' = \alpha$. Thus in distinction to a linear operator, already defined by (I.3.1), an antilinear operator also transforms the complex numbers into their complex conjugates:

$$Ai = -iA. \tag{2.7'}$$

In a scalar-product space $\mathscr{H}$ one can uniquely define for every antilinear operator $A$ (defined everywhere in $\mathscr{H}$ or in a dense subspace thereof) an antiadjoint operator $A^\dagger$ (often simply called the adjoint operator) by

$$(Af, g) = (A^\dagger g, f) = (f, A^\dagger g)^*, \qquad f, g \in \mathscr{H}. \tag{2.8}$$

The following relations hold for a linear operator $L$ and antilinear operators $A_1, A_2$:

$$(A_1 A_2)^\dagger = A_2^\dagger A_1^\dagger, \quad (AL)^\dagger = L^\dagger A^\dagger, \quad (A_1 L A_2)^\dagger = A_2^\dagger L^\dagger A_1^\dagger. \tag{2.9}$$

An antilinear operator is called *antiunitary* if it preserves the norm, i.e., if

$$(Ag, Af) = (A^\dagger Af, g) = (f, g), \quad \text{or } A^\dagger = A^{-1}. \tag{2.10}$$

The product of two antiunitary operators is a unitary operator.

For antilinear operators one has, in analogy to the case for linear operators,

$$\text{Tr}(AW) = \text{Tr}(WA), \qquad \text{Tr}(QAWA^\dagger) = \text{Tr}(A^\dagger QAW), \tag{2.11}$$

where $W$ and $Q$ are Hermitian operators. It is always assumed that the operators are such that all traces that occur in the proof are well defined.〛

If we choose for the representative of the velocity reversal or time reversal, $A_T$, not a unitary operator as we did for the space inversion in Section V.4, but an antiunitary operator, then (2.6) will not lead to a contradiction, but—using (2.7c)—back to (2.5).

We therefore make the following statement: The time-reversed or velocity-reversed state $W^T(t)$ of the state $W(t)$ is obtained from $W(t)$ by

$$W^T(t) = A_T W(t) A_T^\dagger, \tag{2.12}$$

where the time-reversal operator $A_T$ is an antiunitary operator

$$A_T^\dagger A_T = 1 \tag{2.13}$$

that fulfills the defining relations (2.2)–(2.4).[1]

In addition $A_T$ fulfills

$$A_T^2 = \epsilon 1, \quad \text{where } \epsilon = +1 \text{ or } -1. \tag{2.14}$$

Furthermore we require that the energy and transition operators commute with the time-reversal operator $A_T$:[2]

$$A_T K - K A_T = 0, \qquad A_T H - H A_T = 0, \qquad A_T T - T A_T = 0, \tag{2.15}$$

and that the $S$-operator fulfill

$$A_T^\dagger S A_T = S^\dagger. \tag{2.15'}$$

(2.15) and (2.15') are the statement of time reversal invariance.

The justification of this statement consists of two parts. The first and more difficult part is to show that the operation of time reversal is represented by a semiunitary operator [i.e., an operator fulfilling (2.7a), (2.7b), (2.13)]. In our heuristic arguments above, we had simply assumed that—in resemblance to the basic assumption I—time reversal is represented by something like a linear operator. The justification of this is, essentially, the celebrated Wigner theorem and applies to a much wider class of transformations than that of time reversal. This part is briefly discussed in the appendix to this section.

The second part in the justification of the above statement is to establish the particular properties of the time reversal operator expressed by the relations (2.2), (2.3), (2.4), (2.15).

From the physical interpretation of the time-reversal or velocity reversal operation, the following requirement results: If the expectation value of the momentum in the state $W(t)$ is $\bar{p}$, then the expectation value of the momentum in the state $W^T(t)$ should be $-\bar{p}$. This means

$$\text{Tr}(P_i W(t)) = -\text{Tr}(P_i W^T(t)) = -\text{Tr}(P_i A_T W(t) A_T^\dagger) = -\text{Tr}(A_T^\dagger P_i A_T W(t)). \tag{2.16}$$

As this is to be fulfilled for any state $W(t)$ and its corresponding time-reversal state $W^T(t)$, it follows that

$$P_i = -A_T^\dagger P_i A_T,$$

from which (2.3) is obtained. Equation (2.2) follows in the same way. Equation (2.4) follows from (2.3) and (2.2), as mentioned above. It has also already been mentioned that from (2.6) the antilinearity of $A_T$ follows.

---

[1] If there exist additional independent observables that cannot be expressed in terms of $P_i$, $Q_i$, and $J_i$, then their relation to $A_T$ also has to be specified.

[2] So far, this requirement has been found to hold except for decaying $K^0$ mesons, and there may be a superweak part of the Hamiltonian that does not fulfill (2.15).

Executing the operation of time reversal two times on all states,

$$W^{TT}(t) = A_T W^T(t) A_T^\dagger = A_T A_T W(t) A_A^\dagger A_A^\dagger, \tag{2.17}$$

should lead to the original state. From this requirement it follows that

$$A_T A_T = \epsilon 1 \quad \text{with } \epsilon\epsilon^* = 1. \tag{2.18}$$

But as $A_T A_T^2 = A_T A_T A_T = A_T^2 A_T$ and consequently $A_T \epsilon = \epsilon A_T$, it follows from the antilinearity that $\epsilon$ must be real. From this (2.14) follows.

Whether one has to choose for the time-reversal operator an $A_T$ fulfilling (2.14) with $\epsilon = +1$ or with $\epsilon = -1$ has—at least at the present time—to be decided by experience. The following choice has proved to be without contradiction to experimental observations:

$$\epsilon = (-1)^{2s}, \tag{2.19}$$

where $s$ is the spin of the physical system.[2a]

The requirement (2.15) is empirical in nature: No interaction has been found (except for the abovementioned $K^0$ system, which we shall ignore here) that violates time reversal invariance as stated by (2.15). Also, if the Heisenberg equation of motion (XII.1.24) with (XII.1.26) is to be fulfilled by the position operator, then it follows from (2.2) and (2.3) and the antilinearity of $A_T$ that $H$ must commute with $A_T$.

Equation (2.15′) is a consequence of (2.15) and the antiunitarity of $A_T$.

Therewith we have justified the above statement.

The state $W$ and its $A_T$-transformed state $W^T$ may be considered as different states of the same physical system. It is, however, more practical to restrict the scope of the physical system and to consider all the $A_T$-transformed states of all the states of a physical system as belonging to a new system, the $A_T$-transformed system of the original system. To obtain an idea about the property of the $A_T$-transformed system, let us consider a particular state:

$$W^T(t) = A_T W(t) A_T^\dagger.$$

From the time-development axiom V—in particular, from (XII.1.15) and (XII.1.16)—it follows that

$$W^T(t) = A_T e^{-itH} W_0 e^{itH} A_T^\dagger = e^{+itH} A_T W_0 A_T^\dagger e^{-itH}$$
$$= e^{-i(-t)H} W_0^T e^{+i(-t)H}. \tag{2.20}$$

Thus the $A_T$-transformed system develops forward in time in the same way as the original system would develop backward: $W(-t) = e^{-i(-t)H} W_0 e^{i(-t)H}$. If an original scattering system develops from a particular in-state $W^{in}$ into a particular out-state $W^{out}$, then the time-reversed system develops from an in-state $(W^T)^{in}$ agreeing with the out-state of the original system $(W^T)^{in} = W^{out}$ into the out-state $(W^T)^{out} = W^{in}$.

[2a] E. P. Wigner, in F. Gürsey (ed): *Group Theoretical Concepts and Methods in Elementary Particle Physics*. Gordon and Breach, New York, 1964, p. 37.

# Appendix to Section XIX.2

In this Appendix we give a brief description of the conditions and arguments that lead to the statement that the transformation of a physical system is described by a semiunitary operator. We choose here the transformation of time reversal.

Let $W^T(t)$ denote the state that is obtained from the state $W(t)$ by the operation of time reversal $\mathscr{T}$:

$$\mathscr{T}: W(t) \to W^T(t).$$

It is natural to require that:

1. If two states have no property in common, then the $\mathscr{T}$-transform of these two states should also have no property in common.
2. If one state has a more specific property than a second state, then its $\mathscr{T}$-transform should also have a more specific property than the $\mathscr{T}$-transform of the second state.

Let us restrict ourselves to states that are described by projection operators $\Lambda_1$ and $\Lambda_2$. Then requirement 1 means:

$$\Lambda_1 \Lambda_2 = 0 \quad \Rightarrow \quad \Lambda_1^T \Lambda_2^T = 0,$$

or

$$\Lambda_1 \mathscr{H} \perp \Lambda_2 \mathscr{H} \Rightarrow \Lambda_1^T \mathscr{H} \perp \Lambda_2^T \mathscr{H}$$

(orthogonal properties go into orthogonal properties). Requirement 2 means:

$$\Lambda_1 \Lambda_2 = \Lambda_1 \quad \Rightarrow \quad \Lambda_1^T \Lambda_2^T = \Lambda_1^T.$$

$\Lambda_1$ is the state with the more specific property than $\Lambda_2$. (As an example, $\Lambda_2$ may be the state that has angular momentum $j = 2$, and $\Lambda_1$ the state that has angular momentum $j = 2$ and angular-momentum component $j_3 = +2$.)
It is further natural to require that:

3. Time reversal is a one-to-one mapping in the set of states.

From these three requirements it follows by a general mathematical theorem[3] that time reversal $\mathscr{T}$ is described by a semilinear operator $A_T$:

$$\mathscr{T}: W(t) \to W^T(t) = A_T W(t) A_T^\dagger.$$

It is further natural to require that:

4. Overall probability should be preserved by the operation of time reversal. This means that for *all* $W$ and their corresponding $W^T$

$$\mathrm{Tr}\, W = \mathrm{Tr}\, W^T.$$

From this follows (2.13).

[3] E. Artin, *Geometric Algebra*, Academic Press, New York, 1951, p. 88.

Thus, the operation of time reversal is represented by a semiunitary (unitary or antiunitary) operator.

So far no specific property of the time-reversal operation has been used, and the above arguments apply to any transformation of the physical states with the properties 1, 2, 3, and 4. Thus every such transformation is represented by a semiunitary operator which is in general determined up to a phase factor. The celebrated Wigner theorem,[4] which states that a symmetry operation of a quantum physical system is induced by a unitary or antiunitary operator, is contained in the above statement.

## XIX.3   Time-Reversal Invariance and the Properties of the $S$-Matrix

Before we can apply the assumption of time-reversal invariance to find the properties of the $S$-matrix and $T$-matrix elements, we have to determine the action of $A_T$ on the basis vectors. This is done in the same way as it was for the parity operator $U_P$ in Section V.4. For the momentum eigenvector $|\mathbf{p}\eta\rangle$ it follows from (2.14) that

$$P_i(A_T|\mathbf{p}\eta\rangle) = -p_i(A_T|\mathbf{p}\eta\rangle). \tag{3.1}$$

Consequently, $(A_T|\mathbf{p}\eta\rangle)$ is also an eigenvector of $P_i$ but with eigenvalue $-p_i$. If there are no additional eigenvectors and $\mathbf{p}$ itself labels the basis system completely, then it follows from (3.1) that

$$A_T|\mathbf{p}\rangle = \alpha(\mathbf{p})|-\mathbf{p}\rangle,$$

where $\alpha$ is a phase factor, which, because of

$$A_T^2|\mathbf{p}\rangle = \epsilon 1|\mathbf{p}\rangle = \alpha(\mathbf{p})A_T|-\mathbf{p}\rangle = \alpha(\mathbf{p})\alpha(-\mathbf{p})|\mathbf{p}\rangle,$$

fulfills

$$\alpha^2 = \epsilon \quad (= +1 \text{ for the spinless case}) \tag{3.2}$$

and can be shown to be independent of $\mathbf{p}$.

For the angular-momentum eigenvectors $|Ell_3\rangle$ it follows in the same way, from (2.4) and (2.14), that

$$A_T|Ell_3\rangle = \alpha|El - l_3\rangle. \tag{3.3}$$

If there are additional quantum numbers $\eta$ (internal quantum numbers, channel labels), then the action of $A_T$ on the basis vectors $|\ldots\eta\rangle$ depends upon the relation between $A_T$ and $\eta^{\mathrm{op}}$. Let us assume that $\eta^{\mathrm{op}}$ does not con-

---

[4] Proofs of the Wigner theorem can be found in Ludwig, (1954, p. 101); U. Uhlhorn, *Ark. Fys.* **23**, 307 (1963); L. O'Raifeartaigh, G. Rasche, *Ann. of Phys.* **2b**, 155 (1963); V. Bargmann, *Ann. of Math.* **59**, 1 (1954). See also J. M. Jauch, in *Group Theory and Its Applications* (Ed. E. M. Loebl), Academic Press, New York, 1968, p. 131, where further references are given.

tain any internal angular momenta. Let us further assume that all $\eta^{\mathrm{op}}$ commute with $A_T$:

$$[A_T, \eta^{\mathrm{op}}] = 0. \tag{3.4}$$

This will be the case if $\eta^{\mathrm{op}}$ stands for the internal-energy operator of the target, for instance. Then

$$A_T|\mathbf{p}\eta\rangle = \alpha(\eta)|-\mathbf{p}\eta\rangle \quad \text{and} \quad A_T|Ell_3\eta\rangle = \alpha(\eta)|El - l_3\eta\rangle, \tag{3.5}$$

where $\alpha = \alpha(\eta)$ may depend upon $\eta$ and perhaps also upon $l$.

Let us now consider the $S$-matrix element in the angular-momentum basis (XVI.1.42) and the reduced $S$-matrix elements $\langle\eta\|S_l(E)\|\eta'\rangle$ defined by (XVI.1.46) and (XVI.1.44). One can either use (2.14) for the $T$-operator together with (3.5) and the definition of the $S$-matrix given by (XVI.1.42), or one can use (2.15) for the $S$-operator directly together with (3.15), to calculate, using (2.8) and (3.5),

$$\begin{aligned}
\langle Ell_3\eta|A_T^\dagger SA_T|E'l'l_3'\eta'\rangle &= (A_T|Ell_3\eta\rangle, SA_T|E'l'l_3'\eta'\rangle)^* \\
&= \alpha(\eta)\alpha^*(\eta')\langle El - l_3\eta|S|E'l' - l_3'\eta'\rangle^* \\
&= \alpha(\eta)\alpha^*(\eta')\langle E'l'l_3'\eta'|S|El - l_3\eta\rangle.
\end{aligned} \tag{3.6}$$

On the other hand, the left-hand side according to (2.15) is equal to

$$\begin{aligned}
\langle Ell_3\eta|S^\dagger|E'l'l_3'\eta'\rangle &= (S|Ell_3\eta\rangle, |E'l'l_3'\eta'\rangle) \\
&= \langle E'l'l_3'\eta'|S|Ell_3\eta\rangle^*.
\end{aligned} \tag{3.7}$$

Thus

$$\langle E'l'l_3'\eta'|S|Ell_3\eta\rangle^* = \langle E'l' - l_3'\eta'|S|El - l_3\eta\rangle\alpha(\eta)\alpha^*(\eta') \tag{3.8}$$

or also, equating the complex conjugate of (3.7) and (3.8),

$$\langle E'l'l_3'\eta'|S|Ell_3\eta\rangle = \langle El - l_3\eta|S|E'l' - l_3'\eta'\rangle\alpha(\eta)\alpha^*(\eta') \tag{3.9}$$

In order to determine the phase factors, we calculate

$$\langle Ell_3\eta|A_T^\dagger A_T|E'l'l_3'\eta'\rangle = \alpha(\eta)\alpha^*(\eta')\langle Ell_3\eta|E'l'l_3'\eta'\rangle^*. \tag{3.10}$$

On the other hand, because of (2.12),

$$\langle Ell_3\eta|A_T^\dagger A_T|E'l'l_3'\eta'\rangle = \langle Ell_3\eta|E'l'l_3'\eta'\rangle = \rho^{-1}(E)\delta(E - E')\delta_{ll'}\delta_{l_3 l_3'}\delta_{\eta\eta'} \tag{3.11}$$

if the generalized eigenvectors are "normalized" in the usual way with real $\rho(E)$. Consequently, comparison of (3.10) and (3.11) shows that we require

$$\alpha(\eta)\alpha^*(\eta') = 1. \tag{3.12}$$

Equations (3.8) and (3.9) with (3.12) are called the *reciprocity theorem*[5] (for the spinless case). This relation is quite distinct from the corresponding

---

[5] The reciprocity theorem is also called the principle of detailed balance or the principle of microreversability.

relations—e.g., (1.7) or (1.8)—following from parity invariance, as it relates two different processes with each other. The left-hand side of (3.9) describes the process $\eta \to \eta'$, whereas the right-hand side describes the process $\eta' \to \eta$. The reciprocity theorem then states that these two processes have the same transition probability. The measurement of these transition probabilities provides a possibility of checking whether time-reversal invariance is fulfilled.

We now want to determine the consequences of time-reversal invariance upon the $l$th partial reduced $S$-matrix element $\langle \eta' \| S_l(E) \| \eta \rangle$ defined by (XVI.1.45) and (XVI.1.43). Inserting these into (3.8), we obtain

$$\langle \eta' \| S_l(E) \| \eta \rangle^* = \langle \eta' \| S_l(E) \| \eta \rangle; \tag{3.13}$$

inserting them into (3.9),

$$\langle \eta' \| S_l(E) \| \eta \rangle = \langle \eta \| S_l(E) \| \eta' \rangle \tag{3.14}$$

Thus the $l$th partial reduced $S$-matrix element $\langle \eta' \| S_l(E) \| \eta \rangle$ is not only unitary [cf. XVI.2.4)] but also real, i.e., it is a symmetric matrix. We shall make use of this fact in the following section.

For an experimental check of the reciprocity theorem we obtain the consequences of (3.14) for the partial cross section. Inserting (3.14) into (XVI.1.48), we obtain for the partial-wave amplitudes (XVI.1.34) the following relation:

$$\sqrt{p'(E)p}\, T_{l\eta}^{\eta'}(E) = \sqrt{p(E')p'}\, T_{l\eta'}^{\eta}(E'). \tag{3.15}$$

And inserting this into (XVI.1.37), one obtains the following relation between the partial cross sections $\sigma_l^{\eta' \leftarrow \eta}$ and $\sigma_l^{\eta \leftarrow \eta'}$ describing the processes $\eta \to \eta'$ and $\eta' \to \eta$, respectively:

$$\sigma_l^{\eta \leftarrow \eta'} = \sigma_l^{\eta' \leftarrow \eta}\, \frac{p^2}{p'^2}. \tag{3.16}$$

An analogous relation holds for the total and differential cross sections. Equation (3.16) has been derived for the case of spinless particles. If projectile and target for the system $\eta$ have the spins $s$ and $s_T$, and for the system $\eta'$ the spins $s'$ and $s'_T$, then (3.16) is replaced by

$$\sigma^{\eta \leftarrow \eta'} = \sigma^{\eta' \leftarrow \eta}\, \frac{p^2}{p'^2}\, \frac{(2s + 1)(2s_T + 1)}{(2s' + 1)(2s'_T + 1)}. \tag{3.16'}$$

For a check of the principle of detailed balance (3.16), we consider the experimental data of Figure 3.1.

This figure depicts the counting rate (which is proportional to the cross section) as a function of the energy for various processes. Parts B and C

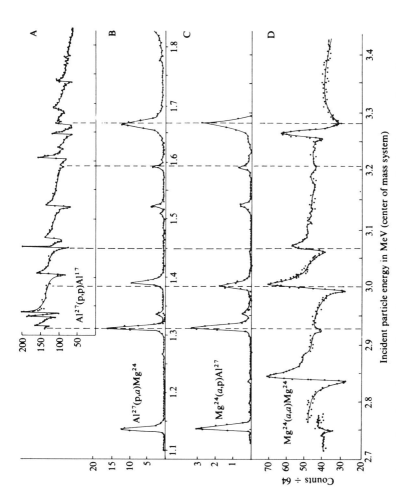

Figure 3.1   Relative cross section as function of energy for various scattering processes: A, elastic scattering of protons on aluminum. B, reaction $Al^{27} + p \rightarrow \alpha + Mg^{24}$. C, reaction $Mg^{24} + \alpha \rightarrow p + Al^{27}$. D, elastic scattering of $\alpha$-particles on magnesium. [From S. G. Kaufmann et al., *Phys. Rev.* **88**, 673 (1952), with permission.]

are for the processes

$$Al^{27} + p \rightarrow \alpha + Mg^{24},$$

$$Mg^{24} + \alpha \rightarrow p + Al^{27},$$

which are related by the principle of detailed balance, and (3.16) should apply, with $\eta$ characterizing the $Al^{27}$-proton system and $\eta'$ characterizing the $Mg^{24}$-$\alpha$ system. The energy scale has been appropriately adjusted, and the curves should agree with each other (except for a factor). The data were taken from two different experiments with different resolution; still, the agreement is convincing. The various bumps correspond to resonances of the $Si^{28}$ system, and we shall discuss this, as well as the comparison with the curves in parts A and D, in the following chapter.

## Problems

1.   Show that the product of two anti-linear operators is a linear operator and that the product of an antilinear operator and a linear operator is anti-linear.

2.   Show that as a consequence of the antiunitarity the time reversal operator must fulfill $A_T^2 = \epsilon$, where $\epsilon = +1$ or $\epsilon = -1$.

3.   Show that from the definition of the Møller wave operators (XV.3.7±) it follows that $\Omega^{\pm} = A_T^{\dagger} \Omega^{\mp} A_T$ if $H$ and $K$ commute with $A_T$. Use this to show that (2.15) is a consequence of (2.14) if the $S$-operator is defined in terms of the energy operator.

4.   Prove that the time-reversal operator transforms the exact energy eigenvectors $|\mathbf{p}, \eta^{+}\rangle$ into the exact energy eigenvectors $|-\mathbf{p}\eta'^{-}\rangle$. (The label $+$ or $-$ refers to the definition according to the Lippman–Schwinger equation.)

5.   Show that the time-reversal operator transforms an out-state into an in-state and vice versa.

6.   Consider a multichannel system of spinless particles. Show that the transformation property under $U_P$ and $A_T$ implies that the $S$-matrix in the momentum basis is symmetric.

7.   Show that from the Heisenberg equation of motion (XII.1.24) for the position operator, $A = Q_i$, with (XII.1.26) and the antilinearity of $A_T$ follows that $H$ commutes with $A_T$.

# CHAPTER XX

# Resonances in Multichannel Systems

In the present chapter we discuss resonances in scattering processes from an initial state into several possible final states, where the internal quantum numbers of the resonances are, in general, different from the quantum numbers of the initial state and the final states. In Section XX.2 we first discuss, in detail, the case of a single multichannel resonance and then the case of a double multichannel resonance (which occurs if there are two resonances with different internal quantum numbers in the same partial wave, with both the resonances coupling to initial and final states). In Section XX.3 the Argand diagrams for inelastic resonances are described.

## XX.1 Introduction

In the discussion in Chapter XVIII we assumed that there is only one channel open in the scattering experiment and that it is this particular channel to which the resonance phenomenon is restricted. Then there is only one reduced $S$-matrix element $\langle \eta \| S_l(E) \| \eta \rangle$ that is different from zero, and it is this reduced matrix element that resonates. This is the case of elastic scattering.

$$a + T \rightarrow R \rightarrow a + T. \tag{1.1}$$

$\eta$ are the quantum numbers of the initial state $(a, T)$, of the final state $(a, T)$, *and* of the resonance $R$.

This is, of course, a very special situation, which in general will not be fulfilled. When the energy is high enough, so that several final channels are open, a resonance can occur not only in the elastic scattering process described by $S_l(E) = \langle \eta_A \| S_l(E) \| \eta_A \rangle$ but also in the inelastic processes described by $\langle \eta \| S_l(E) \| \eta_A \rangle$ with $\eta \neq \eta_A$. In order that one particular resonance shows up in several channels this resonance state must have nonzero transition matrix elements to *all* initial and final states with quantum numbers $\eta$, $\eta'$, $\eta''$. If one also wants to assign internal quantum numbers, say $\kappa$, to the resonance state, then these must be different than the quantum numbers $\eta$; the corresponding observables $\kappa^{op}$ and $\eta^{op}$ cannot commute.

It may be a matter of taste and convenience whether one characterizes these resonance states by internal quantum numbers $\kappa = (\kappa_1 \, \kappa_2 \cdots \kappa_n)$ and considers the resonance energy as a derived quantity $E_R = E(\kappa_R)$, $\kappa_R = (\kappa_{R1} \, \kappa_{R2} \cdots \kappa_{Rn})$ determined by the internal quantum numbers, or whether one characterizes the resonance state directly by the resonance energy $E_R$. The former view is prevailing in particle physics, the latter in nuclear physics. We shall follow here the former view, assuming that the resonance state is characterized in addition to the angular momentum by a set of "internal" quantum numbers $\kappa$, because this allows a very simple derivation of multichannel resonance formulas from the results of Chapter XVIII.

## XX.2 Single and Double Resonances

The reduced matrix elements of the $S$-matrix $\langle \eta \| S_l(E) \| \eta' \rangle$ form a unitary matrix, according to (XVI.2.3). According to a well-known theorem of linear algebra,[1] for any unitary matrix there exists another unitary matrix that diagonalizes it, and the nonzero elements of the diagonal matrix (the eigenvalues) are of absolute value one. Thus for any value of $E$, there exists a unitary matrix $\langle \kappa \| \eta \rangle$ (with an adjoint $\langle \eta \| \kappa \rangle = \overline{\langle \kappa \| \eta \rangle}$),

$$\langle \kappa | \kappa' \rangle = \delta_{\kappa \kappa'} = \sum_{\eta} \langle \kappa \| \eta \rangle \langle \eta \| \kappa' \rangle, \qquad \delta_{\eta' \eta} = \langle \eta | \eta \rangle = \sum_{\kappa} \langle \eta' \| \kappa \rangle \langle \kappa \| \eta \rangle$$

(2.1)

such that

$$\delta_{\kappa \kappa'} e^{i 2 \delta_l^{(\kappa)}(E)} = \sum_{\eta \eta'} \langle \kappa \| \eta \rangle \langle \eta \| S_l(E) \| \eta' \rangle \langle \eta' \| \kappa' \rangle. \qquad (2.2)$$

$\delta_l^{\kappa}(E)$ are called the eigen-phase-shifts or eigenphases. The matrix $\langle \kappa \| \eta \rangle$ may be different for different values of $E$ and, therefore, in general, a function of the scattering energy. From the time-reversal invariance it follows that the reduced $S$-matrix is real [Equation (XIX.3.14)]. From the general theorems[1] of linear algebra it follows that for a real unitary matrix there exists an orthogonal matrix that diagonalizes it. Thus, the $\langle \kappa \| \eta \rangle$ fulfill not only (2.1) but also

$$\langle \kappa \| \eta \rangle = \langle \eta \| \kappa \rangle = \langle \kappa \| \eta \rangle^*. \qquad (2.1a)$$

---

[1] See, e.g., A. Lichnerowicz (1967) or V. I. Smirnov (1964, Vol. III, Part 1, Section 42).

Inverting (2.2) with the help of (2.1), we obtain

$$\langle \eta \| S_l(E) \| \eta' \rangle = \sum_\kappa e^{i2\delta_l^{(\kappa)}(E)} \langle \eta \| \kappa \rangle \langle \kappa \| \eta' \rangle. \tag{2.3}$$

To understand the physics behind this mathematical transformation of the reduced $S$ matrix into diagonal form, let us consider the basis vectors that were used for the space of the initial states $(a, T)$:

$$| Ell_3 \eta \rangle = | Ell_3 \rangle \otimes | \eta \rangle.$$

The orthogonal matrix $\langle \eta \| \kappa \rangle$ transforms this basis system into a new basis system

$$| Ell_3 \kappa \rangle = \sum_\eta | Ell_3 \eta \rangle \langle \eta \| \kappa \rangle \tag{2.4}$$

with the quantum numbers $Ell_3$ unaffected but with new internal quantum numbers $\kappa$. The observables whose eigenvalues are $\kappa$ do not commute with the observables whose eigenvalues are $\eta$:

$$[\kappa^{op}, \eta^{op}] \neq 0. \tag{2.5}$$

The quantum numbers $\eta$ were chosen so that the initial state $(a, T)$ is characterized by a definite value of $\eta$: $\eta = \eta_A = \eta(a, T)$. Now it may be that the quasistationary state $R$ that is formed in the process $a + T \to R \to a + T$ does not have a definite value of the quantum numbers $\eta$ but a definite value of the quantum numbers $\kappa$, say $\kappa = \kappa_R$. Then we can repeat the same considerations as in Sections XVIII.2 and XVIII.6—this time not for the internal quantum numbers $\eta$, but for the internal quantum numbers $\kappa$, in particular for that value $\kappa_R$ of the quantum numbers $\kappa$ that constitutes the internal quantum numbers of $R$. The time delay $t_D^{(\kappa_R)}$ and phase shift $\delta_l^{(\kappa_R)}$ are then characterized by this quantum number in the same way as in Chapter XVIII they are characterized by the quantum numbers $\eta_A$ of the initial (and final) projectile-target system $(a, T)$. Thus it is not for the quantum numbers $\eta_A$ of the initial state $(a, T)$ that (XVIII.6.23) holds, but for the quantum numbers $\kappa_R$ of the resonance state $R$, which now has quantum numbers that are incompatible with the quantum numbers $\eta_A$, i.e., they fulfill (2.5). Thus instead of $S_l = \langle \eta_A \| S_l(E) \| \eta_A \rangle$, we now write Equation (XVIII.6.23) for $\langle \kappa_R \| S_l(E) \| \kappa_R \rangle$:

$$\langle \kappa_R \| S_L(E) \| \kappa_R \rangle = e^{i2\delta_l^R(\kappa_R)} e^{i2\gamma_l(\kappa_R)} = e^{i2\delta_l(\kappa_R)}$$

$$= \left( 1 + \frac{i\Gamma_R}{E_R - E - i\Gamma_R/2} \right) e^{i2\gamma_l(\kappa_R)}. \tag{2.6}$$

$\delta_l^R(\kappa_R)$ here denotes the resonant phase shift, and $\gamma_l(\kappa_R)$ the background phase shift, so that the phase shift is again written

$$\delta_l(\kappa_R) = \delta_l^R(\kappa_R) + \gamma_l(\kappa_R).$$

The reduced $S$-matrix elements $\langle \eta \| S_l(E) \| \eta' \rangle$ between the (reduced) basis vectors $\| \eta \rangle$ have then to be expressed in terms of the reduced $S$-matrix elements $\langle \kappa \| S_l(E) \| \kappa' \rangle$ between the (reduced) basis vectors $\| \kappa \rangle$, using the transformation

$$|\eta\rangle = \sum_\kappa |\kappa\rangle \langle \kappa \| \eta\rangle, \qquad (2.4')$$

which leads to (2.3) with

$$\delta_{\kappa\kappa'} e^{i2\delta_l^{(\kappa)}} = \langle \kappa \| S_l \| \kappa' \rangle.$$

The matrix element $\langle \kappa \| \eta \rangle$ is the number that characterizes the strength of the coupling between the states with the internal quantum numbers $\eta$ and $\kappa$. As mentioned above, they depend in general upon the energy $E$.

It may of course be that a quasistationary state is formed not only for one value $\kappa_R$ of the quantum numbers $\kappa$, but also for other values $\kappa = \kappa_{X_1}, \kappa_{X_2}, \ldots$. This may even happen at the same value $l$ of the angular momentum and even at close energies $E_R$ and $E_{X_1}$. We assume for the sake of definiteness that there are two values $\kappa_R$ and $\kappa_X$ for which quasistationary states are formed with angular momentum $l$, and their energies and widths are $(E_R \Gamma_R)$ and $(E_X \Gamma_X)$, respectively. Then in addition to (2.6) one has also

$$\langle \kappa_X \| S_l(E) \| \kappa_X \rangle = e^{i2\delta_l(\kappa_X)} = \left(1 + \frac{i\Gamma_X}{E_X - E - i\Gamma_X/2}\right) e^{i2\gamma_l(\kappa_X)}. \qquad (2.7)$$

We also assume that all the other $S$-matrix elements are nonresonant, i.e.,

$$\langle \kappa \| S_l(E) \| \kappa \rangle = e^{i2\delta_l(\kappa)} = e^{i2\gamma_l(\kappa)} \quad \text{for } \kappa \neq \kappa_R, \kappa_X, \qquad (2.8)$$

where $\gamma_l(\kappa)$ are the background phase shifts, which are functions of energy $E$. One may be tempted to assume that the background phase shifts in the resonant and non-resonant channels are slowly varying functions of the energy. This is, however, in general not the case[1a] and we shall not make use of such an assumption here.

We shall call the above-described object a double resonance or double multichannel resonance of quantum numbers $(\kappa_R, \kappa_X)$. Thus the formation experiment in the scattering process

$$a + T \rightarrow a' + T$$

can go two ways:

$$(a + T)_{\text{with q.n. } \eta_A} \quad \overbrace{\qquad \begin{array}{c} \rightarrow R_{\text{with q.n. } \kappa_R} \\ \rightarrow X_{\text{with q.n. } \kappa_X} \end{array} \qquad} \quad \rightarrow (T' + a')_{\text{with q.n. } \eta}.$$

---

[1a] These rapid energy variations follow from Wigner's eigenphase repulsion theorem: The eigenphases $\delta_l(\kappa)$ do not cross (modulo $\pi$). As a consequence, near the resonance energy in the channel $\kappa_R$ the eigenphases in all other channels $\kappa$ must also vary. This is assured by the existence of branch cuts in the individual eigenphases (and the corresponding transition matrix elements $\langle \kappa \| \eta \rangle$ which are also functions of $E$) which are in the complex energy plane much closer to the real axis than the resonance pole. These branch cuts do not occur in the complete $S$ matrix and therefore do not have physical significance. The occurence of these branch cuts in the $\langle \kappa \| \eta \rangle$ and $\gamma_l(\kappa)$ must be just such that all these branch cuts cancel out in the background term (2.16), which can be a slowly varying function of energy. A discussion of these singularities of the individual eigenphases is given in C. J. Goebel, K. W. McVoy: Phys. Rev. **164**, 1932 (1967); H. A. Weidenmuller: Phys. Letters **24B**, 441 (1967); and R. H. Dalitz, R. G. Moorhouse: Proc. Roy. Soc. A318, 279 (1970).

Figure 2.1

In the case of elastic scattering $\eta = \eta_A$, $(T' = T, a' = a)$, in the case of a reaction $\eta = r \neq \eta_A$. If one wants to illustrate this by a diagram alalogous to Figure XVIII.1.2, one would draw Figure 2.1.

The $S$-matrix for the double multichannel resonance is obtained by inserting (2.6), (2.7) and (2.8) into (2.3):[2]

$$\langle \eta \| S_l(E) \| \eta' \rangle = \sum_\kappa \langle \eta \| \kappa \rangle \langle \kappa \| \eta' \rangle e^{i2\gamma(\kappa)} + \frac{i\Gamma_R \langle \eta \| \kappa_R \rangle \langle \kappa_R \| \eta' \rangle e^{i2\gamma(\kappa_R)}}{E_R - E - i\Gamma_R/2}$$

$$+ \frac{i\Gamma_Z \langle \eta \| \kappa_X \rangle \langle \kappa_X \| \eta' \rangle e^{i2\gamma(\kappa_X)}}{E_X - E - i\Gamma_X/2} \tag{2.9}$$

The sum over $\kappa$ extends over all channels including the resonating channels $\kappa_R$ and $\kappa_X$.

The amplitude for elastic scattering connected to the $S$ matrix by (XVI.1.48), or (XVI.2.8), is obtained from (2.9) with $\eta = \eta' = \eta_A$ after a short calculation using (2.1):

$$pT_l(p) = e^{i2\gamma(\kappa_R)} \left( \frac{\frac{1}{2}\Gamma_R |\langle \eta_A \| \kappa_R \rangle|^2}{E_R - E - i\Gamma_R/2} + \frac{\frac{1}{2}\Gamma_X |\langle \eta_A \| \kappa_X \rangle|^2 e^{i2(\gamma(\kappa_X) - \gamma(\kappa_R))}}{E_X - E - i\Gamma_X/2} \right)$$

$$+ \sum_\kappa \langle \eta_A \| \kappa \rangle \langle \kappa \| \eta_A \rangle e^{i\gamma(\kappa)} \sin \gamma(\kappa). \tag{2.10}$$

In the case of inelastic scattering from the initial state with quantum numbers $\eta' = \eta_A$ into the final state with quantum numbers $\eta = r$, the scattering amplitude (XVI.1.48) is obtained from (2.9) as

$$\sqrt{p_r p} T_{r,l}^{(A \to r)}(p) = e^{i2\gamma(\kappa_R)} \left( \frac{\frac{1}{2}\Gamma_R \langle r \| \kappa_R \rangle \langle \kappa_R \| \eta_A \rangle}{E_R - E - i\Gamma_R/2} \right.$$

$$\left. + \frac{\frac{1}{2}\Gamma_X \langle r \| \kappa_X \rangle \langle \kappa_X \| \eta_A \rangle e^{i2(\gamma(\kappa_X) - \gamma(\kappa_R))}}{E_X - E - i\Gamma_X/2} \right)$$

$$+ \sum_\kappa \langle r \| \kappa \rangle \langle \kappa \| \eta_A \rangle e^{i\gamma(\kappa)} \sin \gamma(\kappa). \tag{2.11}$$

Equations (2.9)–(2.11) are approximations to the same extent to which (2.6) and (2.7) and (XVIII.6.23) are approximations. They have been derived here from the ordinary Breit–Wigner approximation and the assumption that the resonance (or quasistationary state) has internal quantum numbers $\kappa$

---

[2] We omit the index $l$ in the phase shift $\gamma_l$ and $\delta_l$ whenever possible. $\gamma$ then always denotes the background phase of that partial wave in which the resonance occurs.

incompatible with the internal quantum numbers $\eta$ of the initial and final state. Due to this derivation (2.9)–(2.11) contain the individual eigenphases $\gamma(\kappa)$ though these eigenphases belonging to the intermediate quantum numbers $\kappa$ may be unphysical quantities. The physically observable quantities are the resonance terms and the total background term (summed over all $\kappa$) and only these quantities should be compared with the experimental data.

## Single Multichannel Resonance

Before we continue the discussion of a double resonance, we treat the simpler case that there is only one resonance with quantum number $\kappa_R$. Such a resonance we shall call a single multichannel resonance. This case is obtained from the above formulas (2.10) and (2.11) by omitting the second term in the bracket.

It is customary to introduce the following notation:

$$\Gamma_\eta^{1/2} = \langle \eta \| \kappa_R \rangle \Gamma^{1/2}, \qquad \Gamma = \Gamma_R. \tag{2.12}$$

At this point we use time-reversal invariance. If time-reversal invariance holds, then the unitary matrix $\langle \eta \| \kappa \rangle$ is orthogonal, i.e., the matrix elements are real [Equation (2.1a)]. Thus for time-reversal-invariant interactions $\Gamma_\eta^{1/2}$ are real.[3]

Because of (2.1), the $\Gamma_\eta$ fulfill

$$\sum \Gamma_\eta = \Gamma = \Gamma_R. \tag{2.13}$$

$\Gamma_\eta$ is called the partial width for the channel $\eta$. With this notation the elastic scattering amplitude for a single resonance of quantum number $\kappa_R$ is written

$$p T_l(p) = e^{i2\gamma(\kappa_R)} \frac{\Gamma_{\eta_A}/2}{E_R - E - i\Gamma/2} + b(\eta_A), \tag{2.14}$$

and the reaction amplitude is given by:

$$\sqrt{p_r p}\, T_{r,l}(p) = e^{i2\gamma(\kappa_R)} \frac{\frac{1}{2}\Gamma_r^{1/2}\Gamma_{\eta_A}^{1/2}}{E_R - E - i\Gamma 2} + b(r), \tag{2.15}$$

where

$$b(\eta) = \sum_\kappa \langle \eta \| \kappa \rangle \langle \kappa \| \eta_A \rangle e^{i\gamma(\kappa)} \sin \gamma(\kappa). \tag{2.16}$$

which can be a slowly varying function of energy even if the individual $\gamma(\kappa)$ vary rapidly. The representation (2.14), (2.15) is called the generalized Breit–Wigner approximation. These expressions are very similar to the corresponding expression (XVIII.6.26) for the one-channel resonance, except that the height of the multichannel resonance is smaller by the factor

---

[3] If time-reversal invariance does not hold, $\Gamma_\eta^{1/2}$ is complex and one defines $\Gamma_\eta$ by $\Gamma_\eta = |\Gamma_\eta^{1/2}|^2$. We shall not discuss this case here.

$\langle \eta_A \| \kappa_R \rangle^2$ or $\langle \eta_A \| \kappa_R \rangle \langle r \| \kappa_R \rangle$. The first factor represent the transition amplitude (square root of the transition probability) for a transition from a state with quantum numbers $\eta_A$ to a state with quantum numbers $\kappa_R$ and back to a state with quantum numbers $\eta_A$; the second factor represents the transition amplitude from $\eta_A$ to $\kappa_R$ and from $\kappa_R$ to $r$. Thus the smaller height is due to the inelasticity, i.e., the fact that the resonance contributes not only to the channel $\eta_A$ or $r$, but also to all the other channels $\eta \neq \eta_A$ or $\eta \neq r$.

The cross sections corresponding to the amplitudes (2.15) will then not look much different from the expressions for the cross sections given in Sections XVIII.6 and XVIII.9. Using the abbreviations:

$$\epsilon = \frac{E_R - E}{\Gamma/2}, \tag{2.17}$$

$$R = R^\eta = \langle \eta \| \kappa_R \rangle \langle \kappa_R \| \eta_A \rangle, \tag{2.18}$$

where $\kappa_R$ are the quantum numbers of the resonance $R$, and $\eta_A$ the quantum numbers of the initial state $A = (a, T)$, one can write the $l$th partial cross section $\sigma_l^{(A \to \eta)}$ using (2.14) for $\eta = \eta_A$ and (2.15) for $\eta = r$ with (2.16):

$$\sigma_l^{(A \to \eta)} = 4\pi(2l + 1) \frac{1}{p^2} \left| \sum_\kappa \langle \eta \| \kappa \rangle \langle \kappa \| \eta_A \rangle e^{i\gamma(\kappa)} \sin \gamma(\kappa) + e^{i 2\gamma(\kappa)} \frac{R}{\epsilon - i} \right|^2 \tag{2.19}$$

After some calculation this can be written as

$$\sigma_l^{(A \to \eta)} = 4\pi(2l + 1) \frac{1}{p^2} \left[ |b(\eta)|^2 \right.$$
$$\left. + \frac{R}{\epsilon^2 + 1} \left( R + \sum_\kappa \langle \eta \| \kappa \rangle \langle \kappa \| \eta_A \rangle \sin^2 \gamma(\kappa)(2\epsilon \cot \gamma(\kappa) - 2) \right) \right], \tag{2.20}$$

where $b(\eta)$ is given by (2.16). This is the elastic cross section for $\eta = \eta_A$ and the reaction cross section for $\eta = r$.

For an energy $E$ far away from the resonance energy $E_R$, i.e., for $|\epsilon| \gg 0$, the second term in the bracket on the right-hand side of (2.20) is very small and vanishes for $|\epsilon| \to \infty$. The cross section is then given by

$$\sigma_l^{A \to \eta} \to \sigma_{(nr)} = 4\pi(2l + 1) \frac{1}{p^2} |b(\eta)|^2 \quad \text{for } |\epsilon| \to \infty. \tag{2.21}$$

Thus we see that (2.20) is very similar to (XVIII.9.14): a single resonance in the many-channel case will appear as a structure above a gradually varying background. The shape of this structure will in general not be the same as the one given in Figure XVIII.9.2 and will depend upon the various background phase shifts $\gamma$. However, if these background phase shifts are, in the neighborhood of the resonance energy $E \approx E_R$, approximately the same modulo $\pi$ for all quantum numbers $\kappa$:

$$\gamma_l(\kappa) = \gamma_l(E) = \gamma(E), \tag{2.22}$$

then one can write (using $\langle \eta_A | \eta_A \rangle = \sum \langle \eta_A \| \kappa \rangle \langle \kappa \| \eta_A \rangle = 1$,

$$\sigma_l^{(A \to \eta)} \to \sigma_{(nr)} + \frac{Rq^2 + 2q\epsilon - 2 + R}{\epsilon^2 + 1} R\sigma_l^{(bg)} \quad \text{for } \eta = \eta_A, \quad (2.23)$$

where $q$ is again defined by

$$q = \cot \gamma_l(E) \quad (2.24)$$

and $\sigma_l^{bg}$ is given by

$$\sigma_l^{(bg)} = 4\pi(2l + 1)\frac{1}{p^2} \sin^2 \gamma. \quad (2.25)$$

Equation (2.23) is, indeed, very similar to (XVIII.9.14) with (XVIII.9.6). The shape of a single resonance in the many-channel case will, therefore, be very similar though not identical to the shape depicted in Figure XVIII.9.2.[3a]

From (2.20) we see that the multichannel resonance shows up in the elastic channel, $\eta = \eta_A$, as well as in all the reaction channels $\eta = r$, at the same energy $E_R$ and with the same width $\Gamma$. The shape of the resonance depends upon the background phase shifts $\gamma(\kappa)$. And even if (2.22) holds, the resonance with quantum number $\kappa_R$ will have different though similar shapes in the various channels $\eta$, because $R$ in (2.23) depends in general upon $\eta$ according to (2.18).

However, if one has the particular case that the background for all quantum numbers $\kappa$—the resonating $\kappa = \kappa_R$ and all the nonresonating $\kappa$—is negligible, then one has

$$\sigma_l^{(R)(A \to \eta)} = 4\pi(2l + 1)\frac{1}{p^2}\frac{\frac{1}{4}\Gamma_\eta \Gamma_{\eta_A}}{(E_R - E)^2 - \Gamma^2/4}. \quad (2.26)$$

Thus, the resonance without background appears in all channels as a Breit–Wigner resonance at the same energy and with the same width, but with different heights depending upon the value of the partial width $\Gamma_\eta$.

Further, as $\eta_A$ can really take any value, the same Breit–Wigner resonance also appears for any initial state. Thus, independent of what the initial and final states of a scattering process are, if only their transition amplitudes to the state with quantum numbers $\kappa_R$ are different from zero, the resonance will appear at the same energy and with the same width. This is, of course, what one must expect of a quasistationary state, which is entirely determined by its "quantum numbers" $(E_R, \Gamma, l, \kappa_R)$ and independent of the experimental conditions.[4]

---

[3a] If all the background eigenphases are equal they can be energy independent cf. footnote on p. 2–4.

[4] As we have remarked already in Section XVIII.6, the notion of the quasistationary state $(E_R, \Gamma, l, \kappa_R)$ has a meaning only up to a certain accuracy. If the energy dependence of $\Gamma$ has to be taken into account, then the apparent position, $E_R^{(observed)}$, may be different for different experiments.

Experimental examples of single multichannel resonances are shown in Figure XIX.3.1. This figure shows the various resonances ("energy levels") of $Si^{28}$ as they occur in the processes

$$A: \quad Al^{27} + p \rightarrow (Si^{28})^* \rightarrow p + Al^{27},$$

$$B: \quad Al^{27} + p \rightarrow (Si^{28})^* \rightarrow \alpha + Mg^{24},$$

$$C: \quad Mg^{24} + \alpha \rightarrow (Si^{28})^* \rightarrow p + Al^{27},$$

$$D: \quad Mg^{24} + \alpha \rightarrow (Si^{28})^* \rightarrow \alpha + Mg^{24}.$$

$(Si^{28})$-resonances are observed at the following energy values:[5] 2.75, 2.93, 2.95, 3.00, 3.14, 3.20, 3.28 MeV. Fine discrepancies between the curves—like the double peak around 3.4 MeV in process B, which is matched only by a single peak in process C—can be explained in terms of differences in the energy resolution for the experiments.

Let $\eta_A$ denote the quantum numbers of the $p$-$Al^{27}$ system, $r$ the quantum numbers of $\alpha$-$Mg^{24}$, and $\kappa_{R_1}, \kappa_{R_2}, \ldots$ the quantum numbers of the various $(Si^{28})$-resonances. The energy values of these resonances are sufficiently far apart that one can consider each separately as a single resonance and need not apply the considerations for a double resonance described below.

We see that in general, each single resonance occurs in all four processes. However, whereas in the reactions B and C they usually occur as pure Breit–Wigner resonance, in the elastic processes A and D they have considerable background phases. For example, the resonance at 3.203 MeV, whose quantum number we may call $\kappa_{R_6}$, occurs in B and C as a bump and in A and D as a dip. This indicates that $\langle \eta_A \| \kappa_{R_6} \rangle$ as well as $\langle r \| \kappa_{R_6} \rangle$ are different from zero, and that the sum over $\kappa$ in (2.20) is negligible for $\eta = r$ but not negligible for $\eta = \eta_A = $ (quantum numbers of the $p$-$Al^{27}$ system) and $\eta = \eta_A = r = $ (quantum numbers of the $\alpha$-$Mg^{24}$ system). A possible explanation for this effect is that the background phases $\gamma(\kappa)$ are not negligible, but approximately independent of $\kappa$, so that for the elastic processes something like (2.23) holds. Under the same assumption one obtains for the inelastic processes, using $\langle \eta | \eta_A \rangle = 0$,

$$\sigma_l^{A \rightarrow \eta} \approx \sigma_{nr} + \frac{(\langle \eta \| \kappa_R \rangle \langle \kappa_R \| \eta_A \rangle)^2}{\epsilon^2 + 1}, \qquad (2.23')$$

so that in a reaction the resonance occurs as a pure Breit–Wigner above a possible background $\sigma_{nr}$.

---

[5] These are the energy values on the $\alpha$-particle energy scale. The difference in the origin of the energy axis for the $\alpha$-particle energy and the proton energy is 1.61 MeV, and the energy scales of the processes A, B and C, D have been adjusted by bringing each pair of resonance peaks of the reaction curves B and C into coincidence.

## Double Multichannel Resonance

We now return to the case of two resonating $l$th partial waves: one with the internal quantum numbers $\kappa_R$ and energy and width $(E_R, \Gamma_R)$, and the other with internal quantum numbers $\kappa_X$ and resonance parameters $(E_X, \Gamma_X)$. We start from (2.10) and (2.11) and consider simultaneously the cases of elastic scattering of (2.10) $\eta = \eta_A$ and of the reaction (2.11), $\eta = r$. We use the notation

$$R = R^{\eta} = \langle \eta \| \kappa_R \rangle \langle \kappa_R \| \eta_A \rangle,$$

$$X = X^{\eta} = \langle \eta \| \kappa_X \rangle \langle \kappa_X \| \eta_A \rangle e^{i2(\gamma(\kappa_X) - \gamma(\kappa_R))}. \tag{2.27}$$

We shall always assume time-reversal invariance (2.1b), so that $R$ is real and the phase of $X$ is entirely determined by the difference of the background phases. Equations (2.10) and (2.11) are then written

$$pT_l(p) = e^{i2\gamma(\kappa_R)} \left( \frac{(\Gamma_R/2)R^{\eta}}{E_R - E - i\Gamma_R/2} + \frac{(\Gamma_X/2)X^{\eta}}{E_X - E - i\Gamma_X/2} + b \right), \tag{2.28}$$

where $b$ is given by (2.16).

To simplify our following discussion we ignore the background term $b$. It will have effects similar to those discussed above for the single resonance and in Section XVIII.9.

The cross section for the double multichannel resonance then becomes (if we assume for the sake of simplicity that $X$ is real):

$$\sigma_l^{(R)} = 4\pi(2l + 1) \frac{1}{p^2} \left( \frac{(\Gamma_R^2/4)R^2}{(E_R - E)^2 + \Gamma_R^2/4} + \frac{(\Gamma_X^2/4)X^2}{(E_X - E)^2 + \Gamma_X^2/4} \right.$$

$$\left. + \frac{(\Gamma_R/4)\Gamma_X 2RX[(E_R - E)(E_X - E) + \Gamma_R/4\Gamma_X]}{[(E_R - E)^2 + (\Gamma_R^2/4)][(E_X - E)^2 + (\Gamma_X^2/4)]} \right). \tag{2.29}$$

Thus the $l$th partial cross section is the sum of two Breit–Wigner cross sections, with different widths and different heights, and a third term (the interference term). A double resonance is characterized by four parameters $(E_R, \Gamma_R; E_X, \Gamma_X)$ and the quantities $R$ and $X$, which depend according to their definition (2.27) upon the final-state quantum numbers $\eta$ and also upon the initial-state quantum numbers $\eta_A$. Thus for one channel $R$ and $X$ may have the same sign, and for another channel they may have opposite signs. *The shape of the cross section of a double resonance, therefore, varies with the channel quantum number.* If in one channel it appears as a double hump, it may show up in another channel as one single broad hump. The most favorable circumstances for a double hump are $R \approx X$. Then the third term is (2.29) has two maxima, one near $E = E_R$ and the other near $E = E_X$, which enhance the humps at these values that come from the two Breit–Wigner cross sections in (2.29). Then the double resonance will appear in the cross section as two maxima with different widths, provided the background can be ignored.

To obtain an idea of the possible shapes of such a double resonance, let us consider some special cases. It is useful to introduce the new variable

$$\mathscr{E} = E + \frac{E_R + E_X}{2} \tag{2.30}$$

and denote

$$\Delta = E_R - E_X. \tag{2.31}$$

Then the cross section formula (2.29) is written as

$$\sigma_l^{(R)} = 4\pi(2l + 1)\frac{1}{p^2}\frac{1}{4}\left(\frac{R^2\Gamma_R^2}{(\Delta/2 - \mathscr{E})^2 + (\Gamma_R/2)^2} + \frac{X^2\Gamma_X^2}{(-\Delta/2 - \mathscr{E})^2 + (\Gamma_X/2)^2}\right.$$

$$\left. + \frac{2RX(\mathscr{E}^2 - (\Delta/2)^2 + \Gamma_R\Gamma_X/4)\Gamma_R\Gamma_X}{[(\Delta/2 - \mathscr{E})^2 + (\Gamma_R/2)^2][(-\Delta/2 - \mathscr{E})^2 + (\Gamma_X/2)^2]}\right). \tag{2.32}$$

The interesting situation arises when $E_R$ is close but not identical to $E_X$ and if the widths of the two resonances have approximately the same magnitude as $\Delta = E_R - E_X$. Therefore we shall consider the case

$$\Gamma_R = \Gamma_X = \Delta. \tag{2.33_1}$$

Note that in this case the third term in (2.32) has a zero and a minimum (if sign $R$ = sign $X$) or maximum (if sign $R$ = $-$sign $X$). With (2.33) one obtains from (2.32)

$$\sigma_l^R = 4\pi(2l + 1)\frac{1}{p^2}\frac{\Gamma^2}{4}\left(\frac{R^2}{(\Delta/2 - \mathscr{E})^2 + (\Delta/2)^2}\right.$$

$$\left. + \frac{X^2}{(-\Delta/2 - \mathscr{E})^2 + (\Delta/2)^2} + \frac{2RX\mathscr{E}^2}{\mathscr{E}^4 + (\Delta^2/2)^2}\right). \tag{2.34}$$

The term in the bracket of (2.34)—i.e., $\sigma_l^R$ except for the factor

$$4\pi(2l + 1)\frac{1}{p^2}\frac{\Gamma^2}{4}$$

—is depicted in Figure 2.2 for various values of $X$ ($R = 1$). Ignoring $e^{i2(\gamma(\kappa_R) - \gamma(\kappa_X))}$ $X$ is connected to the coupling constants according to (2.27) by

$$X = \frac{\langle\eta\|\kappa_X\rangle\langle\kappa_X\|\eta_A\rangle}{\langle\eta\|\kappa_R\rangle\langle\kappa_R\|\eta_A\rangle}.$$

The correspondence between the curves and the values of $X$ is as given in the legend. We see that the cross section may look quite different for different couplings between the intial and final channels and the resonance channel. For $X = 1$ one obtains a double peak with the maxima slightly below $E = E_X$ ($\mathscr{E} = -\Delta$) and slightly above $E = E_R$ ($\mathscr{E} = +\Delta$). For $X = -1$, the cross section has one single broad bump centered around $E = \frac{1}{2}(E_R + E_X)$. For $X = 0$ there is, of course, only one bump at $E = E_R$, representing the resonance $R$, as in this case the resonance $X$ does not couple

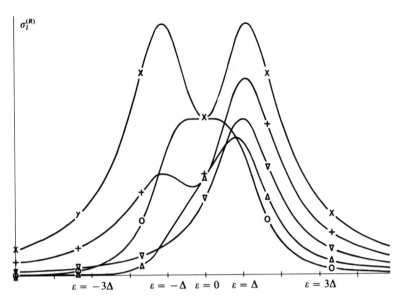

Figure 2.2    Cross section for double resonance: the case $(2.33_1)$ for various values of $X$:

| Symbol: | × | + | ▽ | △ | ○ |
|---|---|---|---|---|---|
| $X$: | 1.0 | 0.5 | 0.0 | −0.5 | −1.0 |

to the channels $\eta$ or $\eta_A$. But also for $X = -0.5$ there is only one single bump, which however does not have the form of a Breit–Wigner resonance. Thus, for different final channels $\eta$, a double resonance may show up in the cross section as a double peak, as a broad single peak with maximum between the two resonance energies, or as a single peak around one of the resonance energies.

The cross sections in Figure 2.2 correspond to the most symmetrical case, $(2.33_1)$. If $\Gamma_X$ or $\Gamma_R$ differ from $\Delta$ the picture changes slightly. Figure 2.3 depicts (2.32) [except for the factor $4\pi(2l + 1)(1/p^2)\Gamma^2$] for the case

$$\Gamma_X = \Delta, \qquad \Gamma_R = \tfrac{1}{2}\Delta. \qquad (2.33_2)$$

The variation in the shape of the cross section with the coupling between the channels is similar to that in Figure 2.2.

In the above considerations we have ignored the background term $b$, so that the above discussions are valid only if (2.16) vanishes. Also we have ignored $\gamma(\kappa_X) - \gamma(\kappa_R)$. From our discussions in Section XVIII.9, and from the footnote on p. 466 we know already that the background phase shifts may be large, so that this idealization may well be inadmissable. Taking the interference between resonances into account will make the complicated features of a double resonance even more complicated. Taking in addition the background from other partial waves and the resolution of the experimental apparatus into account, one concludes that the observation of a

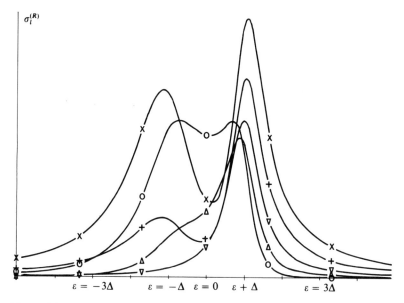

Figure 2.3   Cross section for double resonance: the case $(2.33_2)$.
Symbols as in Figure 2.2.

double multichannel resonance as a double peak in the cross section is an
unlikely possibility. Thus two resonances with different quantum numbers $\kappa_R$
and $\kappa_X$ but with energies $E_R$ and $F_X$, that differ from each other by an amount
comparable with the width and that both couple to the initial and final
channel, will be very hard to observe.

Nonetheless, examples of double resonances have been seen in nuclear
scattering processes. The most prominent examples are the $Be^8$ energy
levels with (angular momentum)$^{parity}$ $= l^P = 2^+$. These resonances have
been observed in the processes

$$He^4 + He^4 \to (Be^8)^* \to He^4 + He^4 \tag{$2.35_1$}$$

and

$$He^4 + He^4 \to (Be^8)^* \to Be^8 + \gamma, \tag{$2.35_2$}$$

at excitation energies of 16.6 and 16.9 MeV of the $Be^8$, with widths of approximately 110 and 80 keV, respectively. We do not want to describe the
details of this system, and shall just establish the connection with the quantities of our formulas (2.28) and (2.29). We ignore the background phases;
then (2.28) can be written

$$pT_l^{A \to \eta_A} = \frac{\frac{1}{2}\Gamma_X \langle \eta_A \| \kappa_X \rangle \langle \kappa_X \| \eta_A \rangle}{E_X - E - i\Gamma_X/2} + \frac{\frac{1}{2}\Gamma_R \langle \eta_A \| \kappa_R \rangle \langle \kappa_R \| \eta_A \rangle}{E_R - E - i\Gamma_R/2} \tag{$2.36_1$}$$

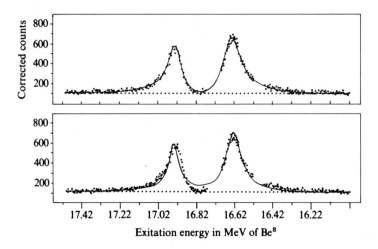

Figure 2.4   Fits of the doublet structure in Be$^8$ in the He$^4$-He$^4$ channel. Note that the energy decreases from left to right. [From W. D. Callender et al.[7] with permission.]

and

$$pT_i^{A\to\eta} = \frac{\frac{1}{2}\Gamma_X\langle\eta\|\kappa_X\rangle\langle\kappa_X\|\eta_A\rangle}{E_X - E - i\Gamma_X/2} + \frac{\frac{1}{2}\Gamma_R\langle\eta\|\kappa_R\rangle\langle\kappa_R\|\eta_A\rangle}{E_R - E - i\Gamma_R/2}. \quad (2.36_2)$$

Here $|\eta_A\rangle$ denotes the internal state of the (He$^4$, He$^4$) system, $|\eta\rangle$ denotes the state of the (Be$^8$, $\gamma$) system, and $|\kappa_X\rangle$ and $|\kappa_R\rangle$ denote the internal state of the Be$^{8*}$ (16.6 MeV) and Be$^{8*}$ (16.9 MeV) respectively, so that (2.36$_i$) is the scattering amplitude for the process (2.35$_i$). Depending upon the ratio $\langle\eta_A\|\kappa_X\rangle^2/\langle\eta_A\|\kappa_R\rangle^2$, the cross section for the process (2.35$_1$) will then have the shape of one of the curves in Figure 2.2 or 2.3. The cross sections have been measured[6,7] and the experimental data points together with a fit to the data are plotted in Figure 2.4.[7]

[6] A. D. Bacher, F. G. Resmini, H. E. Conzett, R. deSwinairski, H. Meiner, J. Ernst, *Phys. Rev. Letters* **29**, 1331 (1972).

[7] W. D. Callender, C. P. Browne, *Phys. Rev.* **2**, 1 (1970). This figure does not give the cross section for the formation process (2.35$_1$) but for the production process

$$B^{10} + d \to \alpha + Be^{8*}$$
$$\phantom{B^{10} + d \to \alpha +}\hookrightarrow \alpha + \alpha$$

$R$ and $X$ in (2.29) are therefore not

$$R = \langle\eta_A\|\kappa_R\rangle\langle\kappa_R\|\eta_A\rangle, \quad X = \langle\eta_A\|\kappa_X\rangle\langle\kappa_X\|\eta_A\rangle,$$

but

$$R = g_R\langle\kappa_R\|\eta_A\rangle, \quad X = g_X\langle\kappa_X\|\eta_A\rangle$$

with

$$g_R \approx g_X.$$

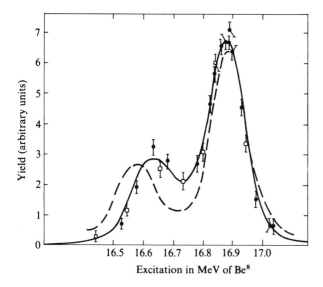

Figure 2.5   Fits of the double structure in $Be^8$ in the $Be^8$-$\gamma$ channel.
[From Nathan et al.[8]: Phys. Rev. Lett. 35, 1137 (1975), with permission.]

The curve in the upper plot shows a fit to the data obtained with an expression (2.29) for a double resonance. The lower plot illustrates the fit with the sum of two Breit–Wigner cross sections,

$$\sigma \sim \frac{\Lambda_1}{(E_X - E)^2 + (\Gamma_X/2)^2} + \frac{A_2}{(E_R - E)^2 + (\Gamma_R/2)^2},\qquad (2.37)$$

which would hold if the states at 16.6 and 16.9 MeV were unrelated single resonances. This fit with two noninterfering Breit–Wigner curves is clearly unacceptable, whereas the fit in the upper plot with a double resonance (two interfering Breit–Wigner curves) is excellent. Thus we conclude that the double-peak structure observed around 16.6 and 16.9 MeV of the $Be^{8*}$ system is a double resonance. Both bumps have approximately the same height, the bump at 16.9 MeV is slightly narrower, and the fitted curve resembles the curve (marked $\times$) in Figure 2.3 for $X/R \approx 1$. Both states $|\kappa_R\rangle$ and $|\kappa_X\rangle$ couple, therefore, with approximately the same strength to the incident channel:

$$\langle \kappa_X \| \eta_A \rangle \approx \langle \kappa_R \| \eta_A \rangle. \qquad (2.38)$$

Figure 2.5 shows a plot of the experimental data[8] for the reaction (2.35$_2$). The solid line is a fit with (2.29); the dashed line is a fit with (2.37). Again the sum of two noninterfering Breit–Wigner curves is clearly ruled out, and a double resonance gives a very good fit. This time, however the curve resembles more the curve (marked $+$) in Figure 2.3 for $X/R \approx 0.5$. From

[8] A. M. Nathan, G. T. Garvey, P. Paul, E. K. Warburton, *Phys. Rev. Letters* **35**, 1137 (1975).

this and (2.38) we conclude that the coupling of the two $|\kappa_R\rangle$ and $|\kappa_X\rangle$ to the outgoing channel $(Be^8, \gamma)$ fulfills the relation

$$|\langle\eta\|\kappa_X\rangle| < |\langle\eta\|\kappa_R\rangle|. \tag{2.39}$$

The experimental data for the $Be^8$ system around 16.6 and 16.9 MeV give a beautiful illustration of the fact that a double resonance may show up in the cross section with a variety of shapes.

## XX.3   Argand Diagrams for Inelastic Resonances

In Section XVIII.8 we discussed the behavior of $pT_l(p)$ in the Argand diagram for the case of purely elastic scattering. In the second part of that section we also described how $T_l(p)$ can be obtained from the experimental data. In the present section we shall give a brief discussion of the behavior of the elastic partial-wave amplitude $pT_l(p)$ in the presence of inelastic processes and of the partial-wave reaction amplitude $\sqrt{p_r p}\,T_{r,l}(p)$. The determination of these amplitudes from the experimental data follows essentially the same procedure as described in the second part of the Section XVIII.8: for the determination of $T_l(p)$ one uses the elastic differential cross sections, and for $T_{r,l}(p)$ the differential cross section for the reaction (initial state $\eta_A$) → (final state $r$).

We will now assume that there is a single multichannel resonance and discuss how this single resonance will show up in the Argand diagrams for the elastic scattering amplitude for the reaction amplitude. We first treat the idealized case without background. The scattering amplitudes are then given according to (2.14) and (2.15) by

$$pT_l^{(R)}(p) = \frac{\frac{1}{2}\Gamma_{\eta_A}}{E_R - E - i\Gamma/2}, \tag{3.1}$$

$$\sqrt{p_r p}\,T_{r,l}^{(R)}(p) = \frac{\frac{1}{2}\Gamma_r^{1/2}\Gamma_\eta^{1/2}}{E_R - E - i\Gamma/2}. \tag{3.2}$$

Here $p$ is the initial momentum and $p_r$ is the final momentum, which is a function of the initial momentum and the difference of internal energies in the initial and final states.[9]

$\Gamma_r$ and $\Gamma_{\eta_A}$ are the partial widths (2.12) for the reaction channel $r$ and for the initial channel $\eta_A$, respectively. $E_R$ and $\Gamma_R = \Gamma$ are the energy and total width of the resonance. We use again the abbreviation

$$\epsilon = (E_R - E)\frac{2}{\Gamma} \tag{3.3}$$

---

[9] For a massive nonrelativistic projectile of mass $m_A$ before and mass $m_b$ after the reaction, $p_r = p_b$ is given by the formula (XVI.1.24).

and also define

$$\sqrt{x_\eta} = \langle \eta \| \kappa_R \rangle = \sqrt{\Gamma_\eta / \Gamma}. \tag{3.4}$$

$\sqrt{x_\eta}$ may be negative.

Equations (3.1) and (3.2) can then be written

$$pT_l^{(R)} = \frac{x_{\eta_A}}{\epsilon - i}, \tag{3.5}$$

$$\sqrt{p_r p} \, T_l^{(R)} = \frac{\sqrt{x_r x_{\eta_A}}}{\epsilon - i}. \tag{3.6}$$

$x_{\eta_A} = \langle \eta_A \| \kappa_R \rangle^2$ is called the elasticity of the resonance and is the transition probability for a transition from a state with quantum numbers $\kappa_R$ into a state with quantum numbers $\eta_A$ at a particular value of $E$. Thus $x_{\eta_A}$ represents the "fraction" of the resonance with internal quantum numbers $\kappa_R$ that is coupled to the elastic channel $\eta_A$. For $x_{\eta_A} = 1$ the resonance couples to the elastic channel only, all $x_\eta$ with $\eta \neq \eta_A$ are zero, and $pT_l^{(R)}$ describes the unitary circle for an elastic resonance as described in Section XVIII.8. The elasticity $x_{\eta_A}$ is related to the inelasticity coefficient $\eta_l(E)$ defined in (XVI.2.10). At the resonance energy this relation is $\eta_l(E_R) = 2x_{\eta_A} - 1$. In contrast to $\eta_l(E)$, which has a strong energy dependence, empirical data show that the elasticity $x_{\eta_A}$ is only very weakly energy-dependent and is therefore a more convenient parameter.

The Argand diagram of (3.5) is depicted in Figure 3.1.[10] It is a circle of diameter $x_{\eta_A}$ lying inside the unitary circle. The resonance energy, $\epsilon = 0$, lies on the top of the circle, and the points $\epsilon = \mp 1$ correspond to $E = E_R \mp \Gamma/2$. One has to distinguish two cases:

(a)  $x_{\eta_A} \geq 0.5$; then the resonance occurs at a phase-shift value

$$\delta_i^{(\eta_A)}(E_R) = \pi/2.$$

(b)  $x_{\eta_A} < 0.5$; then the phase shift at resonance is $\delta^{(\eta_A)}(E_R) = 0$.

However, there is no qualitative difference between these two cases; in both cases the resonant eigen-phase-shift $\delta^{(\kappa_R)}(E)$ goes through $\pi/2$ at $E = E_R$.

The Argand diagram for the resonant reaction amplitude (3.6) is depicted in Figure 3.2. For the diameter of the circle one takes $|\sqrt{x_r x_{\eta_A}}|$. It lies inside the circle of Figure XVI.3.2 for the reaction amplitude, and

$$x_r x_{\eta_A} < \tfrac{1}{4}.$$

If the background is included, the elastic scattering amplitude (2.14) is written

$$pT_l = e^{i2\gamma} \frac{x_{\eta_A}}{\epsilon - i} + b. \tag{3.7}$$

---

[10] Figures 3.1, 3.2, and 3.3 are from Barbaro-Galtieri (1968).

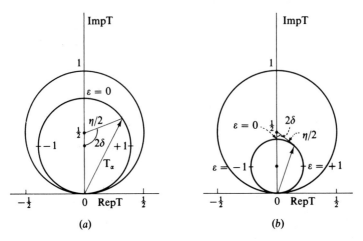

Figure 3.1   The resonant amplitude for the elastic channel is shown for two different values of the elasticity $x_{\eta_A}$: (a) $x_{\eta_A} = 0.75$ and the phase shift for the amplitude goes through 90° at the resonant energy; (b) $x_{\eta_A} = 0.4$ and $\delta = 0°$ at the resonant energy. Notice that the phase of the resonant amplitude in the resonance circle goes through 90° in both cases. [From Barbaro-Galtieri (1968), with permission.]

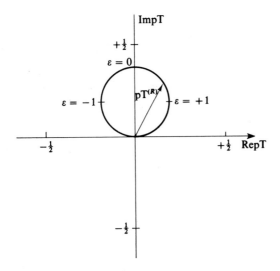

Figure 3.2   The resonant amplitude for scattering into an inelastic channel. [From Barbaro-Galtieri (1968), with permission.]

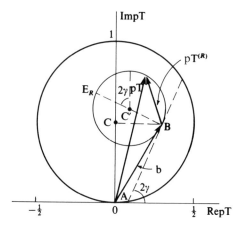

Figure 3.3   Elastic scattering amplitude for the case of a resonance and a background amplitude. If the background is assumed to be constant throughout the resonance region, the amplitude $pT$ is on the circle with center $C'$. In this figure the background amplitude is not purely elastic, but presents some absorption. [From Barbaro-Galtieri (1968), with permission.]

The graphical representation of (3.7) for $\gamma = 0$ is obtained from the graphical representation of (3.5) in Figure 3.1 by displacing the circle by the vector $b$ and starting the resonance circle at the endpoint $B$ of $b$. The resonance energy is then at the top of this resonance circle. The effect of the background phase shift $\gamma$ is then a rotation of the resonance circle by the angle $2\gamma$. The resulting Argand diagram for (3.7) is depicted in Figure 3.3.

The Argand diagrams depicted in the figures of this section and also in the figures of Section XVIII.8 are idealizations. The energy dependence of the background $b(E)$, the background phase shift $\gamma(E)$, and the elasticity distorts the circular behavior of the amplitude. An example of an experimental Argand diagram, obtained from the experimental data by a procedure described in the second part of Section XVIII.8, is depicted in Figure 3.4 below.

The existence and nature of a resonance in the Argand diagram are then determined by comparison with the idealized cases depicted in the above figures. If there is an indication of a circle, one takes this as a sign for a resonance. The resonance energy $E_R$ is then determined as that value of $E$ at which $pT_l(E)$ and the phase shift $\delta_l(E)$ change most rapidly. Around $E_R$ the path $pT_l(E)$ may show as a segment of a circle, and the elasticity $x_{\eta_A}$ is then obtained as the radius of this circle. The total width $\Gamma$ may then be obtained from the energy difference between the points $\frac{1}{4}$ circle to the left and $\frac{1}{4}$ circle to the right of the resonance energy on the circle segment. In this way Argand diagrams have been of great help for displaying resonance behavior and for determining the resonance parameters.

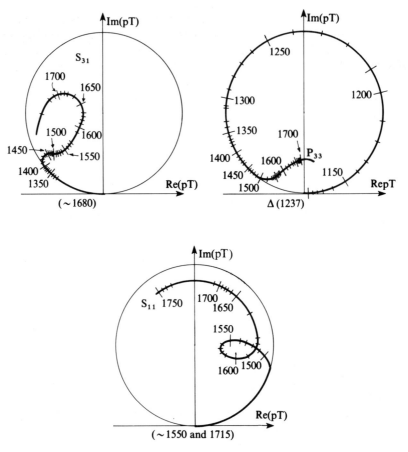

Figure 3.4   Experimental Argand diagrams for some pion-nucleon partial-wave amplitudes. [From A. H. Rosenfeld et al., *Data for Elementary Particle Physics*, 1965, with permission.]

As an illustration, we depict in Figure 3.4 Argand diagrams obtained from the experimental data for some partial-wave amplitudes in $\pi$-$p$ elastic scattering.[11] Each diagram shows the unitary circle and, inside it, $pT_i(E)$ for various partial waves. (For $\pi$-$p$ scattering there are several partial waves for each value of $l$, characterized by isospin $I$ and total angular momentum $j = l \pm \frac{1}{2}$; we cannot discuss these things here, they are also of no relevance for our present discussion.) The numbers at the curves for $pT(E)$ indicate the energy of the $\pi$-$p$ system in MeV $= 10^6$ eV. The second diagram (labeled by $P_{33}$) shows the almost purely elastic resonance $\Delta$ at the energy 1236 MeV. The Argand diagram is almost exactly as described in Figure XVIII.8.1. The third diagram (labeled $S_{11}$) is the Argand diagram of another partial wave with two resonances, one around 1550 MeV and the other around

---

[11] $\pi$ stands for $\pi$-meson and $p$ for proton.

1715 MeV, both inelastic. The first diagram (labeled $S_{31}$) shows another case of an inelastic resonance around 1680 MeV. We see from these examples that the behavior of different partial-wave amplitudes is quite distinct and in general far from the idealized behavior. The partial wave in the second diagram, $P_{33}$, is in fact an exception; all other resonances are inelastic and have a large background term.

# CHAPTER XXI

# The Decay of Unstable Physical Systems

In Section XXI.1 the decaying state is introduced as a resonance for which the production process is ignored. Section XXI.2 gives a heuristic discussion of decay probability (decay rate) and its measurement. Section XXI.3 describes a decaying state by a generalized eigenvector of a Hermitian energy operator with complex eigenvalue. Using this novel concept, the calculation of the decay rate is very simple and is given in Section XXI.4. Section XXI.5 contains a discussion of the partial decay rates and the use of various basis systems for their calculation.

## XXI.1 Introduction

In Chapters XVIII and XX we discussed scattering processes in which an unstable physical system was formed as a quasistationary intermediate state. The measurements were performed at the initial state, which was prepared in the distant past, and at the final state, which was observed when the interaction was no longer in effect and the quasistationary state had ceased to exist. We have already seen in Chapter XX that the property of the quasistationary state is—at least to a high degree of accuracy—independent of the process in which it is observed. Its characteristics are the energy $E_R$, the width $\Gamma$, and the internal quantum numbers $\kappa_R$.

The formation process

$$a + T \to R \to a' + T \tag{1.1}$$

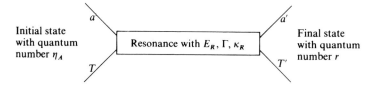

Figure 1.1

in which the quasistationary state is detected is depicted by the diagram in Figure 1.1, and (ignoring background) is described according to (XX.2.15) by the scattering amplitude

$$\sqrt{p_r p} T_{r,l} = \langle r | \kappa_R \rangle \frac{\Gamma/2}{E_R - E - i\Gamma/2} \langle \kappa_R | \eta_A \rangle. \tag{1.2}$$

If $\Gamma$ is sufficiently small and the delay time according to (XVIII.6.36) and (XVIII.6.43) sufficiently large, one may ignore the formation process. One then starts the description at a given time, $t = 0$, with a physical state $W(t = 0) = W^R$, and describes only that part of the process of Figure 1.1 which is depicted by Figure 1.2. $R$ is then considered as a decaying physical system or a decaying state of a physical system.

Examples of decaying states are numerous in physics; "real" stable states are very rare. The radiative transitions of excited atoms or molecules $A^*$ to lower states $A$,

$$A^* \to A + \gamma, \tag{1.3a}$$

are typical examples of decay processes. The unstable physical system had, of course, first to be formed, and this may have happened by a process like

$$\gamma + A \to A^*. \tag{1.3b}$$

But for the excited states of atoms one ignores the mode of formation and starts with an initial state $W^R$ describing the decaying system $A^*$.

The crucial problem is thus the choice of $W^R$ at the arbitrarily chosen initial time $t = 0$. From $W^R$, if the time development is described by a Hamiltonian $H$ according to the basic assumption $V$ of Chapter XII, one can then obtain the state of the decaying system at any later time $t$ by

$$W(t) = e^{-iHt} W^R e^{+iHt}. \tag{1.4}$$

Equation (1.4) is valid if the decaying system is isolated and evolves undisturbed. Whether for a realistic decay process the decaying system really evolves undisturbed is questionable, because it almost unavoidably interacts with its surroundings, especially when the decay process is observed.[1] Therefore the decaying system may be subjected to measurement processes occurring at random times. For example, a decaying particle in a bubble chamber leaves a track of bubbles caused by the interaction of the decaying particle with the environment. Each bubble means a measurement in which

[1] A Beskow, J. S. Nilsson, Arkiv för Fysik **34**, 561 (1967).

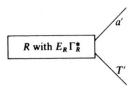

Figure 1.2

the system has been found undecayed. Such a measurement at time $t_1$ changes the state, according to the basic assumption IIIb, into the state

$$W'(t_1) = \Lambda_b W(t_1)\Lambda_b, \qquad (1.5)$$

where $\Lambda_b$ is the projection operator on the space of states of the undecayed system or a subspace of it. Thus in addition to the change of state (1.4) one has to consider the randomly occurring change of state (1.5).

Another problem is the existence of various possibilities for the choice of $W^R$ and their effect on the decay law. A further point that is worth discussing is the detailed decay properties in particular the deviations from the exponential law at small and large times as they follow from less idealised assumptions for $H$ and $W^R$ than those we shall make here. Such finer description will then also require to take the effect of the environment—as described by (1.5)—into account, which will lead back to an exponential decay low.[1a]

All these problems we shall not discuss here.[1a] In the following we shall consider only an approximate description, which will quickly lead to the desired results without regard for the fine points. For small width ($E_R/\Gamma \to \infty$) this will provide a sufficiently accurate description.

## XX.2  Lifetime and Decay Rate

In a decay experiment one has an ensemble of unstable physical systems $R$ in a volume $\mathscr{V}$, which is surrounded by counters that detect the decay products $a'$. The counting rate, i.e., the number of decays per second, $N/T$, is proportional to the number of unstable particles, $N_R = \rho_R \mathscr{V}$. The proportionality constant

$$\lambda = \frac{N/T}{N_R} \qquad (2.1)$$

is called the *initial decay rate*, or often just *decay rate*.

[1a] For a review of these subjects the reader is referred to "Decay Theory of Unstable Quantum Systems" by L. Fonda, G. C. Ghirardi and A. Rimini in Reports on Progress in Physics **41**, 587 (1978). Further methods not discussed in this review can be found in A. P. Grecos: "Solvable Models for Unstable States in Quantum Physics" Advances in Chemical Physics (1978) and A. George. F. Henin, F. Mayné, I. Prigogine: "Particles and Dissipations", Hadron Journal, Vol. 1 (1978).

The number of decays per second, $N/T$, is equal to the rate of decrease of the number of unstable particles $R$, $dN_R/dt$. Therefore, by (2.1),

$$-\frac{dN_R}{dt}(t) = \lambda N_R(t), \tag{2.2}$$

at least at $t = 0$. And if there is no further source or sink of unstable particles $R$ in $\mathscr{V}$, one is led from (2.2) to the exponential decay law

$$N_R(t) = N_R(0)e^{-\lambda t}, \tag{2.3}$$

where $N_R(0)$ is the number of unstable particles $R$ at the time $t = 0$ when the decay process started.

As $N_R(0) = -\int_{t=0}^{\infty} dN_R(t)$, it follows that $-dN_R(t)/N_R(0)$ is the probability of a decay at time $t$ during the time interval $dt$. The decay rate, i.e., the probability per unit time for a decay at $t$, is therefore

$$\frac{d\mathscr{P}^{(cl)}}{dt} \equiv \dot{\mathscr{P}}^{(cl)}(t) = -\frac{1}{N_R(0)}\frac{dN_R(t)}{dt}. \tag{2.4}$$

From (2.2) it then follows that

$$\dot{\mathscr{P}}^{(cl)}(t)|_{t=0} = \lambda, \tag{2.5}$$

which explains the name "initial decay rate" for $\lambda$.

The average lifetime of an unstable particle $R$,

$$\tau = \int_0^{\infty} t\dot{\mathscr{P}}^{(cl)}(t)\, dt, \tag{2.6}$$

is called the *mean life*, or just the *lifetime*. If (2.3) holds, then

$$\tau = \int_0^{\infty} \frac{1}{N_R(0)} N_R(0)e^{-\lambda t}\lambda t\, dt = \frac{1}{\lambda} = \frac{1}{\dot{\mathscr{P}}^{cl}(t=0)}. \tag{2.7}$$

In quantum physics the decaying particles are quantum physical systems and the observed quantities are probabilities. The observables measured on the decaying state $W(t)$ may be the "property" of being in the original resonance state $W^R$. The expectation value of $W^R$ in the state $W(t) = e^{-iHt}W^R e^{iHt}$ represents the probability that $R$ has not decayed. It is called the *nondecay probability*:

$$\mathscr{P}_R(t) = \mathrm{Tr}(W^R W(t)). \tag{2.3a}$$

It corresponds to the classical expression $N_R(t)/N_R(0)$.

Another observable that may be measured is the projection operator $\Lambda$ on the space of physical states of the decay products $(a', T')$, i.e., the property $\Lambda$. The expectation value of $\Lambda$ in the state $W(t)$ represents the probability for observing $a'$ and $T'$. This transition probability from the state $W(t)$ to $\Lambda$,

$$\mathscr{P}(t) = \langle \Lambda \rangle_t = \mathrm{Tr}(\Lambda W(t)), \tag{2.8}$$

is called the *decay probability* of $R$ into $(a', T')$.

The quantum-mechanical decay rate is the transition rate from $W(t)$ into $\Lambda$:

$$\frac{d\mathscr{P}}{dt} = \dot{\mathscr{P}}(t) = \frac{d}{dt} \langle \Lambda \rangle_t.$$

In (XIV.2.18) we have already calculated an expression for it, which we shall use below.

The experimentally measured quantity is the lifetime $\tau$ or the decay rate $\lambda$. For large lifetimes (from 1 sec to 1 year), one measures $N_R(t)$ as a function of $t$ and obtains $\lambda$ from (2.3). For shorter lifetimes down to $10^{-9}$ sec one measures the lifetime by electronic methods. The decay to be studied must be preceded by another event, e.g., the formation of the unstable system, which is used to establish the origin of time. A counter is activated by this event and remains activated for a time $t$. The probability for the decay during this time is, according to (2.4) and (2.3),

$$k \int_0^t dt' \, \lambda e^{-\lambda t'} = k(1 - e^{-\lambda t}) = \frac{\text{Number of times the activated counter detects a decay}}{\text{Number of times the counter is activated}} \tag{2.9}$$

One measures the ratio on the right-hand side as a function of $t$. As $k$ is a constant independent of $t$ [connected with $N_R(0)$ and the detector efficiency], one can calculate $\lambda$ if one knows (2.9) for several values of $t$.

For lifetimes shorter than $10^{-9}$ sec one cannot directly measure $\tau$, but has to resort to indirect methods. As we shall see below, the decay rate is equal to the width. One then obtains the lifetime from the measurement of the width in resonance scattering.

## XXI.3 The Description of a Decaying State and the Exponential Decay Law

We turn now to the quantum-mechanical calculation of the decay probability and decay rate. The first problem we have to face is the appropriate choice of $W(0) = W^R$.

Unstable systems are prepared by a scattering experiment in which the delay time is large. The cross section can then be fitted by a scattering amplitude consisting of a Breit–Wigner resonance term plus background. Intermediate physical states—which, according to Section II.10, are described by elements of $\Phi$—will represent such superpositions. It is often useful and sufficiently accurate to employ a description in which one isolates the intermediate quasistationary system, ignores its mode of formation, and represents it by an idealized decaying state. Such an idealized $W(0) = W^R$—which need not be an operator in $\Phi$—must be chosen as if it was prepared before the decay interaction as an in-state $W^{in}(t)$ which, at $t = 0$, under the influence of the interaction, has aquired a Breit–Wigner energy distribution.

Thus the density matrix[2]

$$\langle E_a \hat{a}^+ | W(0) | E_a \hat{a}'^+ \rangle = \langle E_a \hat{a} | W^{\text{in}}(0) | E_a \hat{a}' \rangle$$
$$= F(E_a - E_R) \rho(E)^{-1} \langle \hat{a} \| W^R \| \hat{a}' \rangle \qquad (3.1)$$

must have a Lorentzian energy distribution function $F(E_a - E_R)$ with the resonance width as the width of the energy distribution:

$$F(E - E_R) = \frac{1}{\pi} \frac{\Gamma/2}{(E_R - E)^2 + (\Gamma/2)^2}. \qquad (3.2)$$

Here we have normalized $F(E - E_R)$ in such a way that[3]

$$F(E - E_R) \to \delta(E - E_R) \quad \text{for } \Gamma/2 \to 0 \qquad (3.3)$$

We shall also normalize the reduced matrix element to one:

$$\sum_{\hat{a}} \langle \hat{a} \| W^R \| \hat{a} \rangle = 1 \qquad (3.4)$$

and discuss the normalization of $W(0) = W^R$. We calculate

$$\text{Tr } W(0) = \sum_a \langle a^+ | W(0) | a^+ \rangle = \sum_{\hat{a}} \int \rho(E) \, dE \langle E \hat{a}^+ | W(0) | E \hat{a}^+ \rangle$$

$$= \sum_{\hat{a}} \int dE \, F(E - E_R) \langle \hat{a} \| W^R \| \hat{a} \rangle$$

$$= \frac{\Gamma}{2\pi} \int_0^\infty dE \, \frac{1}{(E - E_R)^2 + (\Gamma/2)^2}$$

$$= \frac{1}{\pi} \int_{x = -E_R 2/\Gamma}^\infty dx \, \frac{1}{x^2 + 1}$$

$$= \frac{1}{\pi} \left[ \frac{\pi}{2} + \arctan\left( E_R \frac{2}{\Gamma} \right) \right], \qquad (3.5)$$

where the new variable $x$ is

$$x = (E - E_R) \frac{2}{\Gamma}. \qquad (3.6)$$

For the sake of definiteness we have assumed that the spectrum of $H$ and $K$ is continuous and goes from 0 to $\infty$, excluding possible bound states. For physical reason the spectrum of $H$ must be bounded from below, and we have chosen $E_{\min} = 0$. From (3.5) we see that for an infinitely narrow width, $W(0)$ is normalized:

$$\text{Tr } W^R = \text{Tr } W(0) = 1, \qquad (3.5a)$$

---

[2] The first equality of (3.1) has been proven in Appendix XV.A.
[3] See, e.g., Gel'fand and Shilov (1964, Vol. 1, Chapter I, Section 2.1).

because $\arctan(E_R 2/\Gamma) \to \pi/2$ for $E_R 2/\Gamma \to \infty$. As $E_R/\Gamma$ is large but finite, $W(0) = W^R$ is not quite normalized, but the error one makes is small. From now on we shall always work in this approximation, taking

$$E_R/\Gamma \to \infty \quad \text{for } E_R/\Gamma \gg 1 \tag{3.7}$$

and ignore the small error we make.[4] In this approximation the integral over the energy $E$ is taken from $-\infty$ to $+\infty$.

The expectation value of the energy operator in an energy eigenstate $W = |E_0\rangle\langle E_0|$ is $\mathrm{Tr}(H|E_0\rangle\langle E_0|) = E_0$. We shall now calculate the expectation value of $H$ in the state $W(0) = W^R$ given by (3.1) which is to describe a resonance with energy $E_R$ and width $\Gamma$:

$$\mathrm{Tr}(HW^R) = \sum_{\hat{a}} \langle \hat{a} \| W^R \| \hat{a} \rangle \int dE \, \frac{\Gamma}{2\pi} \frac{E}{(E - E_R)^2 + (\Gamma/2)^2}$$

$$= \frac{E_R}{\pi} \int_{x = -E_R\,2/\Gamma}^{\infty} dx \, \frac{1}{x^2 + 1} + \frac{\Gamma}{2} \int_{x = -E_R\,2/\Gamma}^{\infty} dx \, \frac{x}{x^2 + 1}. \tag{3.8}$$

In the limit (3.7) the first integral, according to (3.5), goes to $\pi$, and the second integral goes to zero, so that

$$\mathrm{Tr}(HW^R) = E_R \quad \text{in the limit (3.7).} \tag{3.9}$$

This is the result we desire from comparison with the stationary energy eigenstate.

Equation (3.1) gives only the diagonal matrix elements of the statistical operator $W^R$ for a decaying system. To fix $W^R$ fully we also need to specify its matrix elements between generalized energy eigenvectors of different energy. These are obtained from (3.1), (3.2) and the requirement that $W^R(t)$ describe a decaying state for $t > 0$ rather than a capture state (a state of exponential growth corresponding to a pole in the second energy sheet immediately above the real positive axis).

According to our discussion in Section II.11—in particular, according to (II.11.1)—$W(0)$ should in general be given by

$$W(0) = W^R$$

$$= \sum_{\hat{a}\hat{a}'} \int \rho(E_a)\, dE_a\, \rho(E_a')\, dE_a'\, f(E_a' - E_R) |E_a' \hat{a}'^+\rangle$$

$$\times \langle \hat{a}' \| W^R \| \hat{a} \rangle \langle E_a \hat{a}^+ | f^*(E_a - E_R). \tag{3.10}$$

Hence

$$\langle E\hat{a}'^+ | W^R | E\hat{a}^+ \rangle = |f(E - E_R)|^2 \langle \hat{a}' \| W^R \| a \rangle, \tag{3.11}$$

---

[4] We should remind the reader that every theoretical description is an approximate description and that the Breit–Wigner amplitude for resonance scattering may already be a crude approximation, as remarked in Section XVIII.6. Therefore the choice (3.1), (3.2) may result in greater inaccuracy than the approximation (3.7).

so that by comparison with (3.1) we must have

$$|f(E - E_R)|^2 = \rho(E)^{-1} F(E - E_R) = \frac{\Gamma}{2\pi\rho(E)} \frac{1}{(E - E_R)^2 + (\Gamma/2)^2}. \quad (3.12)$$

Except for an arbitrary factor of modulus 1, there are only two possibilities from (3.12):

$$f(E - E_R) = \sqrt{\frac{\Gamma}{2\pi\rho}} \frac{1}{E_R + i\Gamma/2 - E} \quad (3.13)$$

and

$$f(E - E_R) = \sqrt{\frac{\Gamma}{2\pi\rho}} \frac{1}{E_R - i\Gamma/2 - E}. \quad (3.14)$$

For a decaying state one has to choose (3.14). The property of a state (3.10) with (3.13) is discussed in Problem 1.[4a]

The integral in (3.10) should be extended over the spectrum of $H$; however, in the approximation (3.7) we can take the integral from $-\infty$ to $\infty$. This will include generalized energy eigenvectors that do not belong to the spectrum of $H$; a mathematical remark concerning this point is given below.

To simplify the notation we shall now specify the additional quantum numbers $a$ and then ignore them. As discussed in the preceding chapters, the resonance $R$ will have a definite value of angular momentum $l = l_R$ and internal quantum numbers $\kappa = \kappa_R$. Therefore a choice of angular-momentum eigenvectors

$$|E\hat{a}^+\rangle = |E l l_3 \kappa^+\rangle \quad (3.15)$$

is convenient. The reduced matrix element of $W^R$ in this basis is

$$\langle \kappa l l_3 \| W^R \| \kappa' l' l_3' \rangle = \delta_{\kappa\kappa_R} \delta_{l l_R} \delta_{\kappa'\kappa_R} \delta_{l' l_R} \delta_{l_3 l_3'} \frac{1}{2l + 1} \quad (3.16)$$

for a completely unpolarized state. $W^R$ of (3.10) is then

$$W^R = \frac{1}{2l + 1} \sum_{l_3} \int \rho(E)\, dE \int \rho(E')\, dE'\, f(E' - E_R) |E' l_R l_3 \kappa_R^+\rangle$$
$$\times \langle E l_R l_3 \kappa^+ | f^*(E - E_R), \quad (3.17)$$

which we will write with (3.14) as

$$W^R = \frac{\Gamma}{2\pi} \int_0^\infty \rho^{1/2}(E)\, dE \int_0^\infty \rho^{1/2}(E')\, dE' \frac{1}{E_R - i\Gamma/2 - E'}$$
$$\times |E'^+\rangle\langle E^+| \frac{1}{E_R + i\Gamma/2 - E}, \quad (3.18)$$

ignoring the additional quantum numbers.

---

[4a] It describes a "capture state" and is associated with a pole in the third quadrant of the complex momentum-plane (Figure XVIII.5.3). Whereas the possible values of $E_R + i\Gamma/2$ (position of the poles) depend upon the particular properties (Hamiltonian) of the physical system under consideration, we know already from very general arguments (Section XVIII.5) that with every decaying state $(E_R, \Gamma)$ the physical system must also have a capture state $(E_R, -\Gamma)$.

A proper, stationary energy eigenstate of the observable $H$ with eigenvalue $E_R$ is written as

$$\Lambda_R = |E_R)(E_R|, \qquad (3.19)$$

where

$$H|E_R) = E_R|E_R) \qquad (3.20)$$

and the $|E_R)$ are proper eigenvectors and elements of the space of physical states:

$$(E_R|E_R) = 1 \qquad (|E_R) \in \Phi \subset \mathscr{H} \subset \Phi^\times). \qquad (3.21)$$

As it is important to distinguish between the proper and generalized eigenvectors, we shall here use a separate notation $|E)$ for a proper eigenvector and $|E\rangle$ for a generalized eigenvector (cf. Section II.8). We shall now show that $W^R$ of (3.18) can be written in a form analogous to (3.19).

In order to do this we use a well-known mathematical theorem (the Titchmarsh theorem)[5]
Let $G(u)$ be the limit of a function $G(\omega)$, $\omega = u + iv$, that is analytic in the upper half plane and square-integrable over any line parallel to the real axis:

$$\sup_{v > 0} \int_{-\infty}^{+\infty} |G(u + iv)|^2 \, du < \infty$$

Then

$$G(\omega) = \frac{1}{2\pi i} \int_{-\infty}^{+\infty} \frac{G(\omega')}{\omega' - \omega} \, d\omega' \quad \text{for Im } \omega > 0 \qquad (3.22)$$

and

$$G(\omega) = \frac{1}{\pi i} \mathrm{P} \int \frac{G(\omega')}{\omega' - \omega} \, d\omega' \quad \text{for } \omega \text{ real.} \qquad (3.23)$$

Let us choose a "well-behaved" vector $\phi \in \Phi \subset \mathscr{H} \subset \Phi^\times$ with the property that $\langle {}^+E|\phi \rangle$ is the limit of a function $\langle {}^+\omega|\phi \rangle = G(\omega)$ analytic in the upper half $\omega$-plane that fulfills the condition of the above theorem.[5a] Then, writing $\omega = R_R + i\Gamma/2$, we obtain from (3.23) the connection

$$\sqrt{\rho\left(E_R + i\frac{\Gamma}{2}\right)} \left\langle {}^+E_R + i\frac{\Gamma}{2} \middle| \phi \right\rangle$$

$$= -\frac{1}{2\pi i} \int_{-\infty}^{+\infty} dE \, \rho^{1/2}(E) \langle {}^+E|\phi \rangle \frac{1}{E_R + i\Gamma/2 - E} \qquad (3.24)$$

[5] See, e.g., Nussenzveig (1972, Chapter 1.6).

[5a] This set of well behaved vectors is not dense in $\Phi \subset \mathscr{H}$. In order to obtain all of $\Phi$, one has also to consider those $\phi \in \Phi$ for which $\langle \omega|\phi \rangle$ is analytic in the lower half $\omega$-plane. $\mathscr{H}$ is then given by $\mathscr{H} = \mathscr{H}_+ \oplus \mathscr{H}_-$ where the realization of $\mathscr{H}_+$ ($\mathscr{H}_-$) is the space of Hardy-class functions with respect to the upper (lower) half plane.

and the complex conjugate

$$\left( \left( \sqrt{\rho\left( E_R + i\frac{\Gamma}{2} \right)} \right)^* \left\langle {}^+ E_R + i\frac{\Gamma}{2} \middle| \phi \right\rangle \right)^*$$

$$= \frac{1}{2\pi i} \int_{-\infty}^{+\infty} dE'\, \rho^{1/2}(E') \langle \phi | E'^+ \rangle \frac{1}{E_R - i\Gamma/2 - E'}. \quad (3.25)$$

For the latter we use the notation

$$\left( \sqrt{\rho\left( E_R + i\frac{\Gamma}{2} \right)} \right)^* \left\langle \phi \middle| E_R - i\frac{\Gamma}{2}^+ \right\rangle$$

$$= \left( \sqrt{\rho\left( E_R + \frac{\Gamma}{2} \right)} \left\langle E_R + i\frac{\Gamma}{2} \middle| \phi \right\rangle \right)^*$$

$$= \frac{1}{2\pi i} \int dE'\, \rho^{1/2}(E') \langle\, | E'^+ \rangle \frac{1}{E_R - i\Gamma/2 - E'}. \quad (3.26)$$

If we omit the arbitrary "well-behaved" vector $\phi$, we can define the generalized vectors

$$\left| E_R - i\frac{\Gamma}{2}^+ \right\rangle = \frac{1}{2\pi i} \left( \frac{1}{\sqrt{\rho}} \right)^* \int dE'\, \rho^{1/2}(E') | E'^+ \rangle \frac{1}{E_R - i\Gamma/2 - E'}$$

$$(3.27)$$

and

$$\left\langle E_R + i\frac{\Gamma}{2}^+ \middle| = -\frac{1}{2\pi i} \frac{1}{\sqrt{\rho}} \int dE\, \rho^{1/2}(E) \langle E^+ | \frac{1}{E_R + i\Gamma/2 - E}. \right.$$

$$(3.28)$$

Applying the above theorem to $\langle {}^+E | H | \phi \rangle$ (assuming that $\langle {}^+E | H | \phi \rangle$ also fulfills the conditions of the above theorem), we obtain

$$\left\langle {}^+ E_R + i\frac{\Gamma}{2} \middle| H | \phi \right\rangle$$

$$= -\frac{1}{2\pi i} \frac{1}{\sqrt{\rho}} \int dE\, \rho^{1/2}(E)(E\langle {}^+E | \phi \rangle) \frac{1}{E_R + i\Gamma/2 - E}$$

$$= \left( E_R + i\frac{\Gamma}{2} \right) \left\langle {}^+ E_R + i\frac{\Gamma}{2} \middle| \phi \right\rangle, \quad (3.29)$$

and in the same way

$$\left\langle \phi \middle| H \middle| E_R - i\frac{\Gamma}{2}^+ \right\rangle = \left( E_R - i\frac{\Gamma}{2} \right) \left\langle \phi \middle| E_R - i\frac{\Gamma}{2}^+ \right\rangle \quad (3.30)$$

Thus the $|E_R - i\Gamma/2^+\rangle$ are generalized eigenvectors of $H$ with the eigenvalues $E_R - i\Gamma/2$:

$$H\left|E_R - i\frac{\Gamma}{2}^+\right\rangle = \left(E_R - i\frac{\Gamma}{2}\right)\left|E_R - i\frac{\Gamma}{2}^+\right\rangle, \quad (3.31)$$

$$\left\langle {}^+E_R + i\frac{\Gamma}{2}\right|H = \left(E_R + i\frac{\Gamma}{2}\right)\left\langle {}^+E_R + i\frac{\Gamma}{2}\right|. \quad (3.32)$$

We calculate the norm of this generalized eigenvector $|E_R - i\Gamma/2\rangle$:

$$\left\langle {}^+E_R + i\frac{\Gamma}{2} \middle| E_R - i\frac{\Gamma}{2}^+ \right\rangle$$

$$= \frac{1}{(2\pi)^2}\left|\rho\left(E_R + i\frac{\Gamma}{2}\right)\right|^{-1} \int_{-\infty}^{+\infty} dE' \, \rho^{1/2}(E') \, dE \, \rho^{1/2}(E)$$

$$\times \langle {}^+E | E'^+ \rangle \frac{1}{E_R - i\Gamma/2 - E'} \frac{1}{E_R + i\Gamma/2 - E}$$

$$= \frac{1}{(2\pi)^2}\left|\rho\left(E_R + i\frac{\Gamma}{2}\right)\right|^{-1} \int_{-\infty}^{+\infty} dE \, \frac{1}{(E_R - E)^2 + (\Gamma/2)^2}$$

$$= \frac{1}{2\pi}\left|\rho\left(E_R + i\frac{\Gamma}{2}\right)\right|^{-1} \frac{\Gamma}{2}. \quad (3.33)$$

Though the generalized eigenvectors $|E_R - i\Gamma/2^+\rangle$ are normalizable, the matrix element $\langle E_R + i\Gamma/2|H^n|E_R - i\Gamma^+/2\rangle$ does not in general exist (for $n$ even), as one can easily see by a calculation analogous to (3.33).

Equation (3.33) tempts us to define the "normalized generalized eigenvectors with complex eigenvalue":

$$\left|E_R - i\frac{\Gamma}{2}\right) = (2\pi\Gamma)^{1/2}\left(\left(\rho\left(E_R + i\frac{\Gamma}{2}\right)\right)^{1/2}\right)^* \left|E_R - i\frac{\Gamma}{2}^+\right\rangle, \quad (3.34)$$

$$\left(E_R + i\frac{\Gamma}{2}\right| = (2\pi\Gamma)^{1/2}\left(\rho\left(E_R + i\frac{\Gamma}{2}\right)\right)^{1/2}\left\langle E_R + i\frac{\Gamma}{2}^+\right|. \quad (3.35)$$

The generalized eigenvectors $|E_R - i\Gamma/2^+\rangle$ do not, of course, appear in the spectral resolution

$$\phi = \int \rho(E) dE |E^+\rangle\langle {}^+E|\phi\rangle \quad (\phi \in \phi)$$

for the energy operators (cf. the mathematical note on the nuclear spectral theorem in Section II.8). They are elements of the space $\Phi^\times$ which are not needed in the generalized basis vector expansion of elements $\phi$ of the space of physical states $\Phi$. (As we have seen above, their norm exists and they are even elements of $\mathscr{H}$. However, $\|H^n|E_R - i\Gamma/2)\|^2 = (E_R + i\Gamma/2|H^{2n}|E_R - i\Gamma/2)$ does not exist so that these generalized eigenvectors of $H$ are not in the domain of $H$.[6] That these generalized eigenvectors are elements of $\mathcal{H}$ is of no importance.)

Usually when one constructs the rigged Hilbert space $\Phi \subset \mathcal{H} \subset \Phi^\times$ one tries to do this in such a way that these generalized eigenvectors do not appear in $\Phi^\times$; here we see that they may be useful for the description of decaying states.[6a]

After these mathematical preparations we are ready to rewrite $W^R$ of (3.18). The integration in (3.18) extends over the semibounded spectrum of $H$. The integration in (3.27) and (3.28) extends from $-\infty$ to $+\infty$. Except for the small error of the approximation (3.7), $W^R$ is given by

$$W^R \approx 2\pi\Gamma \sqrt{\rho\left(E_R - i\frac{\Gamma}{2}\right)\rho\left(E_R + i\frac{\Gamma}{2}\right)} \left| E_R - i\frac{\Gamma}{2}^+ \right\rangle\left\langle E_R + i\frac{\Gamma}{2}^+ \right|$$

$$= \left| E_R - i\frac{\Gamma}{2}\right)\left(E_R + i\frac{\Gamma}{2}\right| \tag{3.36}$$

The approximate equality goes over into an exact equality in the limit $E_R 2/\Gamma \to \infty$. Therefore, we define the ideal resonance state by the right-hand side of (3.36). Though this is not exactly a physically preparable state (as in all experimental situations, a resonance appears to be accompanied by a background), we shall from now on use this to describe a decaying state.

For later references we now give the expression for $W^R$ with the additional quantum numbers $\hat{a}$ reinstated. Combining (3.10) and (3.36), one obtains

$$W^R = \sum_{\hat{a}\hat{a}} \left| E_R - i\frac{\Gamma}{2}, \hat{a}'\right)\langle \hat{a}' \| W^R \| \hat{a}\rangle\left(E_R + i\frac{\Gamma}{2}, \hat{a}\right|. \tag{3.36a}$$

Here

$$\left| E_R - i\frac{\Gamma}{2}, \hat{a}'\right) = \left(\left(2\pi\rho\left(E_R - i\frac{\Gamma}{2}\right)\Gamma\right)^{1/2}\right)^* \left| E_R - i\frac{\Gamma}{2}^+ \right\rangle$$

$$= \frac{1}{i}\sqrt{\frac{\Gamma}{2\pi}} \int dE'\, \rho^{1/2}(E') | E'\hat{a}'\rangle \frac{1}{E_R - i\Gamma/2 - E'}. \tag{3.27a}$$

---

[6] In the rigged Hilbert space $\Phi \subset \mathcal{H} \subset \Phi^\times$ the Hermitian observable $H$ is represented by the triplet of operators $H \subset \bar{H} = H^+ \subset H^\times$ where $H$ is essentially self-adjoint (and continuous with respect to the topology in $\Phi$), $\bar{H}$ is the closure of $H$ and $H^\times$ is the adjoint of $H$ in $\Phi^\times$. A generalized eigenvector of $\bar{H}$ is the antilinear functional $F_\omega = |\omega\rangle = |E - i\Gamma/2\rangle$ such that $F_\omega(H\phi) = \langle H\phi|\omega\rangle = \langle\phi|H^\times|\omega\rangle = \omega\langle\phi|\omega\rangle$ for all $\phi \in \Phi$. This with the above "well-behaved" $\phi$ is the precise form of (3.31) (A. Bohm, The Rigged Hilbert Space and Quantum Mechanics [1978], Chapter IV, where the notation $|\bar{\omega}\rangle$ was used for the $|\omega\rangle$ here), and may clarify the ostensibly absurd statement that a generalized eigenvector of $H$ is not in the domain of $\bar{H}$.

[6a] In a different approach generalized eigenvectors with complex eigenvalue have already been used by H. Baumgärtel in *Resonances of Perturbed Self-adjoining Operators and their Eigenfunctionals*, Math. Nachr. 75, 133 (1976).

For example, if the reduced matrix elements of $W^R$ are given by (3.16), corresponding to a resonance with angular momentum $l_R$ and internal quantum number $\kappa_R$, then $W^R$ becomes

$$W^R = \sum_{l_3} \frac{1}{2l+1} \left| E_R - i\frac{\Gamma}{2}, l_R l_3 \kappa_R \right) \left( E + i\frac{\Gamma}{2} l_R l_3 \kappa_R \right|. \quad (3.36b)$$

Equation (3.36) is the form for the statistical operator of a decaying state with resonance parameters $(E_R, \Gamma)$ that is analogous to the form (3.19) for a stationary state. $W^R = |E_R - i\Gamma/2)(E_R - i\Gamma/2|$ is idempotent:

$$W^R W^R = W^R, \quad (3.37)$$

This follows immediately from (3.33).

From the form (3.36) of $W^R$ the exponential decay law for the nondecay amplitude follows immediately. Using (1.4), (3.32), and (3.33), one obtains[6b] for $t > 0$:

$$W(t) = e^{-iHt} \left| E_R - i\frac{\Gamma}{2} \right) \left( E_R + i\frac{\Gamma}{2} \right| e^{iHt}$$

$$= e^{-i(E_R - i\Gamma/2)} e^{i(E_R + i\Gamma/2)} \left| E_R - i\frac{\Gamma}{2} \right) \left( E_R + i\frac{\Gamma}{2} \right|. \quad (3.38)$$

So in the approximation (3.7)

$$W(t) = e^{-\Gamma t} W^R. \quad (3.39)$$

For the nondecay probability (2.3a) we then obtain, with (3.37) and (3.36),

$$\mathscr{P}_R(t) = \text{Tr}(W^R W(t)) = \text{Tr}(W^R W^R) e^{-\Gamma t} = e^{-\Gamma t} \text{Tr } W^R = e^{-\Gamma t}. \quad (3.40)$$

Equations (3.39) and (3.40) have been derived from (3.36), which entails the approximation (3.7).[6c]

Comparing (3.40) with the classical decay law (2.3), we obtain for the initial decay rate and [from (2.7)] for the lifetime

$$\tau = \frac{1}{\lambda} = \frac{1}{\Gamma}. \quad (3.41)$$

Thus the lifetime of a resonance with resonance parameters $(E_R, \Gamma)$ [fulfilling (3.7)], considered as an unstable physical system, is equal to the inverse width.

---

[6b] Equation (3.38) is to be understood as $\langle \phi | W(t) | \phi \rangle = \langle e^{iHt} \phi | \omega \rangle \langle \omega^* | e^{iHt} \phi \rangle$ for every $\phi \in \Phi \cap \mathscr{H}_+$. In a way analogous to (3.36) (3.38) one can define "capture states" of exponential growth using $\phi \in \Phi \cap \mathscr{H}_-$.

[6c] As is well known to specialists, in the usual precise Hilbert space formulation one obtains deviations from the exponential decay law [L. Fonda et al. (1978)]. Deviations for large values of $t$ follows from the condition that the spectrum of $H$ be bounded from below which corresponds to a finite lower limit in the integrals (3.5), (3.8), (3.25), (3.26) and which is not the case if one uses the approximation (3.7). Deviations from the exponential law for small times $t$ follow from the condition that the energy in the decaying state be finite, which corresponds to the condition that the decaying state vector be in the domain of the Hilbert space operator $\bar{H}$ (or even in $\Phi$, the domain of $H$) which is not the case if one uses (3.27a).

## XXI.4 Decay Rate

A very quick but rather abstruse derivation[7] of the initial decay rate $\dot{\mathscr{P}}(0)$ is obtained if one inserts (3.36) into the expression (XIV.2.18), which we write, using (3.36a),

$$\dot{\mathscr{P}}(t=0) = -i \sum_b \sum_{aa'} \int dE\, \rho(E)\, dE'\, \rho(E') \langle b|V|E\hat{a}^+\rangle \langle E'\hat{a}'^+|V|b\rangle$$

$$\sum_{dd'} \left( \frac{1}{E'-E_b-i\epsilon} - \frac{1}{E-E_b+i\epsilon} \right) \langle^+ E\hat{a} \left| E_R - i\frac{\Gamma}{2}, \hat{d}' \right)$$

$$\times \left( E_R + i\frac{\Gamma}{2}, \hat{d} \,\middle|\, E'\hat{a}'\rangle \langle \hat{d}'\| W^R \|\hat{d}\rangle. \tag{4.1}$$

As we have already taken the limit (3.7), we have to take $\epsilon > \Gamma/2$, so the present derivation is only true for narrow widths.

One obtains immediately, using the completeness condition (XIV.2.11b) with (XIV.2.13),

$$\dot{\mathscr{P}}(t=0) = -i \sum_b \sum_{\hat{a}\hat{a}'} \langle b|V \left| E_R - i\frac{\Gamma}{2}, \hat{a}' \right) \left( \hat{a}, E_R + i\frac{\Gamma}{2} \middle| V|b\rangle \langle a'\| W^R \|\hat{a}\rangle \right.$$

$$\times \left( \frac{1}{E_R - E_b - i(\epsilon - \Gamma/2)} - \frac{1}{E_R - E_b + i(\epsilon - \Gamma/2)} \right). \tag{4.2}$$

From this we obtain with (XIV.5.20)

$$\dot{\mathscr{P}}(t=0) = 2\pi \sum_b \sum_{\hat{a}\hat{a}'} \langle b|V \left| E_R - i\frac{\Gamma}{2}, \hat{a}' \right) \left( a, E_R + i\frac{\Gamma}{2} \middle| V|b\rangle \right.$$

$$\times \delta(E_R - E_b) \langle \hat{a}'\| W^R \|\hat{a}\rangle. \tag{4.3}$$

The $T$-matrix $\langle b|V|E_R - i\Gamma/2, \hat{a}\rangle$ is a slowly varying function in the complex energy plane, so that for small values of $\Gamma/2$ one can replace $\langle b|V|E_R - i\Gamma/2, \hat{a}\rangle$ by $\langle b|V|E_R, \hat{a}\rangle$ and thus write for the initial decay rate

$$\dot{\mathscr{P}}(t=0) = 2\pi \sum_b \sum_{\hat{a}\hat{a}'} \delta(E_b - E_R) \langle b|V|E_R, \hat{a}'\rangle (E_R, \hat{a}|V|b\rangle \langle \hat{a}'\| W \|\hat{a}\rangle. \tag{4.4}$$

If the summation extends over all possible decay products, then $\dot{\mathscr{P}}$ is called the total decay rate. For the familiar case of a decaying state with angular momentum $l_R$ and internal quantum numbers $\kappa_R$ [Equation (3.16)], this goes over into

$$\dot{\mathscr{P}}(t=0) = 2\pi \frac{1}{(2l+1)} \sum_{l_3} \sum_b \delta(E_b - E_R)|\langle E_b - E_R)|\langle b|V|E_R, l_R, l_3, \kappa_R)|^2. \tag{4.5}$$

---

[7] A detailed derivation of the initial decay rate is suggested in Problem 2. See also A. Bohm, *Rigged Hilbert Space and Decaying States.* Univ. Texas preprint No. ORO-3992-353 (October) 1978.

Equations (4.4) and (4.5) are formulas used for the calculation of decay rates in all branches of physics. Usually the interaction is weak, so that the Born approximation can be used for the transition matrix element $\langle b|V|E_R l_R l_3 \kappa_R \rangle$. This means $|E_R \hat{a}\rangle$ is replaced by the eigenvector of the free Hamiltonian

$$K = H - V. \tag{4.6}$$

The $|b\rangle$ are already eigenvectors of $K$ [cf. (XIV.2.14)]. So the program for the calculation of the initial decay rate is the following: One finds the eigenstates and generalized eigenvectors of the free-energy operator. These are the energy eigenvectors in the approximation in which the decaying system is considered stable. For instance, in the decay (1.3a), $|E_R, l_R, l_3, \kappa_R\rangle$ will be the proper eigenvector of the energy operator of the atom when the interaction with the radiation is ignored and all energy levels are stable. $|b\rangle$ is the direct product

$$|b\rangle = |E_R^{gr}, l_{gr}, l_3, \kappa_g \rangle \otimes |\gamma\rangle,$$

where the first factor is the ground-state energy eigenvector of the atom, and $|\gamma\rangle$ is a basis vector of the one-photon space. Then one has to conjecture the interaction Hamiltonian $V$ (or the transition operator $T$ if the Born approximation is not satisfactory). For the case of the radiative decays of atoms this can be done to a certain extent by using the correspondence with a classical system; in other situations, e.g., for the decay of elementary particles, one has only very few general principles to limit the possibilities for this conjecture. After $V$ and the solution of the free problem are known, one has to calculate the matrix elements of $V$ between the free eigenvectors and insert it into (4.4) or (4.5).

In this way the transition is considered as a transition from one "stable" state into another, caused by the interaction Hamiltonian $V$ or by the transition operator $T$. The state of the decaying system $W^R$ is then

$$\Lambda^R = \sum_{\hat{a}} |E_R \hat{a}\rangle(\hat{a}E_R|, \tag{4.7}$$

which is orthogonal to the projection operator onto the space of decay products:

$$\Lambda = \sum_b |b\rangle\langle b|, \tag{4.8}$$

$$\Lambda\Lambda^R = 0. \tag{4.9}$$

$\Lambda$ is the projector on the space of physical states of all possible decay products.
Let us then consider

$$I = \Lambda + \Lambda^R \tag{4.10}$$

and its expectation value in the state $W(t)$,

$$\text{Tr}(IW(t)) = \text{Tr}(\Lambda W(t)) + \text{Tr}(\Lambda^R W(t)), \tag{4.11}$$

which represents the probability of finding either the undecayed system or the decay products. At any time, $\text{Tr}(IW(t)) = 1$. $\text{Tr}(\Lambda W(t))$ is the transition probability (2.8), and the nondecay probability (2.3a) is now $\text{Tr}(\Lambda^R W(t))$. Thus (4.11) says:[7a]

$$\mathscr{P}(t) = 1 - \mathscr{P}_R(t) = 1 - e^{-\Gamma t}, \tag{4.12}$$

where we have used (3.40). The initial decay rate is then obtained from (4.12) as

$$\dot{\mathscr{P}}(t = 0) = \Gamma. \tag{4.13}$$

Thus the *total decay rate* of the decaying system is equal to the *total width* of the resonance. Therewith we have obtained the analogue to the classical relation (2.5), but we have also seen under what limitations upon the quantum-mechanical description we have obtained this classical relation.

## XXI.5  Partial Decay Rates

Even more important in practical calculations than the total decay rate are the partial decay rates. In order to obtain a partial decay rate we have to specify the basis vectors $|b\rangle$ in (4.4) or (4.5). For the sake of definiteness we shall give our discussions first in terms of the angular-momentum basis. One can choose

$$|b\rangle = |E_b, \hat{b}\rangle = |E_b, jj_3\eta\rangle \tag{5.1}$$

if the angular momentum commutes with the internal observables:

$$[L_i, \eta^{\text{op}}] = 0. \tag{5.1a}$$

The internal observables $\eta^{\text{op}}$ will in general not agree with the internal observables $\kappa^{\text{op}}$ of which the resonance is an eigenstate with eigenvalue $\kappa_R$. We then write (4.5):

$$\Gamma = 2\pi \sum_{\eta} \int \rho^{\eta}(E)\, dE\, \delta(E - E_R) \sum_{jj_3} \sum_{l_3} \frac{1}{2l+1} |\langle Ejj_3\eta | E_R l_R l_3 \kappa_R\rangle|^2. \tag{5.2}$$

This gives the rate for the decay of $R$ into decay products with any of the internal quantum numbers $\eta$. If the detector detects only decay products with one particular value for $\eta$, then one measures the quantity

$$\Gamma(R \to \eta) = 2\pi \int \rho^{\eta}(E)\, dE\, \delta(E - E_R) \sum_{jj_3} \sum_{l_3} \frac{1}{2l+1} |\langle Ejj_3\eta | V | E_R l_R l_3 \kappa_R\rangle|^2. \tag{5.3}$$

$\Gamma(R \to \eta)$ is called the *partial decay rate* for the decay of $R$ into the state with internal quantum numbers $\eta$ (decay channel $\eta$). Obviously one has

$$\sum_{\eta} \Gamma(R \to \eta) = \Gamma. \tag{5.4}$$

---

[7a] Note that the first part of (4.12) and (4.11) has been derived from (4.10) and (4.9) which cannot be justified if one takes $W^R$ instead of $\Lambda^R$.

Most often $V$ is rotationally symmetric, so that only the term with $j = l_R$ in (5.3) is different from zero.

The weight function $\rho''(E)$ is related to the normalization of the generalized eigenvectors $|E_b, j, j_3 \eta\rangle$:

$$\langle Ejj_3 \eta | E'j'j'_3 \eta' \rangle = \delta_{\eta\eta'} (\rho''(E))^{-1} \delta(E - E') \delta_{jj'} \delta_{j_3 j'_3}. \tag{5.5}$$

Instead of the basis vector (5.1) one may use different basis vectors, e.g., the generalized momentum eigenvectors

$$\langle \mathbf{p}\eta | \mathbf{p}'\eta' \rangle = \delta(\mathbf{p} - \mathbf{p}')\delta_{\eta\eta'} \tag{5.6}$$

if

$$[P_i, \eta^{\mathrm{op}}] = 0. \tag{5.6a}$$

Then one has instead of (5.3)

$$\Gamma(R \to \eta) = 2\pi \int d^3p \, \delta(E(p) - E_R) \sum_{l_3} \frac{1}{2l + 1} |\langle \mathbf{p}\eta | V | E_R l_R l_3 \kappa_R \rangle|^2. \tag{5.7}$$

Instead of the momenta or angular momenta one can choose any other labels

$$|E_b, \hat{b}\rangle = |E_b, \beta, \eta\rangle. \tag{5.8}$$

The only requirement is that

$$[\beta^{\mathrm{op}}, \eta^{\mathrm{op}}] = 0.$$

For the basis vectors that one uses in the decay-rate formula one should always choose the most convenient ones, and which the most convenient ones are depends upon the particular problem.[8] Therefore we now discuss the general case.

Let

$$K, \alpha^{\mathrm{op}}, \kappa^{\mathrm{op}} \tag{5.9}$$

be a complete system of commuting operators for the decaying system, where we have split $\hat{a}^{\mathrm{op}} = (\kappa^{\mathrm{op}}, \alpha^{\mathrm{op}})$ so that the decaying system has a definite value $\kappa = \kappa_R$ for the first set of operators and $\alpha^{\mathrm{op}}$ have not been measured. The reduced matrix element is then

$$\langle \hat{a} \| W^R \| \hat{a}' \rangle = \langle \kappa\alpha \| W^R \| \kappa'\alpha' \rangle = \delta_{\kappa\kappa_R} \delta_{\kappa'\kappa_R} \delta_{\alpha\alpha'} \frac{1}{\dim \mathcal{R}(\kappa_R)} \tag{5.10}$$

where $\mathcal{R}(\kappa_R)$ is the eigenspace of $\kappa^{\mathrm{op}}$ with eigenvalue $\kappa_R$.

Let

$$K, \beta^{\mathrm{op}}, \eta^{\mathrm{op}} \tag{5.11}$$

be a complete system of commuting observables chosen so that the detector detects decay products with definite quantum number $\eta$ and detects states

---

[8] For instance, if it should happen that (5.6a) is not fulfilled but that
$$\hat{P}_i = P_i M^{-1}$$
commutes with $\eta^{\mathrm{op}}$, where $M$ is the mass operator (which is a function of the internal quantum numbers), then one would choose the basis vectors $|\hat{p}, \eta\rangle$, where $\hat{p}_i = p_i/m(\eta)$.

with any values of $\beta$. Then one can ask for the partial decay rate of $R$ into the decay channel $\eta$. Inserting (5.10) into (4.4) and using eigenvectors of (5.9) and (5.11), one obtains

$$\Gamma = \Gamma(R \to \eta), \tag{5.12}$$

where

$$\Gamma(R \to \eta) = 2\pi \int \rho''(E) \, dE \, \delta(E - E_R) \sum_{\beta,\,\alpha} |\langle E\eta\beta | V | E_R \kappa_R \alpha \rangle|^2. \tag{5.13}$$

Here

$$\overline{\sum_{\alpha\beta}} = \frac{1}{\dim \mathscr{R}(\kappa_R)} \sum_{\beta} \sum_{\alpha} \tag{5.14}$$

means summing (or integrating) over all values of the labels for the basis vectors of the space of final states and averaging over all initial quantum numbers. One often states this as "summing over the initial states and averaging over the final states." That is, however, a misleading statement, because the basis vectors $|E_b \beta\eta\rangle$ may have nothing to do with possible final states.

The formula (5.13) for the decay rates has numerous applications in all branches of quantum physics. Whenever it makes sense to speak of a decaying quantum-mechanical system, (5.13) with suitable chosen basis vectors can be applied. The interaction Hamiltonian or transition operator $V$ depends, of course, upon the particular physical system that one considers and is determined by the algebra of observables. To find the right expression for $V$ is one of the tasks of understanding the physical system.

## Problems

1.  Show that the state $W_R$ given by (3.10) with (3.13) corresponds to a state of exponential growth ("capture" state).

2.  In this problem we suggest the calculations of the results in Section XXI.4 without the use of (3.36) but in the approximation (3.7).
(a)  Define the quantity

$$\mathscr{F}_{E_b}(E, E') = \sum_b \sum_{\hat{a}\hat{a}'} \rho(E)\rho(E')\rho(E_b)\langle E_b \hat{b} | V | E \hat{a}^+ \rangle$$

$$\times \langle E'\hat{a}'^+ | V | E_b \hat{b} \rangle \langle \hat{a} \| W^R \| \hat{a}' \rangle (\rho(E_R))^{-1}. \tag{P.0}$$

Extend this function to complex values $\omega_b$, $\omega$, $\omega'$ of the energy and show that, as a consequence of the Hermiticity of $V$ and time-reversal invariance,

$$\mathscr{F}_{\omega_b}(\omega, \omega') = \mathscr{F}_{\bar{\omega}_b}(\bar{\omega}, \bar{\omega}'), \tag{P.1a}$$

$$\mathscr{F}_{\omega_b}(\omega', \omega) = \mathscr{F}_{\omega_b}(\omega, \omega'). \tag{P.1b}$$

Justify

$$\left\langle E_b, \hat{b} | V | E_R + i\frac{\Gamma}{2}, \hat{a}^+ \right\rangle = -\left\langle E_b, \hat{b} | V | E_R - i\frac{\Gamma}{2}, \hat{a}^+ \right\rangle \tag{P.2}$$

using the symmetry relation of the $S$-matrix, (eqn. XVIII.5.10). [(P.2) is needed for the calculations of the following problems].

(b)   Use (3.10) with (3.14) to show that the decay rate (XIV.2.18) can be written as

$$\dot{\mathcal{P}}(t) = \dot{\mathcal{P}}_1(t) + \dot{\mathcal{P}}_2(t) \tag{P.3}$$

with

$$\dot{\mathcal{P}}_1(t) = -i\frac{\Gamma}{2\pi}\iiint dE_b\, dE\, dE'\, \mathcal{F}_{E_b}(E, E')$$

$$\times \frac{e^{-iEt}}{E - E_R + i(\Gamma/2)}\frac{e^{iE't}}{E' - E_R - i(\Gamma/2)}\frac{1}{E' - E_b - i\epsilon}, \tag{P.4}$$

$$\dot{\mathcal{P}}_2(t) = i\frac{\Gamma}{2\pi}\iiint dE_b\, dE\, dE'\, \mathcal{F}_{E_b}(E, E')$$

$$\times \frac{e^{-iEt}}{E - E_R + i(\Gamma/2)}\frac{e^{iE't}}{E' - E_R - i(\Gamma/2)}\frac{1}{E - E_b + i\epsilon}, \tag{P.5}$$

and show

$$\dot{\mathcal{P}}_1(t) = (\dot{\mathcal{P}}_2(t))^*. \tag{P.6}$$

(c)   Use the approximation (3.7) and show, applying the theorem (3.23), that

$$\dot{\mathcal{P}}(t) = 4\pi^2\Gamma e^{-\Gamma t}\, 2\, \mathrm{Re}\, \mathcal{F}_{E_R + i(\Gamma/2) - i\epsilon}\left(E_R - i\frac{\Gamma}{2}, E_R + i\frac{\Gamma}{2}\right). \tag{P.7}$$

(d)   Show that from the condition $\dot{\mathcal{P}}(t \to \infty) = 1$ it follows that

$$4\pi^2\, \mathrm{Re}\, \mathcal{F}_{E_R + i(\Gamma/2) - i\epsilon}\left(E_R - i\frac{\Gamma}{2}, E_R + i\frac{\Gamma}{2}\right) = 1, \tag{P.8}$$

and obtain (4.12) and (4.13).

3.   Assume that the detector has a very good energy resolution $\Delta E_b \ll \Gamma$. Then the decay rate per unit energy as a function of $E_b$ can be measured.

(a)   Using the same method as in Problem 2, show that this decay rate per unit energy is given by

$$\frac{d\dot{\mathcal{P}}}{dE_b}(t = 0) = 2\pi\Gamma\frac{\Gamma}{(E_b - E_R)^2 + (\Gamma/2)^2}\mathcal{F}_{E_b}(E_R - i0, E_b + i0).$$

(b)   Show that for the decaying angular-momentum state (3.16) the decay probability per unit energy is given by

$$\frac{d\dot{\mathcal{P}}(t = 0)}{dE_b} = b(E_b)\sum_b 2\pi\rho^b(E_b)\sum_{l_3}\frac{1}{2l + 1}|\langle E_b\hat{b}|V|E_R l_R l_3 \kappa_R\rangle|^2,$$

where

$$b(E_b) = \frac{\Gamma}{2\pi}\frac{1}{(E_b - E_R)^2 + (\Gamma/2)^2}.$$

$b(E_b)$ is called the natural line width.

Hint: Use (P.2) of problem 2.

# Epilogue

The purpose of a physical theory is fulfilled if it provides a mathematical image of some domain of reality that allows us to relate experimental data and foresee new situations by making mathematical deductions. In this book we have restricted ourselves to this program and have never mentioned any of the further-reaching implications of quantum mechanics. Yet quantum mechanics has affected all scientific and even general human thinking. And when presented in its full generality, as done in this book, it effortlessly reveals the two facts whose lessons reach far beyond the boundaries of physics.

Classical science is based on two assumptions: (1) the deterministic nature of predictions, and (2) the atomistic nature of understanding. Quantum mechanics teaches us to revise both.

Classical theories are deterministic. The laws of classical physics are constructed in such a way that if the initial values of a system's dynamical variables are given, then their precise values can be calculated for any later time. These laws were not simply derived solely from experience, and then accepted as the basis of scientific philosophy and general thinking. Rather, and perhaps to a greater degree, these laws have been extricated from nature because they were in accord with the prevailing philosophical idea that nothing can be without a cause.

Probability statements in classical physics are always associated with insufficient knowledge, i.e., they are statements about the observer's knowledge and not about the physical system, which according to the principles of classical physics can be known to unlimited accuracy. In quantum theory (as described in Section II.6 and elsewhere in the text) one can only say with what probability certain values can be expected, even if one knows the state as well as possible, i.e., even if the system is in a pure state. Thus in quantum theory statements are inherently probabilistic; the occurence of probability functions is not just a consequence of the observer's insufficient knowledge, but a property attributed to the physical systems themselves. Quantum predictions of experimental results are statements of how a microphysical process shows up in the macrophysical domain. These traces of microphysical processes in the macrophysical domain, the only source of human knowledge about such processes, do not obey deterministic laws. Earlier traces of a microphysical process do not determine later traces uniquely, but only probabilistically. Quantum theory teaches us that there are inherent limitations to human knowledge.

The second point, that of the profoundly holistic nature (of the understanding) of quantum physical systems, is not often emphasized, even though it is an obvious consequence of the quantum-mechanical description of physical systems. Although holism has already become rather widely accepted in other disciplines (e.g., psychology), it has been resisted by the physicists, who seem to be influenced by the success of atomism in classical physics. The quantum physical system is a structured whole described by the mathematical structure of an algebra of operators. From the laws of the combination of quantum physical systems (Section III.5), it follows that there are observables—of the combination of the two subsystems (described by $\mathscr{H}_1 \otimes \mathscr{H}_2$) that are incompatible with all observables of either subsystem (described by $\mathscr{H}_1$ or $\mathscr{H}_2$).

Thus, in quantum physics there exist holistic properties that cannot be obtained as combinations of the properties of the subsystems. In this sense the whole is not the sum of the parts.

In the atomistic approach understanding comes from the reduction of the complex system to simpler subsystems by ever finer separations until one comes to the ultimate constituents. In quantum physics the presence of holistic properties prevents this reduction process, and the notion of ultimate constituents loses its meaning. Atomism belongs to classical physics. A quantum physical system such as a molecule cannot be fully understood by dissecting it into nuclei and electrons, although, in the tradition of our scientific heritage, it is tempting to do this. What one arrives at in this way, however, is only the classical analogue of the quantum physical system, as in the Kepler system of proton and electron for the classical analogue of the hydrogen atom. An electron in an atom "is" something different from an electron in a linear accelerator, and the whole picture of the electron can only be displayed by giving its different aspects as they are mathematically described by the various basis systems in the space of physical states.

The visual picture that one usually requires for the process of understanding in is quantum physics not the geometrical picture of the object, but the picture of its image in the space of physical states. The reduction from the more complex to the simpler is performed not on the physical object, leading to simpler consituent objects, but on the space of physical states, leading to the irreducible subspaces for ever simpler structures. At every stage of this reduction one still has a whole picture describing all aspects by the various basis systems of the subspaces, but a simpler one describing a narrower domain of physics.

Dissecting a quantum physical system may destroy it. Therefore a quantum physical system (such as the CO molecule) cannot be understood only atomistically (as a di-atom) but is often more adequately understood holistically and functionally (as a vibrator rotator).

Atomism has been a great achievement of the past, and all technology is based on it. But quantum theory has revealed its limitations and shown that even for simple systems a holistic method is needed also.

# Bibliography

The cited literature has been chosen rather arbitrarily. I have not made a systematic search for the most suitable list of books for further or supplementary reading, and mention just those I happen to have come across. Many of these books I have used myself.

The books and a few review articles are separated into several categories and then listed in alphabetical order; a few comments are added here and there.

## 1 Foundations of Quantum Mechanics

P. A. M. Dirac, *The Principles of Quantum Mechanics*, Clarendon Press, Oxford, 1958 (fourth edition).
   Written by one of the creators of quantum mechanics many years ago, this is still modern and one of the greatest books written on this subject.

J. M. Jauch, *Foundations of Quantum Mechanics*, Addison-Wesley, 1968.
   This text is concerned with the conceptual foundations of quantum mechanics and contains practically no applications. Furthermore, it differs from the present presentation by starting from the lattice structure of quantum mechanics. Still, it is recommended even to those who do not want to learn lattice theory.

G. Ludwig, *Grundlagen der Quantenmechanik*, Springer, Berlin, 1954.

J. von Neumann, *Mathematical Foundations of Quantum Mechanics*, Springer, Berlin, 1932; Princeton University Press, 1955.
   This is the first book written on the Hilbert-space formulation of quantum mechanics developed by its author.

## 2  Scattering Theory

A. I. Baz, Ya. B. Zeldovich, A. M. Peremelov, *Scattering Reactions and Decay in Non-relativistic Quantum Mechanics*, Israel Program for Scientific Translations, Jerusalem, 1969.

L. Fonda, G. C. Ghirardi, A. Rimini, *Decay Theory of Unstable Quantum Systems*, Reports on Progress in Physics, **41**, 587 (1978).

M. L. Goldberger, K. M. Watson, *Collision Theory*, Wiley, New York, 1964.

R. G. Newton, *Scattering Theory of Waves and Particles*. McGraw-Hill, New York, 1966.
   In particular, for the formal theory of scattering this book is highly recommended.

H. M. Nussenzveig, *Causality and Dispersion Relations*, Academic Press, New York, 1972.

A. G. Sitenko, *Lectures in Scattering Theory*, Pergamon Press, 1971.

K. Smith, *The Calculation of Atomic Collision Processes*, Wiley-Interscience, 1971.

John R. Taylor, *Scattering Theory*, Wiley, New York, 1972.
   This is a very clearly written book, which contains more material than the second part of the present book, however, it is written in a somewhat different spirit. It is highly recommended.


## 3  Theory of Angular Momentum and Group Theory

L. C. Biedenharn, J. D. Louck, *Angular Momentum in Quantum Physics*, Addison-Wesley, Publ. Co., 1979.

A. R. Edmonds, *Angular Momentum in Quantum Mechanics*, Princeton University Press, Princeton, 1957.

I. M. Gelfand, R. A. Minlos, Z. Ja. Shapiro, *Representations of the Rotation Group and of the Lorentz Group*, Pergamon Press, New York, 1963.

M. A. Naimark, *Linear Representations of the Lorentz Group*, Pergamon Press, New York, 1964.
   The first part of this book gives an introduction to the theory of group representations and discusses the rotation group in full detail. This is a book written by a mathematician for physicists, very readable, and mathematically rigorous. Some of the material of Chapter I is contained in this book.

M. Hamermesh, *Group Theory*, Addison-Wesley, 1962.

L. Michel, Applications of Group Theory of Quantum Physics: Algebraic Aspects; and

L. O'Raifeartaigh, Unitary Representations of Lie Groups in Quantum Physics.
   Both published in *Group Representations in Mathematics and Physics*, (Ed. V. Bargmann), Springer Lecture Notes in Physics, 1970.
   The two reviews by Michel and O'Raifeartaigh discuss many applications of group theory to quantum physics.

M. E. Rose, *Elementary Theory of Angular Momentum*, Wiley, NewYork, 1957.

## 4 Experimental Subjects

N. L. Alpert, W. E. Keiser, H. A. Szymanski, *Theory and Practice of Infrared Spectroscopy*, Wiley, New York, 1970.

R. P. Bauman, *Absorption Spectroscopy*, Wiley, New York, 1962.

A. Barbaro-Galtieri, Baryon Resonances, in *Advances in Particle Physics*, Vol. 2 (Ed. R. L. Cook, R. E. Marshak) Interscience Publishers, 1968.

G. Herzberg, *Molecular Spectra and Molecular Structure*, D. van Nostrand Company, 1966 (in particular Vol. 1).

This book teaches more than just molecular physics. It is one of the most beautiful books on physics and is highly recommended to every student.

H. S. W. Massey, E. H. S. Burhop, H. B. Gilbody, *Electronic and Ionic Impact Phenomena*, Clarendon Press, Oxford, 1969 (in particular Vol. 1: Collisions of Electrons with Atoms).


## 5 Mathematical Material

A. Bohm, *The Rigged Hilbert Space and Quantum Mechanics*, Springer Lecture Notes in Physics. Vol. 78 (1978).

I. M. Gel'fand and G. P. Shilov, *Generalized Functions*, Vols. 1, 2, 4, Academic Press, 1964.

A. Lichnerowicz, *Linear Algebra and Analysis*. Holden–Day Inc. San Francisco 1967.

K. Maurin, *Hilbert Space Methods*, Polish Scientific Publishers, Warsaw, 1967.

V. I. Smirnov, *A Course of Higher Mathematics*, Pergamon Press, 1964 (in particular Vol. III, Part 1, on linear algebra, and Vol. III, Part 2, on functions of a complex variable).


## 6 History of Quantum Mechanics

Max Jammer, *The Conceptual Development of Quantum Mechanics*, McGraw-Hill, 1966.

F. Hund, *Geschichte der Quantentheorie*, BI Wissenschaftsverlag, 1975.


## 7 Quantum Mechanics, Textbooks and Atomic Physics

G. Baym, *Lectures on Quantum Mechanics*, W. A. Benjamin, New York, 1969.

D. I. Blokhintsev, *Quantum Mechanics*, D. Reidel Publishing Co., 1964.

H. A. Bethe, E. E. Salpeter, *Quantum Mechanics of One- and Two-Electron Atoms*, Springer-Verlag, New York, 1957.

R. H. Dicke, J. W. Wittke, *Introduction to Quantum Mechanics*, Addison-Wesley, 1960.

R. P. Feynman, R. B. Leighton, M. Sands, *The Feynman Lectures on Physics*, Vol. 3, *Quantum Mechanics*, Addison-Wesley, 1965.

S. Gasiorowicz, *Quantum Physics*, Wiley, 1974.

D. T. Gillespie, *A Quantum Mechanics Primer*, Intex Publisher, 1970.

K. Gottfried, *Quantum Mechanics*, W. A. Benjamin, New York, 1966.

L. D. Landau, E. M. Lifshitz, *Quantum Mechanics*, Pergamon Press, 1958.

H. J. Lipkin, *Quantum Mechanics, New Approaches to Selected Topics*, North-Holland, 1973.

E. Merzbacher, *Quantum Mechanics*, Wiley, New York, 1961.

A. Messiah, *Quantum Mechanics*, Vols. 1, 2, Interscience, New York, 1961.

M. Mizushima, *Quantum Mechanics of Atomic Spectra and Atomic Structure*, W. A. Benjamin, New York, 1970.

L. T. Schiff, *Quantum Mechanics*, McGraw-Hill, 1968 (third edition).

## 8  Prerequisites from Classical Physics

V. D. Barger, M. Olsson, *Classical Mechanics: A Modern Perspective*, McGraw-Hill, New York, 1973.

A. O. Barut, *Electrodynamics and Classical Theory of Fields and Particles*, MacMillan, New York 1964.

H. C. Corben, *Classical and Quantum Theories of Spinning Particles*, Holden-Day, San Francisco. 1968.

H. Goldstein, *Classical Mechanics*, Addison-Wesley, Reading, Mass., 1950.

J. D. Jackson, *Classical Electrodynamics*, Wiley, New York, 1975.

F. Rohrlich, *Classical Charged Particles*, Addison-Wesley, Reading, Mass., 1965.

# Index

# Texts and Monographs in Physics

Edited by W. Beiglböck, M. Goldhaber, E. Lieb, and W. Thirring

Texts and Monographs in Physics includes books from any field of physics that might be used as basic texts for advanced training and higher education in physics, especially for lectures and seminars at the graduate level.

---

**Polarized Electrons**
By **J. Kessler**
1976. ix, 223p. 104 illus. cloth

**The Theory of Photons and Electrons**
The Relativistic Quantum Field Theory of Charged Particles with Spin One-Half
Second Expanded Edition
By **J. Jauch** and **F. Rohrlich**
1976. xix, 553p. 55 illus. cloth

**Essential Relativity**
Special, General, and Cosmological
Second Edition
By **W. Rindler**
1977. xv, 284p. 44 illus. cloth

**Inverse Problems in Quantum Scattering Theory**
By **K. Chadan** and **P. Sabatier**
1977. xxii, 344p. 23 illus. cloth

**Quantum Mechanics**
By **A. Böhm**
1979. xvii, 521p. 105 illus. cloth

**Relativistic Particle Physics**
By **H. Pilkuhn**
1979. 320p. approx. 89 illus. cloth

**The Concepts and Logic of Classical Thermodynamics as a Theory of Heat Engines**
Rigourously Constructed upon the Foundation Laid by S. Carnot and F. Reech
By **C. Truesdell** and **S. Bharatha**
1977. xxii, 154p. 15 illus. cloth

**Principles of Advanced Mathematical Physics**
Volume I
By **R.D. Richtmyer**
1978. xv, 400p. 45 illus. cloth

**Foundations of Theoretical Mechanics**
**Part I**: The Inverse Problem in Newtonian Mechanics
By **R.M. Santilli**
1978. 288p. cloth
**Part II**: Generalizations of the Inverse Problem in Newtonian Mechanics
In preparation

**Advanced Quantum Theory and Its Applications Through Feynman Diagrams**
By **M. D. Scadron**
1979. 416p. approx. 78 illus. cloth

# A Springer-Verlag Journal

## Zeitschrift für Physic C — Particles and Fields

*Editors in Chief*    G. Kramer, Hamburg; H. Satz, Bielefeld

*Editors*

K. Fujikawa, Tokyo          J. J. Sakurai, UCLA
K. Gottfried, Cornell       P. Söding, DESY
K. Kajantie, Helsinki       B. Stech, Heidelberg
A. Krzywicki, Orsay         J. Steinberger, CERN
P. Landshoff, Cambridge

*Zeitschrift für Physik* appears in three parts— A: Atoms and Nuclei; B: Condensed Matter and Quanta; C: Particles and Fields. Each part may be ordered separately. Coordinating editor for Zeitschrift für Physik, Parts A, B and C, is O. Haxel, Heidelberg.

*Zeitschrift für Physik C—Particles and Fields* is devoted to the experimental and theoretical investigation of elementary particles. In view of the steadily growing interplay of theory and experiment in this field, particular emphasis is given to a clear and complete presentation of research.

The topics covered include: strong, electromagnetic, and weak interactions of elementary particles, interaction and classification of constituents, and symmetry and unification schemes of different interactions.

---

# Lecture Notes in Physics

*Managing Editor*    **W. Beiglböck**

This series reports on new developments in physical research and teaching— quickly, informally, and at a high level. The type of material considered for publication includes preliminary drafts of original papers and monographs, lectures on a new field or lectures that present a new angle on a classical field, collections of seminar papers, and reports of meetings.

Vol. 81    M.H. MacGregor, **The Nature of the Elementary Particle.** 1978. xxii, 482p.

Vol. 83    **Experimental Methods in Heavy Ion Physics.** Edited by K. Bethge. 1978. v, 251p.

Vol. 84    **Stochastic Processes in Nonequilibrium Systems.** Edited by L. Garrido, P. Seglar, and P.J. Shepard. 1978. xi, 355p

Vol. 88    K. Hutter and A.A.F. Van De Ven, **Field Matter Interactions in Thermoelastic Solids.** 1979. viii, 231p.

Vol. 89    **Microscopic Optical Potentials.** Edited by H. von Geramb. 1979. xi, 481p.

**Springer-Verlag New York Inc.**
175 Fifth Avenue
New York, N.Y. 10010